Its Origin and Control

Second Edition

Kenneth Wark

Cecil F. Warner

PURDUE UNIVERSITY

1817

HARPER & ROW, PUBLISHERS, New York
Cambridge, Hagerstown, Philadelphia, San Francisco,
London, Mexico City, São Paulo, Sydney

Sponsoring Editor: Charlie Dresser
Project Editor: Eleanor Castellano
Production Manager: Marion A. Palen
Compositor: Science Typographers Inc.
Printer and Binder: The Maple Press Company
Art Studio: J & R Technical Services Inc.
Cover design: Mies Hora
(Photo credit: Johnson, De Wys)

AIR POLLUTION: Its Origin and Control, Second Edition

Library of Congress Cataloging in Publication Data
Wark, Kenneth,
Air pollution.

Includes bibliographical references and index.
1. Air-Pollution. I. Warner, Cecil Francis, joint author.
II. Title.
TD883.W28 1981 363.7'392 80-20283
ISBN 0-700-22534-X

Contents

Preface

The first edition of this text has been modified in three broad areas. First, new material has been added in several chapters in order to improve upon the coverage of material. For example, a more thorough discussion of particulate distributions is presented, in conjunction with the material on fractional and overall collection efficiencies. A mathematical analysis of the adsorption wave and its transient behavior also has been included. Secondly, old material has been updated, or removed if no longer appropriate. A major example of this is the complete rewriting of the section on flue gas desulfurization. The chapter on mobile sources has also undergone revision that reflects changes in technology or legislation. Lastly, a considerable number of new problems have been added (and a few old ones removed). This is in keeping with the authors' general philosophy that the reader of this text should be exposed to both the quantitative and qualitative aspects of air pollution control. The data for these new problems are presented in both conventional English units and SI units. SI units are emphasized to a somewhat greater extent than in the first edition. However, a student of air pollution control must be conversant with a large set of diverse engineering units. To minimize this problem, Appendix B contains conversion tables for a number of units commonly used in the field.

It is our intent that the material be suitable for a variety of engineers who wish to gain an introduction to the field of air pollution and its control. An understanding of the fundamentals of thermodynamics is assumed, including some knowledge of chemical equilibrium for ideal-gas mixtures. The required principles of chemical kinetics, important to understand more fully the origin and persistence of numerous pollutants, is presented in the text. A review of some of the basic principles of mass transfer is also given in the text, before presenting the control methods involving absorption.

The new edition continues to present information on four broad areas of interest in the air pollution field: (1) the effects of pollutants on health and welfare, (2) the federal laws and regulations that have been promulgated in an attempt to achieve reasonable ambient air quality, (3) the modeling of atmospheric dispersion of pollutants, and (4) the general and specific approaches to the control of emissions—small- and large-scale, mobile and stationary, combustion and noncombustion. The mechanisms responsible for the effectiveness of a given control device are discussed in some depth. The instrumentation required for the accurate and reliable monitoring of pollutants is covered briefly, since innumerable articles and books are available on the subject in the current literature. Some economic data are presented for comparative purposes. However, inflation has a large effect on construction and operating costs, so a comprehensive list of cost factors is not included.

Similar to the first edition, no attempt has been made to present a complete coverage of the air pollution control field. The space devoted to any particular topic, and the omission of other topics, is mainly a matter of the authors' choice. New methods of control and measurement are constantly being introduced; new or modified laws and regulations are continually being promulgated. Only regular perusal of the current literature will enable one to keep abreast of the developments in the area of air pollution control processes.

Kenneth Wark
Cecil F. Warner

Symbols

LETTER SYMBOLS

A	Cross-sectional area
a	Absorption coefficient for light
	Acceleration
B	Magnetic field intensity
C	Concentration
C_D	Drag coefficient
Coh	Coefficient of haze
c_p	Specific heat at constant pressure
D	Dielectric constant
	Mass diffusivity
D_L	Liquid-phase diffusion coefficient
d	Distance
	Stack diameter
E	Activation energy
	Electric field strength
e	Electric charge
F	Dimensional parameter in plume-rise evaluation
	Force
G	Gas flow rate
g	Gravitational acceleration
g_c	Gravitational constant
H	Effective stack height
	Height of gravity settling chamber
H_t	Height of a transfer unit
h	Actual stack height
	Enthalpy
Δh	Plume rise
I	Light intensity
K	Dust permeability
	Eddy diffusion coefficient
	Mass transfer coefficient
	Scattering ratio
K_c	Cunningham correction factor

K_p	Ideal-gas equilibrium constant
k	Boltzmann's constant
	Mass transfer coefficient
	Rate constant
	Specific heat ratio, c_p/c_v
L	Length
	Liquid flow rate
L_V	Limit of visibility
M	Molar mass (molecular weight)
m	Mass
	Refractive index
MB	million Btu
N	Mass transfer per unit time and unit area
	Number of particles per unit volume
	Number of particles under given size
N_e	Number of effective turns in a cyclone
N_I	Impaction number
N_t	Number of transfer units
n	Exponent in wind velocity profile
	Number of particles per differential size
P	Pressure
	Odor intensity
p	Pressure
Q	Mass emission rate, mass per unit time
	Volume flow rate
Q_h	Heat emission rate, energy per unit time
q	Electric charge
	Emission source strength per unit length
	Quantity of heat transfer
R	Ideal-gas constant
	Radius
	Particulate resistance
Re	Reynolds number, $DV\rho/\mu$
r	Radius (of particle)
S	Filter drag
	Odor concentration
	Separation distance
s	Scattering coefficient for light
T	Temperature
T_{frac}	Fractional light transmission
t	Time
u	Internal energy
	Wind speed
V	Velocity
	Volume
	Voltage

v	Specific volume
W	Mass of adsorbent
	Width
w	Drift velocity
	Logarithm of particle diameter
	Work interaction
X	Liquid-phase mole ratio
x	Distance or coordinate
	Mole fraction
x_S	Stopping distance
Y	Gas-phase mole ratio
y	Distance or direction
	Gas-phase mole fraction
z	Distance or direction

GREEK SYMBOLS

η	Efficiency
θ	Potential temperature
Γ	Adiabatic lapse rate
λ	Wavelength
μm	Microns (micrometers)
μ	Parameter in Gaussian distribution
	Viscosity
ρ	Density
ρ_e	Electrical resistivity
σ	Extinction coefficient
	Standard deviation
ϕ	Angle
	Equivalence ratio
v	Ion velocity

SUBSCRIPTS

a	Atmosphere
C	Carrier gas
G	Gas phase
i	ith species
L	Liquid phase
MM	Mass median
m	Molar value
NM	Number median
OG	Overall gas-phase value
OL	Overall liquid-phase value
S	Solvent
s	Stack
x	Based on mole fraction in liquid phase
y	Based on mole fraction in gas phase

Air Pollution

Chapter 1
Effects and Sources of Air Pollutants

1-1 INTRODUCTION

Air pollution is woven throughout the fabric of our modern life. A by-product of the manner in which we build our cities, air pollution is waste remaining from the ways we produce our goods, transport ourselves and our goods, and generate the energy to heat and light the places where we live, play, and work. The major cause of all air pollution is combustion, and combustion is essential to man. When perfect or theoretical combustion occurs, the hydrogen and carbon in the fuel combine with oxygen from the air to produce heat, light, carbon dioxide, and water vapor. However, impurities in the fuel, poor fuel-to-air ratio, or too high or too low combustion temperatures cause the formation of such side products as carbon monoxide, sulfur oxides, nitrogen oxides, fly ash, and unburned hydrocarbons—all air pollutants.

Air pollution is not a recent phenomenon. King Edward I of England tried to clear the smoky sky over London in 1272 by banning the use of "sea coal." The British Parliament ordered the torturing and hanging of a man who sold and burned the outlawed coal. Under Richard II (1377–1399) and later under Henry V (1413–1422), England took steps to regulate and restrict the use of coal [1]. One of the earliest recorded publications dealing with air pollution is a pamphlet published in 1661 by royal command of Charles II: "Fumifugium; or the Inconvenience of the Air and Smoke of London Dissipated; together with Some Remedies Humbly Proposed," written by John Evelyn, one of the founding members of the Royal Society [1].

The use of coal in the generation of energy was a major factor in the Industrial Revolution, which formed the basis of our current technological society. Unfortunately, intimately associated with the benefits of our technological society is the fouling and degrading of our environment. One of the earliest legal attempts to control air pollution in the United States appears to be an 1895 ordinance making illegal the "showing of visible vapor" as exhaust from steam automobiles.

Such natural processes as forest fires, decaying vegetation, dust storms, and volcanic eruptions have always contaminated the air. Although the total global production of many gases and particulate matter recognized as pollutants is much greater from natural sources than from man-made sources, global distribution and dispersion of those pollutants result in low average concentrations. By precipitation, oxidation, and absorption into the oceans and the soil, the atmosphere can cleanse itself of all known pollutants given sufficient time [2, 3]. On the other hand, man-generated pollutants are usually concentrated in small geographic regions; hence most air pollution is truly man-made. In the United States alone, over 200 million tons of gaseous, solid, and liquid waste products are discharged annually into the atmosphere. Currently the rate at which pollutants are discharged into the atmosphere in highly populated regions at time exceeds the cleansing rate of the atmosphere.

1-2 AIR POLLUTION EPISODES

Although limited air pollution was experienced as early as 1272, it has become a major problem only in relatively recent years, considering man's total history. In December, 1930, a heavily industrialized section of the Meuse Valley, in Belgium, experienced a severe 3-day fog during which hundreds of people became ill and 60 died—more than 10 times the normal number. Shortly afterward, during a thick 9-day fog in January, 1931, 592 people in Manchester and Salford area of England died—again a large jump in the death rate. In 1948, in Donora, Pennsylvania, a small mill town dominated by steel and chemical plants, a 4-day fog made almost half of the 14,000 inhabitants sick. Twenty persons died. Ten years later, Donora residents who had been acutely ill during that episode were found to have a higher rate of sickness and to die at an earlier age than the average for all the townspeople. During a fog in London as far back as 1873, 268 unexpected deaths from bronchitis were reported. It was not until a great fog blanketed London in 1952 that the sinister potential of air pollution became fully apparent. That fog lasted from December 5 to December 8, and 10 days later it was learned that the total number of deaths in Greater London duing that period exceeded the average by 4000. The statistics indicated that almost all those who died unexpectedly had records of bronchitis, emphysema, or heart trouble, and that people in the last category were most vulnerable. Again in January, 1956, 1000 extra deaths in London were blamed on an extended fog. In that year, Parliament passed a Clean Air Act and Britain embarked on a program to reduce the burning of soft coal [4]. The smog conditions of Los Angeles, New York City, Chicago, and other large cities of the United States are widely publicized in today's press.

The misuse of air resources in the USSR is not very different from that in this country. Despite the fact that Russia's current yearly production of cars is one-tenth that of the United States, most Soviet cities experience varying degrees of air pollution. Cities situated in valleys or hilly regions are especially likely to experience dangerous levels of air pollution. In the hilly cities of Armenia, for example, the established health standards for carbon monoxide are often exceeded. Similarly, Magnitogorsk, Alma Ato, and Chelyabinsk, with their metallurgical industries, are frequently covered with a layer of dark blue haze. Like Los Angeles, Tbilisi, the capital of the Republic of Georgia, has smog almost 6 months of the year. Leningrad has 40 percent fewer clear daylight hours than the nearby town of Pavlovsk [5].

1-3 GENERAL NATURE OF AIR POLLUTION PROBLEMS

Only a finite amount of air, land, and water resources exist, and as population increases, the portion available for each person decreases. From the beginning of time until 1900, the population of the world increased to 1.7 billion. By 1974 world population had reached 3.9 billion, and the awesome figure of 7 billion is estimated by the year 2000. The population of the United States has followed a similar trend. In addition, technological advances in the field of agriculture have greatly reduced the number of jobs in rural areas. As in other developed countries, today two-thirds of the population lives in urban areas comprising about 1 percent of the land. Suburban growth and superhighways have made it possible for more people to travel greater distances and thus to converge faster on our cities. Hence an increasing population combined with a high standard of living has led to a drastically intensified output and concentration of air pollutants in localized areas.

COUNTRY	GROSS NATIONAL PRODUCT ($/CAPITA)	COMMERCIAL ENERGY CONSUMPTION (MILLION Btu/CAPITA)	GNP / ENERGY
India	50	5	10
Chile	450	20	22
Japan	600	30	20
USSR	900	70	13
West Germany	1450	85	17
United Kingdom	1500	117	13
Canada	1900	130	15
United States	2850	180	16

During a recent time period in which the population of the United States doubled, there was an eightfold increase in the gross national product (GNP) and an increase in the production of electrical energy by a factor of 13. For developed or developing countries, there appears to be a close relationship between the gross national product per capita (economic level) and energy consumption per capita. This is illustrated by data for selected countries taken from Cook's study of the flow of energy in an industrialized society [6]. All values in the table (see page 3) are approximate.

These data indicate that the quantity of goods and services enjoyed by a citizen is closely related to the quantity of energy consumed (directly or indirectly) by that citizen. In other words, the availability and use of energy is a requisite for a high standard of living. It is striking to translate the American experience into global terms. An increase in energy demand per capita in the developing countries, similar to that experienced in West Germany and the United States, in combination with the increase in global population could result in uncontrolled emissions of air pollutants in catastrophic proportions.

In the past, industry, agriculture, and individual polluters have found it more economical to discharge waste products into the atmosphere than to exercise waste control. In general the organization or activity causing the pollution did not suffer the consequences of the pollution; likewise, those who benefited from a reduction in air pollution resulting from the installation of control equipment did not directly bear the cost of the equipment. In recent years, as the public has become increasingly concerned with environmental problems, air has come to be regarded as a resource within the public domain. Hence air pollution is considered a public problem, a concern not only of those who discharge the pollution, but also of those who may suffer as a result. Thus the laws in some countries now permit an individual or group of private individuals to sue directly an organization or company which is polluting that particular part of the public domain.

The rational control of air pollution rests on four basic assumptions [7].

1. *Air is in the public domain*. Such an assumption is necessary if air pollution is to be treated as a public problem, of concern not only to those who discharge the pollution but also to those who may suffer as a result.

2. *Air pollution is an inevitable concomitant of modern life*. There is a conflict between man's economic and biologic concerns; in the past, this conflict was recognized only after air pollution disasters. We need a systematic development of policies and programs to conserve the atmosphere for its most essential biological function.

3. *Scientific knowledge can be applied to the shaping of public policy*. Information about the sources and effects of air pollution is far from complete, and a great deal of work must be done to develop control devices and methods. Nevertheless, sufficient information is now available to make possible substantial reductions in air pollution levels. Man does not have to abandon either his technology or his life, but he must use his knowledge.

4. *Methods of reducing air pollution must not increase pollution in other sectors of the environment.* Some industries reduce wastes in the air by dissolving them in water and by pouring the polluted water into streams. For example, one proposal to reduce the sulfur dioxide emitted by coal-burning electrical power plants results in the formation of large quantities of either solid or liquid wastes. Such methods are not true solutions to air pollution problems.

1-4 DEFINITION AND GENERAL LISTING OF AIR POLLUTANTS

Before enforceable laws and ordinances can be formulated to control the pollution of the air, the term *air pollution* must be defined. Many definitions have been proposed. One such definition is the following: "Air pollution may be defined as the presence in the outdoor atmosphere of one or more contaminants or combinations thereof in such quantities and of such duration as may be or may tend to be injurious to human, plant, or animal life, or property or which unreasonably interferes with the comfortable enjoyment of life or property or the conduct of business." In a similar vein, the laws of the state of Wisconsin define air pollution as "... the presence in the atmosphere of one or more contaminants in such quantities and of such duration as is or tends to be injurious to human health or welfare, animal or plant life, or property, or would unreasonably interfere with the enjoyment of life or property." Every state in the Union has a similar definition. The Wisconsin law further defines contaminants as "... dust, fumes, mist, liquid, smoke, other particulate matter, vapor, gas, odorous substances, or any combination thereof but shall not include uncombined water vapor."

One method of defining an air pollutant is first to specify the composition of "clean" or "normal" dry atmospheric air and then to classify all other materials or increased amounts of those materials given in the composition of atmospheric air as pollutants if their presence results in damage to human beings, plants, animals, or materials. Table 1-1 lists the chemical composition of dry atmospheric air typically found in rural areas or over the ocean far from land masses. Atmospheric air also contains from 1 to 3 percent by volume water vapor and traces of sulfur dioxide, formaldehyde, iodine, sodium chloride, ammonia, carbon monoxide, methane, and some dust and pollen. Thus, according to the definition presented above, carbon monoxide, hydrocarbon vapors, or ozone in a concentration of greater than 0.04 ppm would be air pollutants.

At the present time neither carbon dioxide nor uncombined water vapor is considered to be a pollutant. This condition could change in the future since the discharge of either substance into the atmosphere in increased quantities may result in a significant change in the global atmospheric temperature. Likewise, certain odors now considered nuisances rather than pollutants may eventually be designated pollutants.

Table 1-1 CHEMICAL COMPOSITION OF DRY ATMOSPHERIC AIR

SUBSTANCE	VOLUME (PERCENT)	CONCENTRATION (PPM)[a]
Nitrogen	78.084 ± 0.004	780,900
Oxygen	20.946 ± 0.002	209,400
Argon	0.934 ± 0.001	9,300
Carbon dioxide	0.033 ± 0.001	315
Neon		18
Helium		5.2
Methane		1.2
Krypton		0.5
Hydrogen		0.5
Xenon		0.08
Nitrogen dioxide		0.02
Ozone		0.01–0.04

SOURCE: *Handbook of Air Pollution*, PHS Publication AP-44 (PB 190-247), 1968 [40].
[a] ppm is an abbreviation of parts per million.

It is common practice to express the quantity of a gaseous pollutant present in the air as parts per million (ppm). Thus

$$\frac{1 \text{ volume of gaseous pollutant}}{10^6 \text{ volumes (pollutant + air)}} = 1 \text{ ppm}$$

$$0.0001 \text{ percent by volume} = 1 \text{ ppm}$$

The mass of a pollutant is expressed as micrograms of pollutant per cubic meter of air. Symbolically,

$$\frac{\text{micrograms}}{\text{cubic meter}} = \mu\text{g/m}^3$$

At 25°C and 101.3 kPa (1 atm) pressure the relationship between parts per million and micrograms per cubic meter is found from

$$\frac{m_{\text{pol}}}{V_{\text{air}}} = \frac{\rho_{\text{pol}} V_{\text{pol}}}{V_{\text{air}}} \frac{P M_{\text{pol}}}{\rho_{\text{pol}} R_u T} = \frac{V_{\text{pol}}}{V_{\text{air}}} \frac{P M_{\text{pol}}}{R_u T}$$

where the pollutant gas is assumed to be an ideal gas, and M_{pol} is the molar mass of the pollutant. If P is taken as 1 atm, T as 298°K, and R_u as 0.08208 atm·m^3/kg mol·°K, then the equation reduces to

$$\frac{m_{\text{pol}}}{V_{\text{air}}} = \frac{V_{\text{pol}}}{V_{\text{air}}} \frac{M_{\text{pol}}}{24.5}$$

where the mass of pollutant per unit volume is now expressed as kg/m^3. Finally, by multiplying the right side by 10^9 to convert the mass to micrograms, and by dividing by 10^6 so that $V_{\text{pol}}/V_{\text{air}}$ can be expressed as parts per million, then the basic relation between $\mu g/m^3$ and ppm at 1 atm and 25°C is

$$\mu g/m^3 = \frac{\text{ppm} \times \text{molecular weight}}{24.5}(10^3) \tag{1-1}$$

For conditions of 1 atm and 0°C (273°K) the constant in the denominator becomes 22.41. Sometimes concentrations are also expressed in parts per billion (ppb) or parts per hundred million (pphm).

The following examples illustrate the calculations required for converting from one system of units to another.

Example 1-1

The exhaust gas from an automobile contains 1.5 percent by volume of carbon monoxide. What is the concentration of CO in milligrams per cubic meter (mg/m^3) at 25°C and 1 atm pressure?

SOLUTION

The conversion of data from percent by volume to parts per million is made by noting that

$$1 \text{ percent by volume} = 1 \text{ percent by moles} = 10^4 \text{ ppm}$$

Hence 1.5 percent by volume is 15,000 ppm of CO. In addition, the molar mass (molecular weight) of CO is 28.0. Substitution of these data into Equation (1-1) yields

$$\mu g/m^3 = \frac{15{,}000\ (28.0)}{24.5} \times 10^3 = 17.1 \times 10^6$$

There are 10^3 micrograms (μg) in 1 milligram (mg). Therefore the concentration of CO in the desired units is

$$\text{CO concentration} = \frac{17.1 \times 10^6(\mu g/m^3)}{10^3(\mu g/mg)} = 17.1 \times 10^3 \text{ mg/m}^3$$

The CO concentration will be greatly reduced when the exhaust gas enters the atmosphere. In contrast to the above value, the federal standard for the ambient atmosphere (based on an 8-hr measurement) is only 10 mg/m^3.

Example 1-2

The average daily concentration of sulfur dioxide is observed to be 415 $\mu g/m^3$ at 25°C and 1 atm at a given locale. What is the concentration of SO_2 in parts per million?

SOLUTION

The molar mass of sulfur dioxide is 64.0. Substitution into Equation (1-1) gives

$$415\ \mu g/m^3 = \frac{(ppm)(64.0)(10^3)}{24.5}$$

Solving for the concentration of SO_2 in parts per million, we find that

$$ppm = \frac{415(24.5)}{64.0 \times 10^3} = 0.159$$

This concentration is above the federal primary ambient standard of 0.14 ppm (based on a 24-hr average value) presented in Table 2-1.

Example 1-3

The ozone concentration is observed to be 118 $\mu g/m^3$ at an urban monitoring station. Does this value exceed the concentration found in standard dry air?

SOLUTION

The molar mass of ozone (O_3) is 48.0. Substituting the appropriate values into Equation (1-1) and solving for the concentration in parts per million, we find that

$$ppm = \frac{118(24.5)}{48(10^3)} = 0.060$$

The concentration of ozone in standard dry air as given in Table 1-1 ranges from 0.01 to 0.04 ppm. Thus the value of 118 $\mu g/m^3$, which corresponds to 0.060 ppm, exceeds the tabulated value.

Example 1-4

The maximum concentration of dust in the atmosphere during a 24-hr period was found to be 0.000130 grain per dry standard cubic foot (gr/dscf). What is the equivalent concentration in micrograms per cubic meter?

SOLUTION

From Appendix B it is found that 1 gr/ft^3 equals 2.29 g/m^3. Hence,

$$\text{concentration} = 0.000130\ gr/ft^3 \left(\frac{2.29\ g/m^3}{1\ gr/ft^3}\right)\left(\frac{10^6\ \mu g}{g}\right)$$

$$= 298\ \mu g/m^3$$

This value is relatively high, and exceeds the 24-hr federal ambient air quality standard of 260 $\mu g/m^3$ for the same time period of measurement for particulate (as shown in Table 2-1).

A general classification of air pollutants is as follows:

1. Particulate matter.
2. Sulfur-containing compounds.
3. Organic compounds.
4. Nitrogen-containing compounds.
5. Carbon monoxide.
6. Halogen compounds.
7. Radioactive compounds.

Particulate matter is frequently divided into subclasses which include fine dust (less than 100 μm in diameter), coarse dust (above 100 μm in diameter), fumes (0.001–1 μm in diameter), and mists (0.1–10 μm in diameter). Fumes are particles formed by condensation, sublimation, or chemical reaction, and sometimes are designated as smoke. Mists are comprised of liquid particles formed by condensation and are fairly large in diameter compared to fumes or smoke. As a larger class, fumes, smoke, mist, and fog are called aerosols.

Items 2 through 6 in the above general list can be grouped into two broad classifications: primary and secondary pollutants. Primary pollutants are those emitted directly from sources, while secondary ones are those formed in the atmosphere by chemical reactions among primary pollutants and chemical species normally found in the atmosphere. Table 1-2 lists some primary and secondary pollutants for classes of substances. Note that carbon dioxide has been placed in parentheses in the table, since CO_2 normally is not considered a pollutant. However, the increase in CO_2 concentration worldwide is the basis for concern regarding its eventual effect. Particulate matter, carbon monoxide, sulfur oxides, oxides of nitrogen, hydrocarbons, photochemical oxidants, asbestos, beryllium, and mercury will be considered in some detail in the following sections.

1-5 PARTICULATE MATTER

Particulate is a term employed to describe dispersed airborne solid and liquid particles larger than single molecules (molecules are approximately

Table 1-2 GENERAL CLASSIFICATION OF GASEOUS AIR POLLUTANTS

CLASS	PRIMARY POLLUTANTS	SECONDARY POLLUTANTS
Sulfur-containing compounds	SO_2, H_2S	SO_3, H_2SO_4, MSO_4 [a]
Organic compounds	C_1–C_5 compounds	Ketones, aldehydes, acids
Nitrogen-containing compounds	NO, NH_3	NO_2, MNO_3 [a]
Oxides of carbon	CO, (CO_2)	(None)
Halogen	HCl, HF	(None)

[a] MSO_4 and MNO_3 designate sulfate and nitrate compounds, respectively.

0.0002 μm in diameter) but smaller than 500 μm (1 μm = 1 micron = 10^{-4} cm). Particles in this size range have a lifetime in suspension varying from a few seconds to several months. Particles less than 0.1 μm undergo random Brownian motions resulting from collision with individual molecules. Particles between 0.1 and 1 μm have settling velocities in still air which, while finite, are small compared with wind velocities. Particles larger than 1 μm have significant but small settling velocities. Particles above approximately 20 μm have large settling velocities and are removed from the air by gravity and other inertial processes. Equations for settling velocities are discussed in more detail in Section 5-5.

A portion of the particles introduced into the atmosphere by man's activity serve as condensation nuclei that influence the formation of clouds, rain, and snow. Some cities display a definite minimum rainfall on Sundays, when particulate concentrations are the lowest. It has been observed that the precipitation at La Porte, Indiana, 30 mi east of Chicago, is significantly larger than for South Bend and Valparaiso, Indiana. These latter two cities are considerably farther east of the region of heavy particulate emissions along the southern shore of Lake Michigan.

1-5-A Visibility and Related Atmospheric Characteristics

One of the most common effects of air pollution is the reduction in visibility resulting from the absorption and scattering of light by airborne liquid and solid materials. Visibility is principally affected by particles that are formed in the atmosphere from gas-phase reactions. Although not visible, carbon dioxide, water vapor, and ozone in increased concentrations change the absorption and transmission characteristics of the atmosphere. Much effort has been expended in recent years to relate the concentration of airborne pollutants to the optical characteristics of various plumes in the atmosphere. Reduction in visibility not only is unpleasing to an individual, but also may have strong psychological effects. In addition, certain safety hazards arise.

SCATTERING AND ABSORPTION OF LIGHT

The *Glossary of Meteorology* [8] gives the following definition of visibility:

> In United States weather observing practice, the greatest distance in a given direction at which it is just possible to see and identify with the unaided eye (a) in the daytime, a prominent dark object against the sky at the horizon, and (b) at night, a known, preferably unfocused moderately intense light source. After visibilities have been determined around the entire horizon circle, they are resolved into a single average value of prevailing visibility for reporting purposes. Visibility is dependent upon both the ability of the eye to distinguish an object because it is in contrast with the background, and the transmission of light through the atmosphere. Changes in contrast with viewing distance occur because (a) additional light is introduced into the sight path or (b) light is lost from the line of sight because of the atmosphere. In both cases the contrast

between the object and the background approaches zero. When the eye can no longer distinguish a difference between the object and the background, the object is said to be beyond the limit of visibility. The aforementioned alteration in contrast is due to the absorption and scattering of light by particles in the atmosphere.

If air molecules were the only factor in the attenuation of light, a visibility of more than 150 mi could be calculated based upon the Rayleigh scattering theory. Smoke in Indianapolis often reduces airport visibility to less than 3 mi and sometimes to nearly 1 mi [9]. According to Blacktin [10], dust particles in a concentration of $2000/\text{cm}^3$ can obscure a mountain at 50 mi, while a concentration of $100{,}000/\text{cm}^3$ can reduce visibility to 1 mi. A concentration of NO_2 of 8 to 10 ppm would probably reduce visibility to about 1 mi. Thus light attenuation by other than Rayleigh scattering must occur. Because of its absorption characteristics, NO_2 causes the sky to appear brownish in color in addition to reducing visibility. A concentration of 0.1 ppm NO_2 would probably not be noticeable, whereas 1.0 ppm could be detected by the eye. The state of California ambient air quality standard for NO_2 was set at 0.25 ppm primarily on the basis of coloration effects.

Consider an object illuminated by a light beam of intensity I, a distance x from an observer. Let the light passing through an incremental distance dx be reduced by absorption and scattering by an amount dI which is proportional to the intensity I. Middleton [11] has shown that

$$dI = -\sigma I \, dx \tag{1-2}$$

where σ is the extinction coefficient and the negative sign indicates that the intensity is reduced along the light path by the atmosphere. Integration over the path length from 0 to d gives (for constant σ)

$$I = I_o \exp(-\sigma d) \tag{1-3}$$

where I is the intensity at d and I_o is the original intensity at $x = 0$.

The extinction coefficient σ includes the effects of both scattering and absorption by gas molecules and aerosols. It is often desirable to separate the effects of scattering and absorption, in which case Equation (1-3) becomes

$$I = I_o \exp[-(a + s)d] \tag{1-4}$$

where a is the absorption coefficient and s is the scattering coefficient.

Thus the fractional transmittance, I/I_o, of light is a function of both the path length d and the interception qualities of the particles. The values of the extinction coefficients are functions of the wavelength of the incident light, particle size and shape, and the optical properties of the particulate. In addition, I/I_o is a strong function of the scattering angle, that is, the angle between the direction of the light and the line of sight of the observer.

The attenuation of light in the atmosphere by scattering is due to particles of a size comparable to the wavelength of the incident light. This is called Mie scattering and is the phenomenon largely responsible for the reduction of atmospheric visibility. Visible solar radiation falls into the range from 0.4 to 0.8 μm, roughly, with a maximum intensity around 0.52 μm. Hence solid and liquid particles in the submicron range from 0.1 to 1 μm are responsible for the decrease in visibility. Interestingly enough, particulate in this same range invades the lungs and can lead to severe health problems. Mie (in Middleton [11]) developed the following expression for the coefficient of scattering:

$$s = NK\pi r^2 \tag{1-5}$$

where N is the number of particles of radius r per unit volume, and K is the scattering area ratio, or the ratio of area of wave front acted upon by the particle to the area of the particle.

Values of K for selected particulate may be found in references 12 and 13. When the particulate matter in the light path is not homogeneous, the value of s is the sum of the individual values of s for the various groups of particles. Thus, for forward scattering,

$$s = \sum_{i=1}^{n} N_i K_i \pi r_i^2 \tag{1-6}$$

For spheres that do not absorb but scatter light, Van de Hulst [13] presents a curve relating K to the term $4\pi r(m-1)/\lambda$, where m is the refraction index of the particulate, r is the particle radius, and λ is the wavelength of the incident light. Figure 1-1 is a reproduction of the data of Van de Hulst. By summarizing a number of measurements of industrial hazes in England, Middleton [11] concluded that the absorption coefficient a is more or less equal in value to the scattering coefficient s.

A relationship between the concentration of particulate in the atmosphere and the visibility can be developed as follows. It is generally assumed that the lower limit of visibility for most people occurs when the light intensity (or contrast) is reduced to 2 percent of the unattenuated light

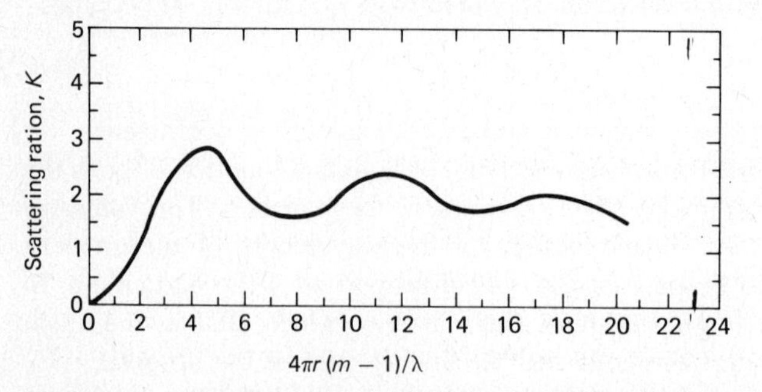

Figure 1-1 Approximate value of the scattering ratio K for nonabsorbing spheres.

beam. (Some experiments show this to be closer to 5 percent.) On this basis the distance d in Equation (1-3) becomes identical to the visibility L_V when 0.02 is substituted for the quantity I/I_o. Hence,

$$\ln\left(\frac{I}{I_o}\right) = -\sigma d$$

$$\ln(0.02) = -\sigma L_V$$

$$L_V = \frac{3.9}{\sigma} \tag{1-7}$$

Note that L_V and σ must be in compatible units. In the preceding derivation it is assumed that the particle density between source and observer is uniform.

If it is further assumed that attenuation is due to scattering only and that the particulate matter is in the form of spheres of the same size and uniformly distributed, an approximation equation for the visibility can be formulated by combining Equations (1-5) and (1-7). As a result,

$$L_V = \frac{5.2\rho_p r}{KC} \tag{1-8}$$

where C is the particulate concentration in the direction of sight, ρ_p is the particle density, and r is its radius. A self-consistent set of units must be used. This expression indicates that visibility is inversely proportional to the scattering coefficient and the concentration of particles in the atmosphere, and in direct proportion to the diameter and specific gravity of the particulate. The greatest range in possible values occurs with the variables C and r. The visibility is greatly affected by concentrations greater than 100 $\mu g/m^3$ and by particles in the 0.1 to 1.0 μm size. Unfortunately, as discussed in Chapter 5, many particulate removal devices are most effective for particle sizes greater than 1 μm. The particulate released to the atmosphere may be small in percent of the overall weight originally in the stack gas before cleaning, but its size is very effective in reducing visibility. An empirical relation [14] between the visual range and the concentration is

$$L_V \approx \frac{1.2 \times 10^3}{C} \tag{1-9}$$

where C is in micrograms per cubic meter and L_V is in kilometers. Data indicate that the actual value of L_V may be a factor of 2 larger or smaller than that calculated from the above expression. However, as a rule of thumb, the product of visibility and concentration in the units expressed for Equation (1-9) is around 1000. Equation (1-9) is an estimate based on typical atmospheric particulate conditions, while Equation (1-8) is the visibility for a given particle size or an appropriate mean particle size. Both equations may be in considerable error when the relative humidity exceeds 70 percent.

Above this value particles tend to develop into fog droplets as a result of hygroscopic tendencies of particulate under high relative humidity conditions. Other empirical relations for scattering coefficients or visibility limits appear in the literature [15].

Example 1-5

Consider oil droplets of 0.6-μm diameter suspended in air and exposed to daytime radiation. The density of the particles is taken to be 0.90 g/cm^3. (a) What is the concentration of particles in micrograms per cubic meter for a visibility of 1 mi, if the value of K, the scattering ratio, is 4.1? (b) What is the concentration of suspended particles with a density of 2.5 g/cm^3 and an effective diameter of 1.0 μm, if the K-value is 2.0 and the visibility is reduced to 5.0 mi?

SOLUTION

(a) The value of K chosen is reasonably large since the particle diameter is quite close to the midrange of visible radiation. Peak emission occurs at 0.52 μm for solar radiation. The concentration of suspended particles necessary for a visibility of 1.0 mi can be determined from Equation (1-8). Substitution of the above values yields

$$C = \frac{5.2\,\rho_p r}{KL_V} = \frac{5.2(0.9)(0.3)}{4.1(1610)} = 2.1 \times 10^{-4}\,\text{g/m}^3 = 210\,\mu\text{g/m}^3$$

In order to achieve a consistent set of units, the distance of 1 mi for L_V has been replaced by 1610 m. (b) The solution in this case is attained in the same fashion as above. Upon substitution of the appropriate data,

$$C = \frac{5.2(2.5)(0.5)}{2(8050)} = 4.04 \times 10^{-4}\,\text{g/m}^3 = 404\,\mu\text{g/m}^3$$

These estimated concentrations required to achieve certain visibilities are fairly typical of dust or aerosol loadings (*loading* is the mass per unit volume of particulate) found in the atmosphere.

COEFFICIENT OF HAZE

The ratio I/I_o discussed in the preceding subsection is one measure of the fractional transmittance of light through the atmosphere. It is a function of the distance of transmission as well as the scattering and absorption effects of the suspended particles. Another measure of reduced light transmission or the possibility of increased soiling due to atmospheric particulate is a unit called the *coefficient of haze*, abbreviated Coh. It is determined by collecting particulate on a clean filter paper and then measuring the decrease in light transmission. The method is standardized according to American Society for Testing and Materials (ASTM) Standard D 1704-16, and the result is expressed as Coh units per 1000 lineal feet of air column.

The Coh unit is measured in terms of the fractional transmittance of light, I/I_o, through the dirty filter paper. By definition,

$$\text{Coh} = 100 \log_{10}\left(\frac{I_o}{I}\right) \tag{1-10}$$

For clean air the value of Coh is 0. It may be shown that 1 Coh unit is associated with a fractional transmittance of 0.977, and 10 Coh units represent a fractional transmittance of 0.794. Hence larger values of the coefficient of haze imply decreased light transmission.

The Coh unit itself is impractical since its value depends upon the quantity of particulate collected on the filter. Increased filtering times lead to increased opacity. To overcome this, the measurement is standardized by reporting the Coh units per equivalent 1000 lineal feet of sample. The actual number of lineal feet of sample during a test equals the volume of the sample drawn through the filter divided by the cross-sectional area of the filter, in appropriate units. Assuming that a constant flow rate is maintained during the test, the actual lineal feet also equal the flow velocity times the time during which particulate is collected. That is,

$$\begin{aligned}\text{number of 1000-ft lengths of sample} &= \frac{\text{ft}^3 \text{ of sample}}{\text{ft}^2 \text{ of filter area}}\,\frac{1}{1000} \\ &= \text{gas velocity (ft/s)} \times \text{time (s)} \times \frac{1}{1000}\end{aligned} \tag{1-11}$$

By combining Equations (1-10) and (1-11) we obtain the relation

$$\begin{aligned}\frac{\text{Coh}}{1000 \text{ ft}} &= \frac{10^5(\text{area of filter})}{\text{volume of sample}}\log_{10}(T_{\text{frac}})^{-1} \\ &= \frac{10^5}{(\text{gas velocity})(\text{time})}\log_{10}(T_{\text{frac}})^{-1}\end{aligned} \tag{1-12}$$

In this last equation the area, volume, velocity, and time must be expressed in square feet, cubic feet, feet per second, and seconds, respectively. A similar equation would be valid if one wished, for example, the Coh units per 1000 m.

Example 1-6

Consider a 70 percent light transmittance after air has passed through a filter at 1.50 ft/s for 1.5 hr. Determine the Coh units per 1000 ft.

SOLUTION

On the basis of Equation (1-12) we find that

$$\frac{\text{Coh}}{1000\text{ ft}} = \frac{10^5}{1.5(1.5)(3600)}\log_{10}\frac{1}{0.70} = 1.91$$

This value represents a moderate amount of air pollution.

Based on experience, the following table provides some rough guidelines for values of Coh/1000 ft versus the degree of air pollution. The reason

Coh/1000 ft	DEGREE OF AIR POLLUTION
0–0.9	Light
1–1.9	Moderate
2–2.9	Heavy
3–3.9	Very heavy
4–4.9	Extremely heavy

these are only rough guidelines can be explained by some further evidence. During one 24-hr period in Las Vegas, Nevada, the particulate loading in the atmosphere due to blowing dust was around 690 $\mu g/m^3$, and the maximum Coh was 2.0. In Yakima, Washington, at some other time the dust loading was 111 $\mu g/m^3$, but the Coh value was greater than 10. In this latter case a great deal of carbonaceous material was present. In comparison, there is an inverse correlation. This discrepancy is due to the two distinct types of particulate present in the two cases. More fundamentally, optical transmittance is a function of particle size and shape, and not of mass.

ATMOSPHERIC ABSORPTION

In addition to a reduction in visibility caused by particulate air pollutants, certain gaseous substances, because of their radiation absorption characteristics, have the potential for causing a deterioration in our environment and in the future may come to be classified as pollutants. The radiation bands for water vapor and carbon dioxide are presented in Figure 1-2, in which the wavelength is given in microns. The wavelength range for visible radiation is from 0.4 to 0.75 μm, and from 1 to 100 μm for infrared. During the normal diurnal cycle solar radiation (primarily short wavelength) passes through the transparent atmosphere and is absorbed by the earth's surface areas. The heated surface areas lose energy by radiating in the infrared. When the atmosphere contains low concentrations of carbon dioxide and water vapor, the quantity of incoming solar energy is approximately equal to the outgoing terrestrial radiation, and the equilibrium temperature of the earth's surface and lower atmosphere is thereby established. With an increase in concentration of either water vapor or carbon dioxide, the atmospheric absorption of the infrared is increased whereas the transmission of short-wave radiation is unchanged. The net result, it has been argued, would be an increase in the average temperature of the earth's surface and lower atmosphere since the quantity of solar radiation during the day would be unchanged while the terrestrial radiation would be reduced. Several authors have predicted that if the current rate of increase in concentration of carbon dioxide in the atmosphere continues, by the year 2000 the surface temperature will have increased approximately 4°F. Further speculation indicates that sufficient ice would melt as a result of the increase in temperature to cause an increase of 60 ft in the level of the oceans. On the basis of the preceding argument, it has been proposed that both water vapor and carbon dioxide should be considered air pollutants.

For the sake of argument it should also be pointed out that increased concentrations of carbon dioxide have been found to stimulate the growth of selected plants. Since plants absorb CO_2 from the atmosphere in their life cycle, the increased growth could cause a corresponding increase in CO_2 absorption, and a new static balance of the CO_2 content in the environment would be established. The new concentration of CO_2 might not be greatly different from the current level.

Another factor affecting the average temperature of the earth is the absorption and reflection of solar radiation caused by particulate matter in the atmosphere. The results of certain studies seem to indicate that the average temperature of the lower atmosphere is dropping and that the reduction in solar energy reaching the surface because of the absorption by particulate is the cause. Thus it can be argued that rather than causing an increase in atmospheric temperature, increased air pollution will cause a reduction in temperature and a return of the ice age.

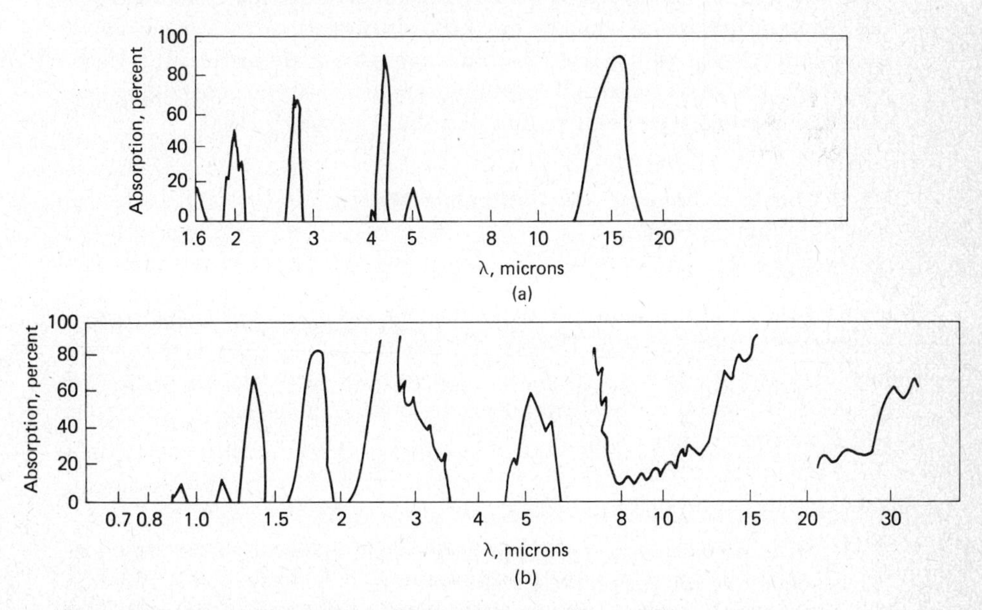

Figure 1-2 Absorption bands of (a) carbon dioxide and (b) water vapor.

1-5-B Effects of Atmospheric Particulate Matter on Materials, Vegetation, and Animals

Airborne particles can be chemically inert or chemically active of themselves; they may be inert but absorb chemically active substances from the atmosphere, or they may combine to form chemically active species. Depending upon its chemical composition and physical state, particulate matter causes wide damage to materials. Particles will soil painted surfaces, clothing, and curtains merely by settling on them. The cost of repainting interior and exterior surfaces of homes and buildings and of cleaning or replacing textiles is estimated at hundreds of millions of dollars annually.

More importantly, particulate matter can cause direct chemical damage either by intrinsic corrosiveness or by the action of corrosive chemicals absorbed or adsorbed, by inert particles emitted into the atmosphere. Metals ordinarily can resist corrosion in dry air alone or even in clean moist air. However, hydroscopic particles commonly found in the atmosphere can corrode metal surfaces with no other pollutants present. Examples of the corrosion of metallic surfaces exposed to industrial atmospheres are well documented [16] and will not be listed here.

Little is known of the effects of particulate matter in general on vegetation. However, several specific substances have been observed to cause some damage. Particles containing fluorides appear to cause some plant damage, and magnesium oxide falling on agricultural soils has resulted in poor plant growth. An animal's health may suffer when the animal feeds on plants covered by toxic particulate. Such toxic compounds may be absorbed into the plant tissues or may remain as a surface contaminant of the plants. Fluorosis in animals has been attributed to their ingestion of vegetation covered with a fluoride-containing particulate matter [16]. Cattle and sheep that have ingested vegetation on which arsenic-containing particles have settled have been victims of arsenic poisoning [16].

1-5-C Effects of Atmospheric Particulate Matter on Human Health

Particulate matter alone or in combination with other pollutants constitutes a very serious health hazard. The pollutants enter the human body mainly via the respiratory system. Damage to the respiratory organs may follow directly, since it has been estimated that over 50 percent of the particles between 0.01 and 0.1 μm which penetrate into the pulmonary compartment will be deposited there [17].

Particulate matter may exert a toxic effect in one or more of the following three ways [16]:

1. The particle may be intrinsically toxic because of its inherent chemical and/or physical characteristics.

2. The particle may interfere with one or more of the mechanisms which normally clear the respiratory tract.
3. The particle may act as a carrier of an absorbed toxic substance.

It is extremely difficult to obtain a direct relationship between exposure to various concentrations of particulate matter and the resulting effects upon human health. The length of time of exposure is important. In some cases it has been observed that exposure to particulate in combination with other pollutants such as SO_2 produces more severe health deterioration than does exposure to each pollutant separately. It is also difficult to reproduce in the laboratory the exact conditions that prevail in the ambient atmosphere.

To date, we have had to rely upon the statistical analysis of such data as increases in hospital admissions, clinic admissions, absences from work and school, and mortality, and the limited data pertaining to the measured concentrations of atmospheric pollutants that prevailed during the subject time periods. Such data do indicate a relationship between increases in particulate concentrations and rises in the number of clinic and hospital visits for upper respiratory infections, cardiac diseases, bronchitis, asthma, pneumonia, emphysema, and the like. Deaths of elderly persons afflicted with respiratory diseases and cardiac conditions also show an increase during

Table 1-3 OBSERVED EFFECTS OF PARTICULATE

CONCENTRATION	MEASUREMENT TIME	EFFECTS
60–180 $\mu g/m^3$	Annual geometric mean with SO_2 and moisture	Acceleration of corrosion of steel and zinc panels
75 $\mu g/m^3$	Annual mean	Ambient air quality standard
150 $\mu g/m^3$	Relative humidity less than 70 percent	Visibility reduced to 5 mi
100–150 $\mu g/m^3$		Direct sunlight reduced one-third
80–100 $\mu g/m^3$	With sulfation levels of 30 $mg/cm^2/month$	Increased death rate of persons over 50 may occur
100–130 $\mu g/m^3$	With $SO_2 > 120\ \mu g/m^3$	Children likely to experience increased incidence of respiratory disease
200 $\mu g/m^3$	24-hr average and $SO_2 > 250\ \mu g/m^3$	Illness of industrial workers may cause an increase in absences from work
260 $\mu g/m^3$	Maximum once in 24 hr	Ambient air quality standard
300 $\mu g/m^3$	24-hr maximum and $SO_2 > 630\ \mu g/m^3$	Chronic bronchitis patients will be likely to suffer acute worsening of symptoms
750 $\mu g/m^3$	24-hr average and $SO_2 > 715\ \mu g/m^3$	Excessive number of deaths and considerable increase in illness may occur

SOURCE: Abridged from data presented in National Air Pollution Control Administration, *Air Quality Criteria for Particulate Matter*, AP-49. Washington, D.C.: HEW, 1969.

periods when the concentration of particulate matter is unusually high for several days. A growing volume of evidence indicates that much of the particulate matter in the atmosphere is carcinogenic in nature [16], especially when combined with cigarette smoking.

A limited quantity of data on the relation between particulate concentration and effects produced is presented in Table 1-3 for illustrative purposes. Also listed in Table 1-3 is the primary air quality standard for particulate as established by the federal government. Presented for comparative purposes, it is the maximum value of particulate concentration permitted in the ambient atmosphere. Values above this are considered to be detrimental to human health.

1-6 CARBON MONOXIDE

Carbon monoxide is a colorless and odorless gas. It is very stable and has a lifetime of 2 to 4 months in the atmosphere. The global emissions of carbon monoxide are large (350 million tons/yr in 1968), of which roughly 20 percent is from man-made sources. Such a quantity would result in an increase of around 0.03 ppm/yr in the ambient concentration. This increase is not observed. Soil fungi [2] may remove a significant amount of the quantity released, and it is generally assumed that CO is oxidized to CO_2 in the atmosphere, although the rate is quite slow. There is some evidence that CO may be chemically active in smog formation.

1-6-A Effects of Carbon Monoxide on Materials and Plants

Carbon monoxide appears to have no detrimental effects on material surfaces. The results of numerous experiments have not shown CO to produce any harmful effects on the higher plant life at concentrations below 100 ppm during exposures for 1 to 3 weeks [18]. Ambient concentrations of CO rarely reach this level even for short periods of time.

1-6-B Effects of Carbon Monoxide on Health

There are many studies that show that high concentrations of carbon monoxide can cause physiological and pathological changes and ultimately death. Carbon monoxide is a poisonous inhalent that deprives the body tissues of necessary oxygen.

Carbon monoxide has long been known to cause death when exposure to a high concentration (>750 ppm) is encountered. The combination of carbon monoxide with hemoglobin leads to carboxyhemoglobin, COHb; the combination of oxygen and hemoglobin leads to oxyhemoglobin, O_2Hb. Hemoglobin has an affinity for CO that is approximately 210 times its affinity for oxygen. That is, the partial pressure of CO required to saturate hemoglobin fully is only $\frac{1}{200}$ to $\frac{1}{250}$ of the partial pressure of oxygen required for complete saturation with oxygen. Exposure to a mixture of both gases leads to equilibrium concentrations of COHb and O_2Hb given by

$$\frac{\mathrm{COHb}}{\mathrm{O_2Hb}} = M\frac{P_{\mathrm{CO}}}{P_{\mathrm{O_2}}} \tag{1-13}$$

where P_{CO} and P_{O_2} represent the partial pressures of CO and O_2 in the inhaled gases, and M is a constant which ranges from 200 to 250 in human blood. Thus the quantity of COHb in the blood is a function of the concentration of CO in the air breathed. Fortunately the formation of COHb in the bloodstream is a reversible process. When exposure is discontinued, the CO that combined with the hemoglobin is spontaneously released, and the blood is cleared of half its carbon monoxide in healthy subjects in 3 to 4 hr [18]. A blood level of 0.4 percent COHb is maintained by the CO produced within the body, independent of external sources.

Example 1-7

Make a rough estimate of the saturation value of COHb in the blood if the CO content in the air breathed is 100 ppm.

SOLUTION

In order to apply Equation (1-13), we assume an M-value of 210. In addition, we assume that the oxygen content of the gas within the lungs is the same as in ambient air, that is, 21 percent or 210,000 ppm. Employing these values we find that

$$\frac{\mathrm{COHb}}{\mathrm{O_2Hb}} = M\frac{P_{\mathrm{CO}}}{P_{\mathrm{O_2}}} = \frac{210(100)}{210{,}000} = 0.1$$

Since the ratio of CO to O_2 in the blood is found to be 1 : 10, the saturation value for carbon monoxide is roughly 9 percent. This is somewhat low, since the oxygen content of incoming air is diluted by the gases which remain in the lungs.

Figure 1-3 relates the percent hemoglobin tied up as COHb as a function of CO concentration in the environment and the exposure time. On the basis of Figure 1-3, for example, exposures of 2 hr at concentrations of 40 ppm result in 2 percent COHb, whereas a concentration of 100 ppm results in approximately 5 percent COHb in healthy male subjects engaged in sedentary activity [18].

The direct effect of COHb is to reduce the oxygen-carrying capacity of the blood. However, a secondary effect is also present. COHb interferes with the release of the oxygen carried by the remaining hemoglobin. This effectively reduces further the oxygen-delivery capacity. Some cases of exposure for 8 hr or more to concentrations of 10 to 15 ppm have caused impaired

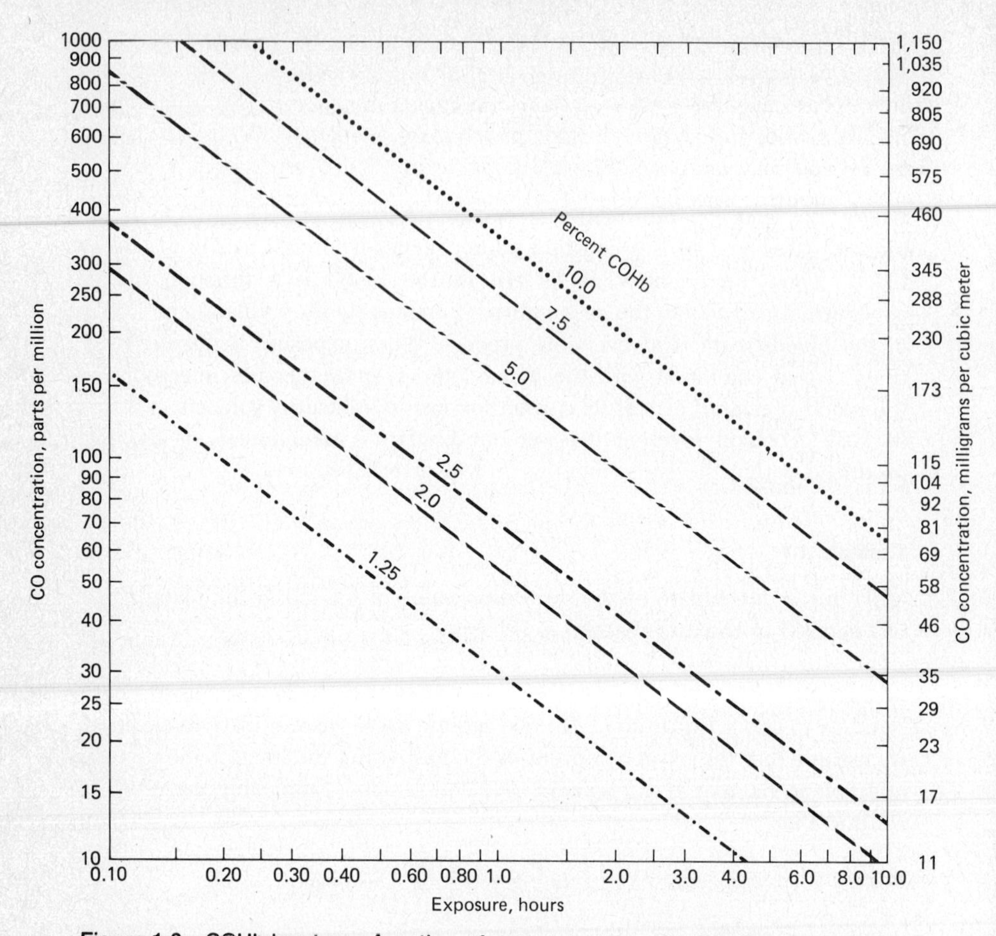

Figure 1-3 COHb level as a function of exposure. (SOURCE: NAPCA. *Air Quality Criteria for Carbon Monoxide*, AP-62, Washington, D.C.: HEW, 1970.)

time-interval discrimination. This concentration range is common during the day along downtown streets. As indicated by Figure 1-3, such exposure would lead to a COHb concentration in the bloodstream of roughly 2.5 percent. Evidence indicates that exposure for 8 hr or more to concentrations of 30 ppm (35 mg/m^3) causes impaired performance on certain psychomotor tests. Such an exposure will produce an equilibrium value of 5 percent COHb in the blood. Exposure to higher concentrations has produced evidence of psychologic stress in patients with heart disease. Evidence has been presented that CO concentrations in the Los Angeles basin are associated with excess mortality [19]. The ambient 24-hr averages for CO are in the range of 10 to 15 ppm for this area. (The vast majority of this CO is due to emissions from gasoline engines.) Many other urban areas have ambient daily averages in the 5- to 20-ppm range. A value of 100 ppm generally is considered an upper limit of safety in industry for healthy persons within certain age ranges for an 8-hr exposure. At 100 ppm, most people experience dizziness, headache, and lassitude. In terms of CO exposure, it should also be noted that cigarette smoke contains 400 to 450 ppm CO. The percent of COHb in blood of cigarette smokers increases as the number of cigarettes smoked per day increases. Blood samples from 29,000 blood donors in U.S. cities, suburbs, and rural areas revealed high concentrations of COHb. Forty-five percent of nonsmokers had blood containing 1.5 percent COHb. The blood of most smokers averaged above 5 percent COHb. Nonsmokers in Chicago averaged 2 percent COHb. One-pack-a-day smokers averaged 6.3 percent, and two-pack-a-day smokers averaged 7.7 percent [20]. Relating either ambient air CO concentration or percent of COHb in the blood to health effects is much more difficult than relating ambient CO concentration to the percent COHb.

There is disagreement as to whether CO is a threshold pollutant. Some experts believe that even small concentrations of CO produce some undesirable effects. Most American scientists believe that carbon monoxide is not a cumulative poison [18]. Chronic poisoning does not occur as a result of long-term exposure to relatively low concentrations. Persons particularly susceptible (those already afflicted with a disease that involves the oxygen capacity of the blood, such as anemia or cardiorespiratory disease) may be affected by CO levels existing on city streets. Exposure to low concentrations of CO (10 to 50 ppm) can impair a person's ability to estimate time intervals and affect his visual acuity with regard to brightness threshold. There is some evidence that such effects on a person driving a car increase his chances of having an accident as his abilities deteriorate [21].

Table 1-4 HEALTH EFFECTS OF CARBON MONOXIDE AND COHb

CARBON MONOXIDE	
ENVIRONMENTAL CONDITION	EFFECTS
9 ppm 8-hr exposure	Ambient air quality standard
50 ppm 6-wk exposure	Structural changes in heart and brain of animals
50 ppm 50-min exposure	Changes in relative brightness threshold and visual acuity
50 ppm 8 to 12-hr exposure nonsmokers	Impaired performance on psychomotor tests
CARBOXYHEMOGLOBIN	
COHb LEVEL (PERCENT)	EFFECTS
< 1.0	No apparent effect
1.0–2.0	Some evidence of effect on behavioral performance
2.0–5.0	Central nervous system effects. Impairment of time-interval discrimination, visual acuity, brightness discrimination, and certain other psychomotor functions
> 5.0	Cardiac and pulmonary functional changes
10.0–80.0	Headaches, fatigue, drowsiness, coma, respiratory failure, death

Table 1-4 lists the health effects attributed to certain CO levels in the environment and COHb levels in the blood. Generally speaking, the COHb level should be less than 2 percent except under extenuating circumstances for short time periods. In the light of these health effects the Environmental Protection Agency's (EPA) air quality standards for CO are: 10 mg/m^3 (9 ppm) for an 8-hr averaging period; 40 mg/m^3 (35 ppm) for a 1-hr averaging period. In the state of California the standard was set in 1969 as 20 ppm for an 8-hr period. The basis for this selection was that community air pollution should not contribute more than 2 percent COHb. In comparison to the primary federal ambient air quality standards listed above, the "significant harm" level defined by EPA for an 8-hr average is 57.5 mg/m^3 (50 ppm). This is roughly equal to a 10 percent carboxyhemoglobin level. The air quality standard of 9 ppm appears to be much lower than the level corresponding to any measured detrimental effect.

1-7 SULFUR OXIDES

Sulfur dioxide and sulfur trioxide are the dominant oxides of sulfur present in the atmosphere. Sulfur dioxide is a nonflammable, nonexplosive, colorless gas that causes a taste sensation at concentrations from 0.3 to 1.0 ppm in air. At concentrations above 3.0 ppm the gas has a pungent, irritating odor. Sulfur dioxide is partly converted to sulfur trioxide or to sulfuric acid and its salts by photochemical or catalytic processes in the atmosphere. Sulfur trioxide and moisture form sulfuric acid. The oxides of sulfur in combination with particulate and moisture produce the most damaging effects attributed to atmospheric air pollution. Unfortunately it has proven difficult to isolate the effects of sulfur dioxide alone. Brief discussions of the effects of sulfurous compounds will be presented in the following sections.

1-7-A Effects of Sulfur Compounds on Visibility and Materials

As pointed out in Section 1-5-A, suspended particulate in the atmosphere reduces the visual range by scattering and absorbing light. Since aerosols of sulfuric acid and other sulfates make up from 5 to 20 percent of the total suspended particulate matter in urban air, they contribute significantly to the reduction in visibility. Investigations [22] indicate that much atmospheric haze is caused by the formation of various aerosols resulting from the photochemical reactions between SO_2, particulate matter, oxides of nitrogen, and hydrocarbons in the atmosphere. In laboratory measurements, mixtures of NO_x and most common hydrocarbons form little or no aerosols when irradiated. However, considerable aerosol formation takes place when mixtures of olefins, NO_2, and SO_2 are irradiated by sunlight. One of the major products of such complex photochemical reactions is light-scattering droplets of sulfuric acid mist.

Measurements indicate that a major fraction of the sulfate in urban air has an effective size of less than 2 μm, with the peak in the size distribution around 0.2 to 0.9 μm. Since the visible wavelength range of the electromagnetic spectrum is roughly from 0.4 to 0.8 μm, the presence of aerosols of this type can cause a pronounced reduction in visibility. Figure 1-4 shows the trend of the influence of SO_2 and relative humidity on the visual range [1]. Although not shown specifically, the combination of increasing SO_2 concentration and increasing relative humidity leads to increasing aerosol concentrations. This in turn is responsible for the decrease in visual range shown in the figure. The typical range of SO_2 concentrations in urban areas, 0.01 to 0.20 ppm, is shown by the cross-hatched region in Figure 1-4. For SO_2 concentrations greater than 0.10 ppm, the limit of visibility becomes significant. Estimates show that a concentration of 0.10 ppm SO_2 with a relative humidity of 50 percent reduces the visibility to about 5 mi [22], as shown in Figure 1-4. It should be noted that when the visibility is less than 5 mi the rate of aircraft landings at major airports must be curtailed.

Sulfur compounds are responsible for the major damage to materials. Unfortunately the relative contribution of each of the individual pollutants is difficult to evaluate precisely. Observations indicate that concentrations of 1 to 2 ppm of SO_2 in the atmosphere cause an increase of 50 to 100 percent in the drying time of a paint film, and higher concentrations—7 to 10 ppm—increase the drying time to 2 to 3 days. The final surface is also less durable when the paint dries in the presence of SO_2. Sulfur oxides seem to have little

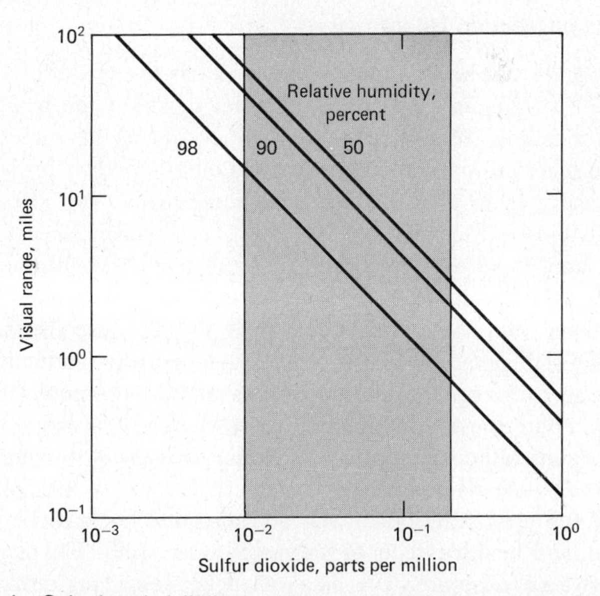

Figure 1-4 Calculated visibility in miles at various sulfur dioxide concentrations and at different relative humidities in New York City.

or no effect on dry, hard paints. Paints containing metal salts react with H_2S. Newer type paints are much more resistant to H_2S.

Sulfur oxides generally accelerate metal corrosion by first forming sulfuric acid either in the atmosphere or on the metal surface. Depending upon the kind of metal exposed as well as the duration of exposure, corrosion rates in urban atmospheres have been observed to be from $1\frac{1}{2}$ to 5 times the rates obtained in rural environments [22]. A threefold reduction in sulfur dioxide in Pittsburgh from 0.15 to 0.05 ppm for the time period 1926 to 1960 caused almost a fourfold reduction in the corrosion rate of zinc. The most detrimental pollutant that contributes to metal corrosion generally is sulfur dioxide. Temperature, and especially relative humidity, also influences corrosion rates significantly. Aluminum is fairly resistant to SO_2 attack. However, at relative humidities greater than 70 percent the corrosion rate is greatly increased [23]. ASTM has reported some long-range studies of atmospheric effects on the tensile strength of aluminum [24]. Exposure in rural areas led to a loss of 1 percent or less in tensile strength over 20 years, while industrial atmospheres led to a 14 to 17 percent loss in the same time period. The literature offers an extensive review of metal corrosion due to different atmospheric conditions through the early 1960s [25]. Sulfurous or sulfuric acids are capable of attacking a wide variety of building materials, including limestone, marble, roofing slate, and mortar. Fairly soluble sulfates are formed which are then leached away by rain. Textiles of nylon, especially nylon hose, are also susceptible to pollutants in the atmosphere. Weakening is apparently caused by SO_2 or by sulfuric acid aerosol.

1-7-B Effects on Human Health

No other pollutant has been studied as intensely as the oxides of sulfur, yet many questions concerning the effects of sulfur dioxide upon health remain unanswered. Because sulfur oxides tend to occur in the same kinds of polluted atmosphere as particulate matter and high humidity, few epidemiologic studies have been able to differentiate adequately the effects of the pollutants.

Various animal species, including man, respond to sulfur dioxide by bronchoconstriction, which may be assessed in terms of a slight increase in airway resistance [22]. Most individuals will show a response to SO_2 at concentrations of 5 ppm and above, and certain sensitive individuals show slight effects at 1 to 2 ppm. Sulfuric acid is a much more potent irritant to man than is sulfur dioxide; therefore, most studies deal with combined sulfurous materials rather than with SO_2 alone. Analysis of numerous epidemiological studies clearly indicate an association between air pollution, as measured by the concentration of SO_2 accompanied by particulate matter and moisture, and health effects of varying severity [22]. This is especially true for short-term exposure. The association between long-term exposure and chronic disease morbidity and mortality is not so clear. Typical con-

Table 1-5 EFFECTS OF SO_2 AT VARIOUS CONCENTRATIONS

CONCENTRATION	EFFECT
0.03 ppm, annual average	1974 air quality standard, chronic plant injury
0.037–0.092 ppm, annual mean	Accompanied by smoke at a concentration of 185 $\mu g/m^3$, increased frequency of respiratory symptoms and lung disease may occur
0.11–0.19 ppm, 24-hr mean	With low particulate level, increased hospital admission of older persons for respiratory diseases may occur. Increased metal corrosion rate
0.19 ppm, 24-hr mean	With low particulate level, increased mortality may occur
0.25 ppm, 24-hr mean	Accompanied by smoke at a concentration of 750 $\mu g/m^3$, increased daily death rate may occur (British data); a sharp rise in illness rates
0.3 ppm, 8 hr	Some trees show injury
0.52 ppm, 24-hr average	Accompanied by particulate, increased mortality may occur

SOURCE: Summarized from data presented in National Air Pollution Control Administration, *Air Quality Criteria for Sulfur Oxides*, AP-50. Washington, D.C.: HEW, 1970.

centration data and associated health effects are presented in Table 1-5. As can be seen from these data, sulfur oxide in combination with particulate and moisture is a potentially serious health hazard (see Table 1-3 also).

A comprehensive review of the physiologic effects of SO_2 and SO_3 on man and animals was made in 1954 by Greenwald [26] and is summarized in reference 1. This report indicates that sulfur dioxide acts as a pungent, suffocating, irritant gas in its effect on the upper respiratory tract under moderate exposure. At concentrations below about 20 ppm, SO_2 produces only acute effects; to date, no chronic or cumulative effects on animals have been reported when exposure levels were moderate and exposure discontinuous. The results of other laboratory tests [1] tend to indicate that concentrations above 1 ppm of pure SO_2 are required before serious or even significant effects may be expected on the health of unimpaired, or, for that matter, impaired individuals.

To illustrate the complexity of determining a satisfactory relationship between human health and the concentration of sulfur dioxide in the atmosphere, we cite the results of two studies. In a study relating the excess mortality in New York City during the years 1963 to 1972 to air pollution as indicated by SO_2 concentration and smoke level, H. Schimmel et al. [27] found that while SO_2 concentrations decreased from a high of 0.219 ppm to 0.03 ppm, mortality did *not* show a corresponding decrease. The smoke level remained approximately the same during the time period. Schimmel et al. [27] also state that studies conducted in England show that mortality and morbidity have decreased with a rather sharp reduction in particulate, whereas the SO_2 levels have not changed very much. In a study of emergency-room visits for asthma over a 4-month period, Goldstein and Block

[28] found no relationship between daily visits to emergency rooms and daily levels of either smoke or SO_2 for residents of Harlem; in Brooklyn, however, they observed a strong correlation between daily visits for asthma and daily levels of SO_2, but not of smoke. The results of these studies support the suggestion that SO_2 and particulate levels should be treated as indices of overall air pollution and not necessarily as agents causing ill health. Another study [29] reports new evidence linking sulfur oxides and suspended particulate to acute respiratory diseases in children. Based on a 1972 survey of children in four New York communities, the study found that exposed children in the 1 to 12 age group experienced a 20 percent greater frequency of acute bronchitis than did children not similarly exposed. The significant increase could not be explained by other factors.

1-7-C Acid Rain

Water droplets formed by condensation in the atmosphere normally should have a pH close to 7. However, the dissolution of atmospheric CO_2 in rainwater tends to lower the pH due to the formation of carbonic acid. This acid is very weak, and at equilibrium with CO_2 the pH of rainwater would lie close to 5.65 in value. By the early 1970s, however, individual pH values from 2 to 6 were measured in various parts of the world, with yearly averages between 4 and 5 [36]. Thus acid concentrations from 10 to 10,000 times greater than expected from natural sources were detected. Measurements in Gander, Newfoundland, since 1975 indicate pH values of 4.0 to 4.5 on a yearly basis in that region [37]. Similar values are valid for the average pH of rainfall in the eastern half of the United States. These low values are due to the transformation of SO_2 and NO_x into acids when they are absorbed in cloud water and raindrops. A typical reaction might be

$$2SO_2 + O_2 \rightarrow 2SO_3$$

$$SO_3 + H_2O \rightarrow 2H^+ + SO_4^+$$

These acid aerosols are then deposited in significant quantities on the surface of land and water masses. Such a phenomenon is called acid rain.

The increasing presence of these acids is attributed to the increase in emissions of SO_2 and NO_x that has occurred since the 1940s through the increased use of fossil fuels in industrialized nations. Data from the eastern United States indicate that roughly 60–70 percent of the acidity is due to sulfuric acid, and 30–40 percent is due to nitric acid [38]. A high proportion of nitric acid derivatives indicates mobile sources are responsible, while a high proportion of sulfuric acid derivatives suggests stationary sources. The concentrations of sulfates and nitrates formed in the atmosphere are frequently low, but the deposition effect is cumulative. The relationship between SO_2 emissions and acidification is well established; the contribution due to NO_x is somewhat less known.

Such pollutants travel in the lowest 2 km of the atmosphere, and are often carried hundreds of kilometers from their source. Their presence has been observed in a qualitative sense by the remarkable increase in summer haze in a number of regions of industrialized nations. Acid rain has affected areas in Switzerland, the southern part of the Scandinavian countries, and especially the northeastern half of North America. In the later case, measurements indicate a severe problem in the Atlantic provinces of Canada and in the Adirondack Mountain region of the United States, for example. Recent data, however, indicates that the problem is increasing in the midwest and southeast parts of the United States. Even major urban areas on the West Coast are affected. The phenomenon has been under active study in Europe since the early 1950s. The problem exists in northern Europe since a major portion of the air pollution produced on the continent is transported in a northeasterly direction toward Scandinavia. Likewise, pollution produced in the central provinces of Canada and in the central and eastern states of the United States is carried in the summer and fall in a northeasterly direction.

Data taken at St. Margarets Bay in eastern Canada indicate that upwards of 25 percent of the suspended particulate was in the form of sulfates [37]. Higher percentages have been measured elsewhere. Although the particulate concentration has been reduced in many industrialized nations due to effective control techniques on sources of particulate, the fraction of the remaining particulate which is in the form of an acid aerosol continues to increase. Conditions in the atmosphere which favor aerosol sulfate formation include poor atmospheric dispersion and ample sunlight. Such conditions also are favorable to ozone formation. The natural acid concentration in rainwater would be less than 10 $\mu g/l$. In Canada in 1975 the acid concentration was around 40 $\mu g/l$ (expressed as H^+). In the Netherlands the average pH fell below 4.0 around 1966 on a yearly basis. A value around 3.7 is possible by 1985 [36]. In Sweden the pH of rainwater is changing by about 0.3 to 0.4 units per decade.

There are several effects of acid rain that are disturbing. First, there is an acidification of natural water sources. This can have a devastating effect on fish life. Trout and salmon are particularly sensitive to a low pH. Reproduction in many fish fails to occur for a pH less than around 5.5 in value. A decrease in plankton and bottom fauna is also observed as the pH is lowered, which reduces the food supply for fish. Secondly, a leaching of nutrients occurs in the soil. This demineralization can lead to loss in productivity of crops and forests, or a change in the natural vegetation. Vegetation itself can be directly damaged, and an increase in corrosion of materials is observed. The severity of damage to soils and bodies of water is partially determined by the minerals in the soil in a given region. Those areas that contain rock such as calcium carbonate or similar minerals are buffered against the onslaught of acid rain. In such regions lakes tend to maintain a pH closer to 5 or 6.

Since there is no way to alter the results of acid rain once it has occurred in a region, the only solution is the control of emissions at the source. Desulfurization of fossil fuels has been considered important in the past in order to prevent local air pollution. It is now becoming more apparent that SO_2 (and NO_x) control is needed to prevent the long-range dispersion of pollutants as well. Thus emission control takes on an international scope, since emissions from one country are affecting the nature of rain in another country. Continued monitoring and evaluation of the situation will be needed if the full extent of the problem is to be realized.

1-8 EFFECTS OF HYDROCARBONS, OXIDES OF NITROGEN, PHOTOCHEMICAL OXIDANTS, ASBESTOS, AND METALS ON MATERIALS AND HEALTH

1-8-A Hydrocarbons

Hydrocarbons do not appear to cause any appreciable corrosive damage to materials. Particulate, or soot made up of unburned hydrocarbons and carbon, soils surfaces. Of all the hydrocarbons, only ethylene has adverse effects on plants at known ambient concentrations. The principal effect of ethylene is to inhibit plant growth [30].

To date, studies of the effects of ambient air concentrations of gaseous hydrocarbons have not demonstrated direct adverse effects upon human health [28]. Studies of the carcinogenicity of certain classes of hydrocarbons do indicate that some cancers appear to be caused by exposure to aromatic hydrocarbons found in soots and tars. Identifiable airborne carcinogens are mostly polynuclear aromatic hydrocarbons [7]. Unburned hydrocarbons in combination with the oxides of nitrogen in the presence of sunlight form photochemical oxidants, components of photochemical smog, that do have adverse effects on human health and on plants.

1-8-B Oxides of Nitrogen

Of the six or seven oxides of nitrogen, nitric oxide (NO) and nitrogen dioxide (NO_2) are important air pollutants. Although N_2O is commonly present in the lower atmosphere (formed by biological action at the earth's surface) it is not considered an air pollutant. Neither NO nor NO_2 causes direct damage to materials; however, NO_2 can react with moisture present in the atmosphere to form nitric acid, which can cause considerable corrosion of metal surfaces. Nitrogen dioxide absorbs visible light and at a concentration of 0.25 ppm will cause appreciable reduction in visibility. Nitrogen dioxide at a concentration of 0.5 ppm for a period of 10 to 12 days has suppressed growth of such plants as pinto beans and tomatoes. Experiments with navel oranges show that yield is reduced by prolonged exposure to NO_2 at concentrations from 0.25 to 1 ppm.

Nitrogen dioxide acts as an acute irritant and in equal concentrations is more injurious than NO. However, at concentrations found in the atmosphere NO_2 is only potentially irritating and potentially related to chronic pulmonary fibrosis [7]. Some increase in bronchitis in children (2 to 3 years old) has been observed at concentrations below 0.01 ppm. In combination with unburned hydrocarbons, the oxides of nitrogen react in the presence of sunlight to form photochemical smog. The chemical reactions involved are discussed in Section 9-5. It is because of this chemical activity that the primary air quality standard for the oxides of nitrogen has been set as 100 $\mu g/m^3$ annual average. The components of photochemical smog which are the most damaging to plants and detrimental to human health are the photochemical oxidants, which we discuss in the following section.

1-8-C Photochemical Oxidants

Oxidizing agents, ozone (O_3), peroxyacetyl nitrate (PAN), peroxybenzoyl nitrate (PBN), and other trace substances which can oxidize the iodide ion of potassium iodide are termed photochemical oxidants. Ozone and PAN are present in the highest concentrations, and the damaging effects of photochemical smog are generally related to the concentrations of these species. The aerosols formed during the chemical reactions that create smog cause a marked reduction in visibility and give the atmosphere a brownish cast. Ozone attacks synthetic rubbers, thereby reducing the life of tires, rubber insulation, and so on. Ozone inhibitors can be built into rubber products. Ozone also attacks the cellulose in textiles, reducing the strength of such items. All of the oxidants cause some fading of fabrics. Oxidants, primarily PAN and PBN, cause severe eye irritation, and in combination with ozone they irritate the nose and throat, cause chest constriction, and at high concentration (3900 $\mu g/m^3$) produce severe coughing and inability to concentrate. Table 1-6 presents a summation of the effects of ozone and photochemical oxidants [31].

1-8-D Asbestos and Metals

Many studies have shown a higher than expected incidence of bronchial cancer among people whose occupations expose them to asbestos. In addition, asbestos has been identified as a causal factor in the development of cancers of the membranes lining the chest and abdomen. Beryllium has also proved dangerous to health, producing both acute and chronic lethal inhalation effects as well as causing damage to the skin and eyes [32]. Most of the cases studied concern occupational exposure. Exposure to metallic mercury vapors may cause injury to the central nervous system and kidneys. And mercury can accumulate in the body system, eventually causing brain damage.

Table 1-6 HEALTH EFFECTS OF OZONE AND PHOTOCHEMICAL OXIDANTS

CONCENTRATION		OZONE	
(ppm)	($\mu g/m^3$)	EXPOSURE	EFFECTS
0.02	40	1 hr	Cracked, stretched rubber
0.03	60	8 hr	Vegetation damage
0.10	200	1 hr	Increased airway resistance
0.30	590	Continuous working hours	Nose and throat irritation, chest constriction
2.00	3900	2 hr	Severe cough

CONCENTRATION		PHOTOCHEMICAL OXIDANTS	
(ppm)	($\mu g/m^3$)	EXPOSURE	EFFECTS
0.05	100	4 hr	Vegetation damage
0.10	200		Eye irritation
0.13	250	Maximum daily	Aggravation of respiratory diseases
0.03		1 hr	Impaired performance of athletes
0.08	160	1-hr maximum	Air quality standard

SOURCE: National Air Pollution Control Administration, *Air Quality Criteria for Photochemical Oxidants*, AP-63. Washington, D.C.: HEW, 1970.

Because of these proven hazards, emission standards were established in 1973 for asbestos, beryllium, and mercury. The standard for asbestos stipulates how asbestos material is to be applied as well as the precautions to be taken in its manufacture and the demolition of buildings. The beryllium standard applies to industrial processes involving beryllium, beryllium ore, and alloys containing more than 25 percent beryllium by weight, and stipulates emission rates from such processes. The mercury standard applies to stationary sources which process mercury ore, recover mercury, and use mercury chlor-alkali cells to produce chlorine gas and alkali metal hydroxide.

1-9 INJURY TO VEGETATION

Damage to plants from air pollution usually occurs in the leaf structure, since the leaf contains the building mechanisms for the entire plant. A leaf typically may be divided into three regions. The epidermis forms a protective layer on the outside. The mesophyll is the center section of the leaf, and contains two layers of cells called the palisade and the spongy parenchyma. In addition, a dense network of veins run throughout the leaf from its base or stem. The veins provide the transport system for water and other chemicals as they are transferred to other portions of the plant. Of particular interest in air pollution studies are the openings through the epidermis into the mesophyll, called stomata. Gases and vapors pass in and out of the leaf structure through the stomata, each of which is surrounded by special guard cells that open or close the stomata.

Among the most frequently encountered gases toxic to vegetation are sulfur dioxide, ozone, PAN, hydrogen fluoride, ethylene, hydrogen chloride, chlorine, hydrogen sulfide, and ammonia.

SULFUR DIOXIDE

The effect on leaves of excessive SO_2 in the atmosphere appears first as a cellular injury to the spongy parenchymal area in the mesophyll, followed by damage to the palisade region. On initial attack the leaf appears water-soaked. Upon drying, a bleached or ivory color appears in the affected areas. Apparently there is a threshold value below which the leaf is capable of consuming the gas without injury. One threshold value cited is 0.3 ppm (785 $\mu g/m^3$) sustained exposure for 8 hr. Concentrations of 0.3 to 0.5 ppm for several days lead to chronic injury to sensitive plants. The SO_2 enters the stomata directly and the plant cells in the mesophyll convert it to sulfite and then into sulfate. Apparently when excessive SO_2 is present the cells are unable to convert sulfite to sulfate fast enough, and disruption of the cell structure begins. Spinach, lettuce, and other leafy vegetables are most sensitive, as are cotton and alfalfa. Pine needles are also affected, with either the needle tip or the whole needle becoming brown and brittle. A considerable amount of literature (over 40 articles in reference 22) deals with plant damage caused by sulfur compounds.

OZONE

Effects of ozone on vegetation were first noted in the late 1950s. It first attacks the palisade region in the mesophyll. The cell structure collapses, and buff or reddish-brown pigmentation (stipple) appears on the upper surface. Conifer needle tips become brown and necrotic. Spinach, pinto beans, tomatoes, and white pines are especially sensitive [33]. A large number of pine trees in some of the western forests appear to be dying as the result of prolonged exposure to photochemical oxidants. The injury threshold has been estimated at about 0.03 ppm (59 $\mu g/m^3$) for 4-hr exposure. Damage to the plants cited above has been observed at concentrations of 0.1 ppm or less for time periods of 1 to 8 hr. A concentration of 0.06 ppm for 3 to 4 hr damages alfalfa. The retarding of citrus growth has also been attributed to ozone.

PAN

Peroxyactyl nitrate attacks the spongy parenchyma cells surrounding the air space into which the stomata open. The principal visible effect is the silvering or bronzing of the lower leaf area. The threshold concentration for injury is estimated at 0.01 ppm for a 6-hr exposure, although petunias have been affected at 0.005 ppm for an 8-hr exposure. The youngest leaves in terms of maturity are most sensitive.

HYDROGEN FLUORIDE

Fluorides act as cumulative poisons to plants. Even when exposed to extremely low concentrations, plants will eventually accumulate enough to injure the leaf tissue. The earliest effect is tip and margin burn. Apparently the fluoride enters through the stomata and is carried toward the tip and margin by normal water flow. Eventually internal cells collapse. Upon collapse and drying out of the cells, the injured area turns deep brown to tan. A narrow red-brown line of dead tissue distinctly separates the necrotic region from the healthy area. It has been found that peaches, grapes, and gladioli are quite sensitive to fluorides. The injury threshold may be as low as 0.1 ppb (0.08 $\mu g/m^3$) over a 4- or 5-wk period. Concentrations of the order of 1 ppb can be significant.

ETHYLENE

Among the common hydrocarbons, ethylene appears to be the only one that causes plant damage at known ambient levels. Concentrations of ethylene from 0.001 to 0.5 ppm have caused damage to sensitive plants. The effects of ethylene include flower droppings and failure of the leaf to open properly. Injury to orchids and to cotton has been established. A general threshold value of 0.05 ppm for 6-hr exposure has been reported as a guideline.

Other gases and vapors such as hydrogen chloride, chlorine, hydrogen sulfide, and ammonia, among others, can cause a dramatic collapse of leaf tissue when the plant is exposed to levels greater than 1 ppm. Fortunately these levels usually are reached only during accidental spills. An excellent listing of the sensitivity of over 200 varieties of plants to 13 different pollutants appears in the literature [34]. The list is incomplete in the sense that each plant has not been evaluated for all pollutants. Information on the injury to vegetation caused by gaseous air pollutants can be found in the federal air quality criteria publications as well as in "Recognition of Air Pollution Injury to Vegetation," published by the Air Pollution Control Association [33].

1-10 SOURCES OF AIR POLLUTANTS

Once selected substances have been carefully studied and declared to be air pollutants on the basis of the evidence available, it is of immediate interest to ascertain the major sources of these substances. The nonviable solid particulates in the atmosphere are generated from the combustion of fuels such as coal and fuel oil in stationary furnaces and from gasoline, diesel fuel, and jet-engine fuel in mobile sources. Manufacturing processes such as grinding, smelting, crushing, and grain milling and drying also contribute to air pollution. The major source of the sulfur oxides is from the combustion of fuels containing sulfur. However, some sulfur compounds are released to the atmosphere from ore processing and manufacturing processes involving sulfuric acid.

Carbon monoxide is generated primarily by incomplete combustion of carbonaceous fuels in automobile engines and space-heating units. Unburned hydrocarbons result from incomplete combustion of fuels and from petroleum refining. A relatively small portion of the hydrocarbons results from other operations such as dry cleaning, the evaporation of industrial coating, and the cleaning of manufactured parts. Oxides of nitrogen are formed in a combustion process when the nitrogen in the air or in the fuel combines with oxygen at elevated temperatures. A very small quantity of the oxides is released from plants employing or manufacturing nitric acid.

The 1977 nationwide emissions from the major sources of air pollution in the United States are presented in Table 1-7. Other published data may show some deviations from this table; however, even though the actual quantities may not be exact, the relative magnitudes are of value in indicating the magnitude of the problem associated with each pollutant. From Table 1-7, we see that transportation was the major source of pollution on a tonnage basis in 1977, followed by industry and the generation of electrical power. In order to improve the quality of our air, we must direct a major effort toward reducing pollution from those sources. The annual emission levels of the five major pollutants estimated for 1970 and 1977 and predicted for 1985 are listed in Table 1-8. Note that predictions indicate an increase in levels for two pollutants, while the other three will decrease.

The damaging effects of air pollutants are not equivalent on an equal-mass basis. That is, a given concentration of SO_2, for example, may be more detrimental to health than an equivalent concentration of carbon monoxide. Thus, while information presented in Tables 1-7 and 1-8 is important, the true impact of the pollutants upon our environment is not given by these data. Some authors [35] have proposed that the quantity of a pollutant should be multiplied by a "tolerance factor" to obtain a combined pollution index as a true measure of its total contribution to air pollution.

Table 1-7 ESTIMATED NATIONWIDE EMISSIONS FOR 1977 (10^6 metric tons/yr)

POLLUTANT	TRANSPORTATION	FUEL COMBUSTION, STATIONARY SOURCES	INDUSTRIAL PROCESSES	SOLID WASTE DISPOSAL	MISCELLANEOUS	TOTAL
CO	85.7	1.2	8.3	2.6	4.9	102.7
SO_x	0.8	22.4	4.2	0.	0.	27.4
NO_x	9.2	13.0	0.7	0.1	0.1	23.1
C_xH_y	11.5	1.5	10.1	0.7	4.5	28.3
Particulate	1.1	4.8	5.4	0.4	0.7	12.4
TOTAL	108.3	42.9	28.7	3.8	10.2	193.9
PERCENT	55.9	22.1	14.8	2.0	5.2	

NOTE: A zero indicates emissions of less than 50,000 metric tons per year.

Table 1-8 ESTIMATED U.S. EMISSIONS FOR 1970 AND 1977 AND PREDICTED FOR 1985 (million tons/yr)

	CO	HC	SO_x	NO_x	PARTICULATE
1970	112.4	32.5	32.8	21.6	24.4
1977	113.0	31.1	30.1	25.4	13.6
1985	45.3	21.4	31.0	22	11.0

SOURCE: 1970 and 1977 data, EPA National Emissions Trends Report, 1977; 1985 projection, DOE Report, *1985 Air Pollution Emissions*.

QUESTIONS

1. Discuss the causes of air pollution.
2. What constitutes an air pollutant?
3. Why are water vapor and carbon dioxide not air pollutants in a conventional context?
4. List the pollutants for which air quality standards have been established.
5. Why are emission standards established for asbestos, beryllium, and mercury when no air quality standard exists?
6. Estimate the visual range in an atmosphere containing small particles having a concentration of 150 $\mu g/m^3$.
7. Explain how the relation given by Equation (1-4) could be employed in the operation of a meter to measure the smoke density of a diesel engine.
8. What are the difficulties encountered in developing a meter to measure the particulate concentration in a plume based on light transmitted through the plume?
9. What are the effects of particulate on health? What particle size range is of major importance in impairing health?
10. What is the effect of CO upon the human body?
11. Upon what basis was the air quality standard for the oxides of nitrogen set?
12. What was the major source of CO in the United States in the 1970s?
13. What was the major source of particulate in the United States in 1977? How has this changed since 1966?
14. What are the four basic assumptions for the rational control of air pollution in the United States?

PROBLEMS

1-1. The air quality standard for carbon monoxide (CO) in the state of California is 10 ppm measured over a 12-hr averaging time. What is the equivalent concentration in milligrams per cubic meter at 25°C?

1-2. The primary air quality standard for sulfur dioxide (SO_2) as an annual average is 80 $\mu g/m^3$. What is the equivalent concentration in parts per million at 25°C?

1-3. A wind storm on a dry summer day increased the particulate concentration in an urban area to a level of 215 $\mu g/m^3$. Determine the equivalent concentration in grains per cubic foot, for conditions of 1 atm and 25°C.

1-4. The ozone, O_3, concentration sometimes reaches a value of 0.15 ppm over a 1-hr period in urban areas with photochemical smog problems. Determine by what percentage this level exceeds the federal ambient standard of 240 $\mu g/m^3$ for the given time period, if the temperature is 25°C.

1-5. The concentration of carbon monoxide (CO) in cigarette smoke reaches levels of 400 ppm or higher. For this particular value determine the percent by volume and the concentration in milligrams per cubic meter, at 25°C and 1 atm.

1-6. The primary air quality standard for NO_x expressed as NO_2 as an annual average is 100 $\mu g/m^3$. What is the equivalent concentration in parts per million at 25°C?

1-7. The particulate concentration in a large city is reported to be 160 $\mu g/m^3$ on a particular day. (a) Convert this value, which is relatively high for an urban atmosphere, to grains per cubic foot. (b) In industrial situations where particulate is being carried by ducts to collection equipment, the dust "loading" may be as high as 10 gr/ft^3. In this situation, what is the equivalent concentration in micrograms per cubic meter?

1-8. The visibility due to scattering only is found to be 2 mi. What percentage of light will pass through a length of 0.2 mi if the limit of visibility is defined as (a) a 98 percent reduction and (b) a 99 percent reduction in the original light intensity?

1-9. If the limit of visibility is defined as the distance when I/I_o reaches 0.02 in value, then determine the percent extinction that occurs in the first (a) 10 percent, (b) 20 percent, and (c) 50 percent of the path length.

1-10. Reconsider Problem 1-9 for the situation where I/I_o is 0.05.

1-11. A uniform dispersion of spheres with a uniform diameter of 0.8 μm attenuates a light source by 92 percent at a test distance of 1000 m. The particle material density is 1.15 g/cm^3 and the particulate concentration in the air is 745 $\mu g/m^3$. Determine (a) the scattering coefficient in m^{-1} if absorption is neglected, (b) the value of the scattering ratio, K, for the substance, and (c) the limit of visibility in kilometers.

1-12. A visibility of 2 mi characterizes an atmosphere containing droplets of oil having a diameter of 0.4 μm and a specific gravity of 0.83. What is the estimated droplet concentration in micrograms per cubic meter if $m = 1.50$ and the middle range of the visible spectrum is used?

1-13. The visibility due to scattering only is found to be 2 mi. What percentage of light will pass through a length of 0.2 mi?

1-14. The atmosphere contains 500 particles/cm^3 whose average diameter is 1 μm and whose density is 1.5 g/cm^3. The scattering area ratio K_s is 2.5. If both the scattering and absorption coefficients are equal, what is the percentage of attenuation for a light path length of 1 km?

1-15. Derive Equation (1-8) for the limit of visibility.

1-16. Consider Equation (1-4) to be modified to the following form: $I/I_o = \exp(-kCd)$, where C is the concentration in parts per million and k is a function of λ and has units of ppm^{-1} $(mi)^{-1}$. Consider $d = 2$ mi, C of $NO_2 = 0.5$ ppm, and $k = 1.05$ at 0.500 μm and 0.18 at 0.600-μm wavelength. (a) What is the percentage attenuation of the signal at these two wavelengths? (b) Explain why NO_2 gas appears brownish yellow in the atmosphere.

1-17. Consider Equation (1-4) to be modified to the following form: $I/I_o = \exp(-kCd)$, where C is the concentration in parts per million, and k is a function of wavelength and has units of ppm^{-1} (mi^{-1}). Consider the situation where d is 1.0 mi and k is 1.05 at 0.500 μm. During a test the value of I/I_o is found to be (a) 0.40, (b) 0.75, and (c) 0.90. What is the concentration of NO_2 in parts per million during the test?

1-18. Dust with a density of 1.4 g/cm^3, a scattering area ratio K of 2.2, and an effective diameter of 1.4 μm has a concentration of 200 $\mu g/m^3$ in the atmosphere. Estimate the limit of visibility in kilometers.

1-19. A dust-laden air stream is passed through an initially clean filter at a flow velocity of 0.5 m/s. After a test period of 4 hr the transmittance is found to be 68 percent. Determine (a) the equivalent path length, in meters, (b) the value of the Coh, and (c) the value of the Coh per 1000 lineal meters.

1-20. Dust is removed from an air stream by passing it through a clean filter. After a test period of 5 hr for a flow velocity of 1.2 ft/s, the transmittance is found to be 84 percent. Determine (a) the equivalent path length, in feet, (b) the value of the Coh, and (c) the value of the Coh per 1000 lineal feet.

1-21. The value of the Coh per 1000 ft is 4.7 for a test which took 2 hr and the transmittance was 65 percent of that of a clear filter sample. (a) What air-velocity was used during the test, in feet per second? (b) What is the equivalent path length of the collected sample in feet?

1-22. Assume that there are 5000 ml of blood in the body and that blood normally contains 20 ml of O_2 per 100 ml of blood. A person is breathing at a volume rate of 3.6 l/min, and the air contains 600 ppm of CO. Determine the time required, in minutes, for the blood to become 40 percent saturated with CO if the blood CO level initially is (a) 0 percent and (b) 5 percent. Assume that all CO inducted into the lungs is absorbed.

1-23. Consider that cigarette smoke contains an average of 450 ppm carbon monoxide. If the average oxygen content in the air in the lungs is 19.0 percent, what percentage of the saturation level would the COHb concentration ultimately reach?

1-24. Assume that there are 4800 ml of blood in the body and that blood normally contains 20 ml of O_2 per 100 ml of blood. A person under heavy exertion is breathing at a volume rate of 4.2 l/min, and the polluted air contains 100 ppm carbon monoxide. Determine the time required, in minutes, for the blood to become 7 percent saturated with CO if the blood CO level initially is (a) 0 percent and (b) 2 percent. Assume that all carbon monoxide inducted into the lungs is absorbed.

1-25. Consider the situation where the gas in the lungs of a human contains 220 ppm of carbon monoxide and the average oxygen content is 17.5 percent. If these concentrations were maintained, what percentage of the saturation level would the COHb concentration eventually reach?

References

1. A. C. Stern, ed. *Air Pollution*. Vol. I, 2d ed. New York: Academic Press, 1968.
2. R. E. Inman et al. "Note on Uptake of CO by Soil Fungi," *J. Air Pollu. Control Assoc*. **21**, no. 10 (1971): 646.

3. A. C. Hill. "Vegetation: A Sink for Atmospheric Pollutants." *J. Air Pollu. Control Assoc.* **21**, no. 6 (1971): 341.
4. *The New Yorker*, April 13, 1968.
5. M. I. Goldman. "The Convergence of Environmental Disruption." *Science* **170** (October 1970): 37–42.
6. E. Cook, "The Flow of Energy in an Industrial Society." *Scientific American* **225** (September 1971): 135–144.
7. American Association for the Advancement of Science. *Air Conservation*. Washington, D.C., 1965.
8. R. E. Huschke, ed. *Glossary of Meteorology*. Boston: American Meteorology Society, 1959.
9. L. A. Schaal. "Meteorology and Air Pollution in Indiana." Proceedings, Seventh Conference on Air Pollution Control, Purdue University, 1968.
10. S. C. Blacktin. *Dust*. Cleveland, Ohio: Sherwood Press, 1934.
11. W. E. K. Middleton. *Vision Through the Atmosphere*. Toronto, Canada: Univ. of Toronto Press, 1952.
12. J. T. Peterson and R. A. Bryson. "Atmospheric Aerosols: Increased Concentrations During the Last Decade." *Science* **162** (1968): 120.
13. H. C. Van de Hulst. *Light Scattering by Small Particles*. New York: Wiley, 1957.
14. R. D. Ross, ed. *Air Pollution and Industry*. New York: Van Nostrand Reinhold, 1972.
15. H. J. Ettinger and G. W. Royer. "Visibility and Mass Concentrations in a Nonurban Environment." *J. Air Pollu. Control Assoc.* **22** (February 1972): 108.
16. NAPCA. *Air Quality Criteria for Particulate Matter*, AP-49. Washington, D.C.: HEW, 1969.
17. J. K. Burchard. "The Significance of Particle Emissions." *J. Air Pollu. Control Assoc.* **24**, no. 12 (1974): 114.
18. NAPCA. *Air Quality Criteria for Carbon Monoxide*, AP-62. Washington, D.C.: HEW, 1970.
19. A. C. Hexter and J. R. Goldsmith. "Carbon Monoxide: Association of Community Air Pollution with Mortality." *Science* **172** (1971): 265.
20. *Science News* **99**, no. 3 (1971).
21. T. Beard and A. Wertheim. *Am. J. Pub. Health* **57** (1967): 2013.
22. NAPCA. *Air Quality Criteria for Sulfur Oxides*, AP-50. Washington, D.C.: HEW, 1970.
23. P. M. Aziz and H. P. Godard. *Corrosion* **15** (1959): 39.
24. C. J. Walton and W. King. *ASTM Spec. Tech. Publ.* **175** (1956): 21.
25. H. C. Muffley. "Influence of Atmospheric Contaminants on Corrosion." Literature Report, Defense Documentation Center, Alexandria, Va., 1963.
26. I. Greenwald. "Effects of Inhalation of Low Concentrations of Sulfur Dioxide on Man and Other Mammals." *Arch. Ind. Hyg., Occupational Med.* **10** (1954): 455.
27. H. Schimmel, T. J. Murawski, and N. Gutfled. "Relations of Pollution to Mortality, New York City, 1963–1972." Paper No. 74-220, Annual Meeting, Air Pollution Control Assoc., Denver, Colo., 1974.
28. I. F. Goldstein and G. Block. "Asthma and Air Pollution in Two Inner City Areas in New York City." *J. Air Pollu. Control Assoc.* **24** (July 1974): 665–670.

29. Paper by the EPA National Environmental Research Center, presented at the American Medical Association Air Pollution Medical Research Conference, San Francisco, Dec. 5–6, 1974.
30. NAPCA. *Air Quality Criteria for Hydrocarbons*, AP-64. Washington, D.C.: HEW, 1970.
31. NAPCA. *Air Quality Criteria for Photochemical Oxidants*, AP-63. Washington, D.C.: HEW, 1970.
32. "News Focus." *J. Air Pollu. Control Assoc.* **23** (May 1973): 398.
33. *Recognition of Air Pollution Injury to Vegetation*. Pittsburgh, Pa.: Air Pollution Control Association, 1970.
34. A. C. Stern, H. C. Wohlers, R. W. Boubel, and W. P. Lowry. *Fundamentals of Air Pollution*. New York: Academic Press, 1973.
35. L. R. Babcock, Jr. "A Combined Pollution Index for Measurement of Total Air Pollution." *J. Air Pollu. Control Assoc.* **20** (October 1970): 653–659.
36. A. J. Vermeulen. "Acid Precipitation in the Netherland." *Environ. Sci. Tech.* **12**, no. 9 (September 1978): 1016–1021.
37. R. W. Shaw. "Acid Precipitation in Atlantic Canada." *Environ. Sci. Tech.* **13**, no. 4 (April 1979): 406–411.
38. N. R. Glass, G. E. Glass, and P. J. Rennie. "Effects of Acid Precipitation." *Environ. Sci. Tech.* **13**, no. 11 (November 1979): 1350–1355.
39. "News Focus". *J. Air Pollu. Control Assoc.* **29**, no. 10 (October 1979): 1074–1075.

Chapter 2
Federal Legislation and Regulatory Trends

2-1 INTRODUCTION

The existence of air pollution and the need for legislation to protect the health and welfare of the general populace are not modern phenomena. In thirteenth-century England the burning of soft coal polluted the atmosphere in urban areas to such an extent that in 1273 England passed a law to reduce air pollution from this source. However, it was not until the 1940s in the United States that efforts to control the degree of air pollution were initiated. Historically, the first measures against air pollution were taken by the state of California, in light of worsening conditions within the state and especially in the Los Angeles basin. It soon became apparent that other urban areas in the country were experiencing a decline in air quality, although frequently for reasons different from those that led specifically to the Los Angeles-type smog. As a result, federal intervention in the overall air pollution area was deemed necessary. The major development of legislative and regulatory acts took place over the period from 1955 to 1970. Without any precedents to rely on (except for some legislation in the water pollution area), legislators had to vary the approach of each succeeding act in order to overcome deficiencies in preceding acts. Nevertheless, by the mid-1970s the philosophical basis for federal regulation of air pollution was well developed. Of course, in some areas the actual regulations are continually changing in the light of new technological and economic developments. The energy reallocation problems of the 1970s also have an impact on the implementation of federally enacted air pollution control laws and regulations.

2-2 THE HISTORY OF FEDERALLY ENACTED LAWS

The early development of air pollution laws by the U.S. Congress is interesting as a classic example of federal versus states' rights. Congress was unwilling to grant the federal government a high degree of control, since air pollution problems were essentially local or regional ones. Although a number of affected regions encompassed two or more states, it was felt that air pollution control should be mandated on the state or local level. Although philosophically sound, this approach proved impractical for a number of reasons. Thus the history of air pollution legislation in this country exhibits increasing federal involvement, especially in terms of regulatory agencies. That history begins in 1955.

2-2-A The Air Pollution Control Act of 1955, Public Law 84-159, July 14, 1955

This act, the first federally enacted legislation, was narrow in scope and potential, primarily because of the federal government's hesitation to encroach on states' rights. The act considers prevention and control of air pollution at its source to be primarily the responsibility of state and local governments. However, it states that the growth and complexity of air pollution phenomena have resulted in increasing dangers to the population. As a result, federal financial assistance and leadership are essential for the development of cooperative federal, state, regional, and local programs to control and prevent air pollution.

The 1955 act initiated the following action:

1. Research on the effects of air pollution by the Public Health Service.
2. Provision for technical assistance to the states by the federal government.
3. Training of individuals in the area of air pollution.
4. In-house and out-of-house research on air pollution control.

Although this act was a small step, it was a necessary one toward the identification of air pollution sources and the analysis of pollution effects, and toward the effective legislation and enforcement by regulatory agencies which was developed over the next 15 years. It might also be noted that the English Clean Air Act was enacted shortly after this, in 1956. The impetus for this English act was the severe pollution episode in London of December, 1952, which was followed by a similar fog and inversion in January, 1956.

2-2-B Air Pollution Control Act Amendments of 1960, Public Law 86-493, June 6, 1960, and Amendments of 1962, Public Law 87-761, October 9, 1962

Although it was not until 1963 that a major change was made in the overall federal program for air pollution control and abatement, a minor but important amendment to the 1955 act was passed by Congress in 1960. In light of worsening conditions in urban areas due to an increasing contribution to air pollution by mobile sources, Congress directed the surgeon general to conduct a thorough study of motor vehicle exhausts in terms of their effects on human health. This study was completed in 1962 and reported to the Congress [1]. As a result, the act of 1955 was further amended in October, 1962. This amendment required the surgeon general to conduct studies relating the effects of motor vehicle exhausts to health. A more formal process for a continual review of the motor vehicle pollution problem was included in the 1963 act.

2-2-C The Clean Air Act of 1963, Public Law 88-206, December, 1963

Although not a landmark piece of legislation, the Clean Air Act of 1963 did provide for the first time for federal financial aid for research and technical assistance. Legislators continued to voice great concern about states' rights, and it was not clear at this moment of history how the responsibilities for air pollution control should be divided among federal, state, and local governments. Only in the case of interstate commerce was it evident that federal jurisdiction was necessary, on the basis of previous federal legislation in other areas. As a result, the act encouraged state, regional, and local programs for the control and abatement of air pollution, while reserving federal authority to intervene in interstate conflicts. Thus the classical three-tier system of government was preserved. At the same time the act recognized the need for state and local involvement in the implementation and enforcement of effective programs. At this particular time only 13 states had passed laws providing for statewide control of air pollution.

Specifically, the 1963 act provided for:

1. Acceleration in the research and training program.
2. Matching grants to state and local agencies for air pollution regulatory control programs, with the federal government paying two-thirds to three-quarters of the costs.
3. Developing air quality criteria.
4. Initiating efforts to control air pollution from all federal facilities.
5. Federal authority to abate interstate air pollution.
6. Encouraging efforts on the part of automotive companies and the fuel industries to prevent pollution.

In item 5, the federal enforcement authority was strictly limited to situations where the pollution from one state might endanger public health or welfare in another state. Two other provisions of the act were noteworthy in that they led to important subsequent legislation in the air pollution acts of 1967 and 1970: (1) development of air quality criteria which were to be used as guides in setting air quality standards and emission standards, and (2) research authority to develop methods for sulfur removal from fuels.

The Clean Air Act of 1963 provided for a formal process for reviewing the status of the motor vehicle pollution problem. A technical committee was to be formed with representatives from the department of Health, Education, and Welfare (HEW), the automotive industry, the manufacturers of control devices, and the producers of motor fuels. The committee was instructed to monitor the progress toward control, and to indicate those specific areas in which added research and development were essential. These initial investigations of mobile sources led to the next federal legislation, which dealt with motor vehicles.

2-2-D Motor Vehicle Air Pollution Control Act of 1965, Public Law 89-272, October 20, 1965

In the year following the 1963 act, extensive hearings were held in Congress on the control of automotive emissions. In a summary of the hearings [2] we find the following statement:

> Automotive exhaust was cited [in the 1963 hearings] as responsible for some 50 percent of the national air pollution problems. It is, in many respects, the most important and critical source of air pollution, and it is, beyond question, increasing in seriousness despite preliminary and isolated efforts to control it.

The result of these hearings was the Motor Vehicle Air Pollution Control Act of 1965, which in reality was an amendment to the Clean Air Act of 1963. The 1965 act formally recognized the technical and economic feasibility of setting automotive emission standards. It also recognized that local control would prove impractical, and a great deal of confusion would arise from different standards for the 50 states. Hence national standards should be set for automotive exhaust. On the basis of the hearings, it was deemed feasible to apply the then-current California state emission standards for hydrocarbons and carbon monoxide on a national basis. The automotive industry at that time disputed the practicality of national emission standards, but agreed to meet them with 1968 model cars. The 1965 act also stipulated that controls would be tightened as technological advances became available in conjunction with reasonable costs. The state of California was given waivers on the motor vehicle requirements, in order for it to promulgate standards more appropriate to the unique air pollution problems of that region.

Aside from the motor vehicle emission requirements, the Amended Clean Air Act of 1965 also gave the secretary of HEW the authority to intervene in intrastate air pollution problems of "substantial significance."

2-2-E The Air Quality Act of 1967, Public Law 90-148, November 21, 1967

By early 1967 the controversial topic of national emission standards, regionally implemented, versus regional standards regionally implemented was under heavy debate in Congress. National emission standards, of course, infringed upon states' rights. On January 30, 1967, President Johnson proposed national emission standards for industries which "contribute heavily" to air pollution [3]. However, the Senate Committee on Public Works felt that other considerations offset the arguments for national emission standards. The resulting Muskie bill rejected national standards, and attempted to implement a federal control program within the context of traditional governmental institutions. Subsequent success would depend upon the degree of cooperation among various governmental levels. Nevertheless, a 2-year study on the concept of national emission standards for stationary sources was provided for by the act, and this study served as a basis for 1970 legislative action.

Among the provisions of the 1967 Air Quality Act were:

1. Establishment of eight specific areas in the United States on the basis of common meteorology, topography, and climate. (The purpose of this geographical subdivision is not clear, and it never served any useful purpose.)
2. Designation of "air quality control regions" (AQCRs) within the United States, either inter- or intrastate. Within an AQCR an evaluation was to be conducted to determine the nature and extent of the pollution problem, including items such as meteorological data, ambient air quality data, and an emission inventory.
3. Development and issuance of air quality criteria (AQCs) for specific pollutants that have identifiable effects on human health and welfare.
4. Development and issuance of information on recommended air pollution control techniques (which led to the evolution of air pollution control technique reports). On the basis of this information, the federal government could recommend to state and local control agencies the technologies available to achieve the levels of air quality suggested in the air quality criteria reports. Such recommendations would require the initiation of joint industry-government research to develop and demonstrate the technology.
5. Requirement on a fixed time schedule for state and local agencies to establish air quality standards consistent with air quality criteria. The states were allowed to set higher standards than recommended in the criteria reports. If a state did not act, the secretary of HEW had the authority to establish air quality standards for each air quality region. The strict timetable included 90 days to file a letter of intent after receiving any air quality criteria and control techniques for a specific pollutant, adoption of air quality standards within 180 days after public hearings, and implementation of standards within another 180 days. Such a procedure was to be followed for each pollutant. If a state did not effectively enforce the standards set within 180 days notice by HEW, the U.S. Attorney

General could bring suit against the state. Thus states were given primary responsibility for action, but a very strong federal fallback authority was provided.

In the 2 years that followed enactment of the Air Quality Act of 1967, the federal program was not implemented according to the required time schedule. Only a limited number (108) of AQCRs were designed by 1970. The National Air Pollution Control Administration (NAPCA), which directed federal surveillance of the overall program, was understaffed. In addition, the procedure of hearings, and so forth, for setting up AQCRs proved to be too complex. Consequently, both the President and the Congress proposed new legislation early in 1970.

2-2-F The Clean Air Amendments Act of 1970, Public Law 91-604, December 31, 1970

New Year's Day, 1971, heralded the beginning of a new environmental decade, since the new year came in with an air pollution control law. On the preceding day President Nixon signed the bill known as the Clean Air Amendments Act of 1970. These amendments extended the geographical coverage of the federal program aimed at the prevention, control, and abatement of air pollution from stationary and mobile sources. All administrative functions formerly assigned to the secretary of HEW were transferred to the administrator of the newly created Environmental Protection Agency (EPA). Industry, federal agencies, and state administrators were faced with new requirements and new deadlines. The major goal of the act was the achievement of clean air throughout the nation by mid-decade, that is, July, 1975. The major provisions of the act were the following:

1. Additional research efforts were requested. A portion of $350 million to be spent over a 3-year period was earmarked for research on fuels at stationary sources. Special emphasis was given to (a) methods of cleaning fuels prior to combustion, rather than flue-gas cleaning techniques, (b) improved combustion techniques, and (c) methods for producing new or synthetic fuels which have a lower potential for creating polluted emissions. Authorization totaling $45 million for fundamental air pollution studies was contained in the legislation. A sum of $15 million was assigned to research on health and welfare effects of air pollutants, especially long-term effects on highly susceptible individuals. A second study for $30 million would assess the cause and effects of noise pollution. Thus this act recognized for the first time the need to study the increased noise levels that were occurring in urban areas. An Office of Noise Pollution and Abatement was to be established within the framework of EPA. Studies made by this office would lead to recommendations for appropriate future legislation.

2. Additional state and regional grant programs were authorized, and matching grants established for implementing standards.

3. *National ambient air quality standards*—including primary standards for the protection of public welfare—were to be set by EPA. The primary ambient air quality standards were to be set at values the "attainment and maintenance of which in the judgment of the administrator, based on such criteria and allowing an adequate margin of safety, are requisite to protect the public health."

4. The designation of air quality control regions was to be completed. Inclusion of all areas within a state in air quality control regions was required. Existing AQCRs were preserved, and a mechanism was provided for designating new ones. As a result of this action, a total of 235 AQCRs were designated by March 31, 1971. Ninety percent of these were intrastate. By mid-1975, the number of AQCRs totaled 247.

5. Implementation plans to meet standards were to fit a given timetable, but they would continue to be initiated by the states. Establishment of statewide plans for implementation designed to achieve primary or public health standards within 3 years was required. Section 107 of the 1970 act states "each state shall have the primary responsibility for assuring air quality within the entire geographic area comprising each state by submitting an implementation plan for such state which shall specify the manner in which national primary and secondary ambient air quality standards will be achieved and maintained within each air quality control region in each state." Although the methods used in each AQCR could be different, all methods had to result in meeting the standards. National ambient air quality standards and implementation plans were to be established for SO_2, particulate matter, carbon monoxide, hydrocarbons, and oxidants, as well as nitrogen oxides, lead, polynuclear organic matter, fluorides, and odors.

6. Standards of performance for *new* stationary sources were to be required. Each state was to implement and enforce the standard of performance. Thus the act explicitly called for retention of state authority for control of air pollution emissions. Some 19 industries were faced with the control of 14 selected agents which generally are specific to these industries. These agents include arsenic, chlorine gas, hydrogen chloride, copper, manganese, nickel, vanadium, zinc, barium, boron, chromium, selenium, pesticides, and radioactive substances. Before a new stationary source could begin operation, state or federal inspectors were required to certify that the controls would function. (These standards more than likely will change in the future.) The new stationary sources had to remain in compliance throughout the lifetime of the plant.

7. National emission standards for hazardous air pollutants were to be set and would apply to existing as well as new plants. The hazardous agents included lead, mercury, cadmium, and asbestos. Exemptions were possible.

8. Industry was required to monitor and maintain emission records and make them available to EPA officials. This provision included a right of entry to examine records.

9. Imposed fines and criminal penalties for violation of implementation plans, emission standards, performance standards, and the like, were tougher

than under earlier law. Conviction for a knowing violation was made subject to a penalty of $25,000 per day or imprisonment for 1 year, or both. For a second knowing violation the penalty was increased to $50,000 per day of the violation, imprisonment for 2 years, or both.

10. *Automobile emission standards* were to be generally set at a 90 percent reduction of the partially controlled 1970 levels. This portion of the act was and remains quite controversial, since levels of performance are set by Congress itself. An equally valid approach would have been to issue to EPA the right to set standards in the light of on-going technological and economic improvements in equipment, this providing some flexibility in the law. The EPA administrator was provided with the power to grant a single 1-year extension if evidence indicated that this would be prudent. With the hindsight gained in the 9 years since the signing of the act (at the time of this writing), we can conclude that a more flexible provision in the law originally might have saved both industry and the public a great deal of consternation. Now, with the additional factor of shortages of acceptable forms of energy, the original law has had to be amended several times to provide the requisite flexibility.

11. The development of low-emission vehicles was to be encouraged with appropriations for research. A sum of $89.1 million was authorized for a 6-year program (1970–1975) to develop a low-emission alternative to the current internal combustion engine.

12. *Aircraft emission standards* were to be developed by EPA.

13. Citizens' suits were permitted "against any person, including the United States, alleged to be in violation of emission standards or an order issued by the administrator." Suits could also be brought against the administrator, but only when he failed to act in cases where the law specified that he must. The act also prohibited the federal government from signing contracts with any company which was violating the air quality law.

Hence the Clean Air Amendments of 1970 provided for the first time (a) national ambient air quality standards, (b) national emission standards for stationary sources, and (c) standards of performance for existing plants. It also initiated the study of aircraft emission. Outside the area of air pollution, the cause and effects of noise pollution were to be studied, and a report submitted to Congress. With regard to automotive emissions, a 90 percent control of hydrocarbons and carbon monoxide was to be achieved by 1975-model cars, and a 90 percent reduction of NO_x emissions by the following model year. These deadlines have already been altered by 2 years, and interim values for permitted emissions are in effect (as of early 1975). (A more detailed discussion appears in Chapter 10.) Thus some of the goals set by the law have already been modified.

2-2-G The Clean Air Amendments Act of 1977, Public Law 95-95, August 7, 1977

The Clean Air Act Amendments of 1977 retain the fundamental approach to air pollution control established in 1970, notwithstanding new provisions requiring thorough review of both ambient and emission standards as well as alternative pollution control strategies. By 1977 most areas of the country had still not attained the National Ambient Air Quality Standards (NAAQS) for at least one pollutant. For those areas which have not attained NAAQS, so-called "nonattainment areas," states must have an approved state implementation plan (SIP) revision by July 1, 1979, which provides for attainment of primary NAAQS by December 31, 1982. This requirement is a precondition for the construction or modification of major emission sources in nonattainment areas after June 30, 1979. If despite implementation of all "reasonably available measures" a state cannot attain primary standards for carbon monoxides or photochemical oxidants in timely fashion, it must submit a second SIP revision by December 31, 1982 which provides for attainment by December 31, 1987. All plan revisions must prior to attainment provide for "reasonable further progress" toward attainment in terms of annual incremental reductions in emissions.

For those areas that are cleaner than required by NAAQS, the SIP must include an elaborate program to prevent the significant deterioration of air quality. All such nondegradation area must be designated Class I, II, or III, depending upon the degree of deterioration which is to be allowed and limits are assigned to increases of pollution concentrations for each classification. The classifications are:

Class I: Areas where almost no change from current air quality will be allowed.

Class II: Areas where moderate change will be allowed, but where stringent air quality constraints are desirable.

Class III: Areas where substantial industrial growth will be allowed and where the increase in concentration of pollutants up to the federal standards will be insignificant.

Congress specified which of these areas must be protected by the most stringent Class I designation. All other areas were initially Class II, with the states generally free to redesignate them as Class I or III. Congress also specified the maximum allowable incremental increases in concentrations of sulfur dioxide and particulate for each classification and gave EPA 2 years to originate comparable increments for hydrocarbons, carbon monoxide, photochemical oxidents, and nitrogen oxides.

In both "nonattainment" and "nondegradation" areas, major stationary sources may by constructed only by permit and must, at the very least, meet the new source performance standards prescribed by law. The implication could be drawn that no industrial growth could occur in the nonattainment areas. However, there needed to be a way out of this dilemma, and the solution appeared in the Federal Register of December 21, 1976 as an interpretative ruling on the preconstruction review requirements for all new or modified stationary sources of air pollution. This ruling has since come to be known as the "emission offset policy" or simply the "offset policy." The policy says that a new source can locate in a nonattainment area if its emissions are more than offset by concurrent emission reductions from existing sources in that area. Certain stringent conditions must be met however. These conditions are "designed to ensure that the new sources emissions will be controlled to the greatest degree possible; that more than equivalent offsetting emission reductions (emission offsets) will be obtained from existing sources; and that there will be progress toward achievement of the NAAQS."

The 1977 act also provides for the "banking" of offsets. That is, if the offsets achieved are considerably greater than the new source's emissions, a portion of this "excess" emission reductions can be "banked" by the source for use in future growth. The states, if they choose to allow "banking," become the bankers.

Modifications to the emission standards for vehicles are discussed in Chapter 10. Whether the promise of clean air is made a reality within a reasonable time depends strongly on fully manning, funding, and implementing the law. Additional amendments probably will be necessary, as in the past, in the light of technological and economic advances, as well as in terms of the overall energy requirements of the nation.

2-3 AIR QUALITY CRITERIA AND AMBIENT AIR QUALITY EMISSION STANDARDS

The third item discussed under the Air Quality Act of 1967 (see Section 2-2-E) dealt with the development and issuance of air quality criteria (AQC). This action (originally outlined in the 1963 act) was taken to aid states in developing air quality standards for the ambient atmosphere. The 1970 amendments reaffirmed the need for such criteria. Basically, air quality criteria are expressions of the latest scientific knowledge based on the experience of experts. They indicate qualitatively and quantitatively the relationship between various levels of exposure to pollutants and the short- and long-term effects on health and welfare [4]. Air quality criteria are descriptive, in that they describe effects that can be expected to occur when pollutant levels reach or exceed specific figures for a specific time period. They should delineate the effects from combinations of contaminants, as well as from individual pollutants. Economic and technical considerations are not relevant to the establishment of AQCs. It is the long-term effects which air quality criteria must clearly delineate, if control efforts are to be adequate in overcoming chronic exposure. A number of air quality criteria have been established by EPA [5, 6, 7, 8, 9, 10].

Air quality criteria are an essential step in providing a quantitative basis for air quality standards. Standards, unlike criteria, are prescriptive. They prescribe the pollutant levels that cannot be legally exceeded during a specific time period in a specific geographical region. In accord with the 1967 act, ambient air quality standards were to be left to the discretion of state and local authorities. The 1970 act required federal promulgation of national primary and secondary standards. Within 90 days of the issuance of air quality criteria for a given pollutant, the EPA administrator was required to promulgate standards. Such standards were to be established equitably in terms of the social, political, technological, and economic aspects of the problem, and would be subject to revision as these aspects changed in relation to each other over a period of time. (We might note here the necessity for public education with regard to the price tags associated with achieving various air quality levels. Cost effectiveness is as important a goal when setting air quality standards as when designing any free-enterprise operation.)

The purpose of primary standards, according to the 1970 act, was immediate protection of the public health. As such, primary standards were to be achieved regardless of cost and within a specified time limit. Secondary standards were to protect the public from known or anticipated adverse effects. The time schedule for their achievement was to be determined by state and local governments. Both primary and secondary standards had to be consistent with air quality criteria. In addition, standards had to prevent

Table 2-1 FEDERAL AMBIENT AIR QUALITY STANDARDS[a]

POLLUTANT	AVERAGING TIME	PRIMARY STANDARD	SECONDARY STANDARD	MEASUREMENT METHOD
Carbon monoxide	8 hr	10 mg/m^3 (9 ppm)	Same	Nondispersive infrared spectroscopy
	1 hr	40 mg/m^3 (35 ppm)	Same	
Nitrogen dioxide	Annual average	100 μg/m^3 (0.05 ppm)	Same	Colorimetric using NaOH
Sulfur dioxide	Annual average	80 μg/m^3 (0.03 ppm)		Pararosaniline method
	24 hr	365 μg/m^3 (0.14 ppm)		
	3 hr		1300 μg/m^3 (0.5 ppm)	
Suspended particulate matter	Annual geometric mean	75 μg/m^3	60 μg/m^3	High-volume sampling
	24 hr	260 μg/m^3	150 μg/m^3	
Hydrocarbons (corrected for methane)	3 hr (6–9 a.m.)	160 μg/m^3 (0.24 ppm)	Same	Flame ionization detector using gas chromatography
Ozone	1 hr	240 μg/m^3 (0.12 ppm)	Same	Chemiluminescent method
Lead	3 months	1.5 μg/m^3	Same	

SOURCE: *Federal Register* **36**, no. 84, part II, April 30, 1971, pp. 8186–8201 [11]; **43**, September, 1978, p. 46246.

[a] Standards, other than those based on annual average or annual geometric average, are not to be exceeded more than once a year.

the continuing deterioration of air quality in any portion of an air quality control region. This stipulation has caused much controversy. Some interpret it to mean that a region cannot backslide in air quality, even though the present quality may be far superior to the federal standard. Since new industrial or commercial operations in the region cannot be expected to contribute zero pollution, a strict interpretation of this stipulation would stifle economic growth in a region yet fail to press for a decrease in contaminant emissions from sources already in the region.

Ambient air quality standards have already been set by the EPA administrator for six pollutants, as of April, 1971. Table 2-1 summarizes these standards. Note that the value of a standard depends upon the time period over which the measurement is averaged. The suggested method of measurement is also indicated. Except for those values which are annual mean values, the standards listed are not to be exceeded more than once a year.

2-4 NATIONAL EMISSION OR PERFORMANCE STANDARDS

Emission standards essentially place a limit on the amount or concentration of a contaminant that may be emitted from a source. In order to maintain or improve the existing ambient air quality within a region to comply with federal or state air quality standards, it is frequently necessary for certain industries to be regulated by emission standards promulgated by the federal or state government. These performance standards usually reflect the maximum degree of emission control deemed achievable with the present technology in that industry (or borrowed from another industry). A technological development is only considered available, however, if no undue economic hardship will be forced upon industrial plants that are required to install it. As a consequence, emission standards based on the "best practical means" could become more stringent as the state of the art of control equipment improves with time.

The establishment of emission standards for a region is not a simple task. A number of factors must be taken into account.

1. Of major concern is the availability of technology which would be appropriate for the cleanup of a given type of industry. In some cases (the removal of SO_2 from power-plant stack gas, for example) a technological development may merely show promise of availability and a decision must be made on the basis of that promise. In such cases it is probably best not to make a firm commitment until it is apparent that alternate choices are not or will not be available in the near future and a delay will endanger a portion of the population in that region.

2. Monitoring stations must be available to measure (a) the actual industrial emissions for which control is sought and (b) the ambient air quality so that the effectiveness of the standards can be established.

3. Regulatory agencies must be organized to cope with the measurement and enforcement of the standards. Emission standards are worthless unless local authorities are given the power to enforce them and to penalize offenders. And if local authorities run into cases that are too difficult, provision must be made for federal intervention.

4. Since different pollutants may be present in the control region, the synergistic effects of those particular agents should be clearly established. Setting emission standards for individual pollutants on the basis of their individual effects may not be sufficient in some cases.

5. In many urban areas it may be necessary to prepare a diffusion model that predicts with reasonable accuracy the effects of curtailment of various emissions on the ambient air quality. Among other information, such a model should include the effects of atmospheric chemical reactions, the meteorology and topology of the region, and the location and emission rates of known sources. The defense of local emission standards may prove difficult without such models.

6. Finally, in conjunction with item 5, every effort should be made to obtain a reasonably accurate estimate of future growth or decline of industry and population within the region. The addition of several emission sources in a localized area may severely affect the emission rates required for presently operating industries. Hence knowledge of growth factors is an extremely important consideration in the determination of emission standards.

In a number of instances it has been imperative that the federal government set the basic standards in certain industries, without state interference. For example, it would not be reasonable for each of the 50 states to set its own standards for the automobiles brought into and sold in that state, for economic reasons alone. Hence the federal government set the standards (with the exception of the state of California) for emissions of certain pollutants from these mobile sources (see Chapter 10 for details on the history of these standards). Likewise, it would not be economically feasible for the manufacturers of aircraft engines to meet different standards for the hundreds of different airports across the nation. Thus federal emission standards have been set for engines manufactured after 1978, with retrofitting required in some cases to reduce certain pollutants from engines produced before that time (see Chapter 10 for details on aircraft emission standards).

The establishment of federal standards for automotive and aircraft engines is a sensible approach, since the manufacturers are few in number but the products influence air quality in all urban areas. Another area where the case for strong federal standards can be strongly defended encompasses those industries in which plants are numerous, are spread geographically across the nation, and provide a basic commodity essential for the development of the country. If the standards were set by the states, unfair economic advantages might develop. States would compete with each other in setting low standards in order to attract these basic industries. As a consequence, federal emission or performance standards have been set for a number of industrial processes to date. These are "new-source" performance standards, and pertain to new stationary sources where a broad category of source might contribute significantly to endangering the public health. Interestingly, the standards apply also when modifications are undertaken to existing facilities. Modifications in this case are defined as both operational and physical changes which either increase emission rates or initiate new emissions from the plant.

On December 23, 1971, final standards were published for certain plants commencing construction or modification after August 17, 1971. The industries involved are steam generators, Portland cement plants, incinerators, nitric acid plants, and sulfuric acid plants. These emission standards are summarized in Table 2-2. A second set of new-source performance standards proposed by EPA appeared in the *Federal Register* on June 11, 1973, and were promulgated within 90 days. These cover asphalt concrete plants, petroleum refineries, storage vessels for petroleum refineries, storage vessels for petroleum liquids, secondary lead smelters, secondary brass and bronze ingot production plants, iron and steel plants, and sewage treatment plant incinerators. A summary of these regulations appears in Table 2-3.

Table 2-2 STANDARDS OF PERFORMANCE FOR NEW STATIONARY SOURCES, EFFECTIVE DECEMBER 23, 1971[a]

1. Steam electric power plants. The original regulations, promulgated for fossil fuel-fired units on December 23, 1971, have been significantly revised as of February 6, 1980. The new regulations apply to units of more than 73 MW (250 million Btu/hr) of heat input, for which construction began after September 18, 1978. Regulations are directed toward the control of SO_2, particulates, and NO_x. See Table 2-6 for information.
2. Portland cement plants. Particulate matter standards for kilns are a maximum 2-hr average emission of 0.30 lb/ton of feed (0.15 kg/metric ton) and an opacity not greater than 20 percent (except when uncombined water is present). For clinker coolers, opacity must be 10 percent or less and the maximum 2-hr average emission is 0.10 lb/ton of feed (0.050 kg/metric ton).
3. Solid waste incinerators. Established for new incinerators with a charging rate in excess of 50 tons/day. Particulate emission standard is a maximum 2-hr average concentration of 0.08 gr/dscf (0.18 g/m^3) corrected to 12 percent carbon dioxide.
4. Nitric acid plants. Standard is a maximum 2-hr average nitrogen oxide emission of 3 lb/ton of acid produced (1.5 kg/metric ton), expressed as nitrogen dioxide. Applicable to any unit producing 30 to 70 percent nitric acid by either pressure or atmospheric pressure processes.
5. Sulfuric acid plants. Applies to plants employing the contact process. Standard is a maximum 2-hr average emission of SO_2 of 4 lb/ton of acid produced (2 kg/metric ton). An acid mist standard, a maximum 2-hr average emission of 0.15 lb/ton of acid produced (0.75 kg/metric ton), also is established.

SOURCE: EPA, "Standards for Performance for New Stationary Sources," *Federal Register* **36**, December, 1971, pp. 24876-24895 [12].

[a] Standards apply to the upgrading or modernization of existing plants as well as new construction.

Table 2-3 STANDARDS OF PERFORMANCE FOR NEW STATIONARY SOURCES, EFFECTIVE AUGUST, 1973

1. Asphalt concrete plants. No discharge of particulate matter in gas effluent in excess of 70 mg/m^3 (dry normal) (or 0.031 gr/dscf). Opacity limited to 10 percent or less, except for 2 min in any hour.
2. Petroleum refineries. Particulate from fluid catalytic cracking unit catalyst regenerator may not exceed 50 mg/m^3 (0.022 gr/dscf). Opacity restricted to 20 percent or less except for 3 min in any hour. If auxiliary fuels are burned in an incinerator waste-heat boiler, particulate in excess of above shall not exceed 0.043 kg/million kJ (0.10 lb/million Btu). In addition, no discharge of CO in excess of 0.050 percent by volume, and no release of H_2S in excess of 230 mg/m^3 (dry normal) (0.10 gr/dscf).
3. Storage vessels for petroleum liquids. Stored liquids may not have a true vapor pressure which is (a) 78 mm Hg (1.52 psia) or less unless the storage vessel is equipped with a conservation vent; (b) in excess of 78 mm Hg but not greater than 570 mm Hg (11.1 psia) unless equipped with a floating roof; (c) in excess of 570 mm Hg unless equipped with a vapor recovery system.
4. Secondary lead smelters. No gases emitted from the blast or reverberatory furnace may contain particulate in excess of 50 mg/m^3 (dry normal) (or 0.022 gr/dscf) or exhibit 20 percent opacity or greater except for 2 min in any hour. No discharge from any pot furnace may exhibit 10 percent opacity or greater except for 2 min in any hour.
5. Secondary brass and bronze ingot production plants. No discharge of particulate in excess of 50 mg/m^3 (dry normal) (or 0.022 gr/dscf) from a reverberatory furnace. Discharges from reverberatory, blast, or electric furnaces may not exhibit 10 percent opacity or greater except for 2 min in any hour.
6. Iron and steel plants. Particulate discharges may not exceed 50 mg/m^3 (or 0.022 gr/dscf) from basic oxygen furnaces, and the opacity must be 10 percent or less except for 2 min in any hour.
7. Sewage treatment plant incinerators. No discharge permitted for particulates in excess of 70 mg/m^3 (dry normal) (or 0.031 gr/dscf) or which exhibit 10 percent opacity or greater, except for 2 min in any hour.

The amendments to the 1970 Clean Air Act also provided for national emission standards for hazardous air pollutants. Such pollutants include those which, in the judgment of the EPA administrator, may cause or contribute to serious irreversible or incapacitating reversible illness. Standards have been promulgated for asbestos, beryllium, and mercury [13], as of March 30, 1973. The specific regulations for each of these health hazards are briefly summarized in Table 2-4.

In late 1974, EPA proposed performance standards for primary copper, zinc, and lead smelters. The regulations are to apply chiefly to new plants, but they also cover modifications of existing plants. The proposed particulate standard is a maximum of 50 mg/m^3 (dry) measured at standard conditions, and the SO_2 standard requires emissions of less than 0.65 percent of volume. In addition, the visibility standard is such that the plume opacity cannot be greater than 20 percent for more than 2 min in any hour. New source performance standards for five other categories of stationary air pollution sources were proposed in October, 1974, by EPA. These standards appear in the issues of the *Federal Register* for October 21–24 and are also

Table 2-4 SUMMARY OF NATIONAL EMISSION STANDARDS FOR ASBESTOS, BERYLLIUM, AND MERCURY

1. The asbestos standards prohibit visible emissions from any asbestos mill, from plants that manufacture such products as insulating and fireproofing materials, felt, and floor tile, and from spray-on applications of materials containing more than 1 percent asbestos on a dry basis. The regulations also specify certain precautions to be taken during the demolition of buildings.
2. Beryllium standards stipulate that (a) emissions from stationary sources shall not exceed 10 g of beryllium over a 24-hr period—or an ambient concentration limit of 0.01 $\mu g/m^3$ in the vicinity of the source may be requested from the EPA administrator—and (b) the burning of beryllium and/or beryllium-containing waste is prohibited except in incinerators. Incinerators emissions must comply with the above 24-hr standard. Beryllium emission standards are also set for rocket-motor test firings.
3. The mercury standard applies to stationary sources which process mercury ore to recover mercury, and to those processes which use mercury chlor-alkali cells to produce chlorine gas and alkali metal hydroxide. In these cases the emissions are not to exceed 2300 g of mercury in a 24-hr period.

summarized in reference 14. The sources include electric arc furnaces, ferroalloy facilities, phosphate fertilizer industries, primary aluminum plants, and coal preparation plants.

Standards of performance for new stationary sources with heat input less than 250 million Btu/hr and for sources in operation prior to 1972 are established by the various states. Most states have adopted emission standards for particulate and sulfur dioxide which vary according to the heat input of the source as depicted in Figure 2-1. Representative values of a, b, c, d, e, and f for selected states are presented in Table 2-5. The values presented in Table 2-5 were chosen by the state agencies to support their

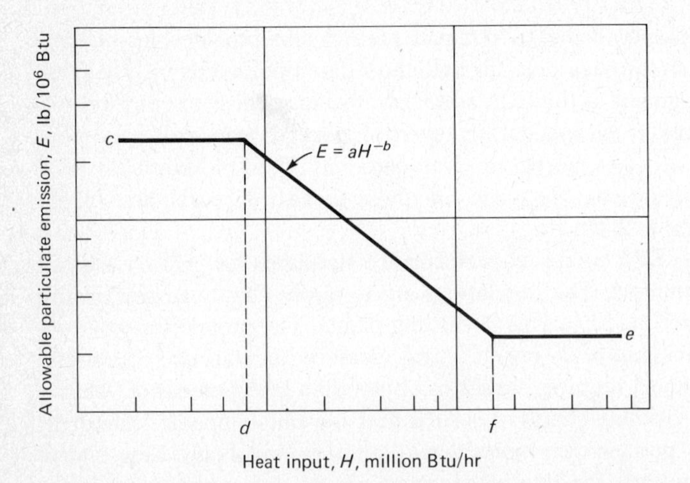

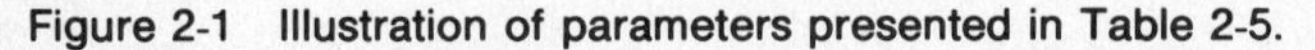
Figure 2-1 Illustration of parameters presented in Table 2-5.

respective implementation plans so as to attain compliance with the Federal ambient air quality standards. The values are subject to change as both the Federal standards and the state implementation plans are changed.

Note that for sources having heat inputs less than the value of (d), the allowable emission rate is constant at the value of (c) and for sources larger than the value of (f) the allowable emission rate is constant at the value of (e). The following example for the state of Indiana illustrates the application of the authorized procedure.

Example 2-1

Determine the standard of performance for particulate matter for a new industrial power plant generating 10,000 kW and burning Indiana coal.

SOLUTION

Assuming a thermal efficiency of 25 percent, and noting that η_{th} = net power/heat input rate, the heat input is calculated as follows.

$$H = \frac{10{,}000 \text{ kW}}{0.25} 3413 \text{ Btu/kWhr} = 136.5\, M \text{ Btu/hr}$$

The value of H expressed in MBtu/hr is larger than the value of d in Table 2-5 for Indiana, therefore the equation shown in Figure 2-1 must be used, namely,

$$E = aH^{-b}$$

From Table 2-5 values of a and b are obtained. Hence

$$E = (0.865)(136.5)^{-0.159} = 0.396 \text{ lb}/10^6 \text{ Btu}$$

Similar relationships are available for obtaining allowable emission rates for SO_2. In all cases the pertinent state regulations should be consulted for specific emission standards. It is important to note that when emissions are related to heat release, as by an equation of the type $E = aH^{-b}$, then an emission standard cannot be met by simply diluting the exhaust stream with fresh air in order to reduce the exit concentration.

In February, 1980, the EPA in the United States promulgated new regulations for the control of SO_2, particulates, and NO_x from electric utility steam generating units. These regulations made significant changes in those standards passed in December, 1971. For example, the 1971 regulation for the control of SO_2 merely stated a maximum emission rate for gaseous, liquid, and solid fuels. The 1980 regulation for coal-fired units states a minimum reduction of potential emissions as well as a maximum allowable emission rate (Figure 2-2). In addition, sulfur content and heating value of the fuel are also considered. A brief summary of the important points made in these regulations is presented in Table 2-6.

Table 2-5 EMISSION STANDARDS FOR PARTICULATE MATTER FROM FUEL BURNING EQUIPMENT[a] (15)

	PARAMETERS FOR COMPUTING ALLOWABLE EMISSION RATE					
	a	*b*	*c*	*d*	*e*	*f*
STATE AND SOURCE						
Colorado	0.5	0.26	0.50	10	0.10	500
Idaho	1.026	0.233	0.60	10	0.12	10,000
Illinois	5.18	0.715	1.00	10	0.10	250
Indiana	0.865	0.159	0.60	20	0.30	10,000
Pennsylvania	3.6	0.56	0.40	50	0.10	600
NEW SOURCE						
Kentucky	1.919	0.535	0.56	10	0.10	250

[a] Values to be used in conjunction with Fig. 2-1.

Table 2-6 NEW SOURCE PERFORMANCE STANDARDS FOR FOSSIL-FUELED ELECTRIC UTILITY STEAM GENERATING UNITS (1980)

The 1980 regulations for steam electric power plants apply to units of more than 73 MW (250 million Btu/hr) of heat input, for which construction began after September 18, 1978. Regulations are directed toward the control of SO_2, particulates, and NO_x. The summary below contains most of the major points in these regulations.

a. SO_2 standard. For a coal-fired unit, the standard requires at least a 90 percent reduction of potential SO_2 emissions and limits the SO_2 emission rate to 1.2 lb/10^6 Btu input, or requires at least a 70 percent reduction and limits the emission rate to 0.6 lb/10^6 Btu. In addition, the standard specifies a unique maximum allowable emission rate and unique minimum reduction of potential emission based on sulfur content and heating value of the coal. The complicated standard is best interpreted by employing Figure 2-2. To determine the standard, a radial line is drawn from the origin (upper left corner of the diagram) to a point on the curvilinear sulfur content-heating value coordinate system. Where the line crosses the "admissible region" boundary fixes the values of minimum reduction and maximum emission rate required by the standard. For gaseous and liquid fuels, SO_2 emissions are limited to 0.20 lb/million Btu heat input (86 g/million kJ).

b. Particulate standard. Emissions are limited to 0.03 lb/million Btu heat input (13 g/million kJ). In addition, the standard limits opacity to 20 percent for a 6-minute average.

c. NO_x standard. These standards vary according to the fuel type.
   1. 0.20 lb/million Btu (86 g/million kJ) heat input for gaseous fuel.
   2. 0.30 lb/million Btu (130 g/million kJ) heat input for liquid fuel, except for shale oil.
   3. 0.50 lb/million Btu (210 g/million kJ) heat input from subbituminous coal, shale oil, and fuels derived from coal.
   4. 0.60 lb/million Btu (260 g/million kJ) heat input from anthracite or bituminous coal.

   The NO_x standards are based on a 30-day rolling average.

SOURCE: EPA, "Standards of Performance for New Stationary Sources; Electric Utility Steam Generating Units," *Federal Register* **45**, February, 1980, pp. 8210–8213.

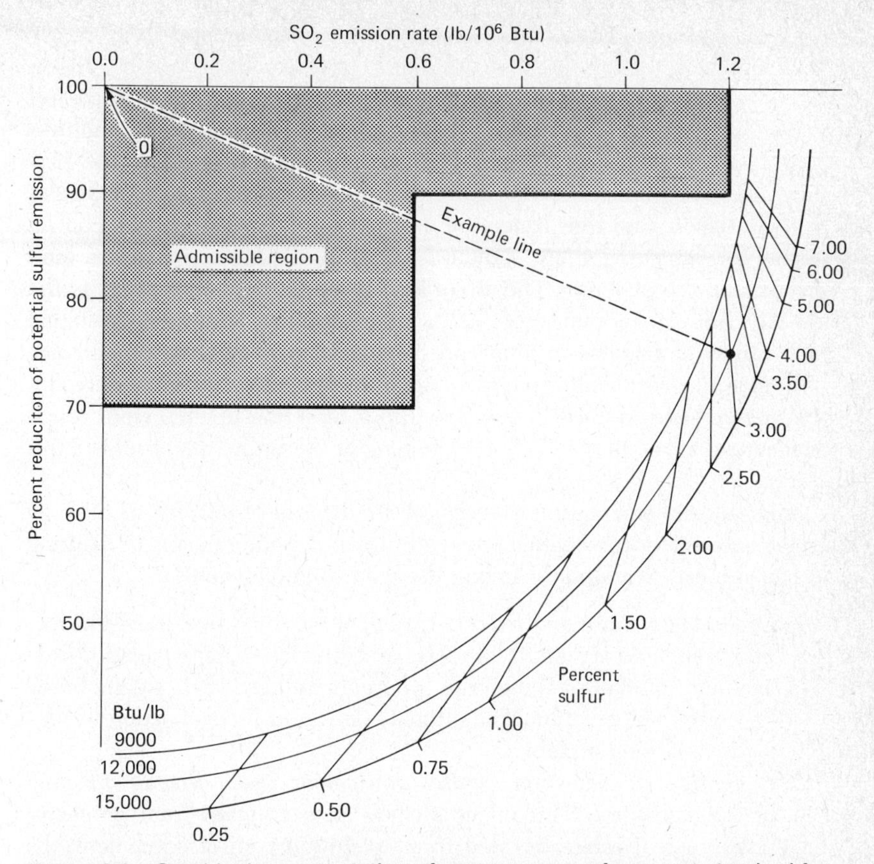

Figure 2-2 Graphical representation of new source performance standard for SO_2 promulgated in 1980 for coal-fired power plants. (Figure devised by J. Molburg, *J. Air Pollu. Control Assoc.* 30 (No. 2), Feb. 1980, 172.)

2-5 IMPLEMENTATION OF AND COMPLIANCE WITH STANDARDS

The Air Pollution Control Acts of 1967 and 1970 provide for the development of plans in each AQCR to implement the control strategy for that region. These plans were required to consider both short-term and long-term requirements for achieving the air quality standards. One approach to regulation frequently advocated by planners is that emissions can be controlled only within the limits of what is technologically and economically feasible at the time. If an industry's compliance with regulations is not deemed feasible, then the industry is granted a variance for a specific time period. During this time period, of course, air pollution is not being reduced, so that such a variance may be considered a license to pollute.

Another approach to regulation holds that emission standards should not be set unless they can be enforced in the present time frame. Such requirements must be economically reasonable, however, and this tends to encourage the setting of a minimum standard for emission controls. Advocates of this approach point out that setting a high standard which reflects emission controls achieved in the better installations in the country (or world) makes it economically impossible for many plants to comply, and thus nothing is achieved by such high standards. In some cases even the best existing control equipment is not sufficient to meet the air quality standards for a given region, and thus it may be necessary to adopt alternative control strategies for achieving these standards. One of these alternatives is land zoning or land-use planning. The use of land zoning is implied strongly in the Clean Air Act Amendments of 1970. By restricting land use within an AQCR, it may be possible to limit emissions in a localized area and enhance the air quality downwind from that area. In the past, unfortunately, industrial parks have frequently developed in a direction upwind from urban or commercial areas. Better land-use planning in the future might relieve this situation.

Once an implementation plan based on the use of current technology that is economically reasonable has been drawn up, then means of assuring compliance must be adopted. Among these are the following:

1. *Fines*. Many state and federal laws now incorporate a provision for fining an industry or its officers, or both, for blatant violations of emission standards. These fines are frequently levied on a daily basis for continued violation. The ultimate deterrent is court action which leads to closing a plant.
2. *Reduction of taxes on capital equipment used directly for air pollution control*. This might include an exemption from property taxes on real estate devoted to air pollution control equipment, as well as exemption from property taxes on the equipment itself. In addition, provision may be made for federal and local governments to grant tax concessions to permit industries to write off some portion (say, one-third) of the purchase price on pollution equipment.
3. *Improved financing arrangements*. For some industries, especially at times when the money market is tight and interest rates high, it may be necessary to provide incentive by offering loans at reduced interest rates. This could be done through the federal government or through private banks.

Another approach to gaining industry compliance with air pollution regulations is the use of fees. Other countries use this technique, but it has not been looked upon as a viable solution in the United States. Under this system, the industry is charged a fee commensurate with the degree of pollution it is discharging. The fee system is not an explicit control technique, and it favors those industries which are in a financial position to accept the penalty. In effect, the offending industry pays a continuing fine for not being in compliance with the law, but continues to pollute the atmosphere. Thus the fee system fails to alleviate the problem.

One of the interesting features of the Clean Air Act Amendments of 1970 is the nature of the citizens' suits made possible by the law. Prior to this law it was necessary for a citizen to be able to prove that he was directly affected, in terms of health or welfare, by an air pollution incident. The suit could not be brought in terms of the effects of the contaminant on someone else, or on the population in general. The 1970 amendments made "class" suits possible in the field of air pollution, and direct proof of injury to the plaintiff is not necessary. This broader interpretation makes the defense against such charges more difficult, and as a result places a greater burden on a given industry to comply with the standards.

QUESTIONS

1. What basic changes in political concepts took place during the development of federal air pollution control legislation between 1955 and 1977?
2. Why is it advantageous to have federal emission standards for vehicles?
3. Discuss why AQCRs are more advantageous than relying upon the separate states to establish air quality standards?
4. Outline the steps that must be taken in the establishment of (a) air quality standards and (b) emission standards.
5. What is the purpose of air quality criteria documents? Why should they be published before the establishment of air quality standards?
6. The SO_x level of one section of a state is much below the federal air quality standard. A large electric utility has proposed to build a coal-burning plant in the region without any SO_2 control equipment. Using the most advanced concepts, it has been determined that even with the discharge from the proposed plant the SO_2 level in the region will still be below the federal air quality standard. Should the plant be built? Discuss.
7. As will be discussed in subsequent chapters, the reduction of air pollution requires an increase in energy consumption. Assuming this to be valid, discuss the apparent conflict between realizing a reduction of air pollution and conserving energy sources.
8. What major political concept was overcome in the establishment of federal control of air pollution?
9. Of what value are diffusion models in the establishment of emission standards?
10. What are the differences between primary and secondary air quality standards?
11. Why are there no emission standards for photochemical oxidants?
12. Why is there an air quality standard for hydrocarbons even though it is recognized that most hydrocarbons per se are not detrimental to health?
13. It has been suggested that the only way of achieving the proposed air quality standards in some cities is to prohibit the use of private automobiles in certain areas. Discuss the ramifications of this proposal.
14. There are no air quality standards for asbestos, mercury, and beryllium, although there are emission standards for these materials. Why?
15. Discuss the advantages of obtaining compliance with standards based upon federal law as compared with compliance based on the older concept of public law.

16. What is meant by an emission offset policy?
17. What is the purpose of designating various regions of the country by classes?
18. What is the purpose or advantage of stating allowable emissions rates in terms of heat release rates?

PROBLEMS

2-1. A newly constructed power plant proposes to burn coal with a heating value of (a) 10,500 Btu/lb and (b) 11,500 Btu/lb. What percent sulfur, by weight, can the coal contain if the emission standard of 1.2 lb/10^6 Btu is to be met without a SO_2 control device?

2-2. A newly constructed power plant proposes to burn coal with a sulfur content of (a) 1.8, (b) 2.4, and (c) 3.0 percent by weight. If the plant is to meet the performance standard of 1.2 lb/10^6 Btu for emissions from new plants, what percent SO_2 removal is required? The fuel heating value is 10,750 Btu/lb.

2-3. A power-plant stack gas at 440°C contains (a) 2200 ppm and (b) 1400 ppm of SO_2. If the volume rate of gas emitted is 15,000 m^3/min, what is the SO_2 emission rate in grams per second? The stack pressure is 1.0 bar.

2-4. A power plant burns 150 tons of coal per day, and there are 18.2 lb of flue gas produced per pound of coal. The sulfur content of the coal is 2.0 percent, and the molar mass (molecular weight) of the flue gas is 30.5. If all the sulfur in the coal appears as SO_2 in the flue gas, what is the concentration of SO_2 in the flue gas in parts per million?

2-5. Consider the design of a new 1000-MW steam power plant. The fuels available are natural gas, residual fuel oil, and a low-sulfur coal. The ash content, sulfur content, and heating value of each are listed below:

	% ASH	% SULFUR	HEATING VALUE
Natural gas	0	0	18,500 Btu/lb (975 Btu/dscf)
Residual oil	0	1.4	18,000 Btu/lb
Coal	9	0.8	12,000 Btu/lb

In addition, the emission factors for particulates, NO_x and SO_x for these fuels are

	PARTICULATES	NO_x	SO_x
Natural gas	15	390	0.4
Residual oil	10	104	157S
Coal	16A	20	38S

Coal emission factors are in pounds per ton of coal burned, oil emission factors are in pounds per 1000 gallons burned, and gas emission factors are in pounds per million cubic feet burned. The specific gravity of the residual oil is 0.94, and the thermal efficiency of the plant is taken to be 39 percent. Symbol *A* is the percent ash, and symbol *S* is the percent sulfur in the fuel.

Determine the rate of emissions of (a) particulates, (b) NO_x, and (c) SO_x for each of the fuels in pounds per hour.

2-6. A municipal incinerator consumes 90 tons per day of waste consisting of 30 percent paper and 70 percent garbage by weight. It has been determined that paper produces 105 dscf/lb of gas and garbage produces 120 dscf/lb when burned with 20 percent excess air. Based on emission factors presented in Table 5-5 and new source performance standards presented in Table 2-2, determine if particulate control equipment is required and what percent removal is required.

2-7. A sulfuric acid manufacturing plant operates such that 50 tons of acid are produced with an estimated emission of 60 lb of acid mist. Based on emission standards for new sources, should control equipment be considered, and if so, what percent removal efficiency is required of the control equipment?

2-8. Based on the sulfuric acid mist emission factors presented in Table 5-5, what is the range of percentages of acid mist which must be removed to meet the new source emission standard presented in Table 2-2?

2-9. A cement plant uses 650 lb of limestone feedstock to produce 376 lb of cement. Based upon the emission factor presented in Table 5-5 and the emission standard presented in Table 2-2, what percent of the particulate generated must be removed?

2-10. A solid waste incinerator generates 17 lb of particulate in the flue gas per ton of waste, without particulate control equipment. If 70 dscf of flue gas are produced per pound of waste incinerated, determine the percent of particulate that must be removed by control equipment from the flue gas to satisfy the federal emission standard shown in Table 2-2.

2-11. A new industrial steam generating plant using 20,000 lb steam/hr and burning coal with a heating value of 9500 Btu/lb is to be built in Idaho. If the plant has a thermal efficiency of 25 percent, determine the allowable particulate emission rate in lb particulate/10^6 Btu.

2-12. The sulfur dioxide emission standard for stationary sources having a heat input H is given by the equation, $E = 6.0$ for H less than 23 million Btu/hr, $E = 1.2$ for H greater than 3000 million Btu/hr, and $E = 17.0\ H^{-0.33}$ for values of H in between, where E is in lb $SO_2/10^6$ Btu input. Determine the allowable emission rate E for a 20,000-kW electric plant having a thermal efficiency of 24 percent.

2-13. A power plant elects to burn coal with a sulfur content of (a) 4.25 percent and (b) 2.78 percent by weight. The heating value of the fuel is 11,300 Btu/lb. If the plant is to meet the current emission standard for new plants, what percent SO_2 removal from the stack gases is required? See Table 2-2.

2-14. A power plant burns 200 tons per day of coal containing 3 percent sulfur. (a) Calculate the rate of emission of SO_2, in grams per second, from the power plant's stack. (b) If the volume flow rate of flue gas containing the SO_2 is 250,000 m^3/hr at 150°C and 1.1 bar, and the molar mass of the gas is 28.5, determine the ppm of SO_2 in the flue gas.

2-15. The air quality standard for particulate is 75 $\mu g/m^3$ maximum during an 8-hr period. In a plant without emission control, 16 lb of solid particulate are discharged per ton of solid waste burned. Assume that the air required to burn 1 ton of waste is 8 tons and that the combustion products are discharged at a temperature of 300°F and atmospheric pressure. If the particulate concentration is reduced by a factor of 10^3 from discharge to receptor, what percentage of the particulate must be removed from the combustion gases to meet the air quality standard if no other particulate is discharged in the geographical area?

2-16. Through the use of Figure 2-2, determine the minimum percent reduction of potential sulfur dioxide emission and the maximum permissible SO_2 emission rate for coal with the following heating value and percent sulfur content: (a) 12,000 Btu/lb, 0.75 percent S; (b) 15,000 Btu/lb, 2.0 percent S; (c) 10,500 Btu/lb, 3.5 percent S.

References

1. Surgeon General, U.S. Public Health Service. *Motor Vehicles, Air Pollution, and Health*, House Document 87-489, June, 1962.
2. U.S. Senate, Subcommittee on Air and Water Pollution, Committee on Public Works. *Step Toward Cleaner Air*, October, 1964.
3. *Air Pollution: Message from the President*, House Document 90-47, January 30, 1967.
4. *Staff Report on Air Quality Criteria*. Subcommittee on Air and Water Pollution, Committee on Public Works, U.S. Senate, July, 1968.
5. NAPCA. *Air Quality Criteria for Particulates*, AP-49. Washington, D.C.: HEW, 1969.
6. NAPCA. *Air Quality Criteria for Sulfur Oxides*, AP-50. Washington, D.C.: HEW, 1970.
7. NAPCA. *Air Quality Criteria for Carbon Monoxide*, AP-62. Washington, D.C.: HEW, 1970.
8. NAPCA. *Air Quality Criteria for Photochemical Oxidants*, AP-63. Washington, D.C.: HEW, March, 1970.
9. NAPCA. *Air Quality Criteria for Hydrocarbons*, AP-64. Washington, D.C.: HEW, 1970.
10. NAPCA. *Air Quality Criteria for Nitrogen Oxides*, AP-84. Washington, D.C.: HEW, January, 1971.
11. *Federal Register* **36**, no. 84, part II, April 30, 1971, pp. 8186–8201.
12. EPA. Standards of Performance for New Stationary Sources, *Federal Register* **36**, December, 1971, pp. 24876–24895.
13. New Focus, *J. Air Pollu. Control Assoc.* **23** (May 1973): 398.
14. New Focus, *J. Air Pollu. Control Assoc.* **24** (December 1974): 1198.
15. S. K. Tabler, "Federal Standards of Performance for New Stationary Sources of Air Pollution." *J. Air Pollu. Control Assoc.* **29** (August 1979): 803.

Chapter 3
Meteorology

3-1 INTRODUCTION

All air pollutants emitted by point and distributed sources are transported, dispersed, or concentrated by meteorological and topographical conditions. The airborne cycle is initiated with the emission of the pollutants, followed by their transport and diffusion through the atmosphere. The cycle is completed when the pollutants are deposited on vegetation, livestock, soil and water surfaces, and other objects, when they are washed out of the atmosphere by rain, or when they escape into space. In some cases the pollutants may be reinserted into the atmosphere by the action of the wind.

In those regions where the topographic and meteorologic conditions are conducive to the accumulation and concentration of pollutants, as in the case of the Los Angeles basin, the pollutants may hasten the deterioration of buildings and adversely affect public health as well as vegetation in the area. During the period of time the pollutants are airborne they may undergo physical and chemical changes. Smog, with the associated eye irritation, is the result of the interaction in the atmosphere of the oxides of nitrogen, selected hydrocarbons, and solar energy. The results of such transformations are not always harmful, however; sometimes they are beneficial, as in the case of some mineral salts that are necessary for plant life. The residence times of various pollutants in the atmosphere will be discussed in the separate chapters treating those pollutants.

In large urban areas pollutants emitted from numerous concentrated sources, as well as distributed sources, are dispersed over the entire geographical area. Any given location within the urban area receives pollutants from the different sources in varying amounts, depending upon prevailing winds, presence of tall buildings, and so on. If the allowable concentration of a selected pollutant at a given location is not to be exceeded, the contributions made by different individual sources must be established.

Thus it is imperative to establish the transport and dispersion patterns for given areas, based on mathematical models of the local atmosphere. After the known emission rate data for the area are supplied to the dispersion model, maps can be drawn of estimated concentrations of various pollutants throughout the region. If the model is successful, the maps should replicate actual data taken at monitoring stations. A successful model then can be used to set emission standards for sources, so that ambient air quality standards may be met. Such models are also important for predicting the influence of new (future) sources on the air quality, and what emission standards must be set for these new sources in order to maintain the desired level of air quality.

The dispersion of a pollutant in the atmosphere is the result of three dominant mechanisms: (1) the general mean air motion that transports the pollutant downwind, (2) the turbulent velocity fluctuations that disperse the pollutant in all directions, and (3) mass diffusion due to concentration gradients. In addition, the general aerodynamic characteristics, such as size, shape, and weight, affect the rate at which the particles of nongaseous pollutants settle to the ground or are buoyed upward. The factors affecting both the wind and atmospheric turbulence will be discussed in the following sections of this chapter. The dispersion of plumes from stacks will be discussed in Chapter 4.

3-2 SOLAR RADIATION

At the upper boundary of the earth's atmosphere the vertical solar radiation, termed the "solar constant," is approximately 8.16 $J/cm^2 \cdot min$. The maximum intensity occurs at wavelengths between 0.4 and 0.8 μm, which essentially is the visible portion of the electromagnetic spectrum. Approximately 42 percent of this energy is either (1) absorbed by the high atmosphere, (2) reflected to space by clouds, (3) back-scattered by the atmosphere, (4) reflected by the earth's surface, or (5) absorbed by water vapor and clouds. Approximately 47 percent of the solar radiation is absorbed by the earth's water and land surfaces. The earth, approximated as a body at roughly 290°K (62°F), radiates long-wavelength radiation with the maximum intensity between 4 and 12 μm (near-infrared region). Much of this radiation is absorbed by the water vapor and carbon dioxide in the atmosphere near the earth's surface. Since both water vapor and carbon dioxide are transparent to most of the solar radiation but absorb the long-wave radiation from the earth's surface, the net effect causes a warming of the atmosphere, depending upon the quantity of H_2O and CO_2 present. This effect has been termed the "greenhouse effect." Some authors have indicated that a 100 percent increase in H_2O and CO_2 content in the atmosphere could result in an increase of between 2° and 4°F in the atmospheric temperature.

On the other hand, one result of industrialization throughout the world is the significant increase in particulates emitted. Particulates in the atmosphere tend to block passage of solar radiation toward the earth's surface. This blocking effect is the opposite of that caused by increased concentrations of CO_2 and water vapor in the atmosphere. That is, a drop in the average atmospheric temperature occurs. It is difficult to predict which of these two factors will have the overriding influence on atmospheric temperatures over the decades ahead, if man-made emissions remain essentially uncontrolled. Nevertheless, a significant change in the average ground temperature in the Northern Hemisphere has been detected, as reported by the National Oceanic and Atmospheric Administration (NOAA) in 1974. From 1890 to 1945 the average temperature rose by nearly 0.9°F, but from 1945 to 1970 the average ground temperature dropped by roughly 0.6°F. It should be noted that the earth's average temperature during the great ice ages was only about 7 degrees lower than during its warmest periods. NOAA has also reported that the amount of sunshine reaching the ground in the continental United States decreased by 1.3 percent between 1964 and 1972. Whether these detectable changes in weather patterns are simply small perturbations that are commonly expected over the centuries, or whether they are due to massive man-made influences, cannot yet be ascertained.

Insolation, or the quantity of solar radiation reaching a unit area of the earth's surface, is a function of many variables. The most important factor is the variation in the angle of incidence, as illustrated in Figure 3-1. In Figure 3-1(a) the increase in surface area receiving the same quantity of insulation in winter compared with summer is illustrated. A similar change is illustrated in Figure 3-1(b) for different geographical locations.

The thickness of the atmosphere and thus the quantity of solar energy absorbed is a function of the time of day, as illustrated in Figure 3-2. The

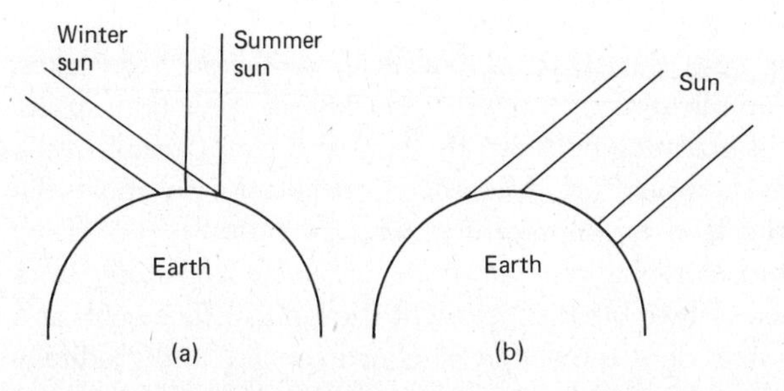

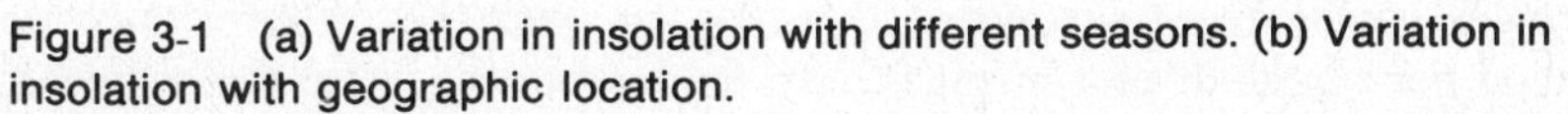

Figure 3-1 (a) Variation in insolation with different seasons. (b) Variation in insolation with geographic location.

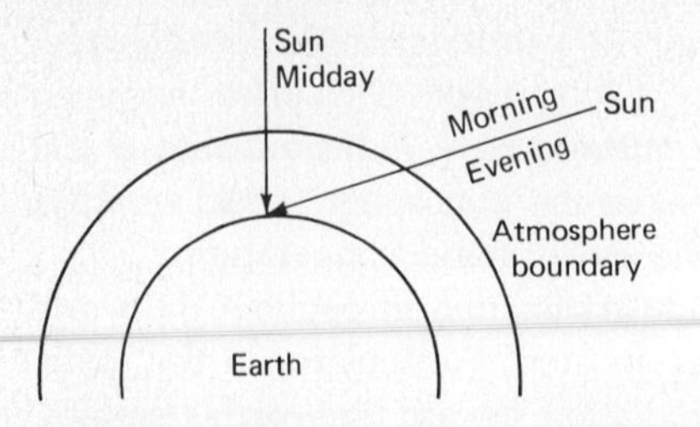

Figure 3-2 Variation in atmospheric thickness traversed by solar radiation with angle of incidence.

sun's rays are tangent to the earth's surface in the morning and evening and approximately perpendicular at noon. The period of insolation for a summer day is approximately twice as long as for a winter day. From the preceding discussion we can see that the actual quantity of solar energy received by a unit surface area on the earth's surface is a complex function of location, season, time of day, and composition of the atmosphere above the surface.

The quantity of incident solar radiation absorbed at the earth's surface is a function of the absorptivity of that surface, that is, whether the surface is earth, rock, water, ice, snow, vegetation, or whatever. Rough bare soil absorbs much more radiation than does ice or highly reflective rock surfaces. The transparency of water extends the thickness of the absorbing layer, and thus more energy is absorbed by a given thickness of water than by the same thickness of opaque land. The fraction of the incident radiation that is reflected by a surface is known as the *albedo* of the surface.

The specific heats of solid earth surface materials are lower than the specific heat of water. Thus, even though the quantity of solar energy absorbed by a unit surface area of land may be the same as that for a unit surface area of water, the resulting temperature increase will be different. Currents in water and the resulting heat transfer by convection cause the energy received at the surface to be transported to a greater depth in water than is the case in rock and soil, where heat transfer by conduction alone occurs. The combination of all the aforementioned effects produces the striking differences between water and land temperatures and hence between marine and continental air temperatures.

The wind patterns near the shore of bodies of water such as lakes, oceans, and bays are complicated by a factor nonexistent over land areas. Differences in the rate of warming between the land and water result in the development of a distinct temperature difference between the air above the water and that above the land by midmorning. The expansion of the rising warmer air over the land causes a general air movement horizontally from the water to the land (sea or lake breeze). At night the land surface cools at a faster rate by radiation than does the water. The air over the land gradually becomes cooler and more dense than the air over the water. Hence the general horizontal air movement is from the land to the water (land breeze).

Since the major large cities are located near large bodies of water, the wind patterns are quite complicated, especially if mountains or large hills are nearby.

3-3 WIND CIRCULATION

The sun, the earth, and the earth's atmosphere form one very large dynamic system. The differential heating of the air gives rise to horizontal pressure gradients, which in turn lead to horizontal motion in the atmosphere. Thus the temperature difference between the atmosphere at the poles and at the equator, and between the atmosphere over the continents and over the oceans, causes the large-scale motions of the air. (Local winds such as lake breezes are caused by local temperature differences.) If the earth were not rotating, air normally would tend to flow directly from high-pressure regions toward regions of low pressure, which in the horizontal usually means from a cold area toward a warm area. In Figure 3-3(a) the flow is shown to be perpendicular to the isobars. The rotation of the earth alters this situation. In addition to the pressure gradient F_p, one must also consider the Coriolis force F_{Cor} arising from the earth's rotation. (The Coriolis force is sometimes termed the horizontal deflection force.) This force accounts for the apparent deflection of a moving air parcel to the right in the Northern Hemisphere, relative to the surface, when the observer faces in the direction of motion of the particle. The Coriolis force in this global system is a function of the velocity of the air parcel, as well as the latitude and the earth's rotational angular speed. It is a maximum at the poles of the earth and zero at the equator. If this vector force is added to the pressure gradient force, then in general the situation is represented by Figure 3-3(b), with the resultant velocity vector at some angle to the isobars. This representation is not one of static equilibrium, however, since the forces are not in balance as drawn. In the upper atmosphere, moreover, air parcels frequently experience relatively little acceleration. Therefore the forces acting on the parcel in this case must essentially be in balance. If we consider only the pressure gradient force and the Coriolis force to be present in the idealized case, the vector representation must appear as in Figure 3-3(c) for parallel isobars. Since the pressure gradient force F_p must be perpendicular to the isobars, the Coriolis force must be parallel to F_p but in the opposite direction, toward the high-pressure region. In addition, the wind velocity and the Coriolis force act at right angles to each other. As a result, the wind must blow parallel to the isobars. Moreover, recall that F_{Cor} acts to the right of the wind velocity in the Northern Hemisphere. Hence the wind must blow so that the low-pressure region is to the left of the direction of motion, as the observer looks downward toward the earth's surface. This idealized or conceptual wind is called the *geostrophic* wind by meteorologists. It is symbolized by V_g in the figure, and it approximates conditions a few hundred meters or more above the surface of the earth. Except in the case of very light winds, the direction

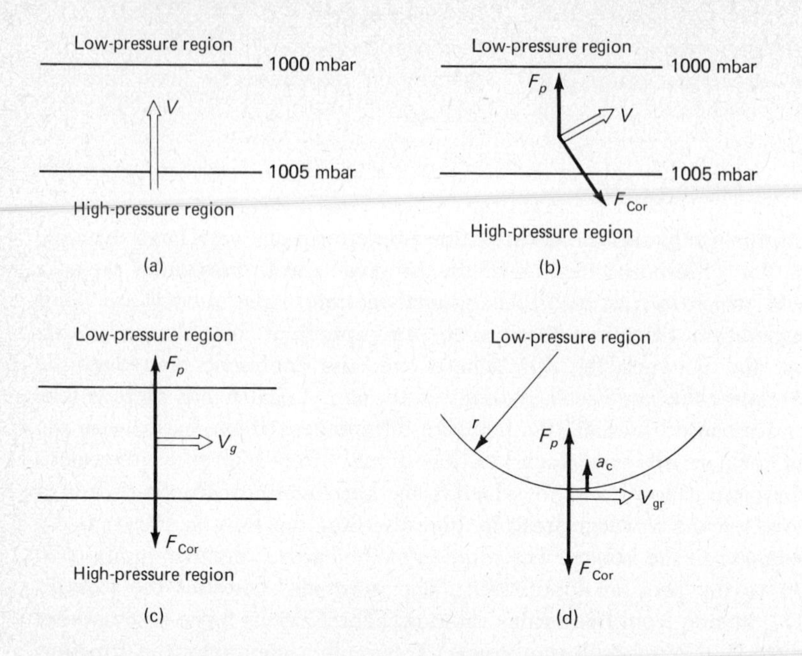

Figure 3-3 **Effect of various forces on wind direction, relative to isobars in the atmosphere. (a) Pressure gradient force only, parallel isobars. (b) Pressure gradient and Coriolis forces, parallel isobars. (c) Pressure and Coriolis forces in balance, parallel isobars. (d) Pressure and Coriolis forces balanced by centripetal acceleration, curved isobar.**

and speed of the actual wind probably do not differ by more than 10 degrees and 20 percent, respectively, from the geostrophic values. For geostrophic flow the isobars coincide with the streamlines of flow.

Another type of wind referred to in meteorology in the *gradient* wind, which is associated with curved isobars. Even though the speed of an air parcel may still be constant, in a curved path the centripetal acceleration a_c must be taken into account. Such curved paths are particularly in evidence around regions of high and low pressure. In the Northern Hemisphere the counterclockwise motion of air around a low-pressure center is termed a *cyclone*; the clockwise motion around a high-pressure center is called an *anticyclone*. Figure 3-3(d) shows the vector diagram for the gradient wind in the vicinity of a low-pressure center. The vector a_c represents the centripetal acceleration inward at a radius of curvature r. As might be anticipated, the gradient velocity, V_{gr}, is a function of the radius of curvature, as well as the pressure gradient, the earth's angular speed, and the latitude. In high- or low-pressure troughs in the atmosphere (where pronounced curvature in the path of an air parcel occurs) the gradient wind velocity is a better approximation than the geostrophic wind velocity to the actual wind condition.

The geostrophic and gradient winds are concepts of practical interest in the absence of a significant frictional force. However, the movement of air near the earth's surface is retarded by frictional effects of the surface roughness. The vertical region between the earth's surface and the upper levels of the atmosphere where the gradient wind concept is valid is called the *planetary boundary layer*. The magnitude of the retardation of wind speed with height and the thickness of the boundary layer are functions of the surface roughness or terrain, as well as of the temperature gradient in the lower atmosphere. The effect of this frictional force, when added to the pressure and Coriolis forces, is to turn the air movement slightly to the left of the gradient wind (when the observer looks downward toward the earth's surface). In effect, the wind is turned at a slight angle toward the low-pressure region. The angular shift is a function of the same variables noted above for the frictional force. The vector diagram in Figure 3-4 illustrates the direction of the resultant wind for straight isobars. The frictional force, F_f, is opposite to the wind direction, and the wind direction must be perpendicular to the Coriolis force vector. Recall that the Coriolis force is proportional to the wind speed. The frictional force directly reduces the wind speed in the boundary layer, thus reducing the Coriolis force, F_{Cor}. The pressure force F_p remains the same, however, so that it no longer is balanced by the Coriolis force as in a geostrophic wind. The result of this imbalance is that the direction of the wind is now across the isobars toward the low-pressure region, rather than parallel to the isobars. Note that the wind speed is now less than the geostrophic wind speed, all other conditions remaining the same.

A steady wind can still exist when the frictional force is included in the analysis. The component of the F_p vector in the wind direction just balances the frictional force, F_f. At the same time, the reduced Coriolis force is just balanced by the remaining component of F_p. An interesting consequence of this is with respect to flow around high- and low-pressure centers, with friction. Figure 3-5(a) shows the vector diagram for circulatory flow close to a high-pressure center. The pressure gradient, Coriolis, and frictional forces are shown, as well as the centripetal acceleration vector. This is merely a superposition of the diagram in Figure 3-4 upon a curved path. The

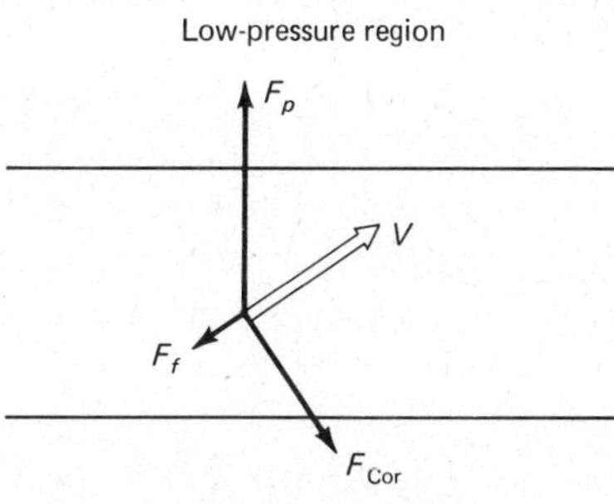

Figure 3-4 Effect of frictional force in the planetary boundary layer on the wind direction.

important thing to note is that the wind velocity is outward from a circle of radius r. Hence flow in the vicinity of a high-pressure center must pass clockwise and with a net flow outward. If a more complete three-dimensional analysis of air flow near a high-pressure center is made, it is found that the flow is downward as well as outward. As a result, air must be brought in from above the center and settle downward to maintain the outward flow. This downward flow is called *subsidence* and is a possible inhibitor of pollutant dispersion in the atmosphere.

Figure 3-5(b) illustrates flow around a low-pressure center. In this case, the wind velocity vector is pointed inward. Thus flow close to a low-pressure center is counterclockwise. Additional analysis reveals that the spiraling flow is upward as well as inward. Hence pollutants in the lower atmosphere will be carried upward and generally will be dispersed over a wide area. In addition, as the air rises it will cool as a result of the decrease in pressure at higher elevations. Water may then condense within the rising air mass. This effect may also have a cleansing action on a polluted atmosphere.

The angular shift of the wind in the planetary boundary layer has an important repercussion on the dispersion pattern of pollutants from stacks. Since the frictional retardation force varies with height above the ground, the amount of angular shift also varies with height. The frictional force in the boundary layer is a maximum at the earth's surface and drops essentially to zero at the top of the boundary layer, where the geostrophic or gradient wind prevails. Hence the angle of displacement of the wind direction due to friction varies from a maximum value near the earth's surface to zero at the top of the boundary layer. As we look downward toward the earth, the wind shifts clockwise with increase in height within the boundary layer. This is shown in Figure 3-6(a) where the angle ϕ in the x-y plane increases in a counterclockwise fashion as the ground level is approached. There are two

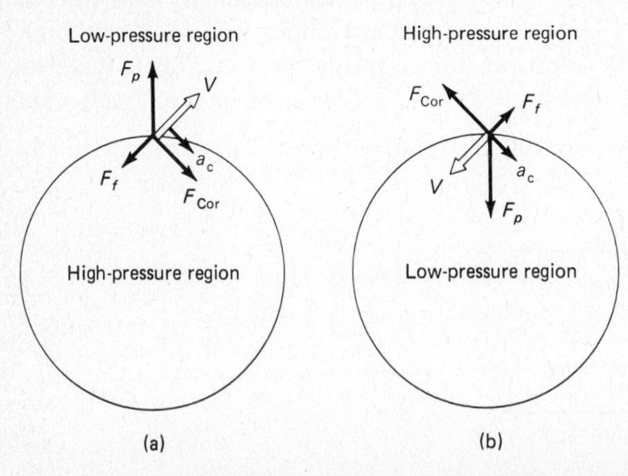

Figure 3-5 Force balances in the vicinity of high- and low-pressure regions. (a) Flow is outward and clockwise around a high-pressure region. (b) Flow is inward and counterclockwise around a low-pressure region.

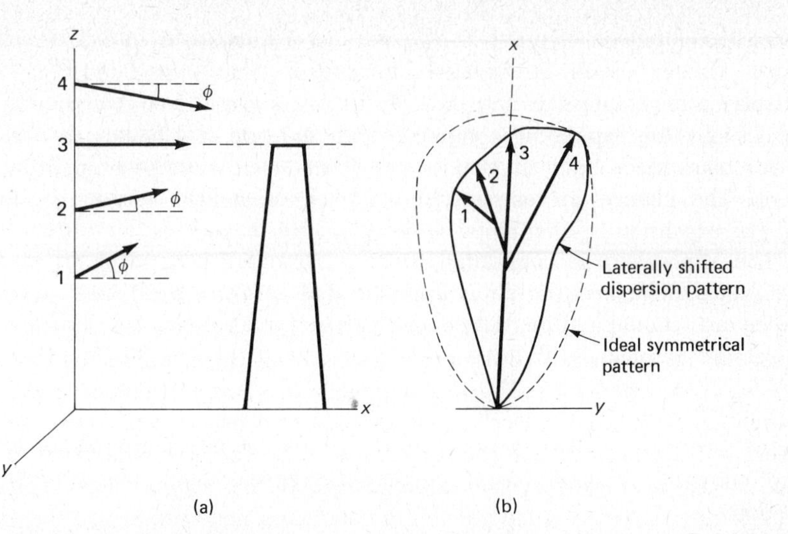

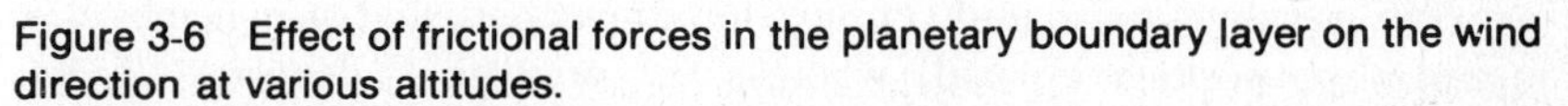

Figure 3-6 Effect of frictional forces in the planetary boundary layer on the wind direction at various altitudes.

ramifications of this shift. First, it is apparent that the wind direction at the base of the stack shown in Figure 3-6(a) may not be a true indication of the direction of travel of pollutants released from the top of a tall stack. Thus wind direction measurements made near the surface of the earth may be somewhat misleading in terms of the actual direction the pollutants take as they leave the stack. In the presence of strong winds (>6 m/s) the change in direction of the velocity u with height is fairly negligible in the lowest 100 m of the planetary boundary layer [13]. However, when the wind speed is less than 6 m/s (13 mi/hr) the change in direction may be appreciable. Values of ϕ in the range of 5 to 15 degrees occur over the ocean, while values from 25 to 45 degrees are typical for low speeds over land masses.

Second, as pollutants are carried downwind they will diffuse outwardly in the y-direction and vertically in the z-direction. As pollutants diffuse vertically in the boundary layer they will encounter a different mainstream wind direction at different heights. Thus the pattern of dispersion downwind frequently will not be symmetrical to the axis of the wind direction at the top of the stack but instead will tend to become skewed. This situation is illustrated in Figure 3-6(b), as one looks downward into the x-y plane. The amount of skewness, as measured by the angle ϕ in Figure 3-6(a), depends upon the relative position of the top of the stack in the boundary layer.

3-4 LAPSE RATE

One of the most important characteristics of the atmosphere is its stability—that is, its tendency to resist vertical motion or to suppress existing turbulence. This tendency directly influences the ability of the atmosphere to disperse pollutants emitted into it from natural or man-made sources. When a small volume of air is displaced upward in the atmosphere, it will encounter a lower pressure and undergo an expansion to a lower temperature. Usually the expansion is rapid enough that we can assume no heat transfer takes place between that parcel of air and the surrounding atmosphere. The change in temperature with elevation that is due to the adiabatic expansion is determined in the following manner.

The atmosphere is considered to be a stationary column of air in a gravitational field, and the air is approximated as a dry ideal gas. In the absence of frictional and inertial effects, a static force balance on a differential element of thickness dz leads to

$$dP = -\rho g\, dz \tag{3-1}$$

where P is the atmospheric pressure, ρ is the atmospheric density (assumed to be constant), g is the local gravitational acceleration, and z is the elevation.

The negative sign results from the convention that the height z is measured as positive upward, whereas the pressure P decreases in this direction.

The first law of thermodynamics for a closed system containing an ideal gas undergoing a quasistatic change of state can be written as

$$\begin{aligned} dq &= du + dw = du + P\,dv \\ &= (dh - P\,dv - v\,dP) + P\,dv = dh - v\,dP \\ &= c_p\,dT - \frac{1}{\rho}\,dP \end{aligned} \tag{3-2}$$

where q is the quantity of heat transfer and c_p is the constant-pressure specific heat.

For an adiabatic process $dq = 0$; thus Equation (3-2) becomes

$$c_p\,dT = \frac{1}{\rho}\,dP \tag{3-3}$$

Substitution of Equation (3-3) into Equation (3-1) and rearrangement gives

$$\left(-\frac{dT}{dz}\right)_{\text{adia}} = \frac{g}{c_p} = \frac{g(k-1)}{kR} \tag{3-4}$$

where k is the specific heat ratio, c_p/c_v. If the change in g and c_p with elevation is assumed to be negligible, then the change in temperature with elevation under adiabatic condtions is a constant, independent of elevation.

The numerical value of dT/dz depends upon the system of units employed. In SI units c_p for dry air equals 1.005 kJ/kg·°C at room temperature and g is 9.806 m/s^2. In English engineering units c_p of dry air equals 0.240 Btu/lb·°F and g is 32.17 ft/s^2. From these data one ascertains that for dry air

$$\left(\frac{dT}{dz}\right)_{\text{dry adia}} = -0.0098°\text{C/m} = -0.0054°\text{F/ft}$$

It is convenient to define a *lapse rate* as the negative of the temperature gradient in the atmosphere. Thus the *dry adiabatic lapse rate*, which is given the special symbol Γ, is

$$\Gamma = \left(\frac{-dT}{dz}\right)_{\text{dry adia}} \cong 1°\text{C}/100 \text{ m} = 5.4°\text{F}/1000 \text{ ft} \qquad \text{(3-5)}$$

The dry adiabatic lapse rate is extremely important in meteorological studies. A comparison of Γ to the actual (environmental) lapse rate in the lower atmosphere indicates to a large degree the stability of the atmosphere. As we shall see below, the degree of stability is a measure of the ability of the atmosphere to disperse pollutants emitted into it. Before examining this effect, it is appropriate to evaluate the international standard atmospheric lapse rate.

Based on meteorological data, an international standard atmosphere has been defined and adopted for comparative purposes. On the average in the middle latitudes the temperature decreases linearly with elevation, until approximately 35,300 ft is attained. The temperature averages 519°R (288°K) at sea level to 393°R (216.7°K) at 35,300 ft (10.8 km). The standard or normal temperature gradient therefore is

$$\left(\frac{dT}{dz}\right)_{\text{std}} = (393 - 519)°\text{F}/35{,}300 \text{ ft} = -0.00357°\text{F/ft} = -0.0066°\text{C/m}$$

The normal or standard lapse rate based on international convention hence is 0.66°C/100 m or 3.6°F/1000 ft. The temperature profile for the standard atmosphere is compared with the adiabatic temperature profile in Figure 3-7.

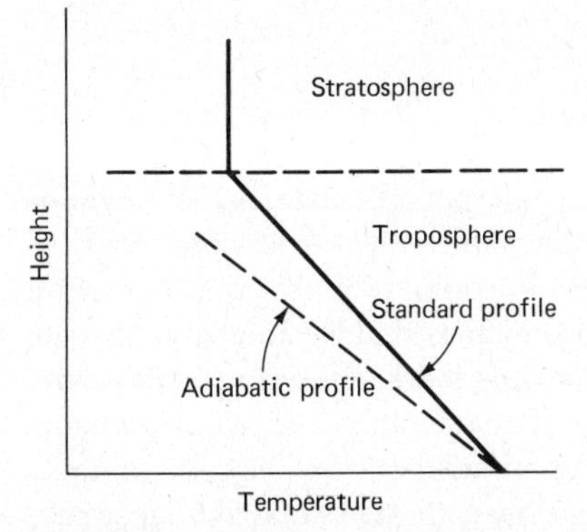

Figure 3-7 Temperature profile of the standard atmosphere in comparison to the adiabatic temperature gradient.

Finally, the saturated (wet) adiabatic lapse rate of the atmosphere is of some interest. When the atmosphere is unsaturated with water vapor, the value of c_p of the dry air–water vapor mixture is essentially that for dry air. Thus dT/dz calculated for such a mixture by Equation (3-4) is unaltered for adiabatic expansion. However, if the air is saturated with water vapor to begin with, a temperature gradient much different from $-5.4°F/1000$ ft exists when an air parcel changes elevation adiabatically. Water vapor will condense when the parcel moves to a higher elevation. This condensation process releases the enthalpy of condensation from the water, and this energy tends to raise the temperature of the mixture. At the same time the upward expansion to a lower pressure tends to decrease the temperature. The relative contributions of these two effects on the change in temperature with elevation can be determined from an energy balance on a parcel of air which is saturated with water vapor. Let ω represent the ratio of the mass of water vapor to mass of dry air per volume of atmospheric air. For a differential change $d\omega$ of water to the liquid phase as the air parcel ascends, the energy release to the vapor phase is $h_{fg}\,d\omega$ in terms of energy per unit mass of dry air. For the vapor phase this is equivalent to a heat addition. Thus Equation (3-2) applied to the vapor phase becomes, when combined with Equation (3-1),

$$dq = c_p\,dT - \frac{1}{\rho}\,dP$$

or

$$-h_{fg}\,d\omega = c_p\,dT + g\,dz$$

Rearrangement leads to the desired result, namely,

$$-\left(\frac{dT}{dz}\right)_{\text{sat}} = \frac{g}{c_p} + \frac{h_{fg}}{c_p}\frac{d\omega}{dz}$$

Note that $d\omega/dz$ is negative for a rising air parcel, since water vapor is condensing. Hence the sum of the two terms on the right is less positive than that for dry air. Thus the overall effect is a lowering of the temperature with increase in elevation, but not by nearly as much as the dry adiabatic change. Because the vapor pressure of water increases markedly with temperature, the quantity $d\omega/dz$ is a strong function of temperature. As an approximation, however, the saturated lapse rate for an adiabatic air parcel is roughly 3.4°F/1000 ft, or 0.6°C/100 m. Although the saturated adiabatic lapse rate is sometimes important in predicting the stability of the atmosphere, it will not be discussed further in this text [11].

3-5 STABILITY CONDITIONS

The degree of stability of the atmosphere must be known if we wish to estimate the ability of the atmosphere to disperse pollutants it receives from man-made sources. A stable atmosphere is one which does not exhibit much vertical mixing or motion. As a result, pollutants emitted near the earth's surface tend to remain there. Whether much mixing occurs on a significant scale in the lower atmosphere is primarily dependent upon (1) the temperature gradient and (2) the mechanical turbulence due to shearing action of the wind. The possibility of thermal mixing can be determined by comparison of the actual (environmental) temperature gradient or lapse rate to the adiabatic lapse rate.

Several possible environmental lapse rates together with the dry adiabatic lapse rate are presented in Figure 3-8. When the environmental lapse rate is greater than the dry adiabatic lapse rate, Γ, the atmosphere is said to be *superadiabatic*. (Note that this means the actual temperature gradient is more negative than the dry adiabatic temperature gradient.) Consider point A in Figure 3-8(a) for the superadiabatic case. When a small quantity of air at temperature A is carried rapidly aloft (perhaps by a turbulent fluctuation in the atmosphere), its expansion closely approximates an adiabatic expansion since the rate of heat transfer across the boundary of the air parcel will be slow compared to its rate of vertical movement. Its final state, therefore, might be position B, which lies along the dry adiabatic line. The temperature at this state B is greater than the surrounding air at this height represented by state C, which lies along the environmental temperature gradient line. This small parcel of air is thus less dense than the surrounding air (same pressure but higher temperature), and it tends to continue its upward motion. Should the same quality of air be dispersed downward, it would undergo an adiabatic compression to a temperature E, lower than the surrounding air, which is at a temperature F. Because of its greater density, the small quantity of air would continue its downward motion. The condition just described is *unstable* since any perturbation in the vertical direction tends to be enhanced. Thus any atmosphere having a superadiabatic lapse rate is unstable.

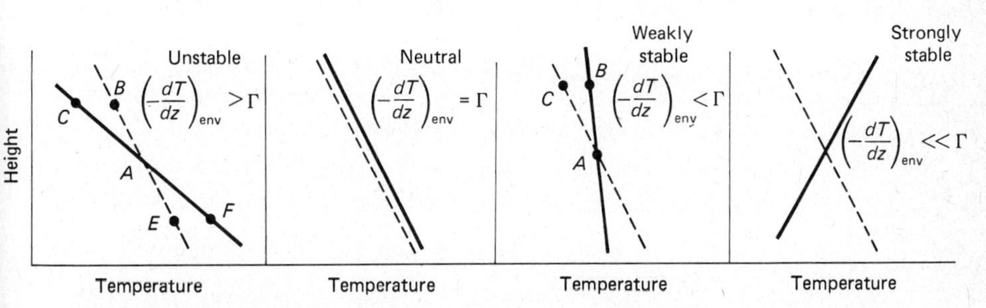

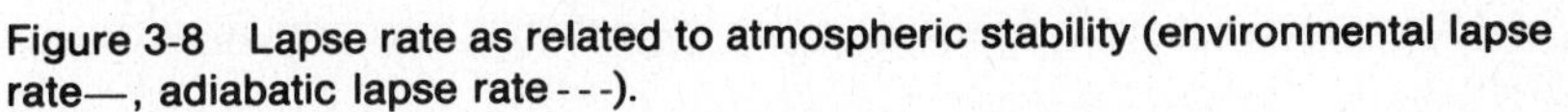
Figure 3-8 Lapse rate as related to atmospheric stability (environmental lapse rate—, adiabatic lapse rate - - -).

When the environmental lapse rate is approximately the same as the dry adiabatic lapse rate [see Figure 3-8(b)], the stability of the atmosphere is said to be *neutral*. Any parcel of air which is carried rapidly upward or downward will have the same temperature as the environment at the new height. Hence there is no tendency for any further vertical movement due to thermal differences and the displaced parcel of air remains in the displaced position. If the environmental lapse rate is less than the dry adiabatic lapse rate (greater temperature gradient), then the atmosphere is called *subadiabatic*. By an argument similar to the one above for the superadiabatic case, it is found that a subadiabatic atmosphere is *stable*. That is, any small air parcel suddenly displaced vertically will tend to return to its original position. As an example, an air element displaced from position A to C in Figure 3-8(c) will be more dense than the environment at B. Hence its tendency will be to sink back toward its original height.

The stability of the atmosphere is also frequently characterized by the potential temperature gradient. The potential temperature θ of dry air is defined as the temperature which a volume of air would have if brought by an adiabatic process from its existing pressure P to a standard pressure P_o of 1000 mbar. This process is illustrated in Figure 3-9 by the path A–B. The equation defining θ is given by

$$\theta = T\left(\frac{P_o}{P}\right)^{(k-1)/k} = T\left(\frac{1000}{P}\right)^{0.288} \tag{3-6}$$

where k is the specific heat ratio and 0.288 is the value of the exponent for dry atmospheric air in the lower atmosphere. In this equation P must be expressed in millibars and T in degrees R or degrees K.

The gradient of the potential temperature may be expressed in terms of the gradient of the environmental temperature and the dry adiabatic lapse

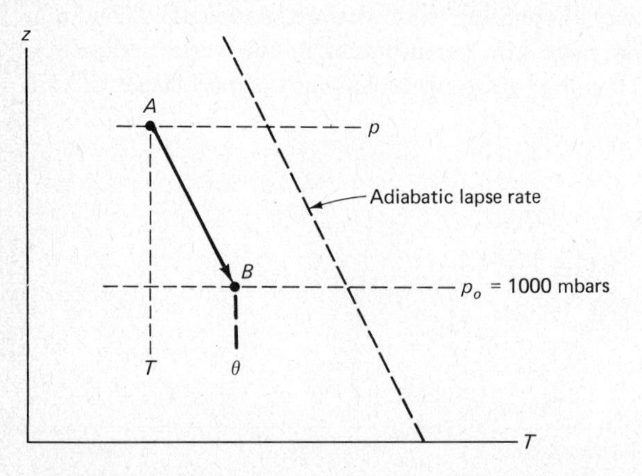

Figure 3-9 **The adiabatic process by which the potential temperature, θ, is defined.**

rate Γ. Logarithmic differentiation of Equation (3-6) with respect to elevation z gives

$$\frac{1}{\theta}\frac{d\theta}{dz} = \frac{1}{T}\left(\frac{dT}{dz}\right)_{\text{env}} - \frac{k-1}{k}\left(\frac{1}{P}\right)\frac{dP}{dz} \tag{3-7}$$

From Equation (3-1),

$$\frac{dP}{dz} = -\rho g$$

The ideal gas equation of state is

$$P = \rho RT \tag{3-8}$$

Substituting Equations (3-1) and (3-8) into Equation (3-7) and simplifying gives

$$\frac{1}{\theta}\frac{d\theta}{dz} = \frac{1}{T}\left[\left(\frac{dT}{dz}\right)_{\text{env}} + \frac{k-1}{k}\left(\frac{g}{R}\right)\right] \tag{3-9}$$

For an ideal gas,

$$\left(\frac{k-1}{k}\right)\frac{1}{R} = \frac{1}{c_p}$$

Therefore

$$\frac{1}{\theta}\frac{d\theta}{dz} = \frac{1}{T}\left[\left(\frac{dT}{dz}\right)_{\text{env}} + \frac{g}{c_p}\right] \tag{3-10}$$

Note that the second term on the right-hand side of Equation (3-10) is the dry adiabatic lapse rate, Γ, as shown by Equation (3-4). Hence the above equation can be written as

$$\frac{1}{\theta}\frac{d\theta}{dz} = \frac{1}{T}\left[\left(\frac{dT}{dz}\right)_{\text{env}} + \Gamma\right] \tag{3-11}$$

Solving Equation (3-11) for $d\theta/dz$ and assuming for small changes in p that $\theta \approx T$, we find that

$$\frac{\Delta\theta}{\Delta z} = \left(\frac{\Delta T}{\Delta z}\right)_{\text{env}} + \Gamma = \left(\frac{\Delta T}{\Delta z}\right)_{\text{env}} - \left(\frac{\Delta T}{\Delta z}\right)_{\text{adia}} \tag{3-12}$$

where $\Delta\theta/\Delta z$ is the potential temperature gradient. Notice that the right-hand side of Equation (3-12) is the difference between the environmental and dry adiabatic temperature gradients. When the temperature decreases with altitude as in the normal or standard atmosphere, $(dT/dz)_{\text{env}}$ will be negative. When $(dT/dz)_{\text{env}}$ is more negative than the dry adiabatic temperature gradient, $\Delta\theta/\Delta z$ will be negative or less than zero. Thus a negative potential temperature gradient is associated with an unstable (superadiabatic) atmosphere. Likewise, when $(dT/dz)_{env}$ is less negative (or even positive) when compared to the dry adiabatic temperature gradient (subadiabatic condition), the potential temperature gradient is positive and the atmosphere is stable. When $\Delta\theta/\Delta z$ is zero, the atmosphere is neutral.

When the temperature increases with altitude, the lapse rate is negative and the atmospheric condition is termed an *inversion* [see Figure 3-8(d)]. It is a condition of strong stability and is characterized by a relatively large positive potential temperature gradient. The effect of an inversion is to reduce the vertical dispersion of pollutants and thus to increase the local concentration. Of the several kinds of inversions, the two most common are those formed by the descent of a layer of air within a high-pressure air mass and those formed by radiation at night from the earth's surface into the local atmosphere. The first type is known as a subsidence inversion. It is an inversion layer whose base normally lies some distance above the earth's surface. This type of inversion is formed by the adiabatic compression and warming of a layer of air as it sinks to lower altitudes in the region of a high-pressure center. For the adiabatic descent of a layer of air which behaves as an ideal gas, Equation (3-3) shows that

$$\left(\frac{dT}{dP}\right)_{adia} = \frac{1}{c_p \rho}$$

Hence the change in temperature at any position within the sinking layer with respect to pressure is a function of the constant pressure specific heat, and the density of the gas at that position. Over a considerable range of temperature the value of c_p for air is essentially constant. However, the density of air at the top of the layer is less than at the bottom, as a result of the change in barometric pressure with altitude. Consequently we note that for the air layer,

$$\left(\frac{dT}{dP}\right)_{top} > \left(\frac{dT}{dP}\right)_{bottom}$$

That is, the top of the layer is warming more rapidly than the bottom of the layer. If the subsidence (general dropping) of the layer persists for sufficient time, a positive temperature gradient will be created within the layer. This gradient fulfills the requirement of an inversions layer. Hence the subsiding air mass acts as a giant lid on the atmosphere below it.

It should be noted that subsidence inversions occur above emission sources, and thus in general do not contribute to short-term pollution problems. However, this type of inversion may persist for several days and greatly contribute to the long-term accumulation of pollutants. The hazardous pollution episodes that have been recorded in the past in major metropolitan regions have frequently been associated with a subsidence inversion. Subsidence inversions are a common feature of the West Coast of the United

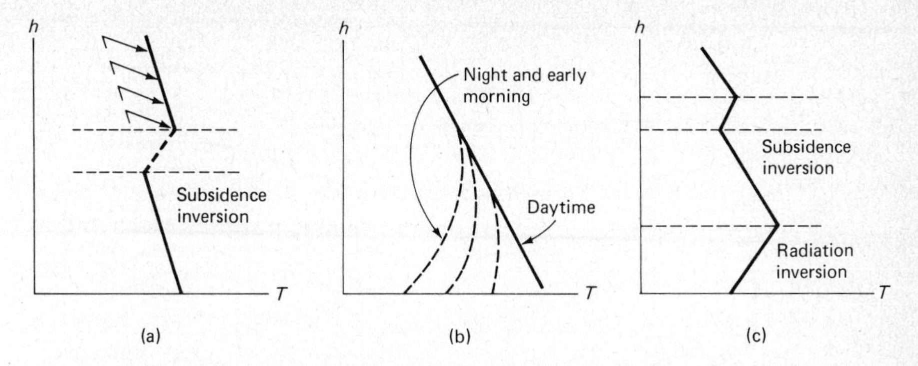

Figure 3-10 **Illustrations of (a) subsidence inversion, (b) radiation inversion, and (c) combination of subsidence and radiation inversions.**

States for approximately 340 days of the year. A typical temperature profile for a subsidence inversion, resulting from the downward flow of air in a high-pressure region, is shown in Figure 3-10(a).

In the second common type of inversion, called a radiation inversion, the surface layers of the atmosphere during the day receive heat by conduction, convection, and radiation from the earth's surface and are warmed. This results in a temperature profile in the lower atmosphere which is represented by a negative temperature gradient. If a clear night follows, then the ground surface radiates heat and quickly cools. The air layers adjacent to the earth's surface are cooled to a temperature below that of the layers of air at higher elevations. Thus the daytime temperature profile becomes inverted, and a stable inversion layer now cloaks the lower atmosphere above the earth's surface. [See Figure 3-10(b).] This type of inversion is strongest just before daylight and during periods of clear skies and light winds. It breaks up as the morning sun heats the ground and reestablishes the moving currents of rising warm air. Figure 3-11 illustrates an actual case of this diurnal shifting between unstable and stable air masses close to the earth's surface. From roughly 6 A.M. to 5 P.M. at this locale, the temperature gradient was negative. However, as sunset approached the gradient reversed itself and remained that way until shortly past sunrise. Note that the inversion layer extended upward for less than 1000 ft. The height of radiation inversions frequently is less than 500 m (1600 ft). Such inversions are suppressed by an appreciable cloud cover and by strong winds. The former prevents to some degree the radiative losses from the earth's surface, and hence the strong cooling of the atmosphere near the surface. Strong winds tend to smooth out any adverse temperature gradient formed by thermal means. Radiation inversions are important in air pollution problems because they occur within the layer of the atmosphere which contains the pollutant sources (unlike subsidence inversions). Furthermore, since radiation inversions are most likely to occur during cloudless and windless nights, there is little likelihood that cleansing by precipitation or flushing by lateral movement will take place.

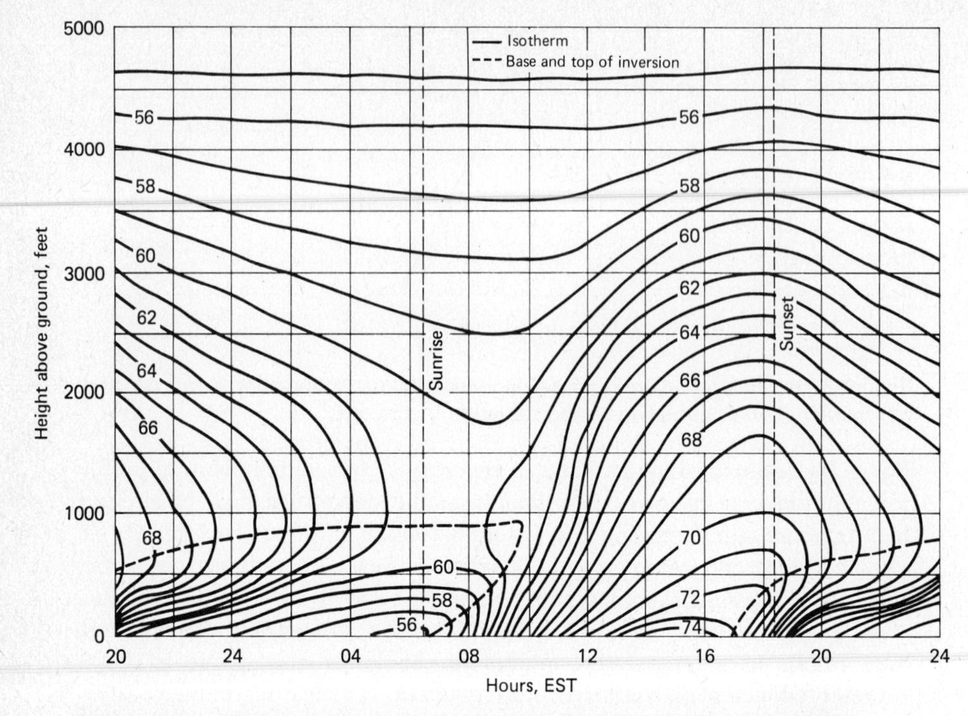

Figure 3-11 Time cross section of average temperature (°F) up to 5000-ft altitude, September, October, 1950, Oak Ridge, Tenn. (SOURCE: U.S. Weather Bureau, *Meteorological Survey of the Oak Ridge Area*. Report ORO-99. Oak Ridge, Tenn.: AEC, 1953.)

It is possible for subsidence and radiation inversions to appear in the atmosphere at the same time. This situation is illustrated in Figure 3-10(c), in terms of a typical temperature profile. The joint occurrence of the two types of inversion leads to a phenomenon called trapping of a stack plume, which will be discussed in a later section. The intensity and duration of an inversion is affected by the season of the year. Fall and winter generally have the longest lasting and the greatest number of inversions. Topography also has an effect upon inversions. For example, cool air flows downward at night along a mountain and can be trapped in the valley by the warm air above it. Until the sun is directly overhead the next day, the valley air may not be heated enough to cause a breakup in the inversion. In Denver, Colorado, in the winter, for example, over half of the inversions last throughout the day [2].

Two other types of localized inversions are possible. One of these is connected with the sea breeze mentioned earlier. The warming of the air in the morning over a land mass will induce the flow of cooler air toward the land from an adjacent large lake or ocean. This results in warmer air aloft and a cooler layer along the surface, or an inversion condition. The other inversion condition is associated with the passage of a warm front across a large land mass. The warm front frequently will tend to override the more dense, cooler air ahead of it, thus creating a localized temperature inversion. A passage of a cold front into and under a warm region ahead of it will lead to a similar situation.

3-6 WIND VELOCITY PROFILE

As we mentioned previously, the movement of air near the earth's surface is retarded by frictional effects proportional to the surface roughness. Thus the nature of the terrain, the location and density of trees, the location and size of lakes, rivers, hills, and buildings produce different wind velocity gradients in the vertical direction. The air layer (called the planetary boundary layer) that is influenced by friction extends from a few hundred meters to several kilometers above the surface of the earth. The depth of this boundary layer is greater for unstable conditions than for stable conditions. Thus pollutants will be dispersed over a greater vertical distance under unstable atmospheric conditions. This leads to a general lessening of pollutant concentration in any given region downwind from a source. However, turbulent fluctuations in an unstable atmosphere may result in instantaneous concentrations which exceed those in a stable atmosphere.

Typical wind-speed profiles during the day and the night are shown in Figure 3-12. Because of a more stable atmospheric condition at night, the profile for the night is usually steeper than that for the day. Note that the wind-speed variation levels out in this case at an approximate height of 600

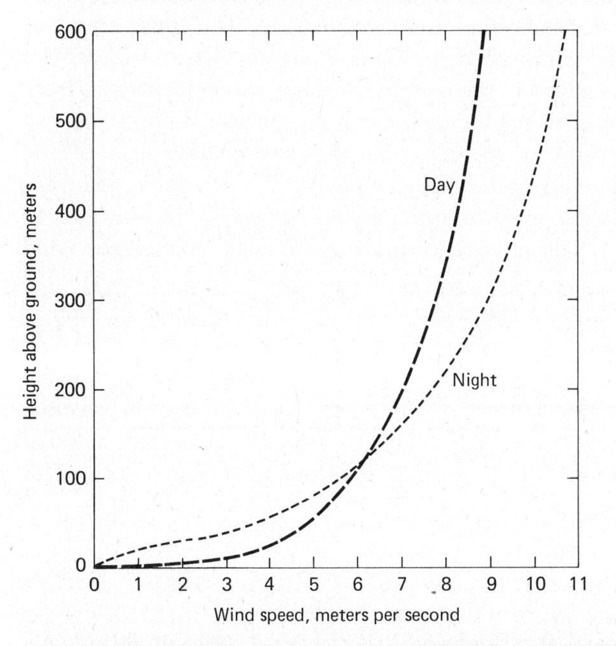

Figure 3-12 Change of wind-speed profile with stability. (SOURCE: D. B. Turner. *Workbook for Atmospheric Dispersion Estimates*. Washington, D.C.: HEW, 1969.)

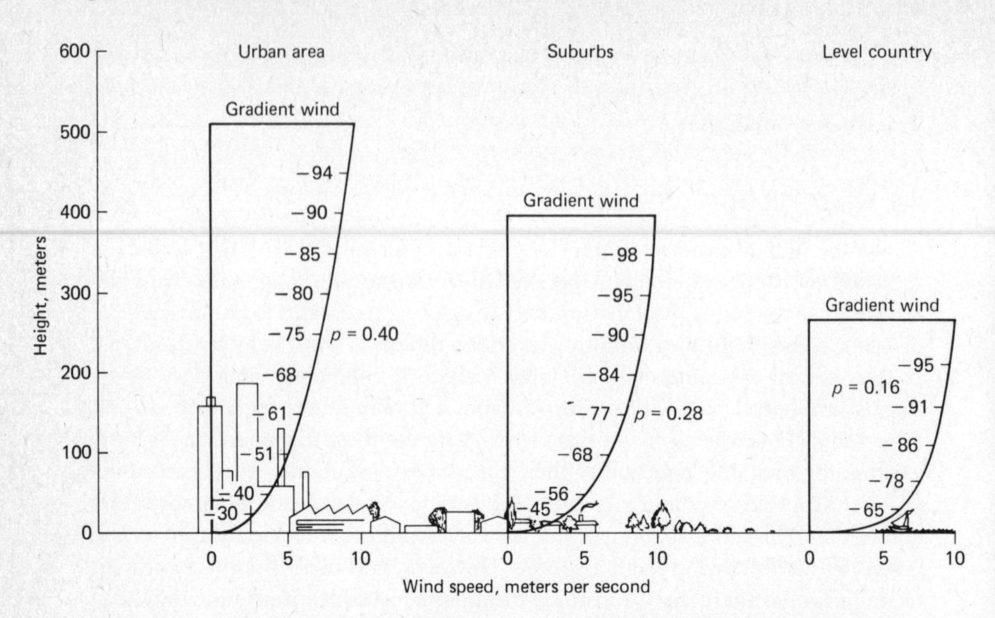

Figure 3-13 Effect of terrain roughness on the wind-speed profile. Values along curves represent percentages of gradient wind value. (SOURCE: D. B. Turner. *Workbook for Atmospheric Dispersion Estimates*. Washington, D.C.: HEW, 1969.)

m. Above this height the frictional effect is negligible, and the wind speed becomes that of the gradient wind discussed earlier. The gross effect of terrain roughness on the wind-speed profile is shown in Figure 3-13. In this particular example the change in the overall boundary layer thickness is from approximately 500 m to 280 m, for decreasing roughness. With decreasing roughness the profile also is steeper near the surface. Because of the appreciable change in wind speed with height in the planetary boundary layer, any wind-speed value must be quoted with respect to the elevation at which it was measured. The international standard height for surface wind measurements is 10 m.

Frequently knowledge of the wind speed at some height other than at the standard is necessary. Numerous efforts have been made to develop suitable analytical expressions relating wind speed to height. Because of the complexity of the phenomena, no completely satisfactory expression is available at this time. However, the power law of Deacon [3] has been found useful for boundary layers up to several hundred meters in depth. This expression is

$$\frac{u}{u_1} = \left(\frac{z}{z_1}\right)^p \tag{3-13}$$

where u is the wind speed at altitude z, u_1 is the wind speed at altitude z_1, and p is the positive exponent, with a value between 0 and 1.

When the environmental lapse rate is approximately the adiabatic value and the terrain is generally level with little surface cover, the value of p to be chosen is approximately $\frac{1}{7}$ [3].

The boundary layer thickness, and thus the wind-speed profile, is a function of the atmospheric stability. Hence the exponent p must vary in relation to the stability characteristics of the atmosphere, as a first approximation. Sutton [4] has suggested that p can be related to a parameter n which is a function of the stability of the atmosphere. This relationship is

$$p = \frac{n}{2 - n} \tag{3-14}$$

The values of n for various stability conditions are given in Table 3-1. Sutton [3] also presents the data of R. Frost relating p to the temperature difference found between elevations of 5 and 400 ft. These data are presented in Table 3-2. The power law expressed by Equation (3-13) may be used to approximate the wind-speed profiles for the various terrains shown in Figure 3-13. It is found reasonably suitable to use p of 0.40 for builtup areas, 0.28 for heavily wooded areas, cities, and suburbs, and 0.16 for flat open country, lakes, and seas, For heights up to several hundred feet above ground level some investigators feel that a logarithmic relation of the type $u = c \ln z$ is more representative of the air layer close to the ground than a power law expression, especially for open terrain.

As will be discussed in Chapter 4, the dispersion of pollutants in the atmosphere is a strong function of the average wind speed at the elevation of the emission source. From the preceding discussion of the variation of wind speed with altitude, we can conclude that the wind speed must be selected

Table 3-1 RELATIONSHIP BETWEEN THE STABILITY PARAMETER n IN EQUATION (3-14) AND THE STABILITY CONDITION OF THE ATMOSPHERE

STABILITY CONDITION	n
Large lapse rate	0.20
Zero or small lapse rate	0.25
Moderate inversion	0.33
Large inversion	0.50

Table 3-2 RELATIONSHIP BETWEEN THE TEMPERATURE DIFFERENCE AND THE PARAMETER p FOR AIR LAYERS FROM 5 TO 400 ft THICK, WHERE $\Delta T = T_{400} - T_5$

ΔT (°F)	p	ΔT (°F)	p
−4 to −2	0.145	2 to 4	0.44
−2.5 to −1.5	0.17	4 to 6	0.53
−2 to 0	0.25	6 to 8	0.63
−1 to 1	0.29	8 to 10	0.72
0 to 2	0.32	10 to 12	0.77

with care, taking into account elevation, nature of the surrounding terrain, and stability of the atmosphere.

3-7 MAXIMUM MIXING DEPTH

The dispersion of pollutants in the lower atmosphere is greatly aided by the convective and turbulent mixing that takes place. The vertical extent to which this mixing takes place varies diurnally, from season to season, and is also affected by topographical features. The greater the vertical extent, the larger the volume of the atmosphere available to dilute the pollutant concentration. Thermal buoyancy effects determine the depth of the convective mixing layer, which is called the *maximum mixing depth* (MMD). Usually data are available as an average for a period of 1 month; consequently the MMD values available are known as mean maximum mixing depths (MMMD).

When an air parcel is heated by solar radiation at the earth's surface, its temperature rises above that of the surrounding air and it acquires buoyancy. The buoyancy acceleration due to this temperature difference can be derived from basic considerations. For a fluid element in the atmosphere in static equilibrium, the governing equation is Equation (3-1), that is,

$$dP = -\rho g\, dz \tag{3-1}$$

Now consider an air parcel in the atmosphere which accelerates upward upon heating. The governing equation for this air parcel is similar to Equation (3-1) except that it contains an inertial term for the acceleration effect. For the air parcel we may write

$$\frac{dV}{dt} = -g - \frac{1}{\rho'}\left(\frac{dP}{dz}\right)$$

where dV/dt is the acceleration of the parcel, ρ' is the heated air parcel density, and ρ in Equation (3-1) is the surrounding unheated air density. No prime has been placed on the quantity dP/dz for the heated air parcel, since the pressure is the same for the heated parcel and the unheated surroundings. Substituting Equation (3-1) into the preceding equation, we find that

$$\frac{dV}{dt} = \left(\frac{\rho - \rho'}{\rho'}\right)g$$

Finally, invoking the ideal gas equation for the heated air parcel and the surrounding atmosphere, which are of equal pressure, we have

$$P = \rho RT = \rho' RT'$$

Replacement of ρ and ρ' in the dV/dt expression by using the ideal gas equations yields

$$\frac{dV}{dt} = \left(\frac{T' - T}{T}\right)g = \text{buoyant acceleration} \tag{3-15}$$

From Equation (3-15) it is seen that an air parcel, after heating, will continue to rise within the local atmosphere until its temperature T' equals the local atmospheric temperature T. At that point the air parcel and its surroundings will be in neutral equilibrium, and that height defines the limit of the convective mixing layer, or the maximum mixing depth. This situation is illustrated in Figure 3-14(a), where the subscript o on the temperature symbols represents the ground-level value. In general the environment has a ground-level temperature of T_o, and its temperature profile is represented by the unbroken line marked $(dT/dz)_{\text{env}}$. The air parcel is heated by solar radiation, for example, to a ground-level temperature T'_o. It expands upward along a dry adiabatic lapse line Γ, shown as a dashed line in the figure. The intersection of these two lines is a measure of the maximum mixing depth, MMD. In very stable air the general atmospheric temperature gradient takes the form shown in Figure 3-14(b). Under such a condition the MMD is considerably lower in height than in the case shown in Figure 3-14(a). Figure 3-14(c) shows the MMD position when an elevated inversion layer is present.

In practice, the MMD is determined with the aid of the temperature profile of the actual atmosphere for several kilometers above the earth's surface. A balloon is sent aloft and temperature values at various altitudes are transmitted back. These are known as radiosonde measurements. These data are plotted versus height. A dry adiabatic temperature line, starting at the normal maximum surface temperature for that month, is also drawn on this plot. The altitude at which the dry adiabatic line intersects the radiosonde measurements is taken as the MMD. Temperature sounding data are usually taken at night, but morning data are also frequently measured.

The values of the MMD are usually lowest at night and increase during the daylight hours. Under a severe inversion at night, the value may be essentially zero, while values up to 2000 and 3000 m are common in the daytime. On a seasonal basis the mean MMD is at a minimum in the winter (December and January) and at a maximum during the early summer (June). It has been noted [5] that extensive episodes of urban air pollution frequently occur when the MMD value is less than 1500 m. Since values less than this are quite common in many urban areas, the potential for air pollution episodes is often high.

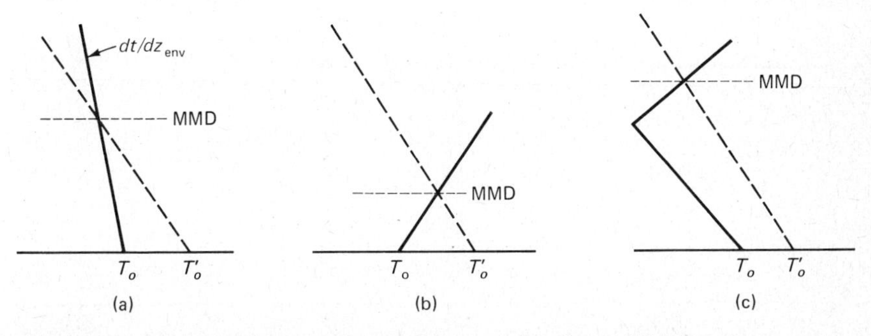

Figure 3-14 Establishment of the maximum mixing depth (MMD) under various atmospheric conditions (adiabatic profile - - -, environmental profile—).

3-8 WIND ROSE

Accurate estimates of the dispersion of pollutants in the atmosphere require a knowledge of the frequency distribution of wind direction as well as wind speed. This type of information varies significantly from city to city, and varies considerably for a given city from month to month. The characteristic patterns of local air movement may be presented in either tabular or graphical form. Table 3-3 is a hypothetical listing of wind speed and direction for an urban area observed at hourly intervals for a period of 1 month (30 days). Data are usually reported at eight primary and eight secondary directions of the compass. The wind speed typically is divided into ranges, as shown in the table. If desired, there are methods for removing bias as well as distributing the data for the calms among the 16 wind directions [5].

Figure 3-15 shows the graphic form of the data in Table 3-3. On polar coordinates the frequencies of various observed wind directions are proportional to the total length of the spokes. The distribution of wind speeds within each direction is indicated by the length of the individual sections of a given spoke. Proceeding outward from the center circle, which represents the percentage of the time calms were observed, the first line segment represents the percentage of time for 0 to 3 mi/hr, the next segment represents 4 to 7 mi/hr, and so on. For most large urban areas such a graph, known as a *wind rose*, will be available for each month of the year, each

Table 3-3 WIND SPEED AND DIRECTION FOR A HYPOTHETICAL CITY FOR A SPECIFIED MONTH

	FREQUENCY OF WIND SPEED OBSERVED AT HOURLY INTERVALS (mi/hr)					
DIRECTION	1–3	4–7	8–12	13–18	19–24	TOTAL
N	3	4	8	4		19
NNE	4	10	3	2		19
NE	8	8	2	2		20
ENE	4	4	7	3		18
E	2	4	3	2		11
ESE	3	3	2	2		10
SE	12	10	15	5		42
SSE	6	17	20	5		48
S	16	24	24	6		70
SSW	7	31	17	3		58
SW	6	48	35	7		96
WSW	5	16	17	7		45
W	6	24	14	6		50
WNW	5	15	14	12		46
NW	4	4	18	28	4	58
NNW	2	8	18	9		37
Calm	73					
TOTALS	164	231	220	101	4	720

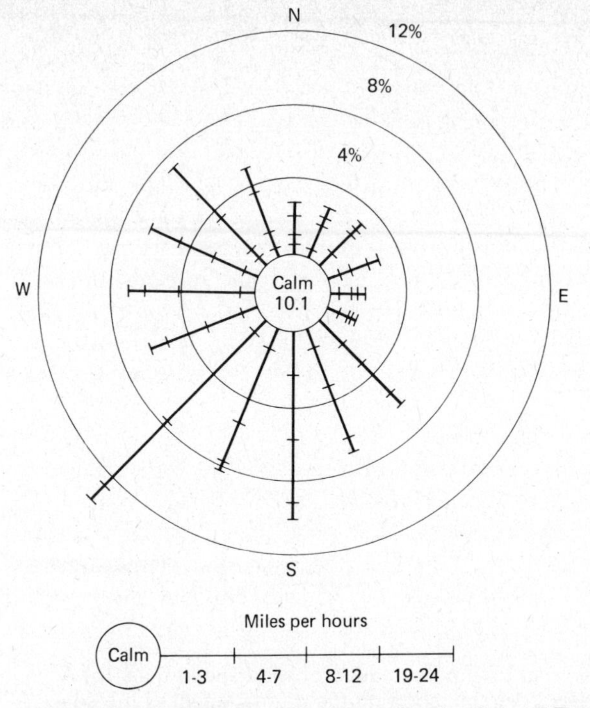

Figure 3-15 Hypothetical wind rose for the data of Table 3-3.

averaged over a number of years in the past. The wind direction indicated in Table 3-3 and Figure 3-15 is the direction from which the wind is coming. Figure 3-16(a) shows an actual wind rose with its typical wind-speed information. The skewness of these particular data should be noted. Such a wind rose would be fairly typical, for example, of data taken at measuring station located in a mountain-valley configuration. The data of Figure 3-16(b), for a different locale, do not show the actual distributions of wind-speed ranges in a given direction, as is noted in Figure 3-16(a). However, the vast difference in the wind rose from day to night is clearly illustrated. In this latter case the data are for New York City, and the shift in the data is due to the sea breeze in that region. For the purpose of predicting pollution dispersion in the local atmosphere, a complete set of wind roses for the year should be available, since data on wind speed and direction may vary widely from month to month.

3-9 TURBULENCE

The dispersion of atmospheric pollutants is accomplished, in general, by two major mechanisms of atmospheric circulation: the average wind speed and the atmospheric turbulence. Atmospheric turbulence is not completely understood. Turbulence in the atmosphere usually includes those fluctuations

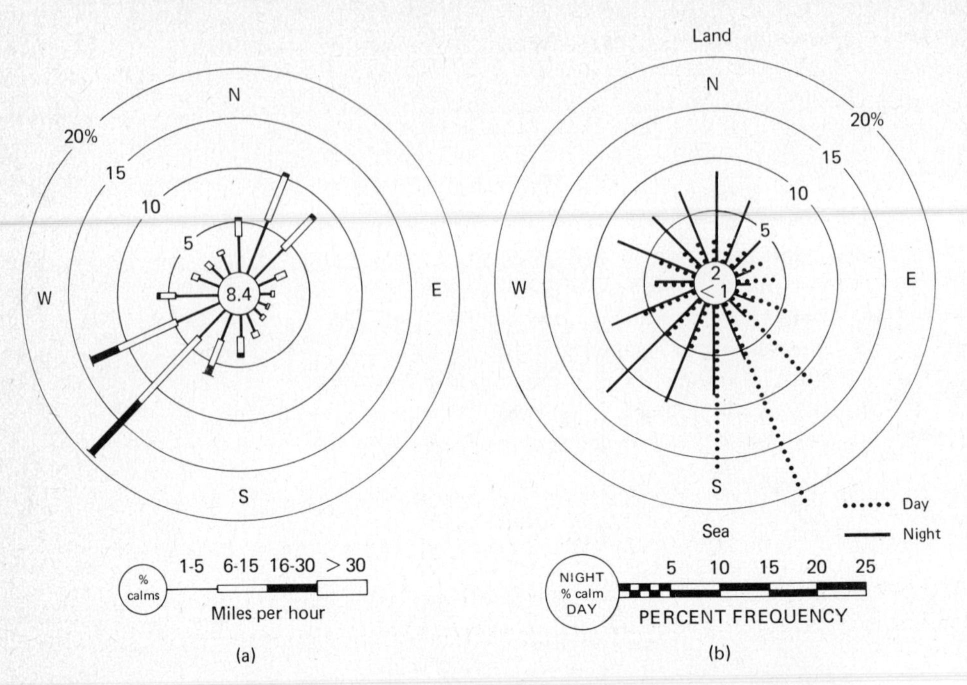

Figure 3-16 (a) A typical wind rose presentation of wind-speed data. (b) A day-night wind rose for New York City showing the diurnal effect of the sea breeze. (SOURCE: D. H. Slade, ed. *Meteorology and Atomic Energy*. Washington, D.C.: AEC, 1968.)

in wind flow which have a frequency of more than 2 cycles/hr. The more important fluctuations have frequencies in the 1 to 0.01 cycle/s range. Atmospheric turbulence is the result of two specific effects: (a) atmospheric heating, which causes natural convection currents ($d\rho/dz$), and (b) "mechanical" turbulence, which results from wind shear effects (du/dz). While both effects are usually present in any given atmospheric condition, either mechanical or thermal (convective) turbulence may prevail over the other. Thermal eddies are prevalent on sunny days when light winds occur and the temperature gradient is highly negative. The period for these cyclic fluctuations would be on the order of minutes. On the other hand, mechanical eddies predominate with neutral stability on windy nights, and the wind fluctuations have periods on the order of seconds. Mechanical turbulence is the result of air movement over the earth's surface and is influenced by the location of buildings and the relative roughness of the terrain.

One of the most useful quantitative descriptions of turbulence is the statistical standard (that is, root mean square) deviation of the wind fluctuations taken over some time period, usually 1 hr. These standard deviations (σ-values) can be converted into estimates of the vertical and horizontal parameters of dispersion equations [6].

3-10 GENERAL CHARACTERISTICS OF STACK PLUMES

As noted in the preceding section, the dispersion of pollutants in the atmosphere is accomplished by two general mechanisms: the average wind speed and the atmospheric turbulence. The effect of the former is simply to carry the pollutants downwind from the source; the latter causes the pollutants to fluctuate from the mainstream concentration in the vertical and cross-wind directions. The two types of turbulence—mechanical and convective—usually occur simultaneously under any atmospheric condition, but in varying ratios to each other. Because of these variations, the general geometric forms of gas plumes emitted from stacks are quite different.

Six classifications of plume behavior are noted in Figure 3-17. In addition to the general variation in geometric shape in the x-z coordinate plane, approximate velocity and temperature profiles are also shown. Gradual transition from one type to another may occur. The *looping* plume shown in Figure 3-17(a) occurs when a high degree of convective turbulence exists. As noted on the figure, the looping plume indicates a superadiabatic lapse rate in the atmosphere, which leads to strong instabilities. The thermal eddies may become large enough to carry a portion of the plume down to ground level for short time periods. Although the large eddies tend to disperse pollutants over a wide region, high concentrations may occur at localized areas on the ground. A looping plume is usually associated with a clear daytime condition accompanied by strong solar heating of the earth's surface and light winds.

A *coning* plume [Figure 3-17(b)] occurs under essentially neutral atmospheric stability, and small-scale mechanical turbulence dominates. Since the thermal heating effect is much lower than in the case of looping plumes, coning occurs when skies are overcast during either the day or night. Winds are typically moderate to strong. The cloud cover prevents incoming solar radiation during the day and outgoing terrestrial radiation at night. The half-angle of the plume, which has the general shape of a cone, has been noted to be roughly 10 degrees. Unlike in looping, in coning the major part of the pollutant concentration is carried fairly far downwind before reaching ground level in significant amounts. This is an especially good condition for estimating pollutant dispersion by the diffusion equations introduced in Chapter 4.

A *fanning* plume occurs in the presence of large negative lapse rates, so that a strong surface inversion takes place to a considerable distance above the stack height. The atmosphere is extremely stable, and mechanical turbulence is suppressed. If the density of the plume is not significantly different from that of the surrounding atmosphere, the plume travels downwind at approximately constant elevation, as shown in Figure 3-17(c). As noted earlier, inversions are characteristic of clear night-time conditions when the earth is cooled by outgoing radiation. When viewed from above, a fanning plume may appear to meander in the horizontal direction as it is

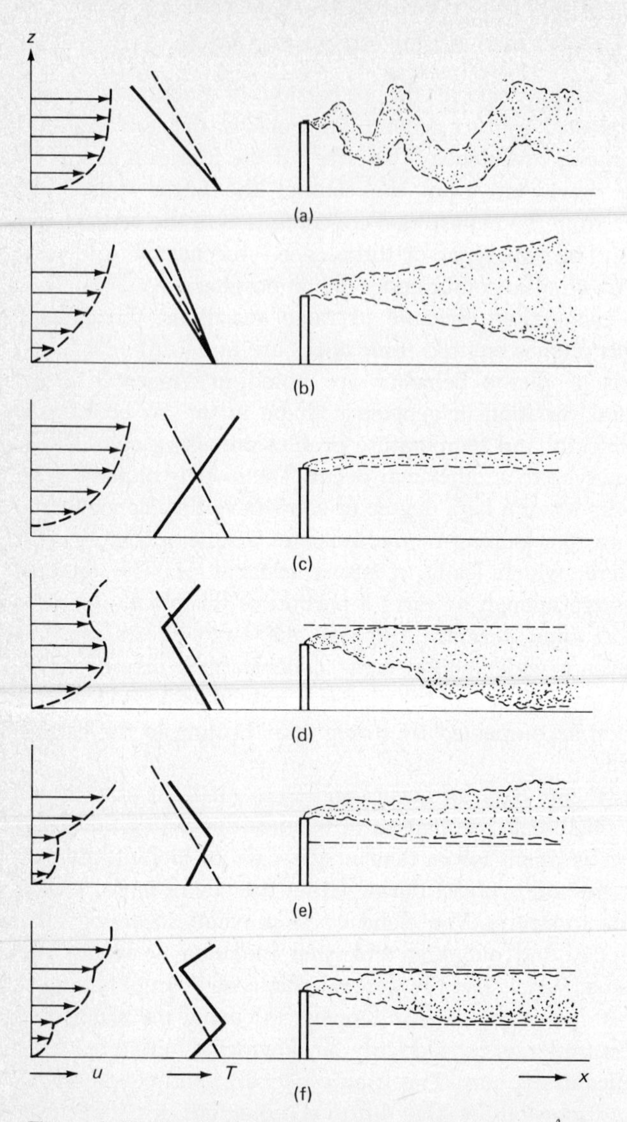

Figure 3-17 Typical velocity profile, temperature profile, and plume shape in the x-z coordinate system for various atmospheric conditions. (Dry adiabatic lapse rate, ---; ambient lapse rate, —.) (a) Looping, strong instability; (b) coning, near neutral stability; (c) fanning, surface inversion; (d) fumigation, aloft inversion; (e) lofting, inversion below stack; (f) trapping, inversion below and above stack height.

carried downwind. As might be anticipated, it is difficult to predict downwind pollutant concentrations when fanning is present. Little pollutant effluent reaches the ground.

Fumigation plumes occur when a stable layer of air lies a short distance above the release point of the plume and an unstable air layer lies below the plume. Figure 3-17(d) illustrates the general profiles when an inversion exists aloft. The temperature profile necessary for fumigation usually arises in early morning, following a night characterized by a stable inversion. The morning sun heats the ground, which in turn leads to the development of a negative temperature gradient from the ground upward. Once the newly formed unstable layer reaches the height of a stack, large concentrations of stack gas (formerly in a fanning profile) will be carried downwind to the surface. Fortunately the conditions for fumigation normally do not last for more than half an hour. However, during that time period relatively high ground-level concentrations will be reached. Fumigation is favored by clear skies and light winds, and is more prevalent in the summer.

The conditions for the *lofting* plume, indicated in Figure 3-17(e), are the inverse of those for the fumigation plume. The inversion layer lies below and the unstable layer lies through and above the plume. This is a favorable situation, since the pollutants are dispersed downwind without any significant ground-level concentrations. While fumigation conditions characterize the early morning after sunrise, lofting conditions prevail in the late afternoon and early evening under clear skies. During the day a negative temperature gradient is set up throughout the lower atmosphere, as a result of solar heating. In the late afternoon radiation from the surface leads to an inversion layer near ground level. As the inversion layer deepens, a lofting plume will change to a fanning plume. Hence lofting is usually a transitional situation. When an inversion exists both below and above stack height, *trapping* results. The diffusion of pollutants is severely restricted to the layer between the two stable regions, as shown in Figure 3-17(f).

The general characteristics of the terrain surrounding a stack and the location and nature of buildings relative to a stack have a marked influence upon the behavior of a plume. Figure 3-18 shows the influence on the wind flow pattern of a bluff structure situated in an otherwise open terrain. Note the "backwash" region of flow just downwind from (behind) the structure. The flow reverses in direction near ground level, as indicated by the velocity profile just behind the building. When a stack is placed upwind from the structure, as in Figure 3-19(a), the important aerodynamic effect of backwash on pollutant dispersion becomes apparent. When the stack height is insufficient, the nature of the flow across the top of the building induces backflow, so that pollutant concentrations are heavy adjacent to the downwind side of the building. The situation worsens as stack height is decreased. A similar condition exists when a stack is situated downwind from a building. Downflow from the building upstream from the stack induces the stack effluent to fall rapidly to the ground downwind from the stack. This is especially true when the building is higher than the stack. Finally, a stack mounted atop or immediately adjacent to a building is shown in Figure 3-19(b). It is apparent that the effluent from the tall stack shown in the figure will not be appreciably affected by the downwash from the building. What would happen if the stack were not much higher than the building is easy to visualize. In such a situation, a good rule of thumb is the following: to prevent the induced flow of stack pollutants toward the ground in the

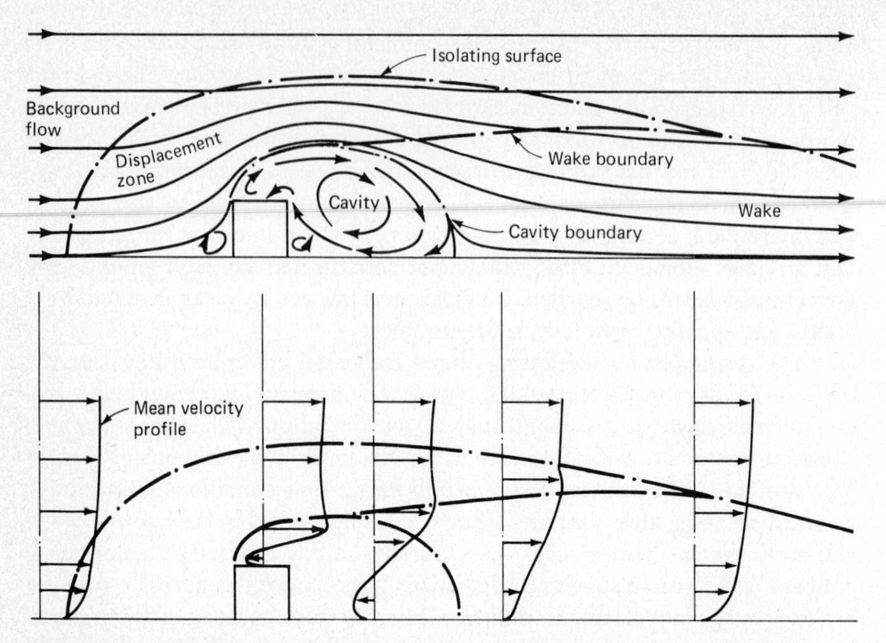

Figure 3-18 General arrangement of flow pattern around a sharp-edged building. (SOURCE: D. H. Slade, ed. *Meteorology and Atomic Energy*. Washington, D.C.: AEC, 1968.)

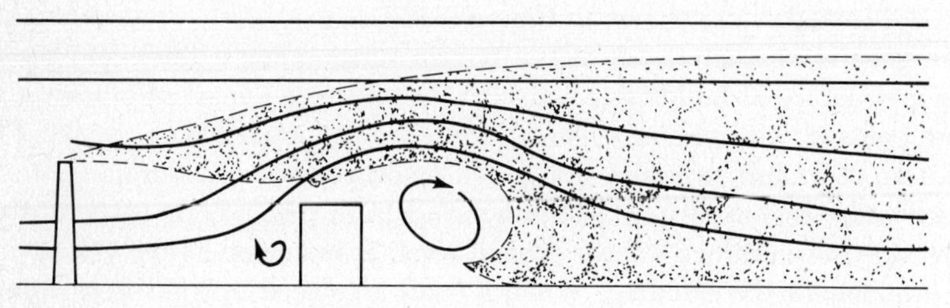

(a)

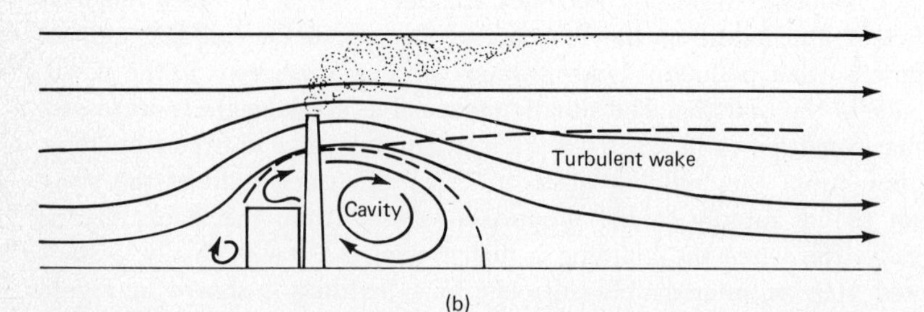

(b)

Figure 3-19 (a) The aerodynamic effect of a building downwind from a stack on the dispersion of gaseous effluent. (b) The effect of a stack mounted atop or immediately adjacent to a building on downwind dispersion. (SOURCE: G. A. Briggs. *Plume Rise*. AEC Critical Review Series, 1969.)

vicinity of a building, the stack height should be at least twice the building height.

3-11 HEAT ISLAND EFFECT

The atmosphere of a large city differs in many ways from that of the countryside. The houses, large office buildings, and factories form an irregular surface that retards the free flow of large air masses and thus retards the winds. Because of the perpendicular building surfaces and ravinelike streets, the city area absorbs more solar energy during the day and retains it for a longer period at night than does an equal area in the country. The city also releases large quantities of particulate into the air. The resulting phenomenon is known as the "heat island." Warm air tends to be concentrated at the center of the city, probably because of the concentration of tall buildings and paved streets. This warm air rises, carrying upward the burden of pollution, then expands and flows outward over the edges of the city. As the air expands, it cools, and thus cooler air at the edges of the city will flow back toward the center of the city near the ground. Thus a self-contained circulatory system is established that can be altered or broken up only by a strong wind. Figure 3-20 illustrates the heat island. The small particulate in the air over the city tends to reduce the solar energy and thus prevent the sun from heating the surface air. Normal vertical air currents are reduced. The air under the city's "dust dome" or "haze hood" on windless days tends to become increasingly polluted.

3-12 GLOBAL CIRCULATION OF POLLUTANTS

From the preceding discussion of meteorology we can conclude that the atmospheric conditions existing at any given location at a specific time are a function of many variables, including terrain, moisture content of the atmosphere, and meteorological conditions. When we consider the quantities of

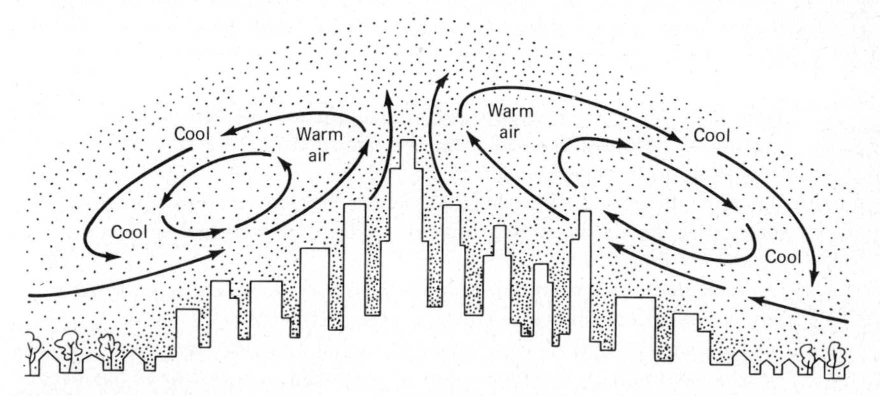

Figure 3-20 Sketch showing the circulation pattern for a city heat island.

the various pollutants released annually, as listed in Chapter 1, the question of their impact upon the global atmosphere assumes major importance. Robinson and Robbins [7] point out that scavenging mechanisms exist in the natural environment for the gaseous emissons CO_2, CO, NO_x and sulfur compounds. Although these scavenging mechanisms are important in controlling long-term accumulations of pollutants in the atmosphere, they do not operate fast enough to provide solutions for local urban air pollution situations. In fact, the existence of an air pollution condition is evidence that the available scavenging mechanism rates have been exceeded by the rates of emission.

In their study, Robinson and Robbins found that while the annual rate of man-generated CO_2 is 1.4×10^{10} tons and the natural generation rate is 10^{12} tons, the CO_2 concentration in the global atmosphere is increasing at a rate of only 0.7 ppm/yr. This increase in atmospheric CO_2 represents about half the CO_2 produced by all combustion sources since 1900; thus a natural scavenging mechanism must exist. Numerous authors [7] have estimated the consumption rate of CO_2 in the biosphere. If vegetation processes were solely responsible for the CO_2 uptake from the atmosphere, its residence time would be 17 years. However, the dominant role of the ocean in the CO_2 cycle has long been recognized. It is estimated that the oceans take up CO_2 at a rate sufficient to give atmospheric CO_2 a mean residence time of 2.5 yr. CO_2 is highly soluble in the ocean, where it is stored or consumed in the growth of marine plants and animals. Robinson and Robbins [7] concluded that the atmospheric residence time of CO_2 is 2 to 4 yr. A condensed summary of the findings of Robinson and Robbins for the selected gaseous pollutants is presented in Table 3-4.

QUESTIONS

1. What are the three dominant mechanisms that account for dispersion of a pollutant into the atmosphere?
2. What forces lead to the geostrophic wind and at what direction does this wind move relative to straight, parallel isobars in the atmosphere?
3. When isobars are curved in the atmosphere, what three forces must be considered to affect an air parcel, and what is the name of the wind resulting from these forces?
4. Describe the flow of air around a high-pressure center in terms of its direction of rotation, its movement perpendicular to the earth's surface, and its motion relative to the center.
5. Describe the flow of air around a low-pressure center in terms of its direction of rotation, its movement perpendicular to the earth's surface, and its motion relative to the center.
6. As one proceeds upward through the planetary boundary layer, what two effects would one observe with respect to the flow of air?
7. Distinguish between the environmental (atmospheric) lapse rate and an adiabatic lapse rate. What are values of the adiabatic lapse rate in the metric and English engineering systems of units?

Table 3-4 GLOBAL SOURCES, CONCENTRATIONS, AND ATMOSPHERIC REACTIONS OF TRACE GASES

			ESTIMATED EMISSIONS (TONS)				
POLLUTANT	MAJOR SOURCES	NATURAL SOURCES	MAN-MADE	NATURAL	ATMOSPHERIC BACKGROUND CONCENTRATION	CALCULATED ATMOSPHERIC RESIDENCE TIME	REMOVAL REACTIONS AND SINKS
SO_2	Combustion of coal and oil	Volcanoes	146×10^6	None	0.2 ppb	4 days	Oxidation to sulfate or after absorption by solid and liquid aerosols
H_2S	Chemical processes, sewage treatment	Volcanoes, biological action in swamp areas	3×10^6	100×10^6	0.2 ppb	2 days	Oxidation to SO_2
CO	Auto exhaust, other combustion	Forest fires, terpene reactions, oceans?	274×10^6	75×10^6	0.1 ppm	3 days	Soil fungi, large sink necessary
NO/NO_2	Combustion	Bacterial action in soil?	53×10^6	NO: 430×10^6 NO_2: 658×10^6	NO: 0.2–2 ppb NO_2: 0.5–4 ppb	5 days	Oxidation to nitrate after aerosol sorption, photochemical reactions
NH_3	Waste treatment	Biological decay	4×10^6	1160×10^6	6–20 ppb	7 days	Formation of ammonium sulfate, oxidation to nitrate
CO_2	Combustion, release from oceans, decay	Biological	1.4×10^{10}	10^{12}	320 ppm	2–4 yr	Absorption in oceans and biologically
HCs	Combustion, chemical processes	Biological	88×10^6	480×10^6	CH_4: 1.5 ppm	CH_4: 16 yr	Photochemical reactions, large sink necessary for methane

SOURCE: E. Robinson and R. C. Robbins. *Sources, Abundance, and Fate of Gaseous Atmospheric Pollutants*. Stanford Research Institute, Report SRI Project PR-6755, February 1968. Supplemental Report, June 1969.

8. How are environmental and adiabatic lapse rates used in conjunction with establishing the degree of stability within the atmosphere?
9. What is meant by superadiabatic, neutral, and subadiabatic conditions in the atmosphere?
10. What is the mechanism that causes a subsidence inversion? What breaks up the inversion?
11. Discuss how an inversion beginning at an elevation of 500 m affects an air pollution problem.
12. Discuss the mechanism that leads to a radiative inversion. What causes it to break up, and at what time does it usually occur?
13. Discuss why it is important to know the maximum mixing depth when calculating the atmospheric concentration of air pollutants.
14. What information is typically available from a wind rose for a given locale?
15. Describe the temperature profiles and the dispersion profiles for plumes known as (a) looping, (b) coning, (c) fanning, (d) fumigation, (e) lofting, and (f) trapping.

PROBLEMS

3-1. On the basis of the average temperature gradient in the situations described, classify the degree of stability of the atmosphere: (a) temperature at ground level is 70°F, temperature at 1500 ft is 80°F; (b) ground-level temperature is 70°F; temperature at 2500 ft is 60°F; (c) ground-level temperature is 60°F, temperature at 1900 ft is 48°F; (d) ground-level temperature is 25°C, temperature at 2000 m is 5°C; (e) ground-level temperature is 30°C, temperature at 500 m is 20°C; (f) ground-level temperature is 25°C, temperature at 700 m is 28°C.

3-2. Determine the potential temperature gradient for each condition described in Problem 3-1.

3-3. Determine the potential temperature for the above ground data quoted in Problem 3-1, if the pressure at those elevations is (a) 970 mbar, (b) 925 mbar, (c) 935 mbar, (d) 820 mbar, (e) 950 mbar, and (f) 930 mbar.

3-4. On the basis of the average temperature gradient in the situations described, classify the degree of stability of the atmosphere: (a) temperature at ground level is 60°F, temperature at 1200 ft is 69°F; (b) temperature at ground level is 55°F, temperature at 900 ft is 51°F; (c) ground-level temperature is 28°C, temperature at 400 m is 20°C; (d) ground-level temperature is 24°C, temperature at 600 m is 18°C.

3-5. For conditions (a) through (d) in Problem 3-4, determine the potential temperature gradient in appropriate units.

3-6. Determine the potential temperature for the above ground-level data cited in Problem 3-4, if the pressure at the given elevations is (a) 980 mbar, (b) 930 mbar, (c) 945 mbar, and (d) 865 mbar.

3-7. The wind has a velocity of 10 mi/hr at 10 m. On a day when the lapse rate is small, what is the velocity at an elevation of 100 m, in miles per hour?

3-8. The ground-level wind velocity at 10 m elevation is 7 mi/hr in a city. What would you estimate the velocity to be at 100-m elevation, in miles per hour?

3-9. Consider that the wind speed u is 2 m/s at a height of 10 m. Estimate the wind speed at heights of (a) 200 m and (b) 300 m for the four stability conditions referred to in Table 3-1.

3-10. Two environmental temperature gradients, $\Delta T/\Delta z$, of (a) $-0.008°F/ft$ and (b) $+0.004°F/ft$ exist at daytime and nighttime, respectively, at a given locale. Using the data of Table 3-2, estimate the ratio of wind speeds at 400 to 5 ft for the two conditions.

3-11. On a given day the ground-level air temperature is 70°F, while the normal maximum surface temperature for that month may be considered to be (a) 90°F and (b) 84°F. At an elevation of 2000 ft the temperature of the air is measured at 65°F. What is the maximum mixing depth, in feet, for the two situations?

3-12. At a given location the ground-level temperature is (a) 60°F and (b) 57°F, while the normal maximum surface temperature for that month is taken to be 80°F. At an elevation of 1000 ft the air temperature is measured at 70°F. What is the maximum mixing depth, in fact, for the two situations?

3-13. At a given location the ground-level air temperature is 18°C, while the normal maximum surface temperature for that month is known to be 30°C. At an elevation of 700 m the temperature is measured at (a) 15°C and (b) 20°C. What is the maximum mixing depth, in meters, for the two cases?

3-14. In a given situation the ground-level air temperature is 15°C, while the normal maximum surface temperature for that month is (a) 26°C and (b) 24°C. At an elevation of 300 m the temperature is found to be 21°C. What is the maximum mixing depth, in meters, for the two cases?

3-15. Assume that the atmospheric lapse rate on a particular day is constant up to 800 m. At ground level P_o is 1020 mbar and T_o is 15°C. A radiosonde measurement indicates that at some elevation z the pressure and temperature are 975 mbar and 11.5°C, respectively. Determine (a) the atmospheric temperature gradient, dT/dz, in degrees Kelvin per meter, (b) the elevation of the observation, in meters, and (c) the potential temperature of the air parcel aloft at elevation z, in degrees Kelvin.

3-16. Solve Problem 3-15 for the following data: P_o is 1010 mbar, T_o is 17.4°C, and at elevation z the pressure and temperature are 960 mbar and 14.7°C, respectively.

3-17. A radiosonde balloon held at an elevation of 30 m recorded at this position a pressure of 1023 mbar and a temperature of 11.0°C. After it was released, it radioed back the following P and T data from positions 2 through 10, respectively: 1012, 9.8; 1000, 12.0; 988, 14.0; 969, 15.0; 909, 13.0; 878, 13.0; 850, 12.6; 725, 1.6; and 700, 0.8, where P is in millibars and T is in degrees Celsius. (a) Estimate the elevation at which each reading listed above was taken, in meters. (b) Plot P and T data on the ordinate of a graph versus elevation on the abscissa, and connect the data point for a given parameter by a series of straight lines.

3-18. Reconsider Problem 3-17 on the basis of the following data. At 30 m the pressure and temperature readings are 1048 mbar and 11.5°C, respectively, and the data at positions 2 through 6 are: 1025, 12.5; 993, 13.2; 927, 12.1; 872, 9.8; 825, 8.4.

3-19. Sketch a temperature-height plot for Oak Ridge, Tennessee, on the basis of Figure 3-11 for the time of (a) 2 A.M, (b) 10 A.M., (c) 2 P.M., and (d) 8 P.M.

References

1. H. H. Lettau and D. A. Haugen. *Handbook of Geophysics*. New York: Macmillan, 1960.
2. *Air Pollution Primer*. New York: National Tuberculosis and Respiratory Disease Association, 1969.
3. O. G. Sutton. *Micrometeorology: A Study of Physical Processes in the Lowest Layers of the Earth's Atmosphere*. New York: McGraw-Hill, 1953.
4. O. G. Sutton. *Quart. J. Roy. Meteorol. Soc.* **73** (1947): 426.
5. G. C. Holzworth. *Monthly Weather Rev*. **92** (1964): 235.
6. H. G. Haughton. "On the Annual Heat Balance of the Northern Hemisphere." *Meteorology* **2** (1954): 7.
7. E. Robinson and R. C. Robbins. *Sources, Abundance, and Fate of Gaseous Atmospheric Pollutants*. Stanford Research Institute, Report SRI Project PR-6755, February 1968. Supplemental Report, June 1969.
8. U.S. Weather Bureau, *Meteorology and Atomic Energy*. Report AECU-3066, AEC, Washington, D.C., 1955.

Chapter 4
Dispersion of Pollutants in the Atmosphere

4-1 INTRODUCTION

The atmospheric dispersion of effluents from vents and stacks depends upon many interrelated factors: for example, the physical and chemical nature of the effluents, the meteorological characteristics of the environment, the location of the stack with relation to obstructions to air motion, and the nature of the terrain downwind from the stack. Several analytical methods have been developed to relate the dispersion of effluents to a selected number of the aforementioned factors; however, none accounts for all of them.

Stack effluents may consist of gases alone, or gases and particulate matter. If the particles are on the order of 20 μm or smaller in diameter, they have such a low settling velocity that they move essentially in the same manner as the gas in which they are immersed. The analytical procedures developed for gas dispersion may be applied to the dispersion of small particles. Large particles, however, cannot be treated in the same way; they have a significant settling velocity which results in a higher ground-level concentration of the solid pollutant closer to the stack than is the case for gases. The deposition of particulates will be discussed in Chapter 5.

To achieve maximum dispersion, the effluents should leave the stack with sufficient momentum and buoyancy that they continue to rise from the stack exit. When there is no wind speed, low-density plumes tend to reach high elevations, and ground concentrations are low. Large particulates and dense gas plumes fall to the ground in the vicinity of the stack. High wind velocities increase the diluting action of the atmosphere, giving rise to lower ground-level concentrations downwind from the stack.

The rise of the majority of hot plumes is caused almost entirely by buoyancy due to the higher temperature of the gases. When the plume is deflected over in the wind, it is diluted along its axis in proportion to the average wind speed u at plume elevation so that the buoyancy is reduced. In stratified air the plume's buoyancy is dissipated as a result of the stability of the surrounding atmosphere which is characterized by the potential temperature gradient. When neutral atmospheric conditions exist, the plume is diffused by turbulence, the intensity of which is a function of ground roughness, height, and most importantly, wind speed.

To prevent downwash of the plume at the stack exit, the gas exit velocity V_s must be sufficiently large. If 98 percent of the wind speed u is equal to or less than 15 m/s, an exit velocity of 20 m/s would protect against downwash 98 percent of the time. Another approximation is expressed by the following ratio:

$$\frac{V_s}{u} > 2$$

That is, downwash from a stack is minimal when the stack gas speed is at least twice as great as the wind speed at the top of the stack.

The ability to predict ambient concentrations of pollutants in urban areas on the basis of dispersion from sources within the region is essential if federal ambient air quality standards are to be attained and maintained, in spite of future industrial and residential growth. Thus mathematical models for estimating the dispersion of pollutants from ground and elevated sources, whether single or grouped, must be developed to simulate the atmospheric process.

4-2 THE EDDY DIFFUSION MODEL

The most comprehensive approach to transport theory is based on the *eddy diffusion model*, which in turn involves the use of the "mixing length" concept. This is the usual starting point in the development of a dispersion model for the atmosphere. The basic equation for this model is mathematically quite complex, but with minor assumptions it may be reduced to the form

$$\frac{dC}{dt} = K_{xx}\left(\frac{\partial^2 C}{\partial x^2}\right) + K_{yy}\left(\frac{\partial^2 C}{\partial y^2}\right) + K_{zz}\left(\frac{\partial^2 C}{\partial z^2}\right) \tag{4-1a}$$

where C is the concentration, t is the time, and the K_{ii} quantities are the eddy diffusion coefficients in the three coordinate directions. This equation is known as the *Fickian diffusion equation*. It is difficult to apply this result, however, to the actual process in the atmosphere. Consequently, the following additional assumptions are commonly made:

1. The concentration of the pollutant emanates from a continuous point source.
2. The process is steady state, that is, $dC/dt = 0$.
3. The major transport direction due to the wind is chosen to lie along the x-axis.
4. The wind speed u is chosen to be constant at any point in the x, y, z coordinate system.
5. The transport of pollutant due to the wind in the x-direction is dominant over the downwind diffusion, that is, $u(dC/dx) \gg K_{xx}(\partial^2 C/\partial x^2)$.

As a result, the Fickian diffusion equation reduces to

$$u\frac{\partial C}{\partial x} = K_{yy}\frac{\partial^2 C}{\partial y^2} + K_{zz}\frac{\partial^2 C}{\partial z^2} \tag{4-1b}$$

where $K_{yy} \neq K_{zz}$. The solution to this equation must also fulfill the following boundary conditions:

1. $C \to \infty$ as $x \to 0$ (large concentration at the point source).
2. $C \to 0$ as $x, y, z \to \infty$ (zero concentration at a great distance from the source).
3. $K_{zz}(\partial C/\partial z) \to 0$ as $z \to 0$ (no diffusion into the surface).
4. $\int_0^\infty \int_{-\infty}^\infty uC(x, y, z)\, dy\, dz = Q, x > 0$ (rate of transport of pollutant downwind is constant and equal to the emission rate Q of the pollutant at the source).

Lowry and Boubel [1] give the following approximate solution to the above equation.

$$C(x, y, z) = \frac{Q}{4\pi r(K_{yy}K_{zz})^{1/2}} \exp\left[\frac{-u}{4x}\left(\frac{y^2}{K_{yy}} + \frac{z^2}{K_{zz}}\right)\right] \tag{4-2}$$

where $r^2 = x^2 + y^2 + z^2$. Unfortunately, Equation (4-2) shows two serious departures when compared to experimental evidence for center-line concentrations. Along the center line at ground level Equation (4-2) reduces to

$$C(x, 0, 0) = \frac{Q}{4\pi x(K_{yy}K_{zz})^{1/2}}$$

Therefore the approximate solution to the simplified theoretical equation indicates that the ground-level value of C along the center line of the plume is inversely proportional to x and independent of the wind speed u. Experimental observations indicate that C is inversely proportional to $(ux^{1.76})$. Improved solutions to the eddy diffusion model are currently being sought, to improve its agreement with observations. Among techniques being tested is one involving the use of an analog-digital hybrid computer [2]. Until the acceptance and ready availability of more sophisticated methods [3], other models must be considered. In seeking other models, the format of Equation (4-2) is helpful. This equation indicates that away from the center line the concentration decays exponentially in both the y- and z-directions. Mathematically this means that C in the cross-wind and vertical directions may be "normally" distributed. In addition, the decrease in the value of C in the x direction is largely dependent upon the values of K_{zz} and K_{yy}. Consequently, any other model should show heavy dependence upon the diffusion coefficients as well. The most widely accepted model at the present time is the Gaussian plume model. It does exhibit the "normal" distribution suggested by Equation (4-2), and it does require extensive information, in an indirect manner, on the mass diffusion coefficients in the y- and z-directions.

4-3 THE GAUSSIAN OR NORMAL DISTRIBUTION

In Section 4-4 we develop a model for estimating the concentration of gaseous pollutants downwind from a source. Although several basic approaches to the problem are possible, usually a number of simplifying assumptions are necessary in any case to obtain a mathematically tractable solution. As a result, all these theories tend to lead to the same distribution function for the pollutant concentration, that is, a *Gaussian* distribution function. To understand the significance of this type of distribution function in the context of air pollution, it is useful to review some of the general characteristics of thc Gaussian or normal distribution.

A variable x is said to be normally distributed if the density function $f(x)$ fulfills the relation

$$f(x) = \frac{1}{\sigma(2\pi)^{1/2}} \exp\left[\frac{-(x-\mu)^2}{2\sigma^2}\right] \tag{4-3}$$

where μ is any real number and σ is any real number with a value greater than zero. The quantity σ is known as the standard deviation. The nature of this function is more easily grasped by reference to Figure 4-1. The value of $f(x)$ is the vertical height above the horizontal axis. The value of μ sets the location of the maximum value of $f(x)$ on the x-axis, and the curve is symmetrical with respect to the position of μ. When $\mu = 0$, the curve is symmetrical around the $x = 0$ axis. Hence μ simply shifts the position of the overall distribution curve with respect to $x = 0$, as noted for the case where $\mu = -2$.

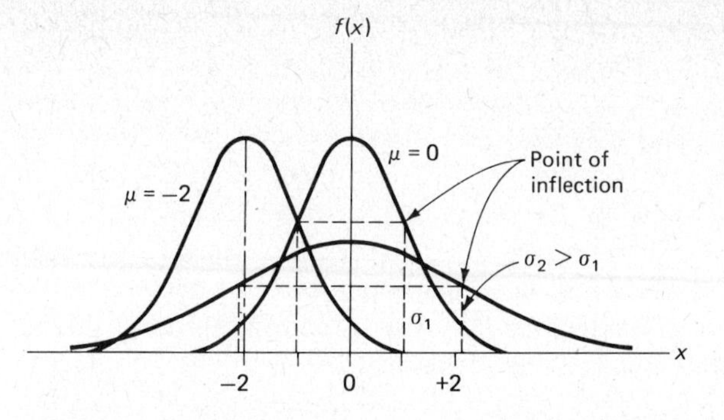

Figure 4-1 The Gaussian or normal distribution function for different values of μ and σ.

The normal or Gaussian distribution function represented by Equation (4-3) is in a normalized form. That is, the area under the curve has a value of unity. The role of σ is to broaden or sharpen the shape of the curve, while still retaining a unit area under the curve. The standard deviation, σ, is a measure of the position of the point of inflection on either side of the curve. When σ increases, as shown by the case of the σ_2 versus σ_1 curves centered at $x = 0$ in Figure 4-1, the maximum value of $f(x)$ decreases but $f(x)$ retains a significant value over a wider range around the major axis. This is necessary, of course, if the area under both of the curves centered at $x = 0$ is to be the same. In general, over 68 percent of the area under the curve lies between $+\sigma$ and $-\sigma$, and over 95 percent lies between $\pm 2\sigma$. This increased spread in the distribution function as σ increases has an important physical significance in the atmospheric dispersion of pollutants.

It is important to keep in mind the role of μ and σ in determining the general position and shape of the Gaussian distribution function as we develop the atmospheric dispersion equations for various situations. In general these dispersion equations will take on the format of a double Gaussian distribution. A double Gaussian distribution in two coordinate directions, such as y and z, is simply the product of the single Gaussian distributions in each of the coordinate directions. Hence,

$$f(y,z) = \frac{1}{2\pi\sigma_y\sigma_z}\exp\left[\frac{-(y-\mu_y)^2}{2\sigma_y{}^2} + \frac{-(z-\mu_z)^2}{2\sigma_z{}^2}\right] \tag{4-4}$$

where σ_y, σ_z, μ_y, and μ_z have essentially the same interpretation as that for the single Gaussian distribution. We will need this expression for comparative purposes in the next section.

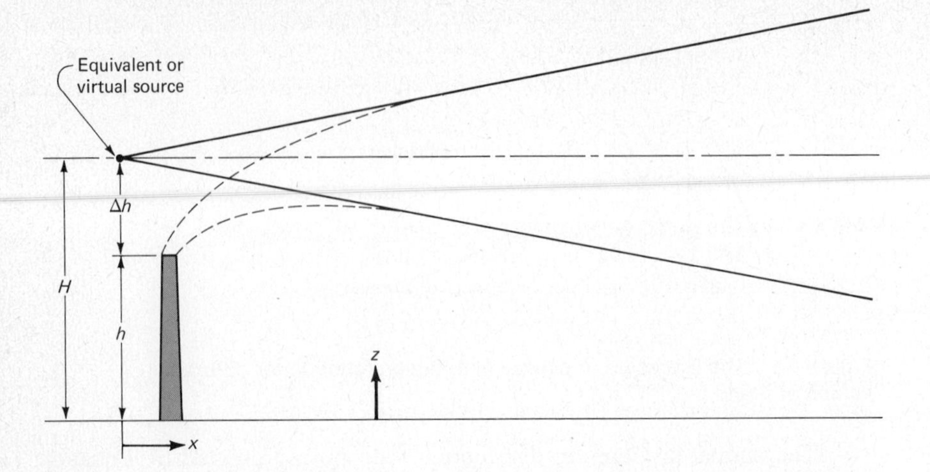

Figure 4-2 A dispersion model with virtual source at an effective stack height *H*.

4-4 THE GAUSSIAN DISPERSION MODEL

A mathematical model of atmospheric dispersion must attempt to simulate the gross behavior of plumes emitted from ground-level or stack-height sources. For localized point sources such as a stack, the general appearance of the plume might be represented by the schematic shown in Figure 4-2. Although the plume originates at a stack height h, it rises an additional height Δh, owing to the buoyancy of the hot gases and the momentum of the gases leaving the stack vertically with a velocity V_s. Consequently, for practical purposes the plume appears as if it originated as a point source at an equivalent stack height $H = h + \Delta h$. The point source also lies somewhat back along the center line from the stack position at $x = 0$.

One possible model for the physical situation shown in Figure 4-2 is developed in the appendix to this chapter. It is based on the mass diffusion of pollutant in the y- and z-directions as a fluid element is carried downwind in the x-direction with a wind speed u. The necessary assumptions for the model are enumerated in the appendix to this chapter. In summary, they include steady state, negligible mass diffusion in the x-direction, a constant wind speed u at all positions, and constant mass diffusivities D_x, D_y, and D_z in the respective coordinate directions. It is also common to neglect the distance from the equivalent or virtual source to the actual stack position. Hence the point source appears to be situated at $x = 0$ and at a height H.

In the appendix to this chapter it is shown that one appropriate representation of the concentration profile downwind from a point source is given by the general equation

$$C = Kx^{-1}\exp\left[-\left(\frac{y^2}{D_y} + \frac{z^2}{D_z}\right)\frac{u}{4x}\right] \tag{4-5}$$

where K is an arbitrary constant whose value is determined by the boundary conditions on the specific atmospheric problem. The evaluation of K for several specific situations is also noted in the appendix. Only the results of those evaluations will be quoted below.

4-4-A Point Source at Ground Level

For a point source at ground level the proper expression for K is

$$K = \frac{Q}{2\pi(D_y D_z)^{1/2}} \tag{4-6}$$

where Q is the strength of the emission source, that is, mass emitted per unit time. By substitution of Equation (4-6) into Equation (4-5) we find that the concentration of a pollutant emitted from a point source at ground level is modeled by the expression

$$C(x,y,z) = \frac{Q}{2\pi x(D_y D_z)^{1/2}} \exp\left[-\left(\frac{y^2}{D_y} + \frac{z^2}{D_z}\right)\frac{u}{4x}\right] \tag{4-7}$$

This equation has the format of the double Gaussian or normal distribution as expressed by Equation (4-4). Since for a ground-level source the maximum concentration in the y- and z-directions should occur along the center line at ground level, the values of μ_y and μ_z in Equation (4-4) are zero for this physical situation. Hence Equation (4-4) reduces to the form

$$f(y,z) = \frac{1}{2\pi\sigma_y\sigma_z} \exp\left(\frac{-y^2}{2\sigma_z{}^2} + \frac{-z^2}{2\sigma_z{}^2}\right)$$

It has been found convenient to reorganize Equation (4-7) into a form similar to the above expression. In order to do this we make the following definitions:

$$\sigma_y{}^2 \equiv \frac{2D_y x}{u} \quad \text{and} \quad \sigma_z{}^2 \equiv \frac{2D_z x}{u} \tag{4-8}$$

Substitution of these two definitions into Equation (4-7) leads to the following relationship for the concentration downwind from a ground-level point source:

$$C(x,y,z) = \frac{Q}{\pi u \sigma_y \sigma_z} \exp\left[-\frac{1}{2}\left(\frac{y^2}{\sigma_y{}^2} + \frac{z^2}{\sigma_z{}^2}\right)\right] \tag{4-9}$$

When Equation (4-9) is rearranged so that the left side is equal to $Cu/2Q$, then the right side will have the identical format of $f(y,z)$ described above, which is a double Gaussian type. The units on the gaseous concentration C

are determined by the units used to express the quantities Q, u, σ_y, and σ_z. In the technical literature σ_y and σ_z usually are given in meters, and u in meters per second. If C is desired in micrograms per cubic meter, then the emission rate Q must be expressed in micrograms per second. If y and z are taken to be zero, then Equation (4-9) reduces to

$$C(x,0,0) = \frac{Q}{\pi u \sigma_y \sigma_z} \tag{4-10}$$

This equation applies to the ground level, center-line concentration from a point source at ground level.

4-4-B Point Source at Elevation *H* Above the Ground, with Reflection

For emission from a stack with effective height H the exponential term in Equation (4-9) containing z^2 must be altered. From the section describing a Gaussian distribution recall that the entire curve is shifted a distance μ_z from the zero axis of z by writing the exponential term in the form $\exp\{-\frac{1}{2}[(z-\mu_z)/\sigma_z]^2\}$. For an elevated source this is equivalent to replacing z in Equation (4-9) by $(z-H)$. This substitution cannot be made directly in Equation (4-9), since the method of evaluating K in Equation (4-5) is also altered. This reevaluation of K is discussed in the appendix to this chapter. The result is that K has one-half the value found previously by Equation (4-6). Likewise, the coefficient in front of the exponential terms is now one-half its former value for the ground source. Consequently, for an elevated point source of a gaseous pollutant, *without reflection*,

$$C = \frac{Q}{2\pi u \sigma_y \sigma_z} \exp\left\{-\frac{1}{2}\left[\frac{y^2}{\sigma_y{}^2} + \frac{(z-H)^2}{\sigma_z{}^2}\right]\right\} \tag{4-11}$$

The restriction "without reflection" is extremely important. The above equation is an appropriate expression for the concentration in the downwind direction up to the point in the x-direction where the concentration at ground level $(z=0)$ is significant. Then appreciable "reflection" of the gaseous pollutant will occur by diffusion back into the atmosphere from ground level. Such a model assumes that the earth's surface is not a sink for a pollutant.

It is a relatively simple task to modify the preceding equation to account for reflection of a gaseous pollutant back into the atmosphere, once it has reached ground level. By referral to Figure 4-3, we see that reflection at some distance x is mathematically equivalent to having a mirror image of the source at $-H$. The shaded area beyond position I on the diagram indicates the region of the atmosphere in which the concentration will increase over that normally supplied by the source at H. This increased concentration is determined mathematically by linear superposition of two

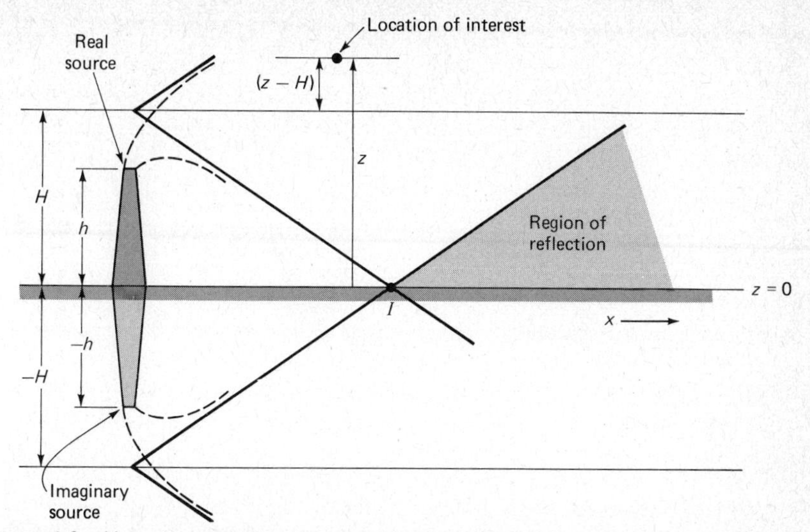

Figure 4-3 Use of an imaginary source to describe mathematically gaseous reflection at surface of the earth.

Gaussian-type concentration curves, one centered at H and the other at $-H$. This is equivalent to adding together two equations like Equation (4-11). However, one equation contains a $(z+H)$ term, rather than a $(z-H)$ term. As a result, the concentration equation for an elevated source *with reflection* becomes

$$C(x,y,z) = \frac{Q}{2\pi u\sigma_y\sigma_z}\left[\exp -\left(\frac{y^2}{2\sigma_y{}^2}\right)\right]\left\{\exp\left[\frac{-(z-H)^2}{2\sigma_z{}^2}\right] + \exp\left[\frac{-(z+H)^2}{2\sigma_z{}^2}\right]\right\} \tag{4-12}$$

The effect of ground reflection on the pollutant concentration above ground level is shown in Figure 4-4. At position I the two Gaussian-type curves predict essentially no overlap in concentration, but at positions downwind from I the overlap will become significant and increase as x increases. At position J downwind the overlap is appreciable. By adding that portion of the lower curve which extends above ground level $(z=0)$ to the original upper curve, we find that the upper concentration curve is altered by the addition of the shaded area shown. At some distance K further downwind from J, the shaded contribution due to reflection might lead to the profile shown at K in Figure 4-4. Obviously the effect of ground-level reflection is to increase the ground-level concentration well above that anticipated without reflection.

Another equation of significance when considering ground-level reflection is that representing the concentration at ground level. In this case $z=0$,

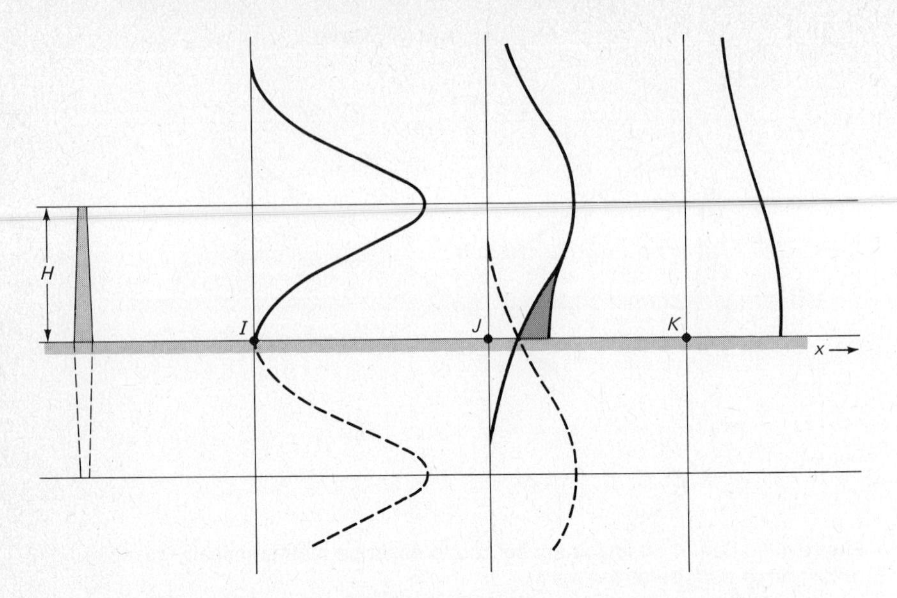

Figure 4-4 Effect of ground reflection on pollutant concentration downwind.

and Equation (4-12) reduces to

$$C(x, y, 0) = \frac{Q}{\pi u \sigma_y \sigma_z} \exp\left(\frac{-H^2}{2\sigma_z^{\,2}}\right) \exp\left(\frac{-y^2}{2\sigma_y^{\,2}}\right) \tag{4-13}$$

If the center-line, ground-level concentration estimate is desired, the last exponential term becomes unity.

A typical concentration profile in the z-direction at a given value of x and a profile in the x-direction along the center line at ground level are superimposed on a schematic of the diffusion process from an elevated stack in Figure 4-5. Note that the Gaussian distribution in the z-direction is centered at the effective stack height H. Also, the center-line concentration downwind maximizes at some value of x and then falls off at increasing values of x. A profile similar to that in the z-direction would also be valid in the y-direction. However, the sharpness of the Gaussian distributions in the y- and z-directions could be quite different, since the values of σ_y and σ_z at a given x are found to be significantly different.

4-5 EVALUATION OF THE STANDARD DEVIATIONS

Several equations were developed in the preceding section for estimating the concentration downwind resulting from a continuous plume. (The preceding equations are not valid for instantaneous or intermittent puffs of pollutants from a point source.) Besides physical data such as the coordinates x, y, and z, the emission strength Q, and the effective height H of the plume center line, it is necessary to have values of u, σ_y, and σ_z.

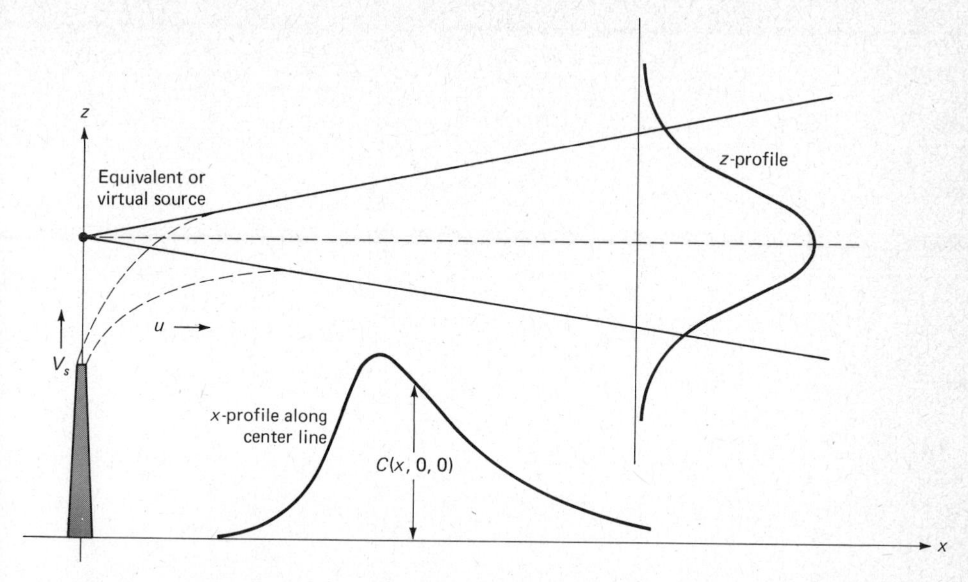

Figure 4-5 Concentration profiles along the center line in the *x*-direction and in the *z*-direction.

It has been pointed out in the preceding chapter that the wind speed, u, is a function of height, z. The typical variation of u with z is given by Equation (3-13), which includes a factor to adjust for various stability conditions in the atmosphere. The appropriate value of u to use in the dispersion equations is the mean value taken through the plume [4]. In most cases it would be impossible to determine the mean, since sufficient atmospheric data would not be available. In lieu of this, the average wind speed at the top of the stack is commonly used. Since in most cases not even this value is known, the measured meteorological value at 10 m is used in conjunction with Equation (3-13) to estimate the wind speed at the stack height.

The values of σ_y and σ_z have already been shown to be related to the diffusion coefficients or mass diffusivities of a gas through another media in the y- and z-directions. As might be anticipated from the physical description of the diffusion problem, the horizontal and vertical deviations, σ_y and σ_z, are a function of the downwind position x as well as the atmospheric stability conditions. Many experimental measurements in the atmosphere have led to an evaluation and correlation of σ_y- and σ_z-values. There are several sets of charts for these two parameters, and the range of stability conditions covered in the different sets do not normally coincide. One widely accepted set of charts is presented as Figures 4-6 and 4-7, as prepared by Turner [4]. These correlations are based on the following restrictions:

1. The concentrations estimated from the use of these charts should correspond to a sampling time of approximately 10 min.

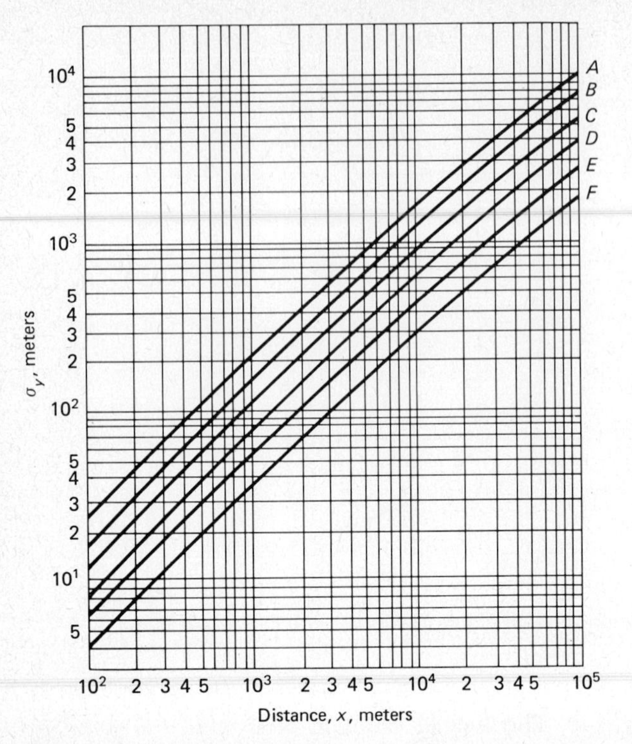

Figure 4-6 Standard deviation, σ *y*, in the crosswind direction as a function of distance downwind. (SOURCE: D. B. Turner. *Workbook of Atmospheric Dispersion Estimates*. Washington, D.C.: HEW, 1969.)

2. The horizontal and vertical deviations are based on a terrain representative of open country.
3. The estimated concentrations more nearly represent only the lowest several hundred meters of the atmosphere.

As noted by Turner, and indicated by the dashed lines over most of the range of the σ_z chart, the σ_z-values are more in doubt than the σ_y-values. This is especially true for downwind distances of more than 1 km. In several cases, such as for neutral to moderately unstable atmospheric conditions and distances out to a few kilometers, the center-line, ground-level concentrations based on these charts should be within a factor of 2 or 3 of actual values.

Turner has also prepared a listing of atmospheric conditions which aids in determining which of the six stability classes (A through F) appearing on the σ charts is appropriate. Table 4-1 shows this key to the various stability categories, and the following comments may help explain the use of this table.

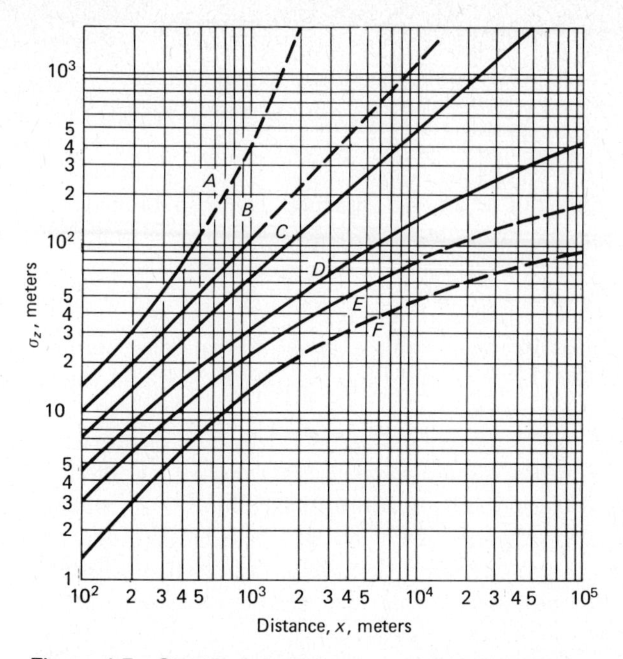

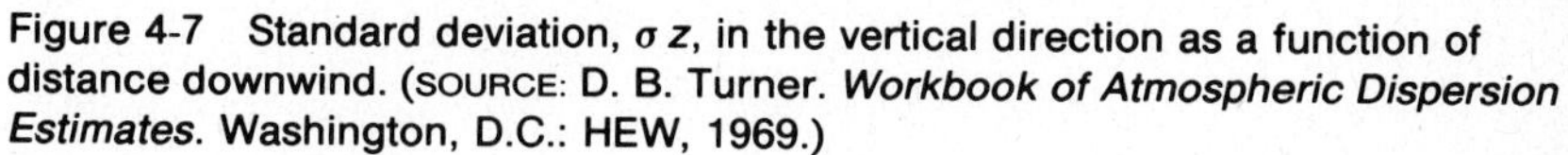

Figure 4-7 Standard deviation, σz, in the vertical direction as a function of distance downwind. (SOURCE: D. B. Turner. *Workbook of Atmospheric Dispersion Estimates*. Washington, D.C.: HEW, 1969.)

Table 4-1 KEY TO STABILITY CATEGORIES

SURFACE WIND SPEED AT 10 m (m/s)	DAY			NIGHT	
	INCOMING SOLAR RADIATION			CLOUD COVER	
	STRONG	MODERATE	SLIGHT	MOSTLY OVERCAST	MOSTLY CLEAR
CLASS[a]	(1)	(2)	(3)	(4)	(5)
<2	*A*	*A–B*	*B*	*E*	*F*
2–3	*A–B*	*B*	*C*	*E*	*F*
3–5	*B*	*B–C*	*C*	*D*	*E*
5–6	*C*	*C–D*	*D*	*D*	*D*
>6	*C*	*D*	*D*	*D*	*D*

SOURCE: D. B. Turner. *Workbook of Atmospheric Dispersion Estimates*. Washington, D.C.: HEW, 1969.

[a] The neutral class, *D*, should be assumed for overcast conditions during day or night. Class *A* is the most unstable and class *F* is the most stable, with class *B* moderately unstable and class *E* slightly stable.

The following items refer to the classes numbered in Table 4-1.

1. Clear skies, solar altitude greater than 60 degrees above the horizontal, typical of a sunny summer afternoon. Very convective atmosphere.
2. Summer day with a few broken clouds.
3. Typical of a sunny fall afternoon, summer day with broken low clouds, or summer day with clear skies and solar altitude from only 15 to 35 degrees above horizontal.
4. Can also be used for a winter day.

When estimating gaseous dispersion from a given source, one normally would choose that stability class typical of the region which would lead to the worst possible pollution episode.

Due to the reduced size of Figures 4-6 and 4-7, it is difficult to read σ_y and σ_z values from these graphs without some loss in accuracy. On the basis of the original source, Table 4-2 lists σ_y and σ_z values for the six stability classes for some arbitrarily chosen downwind distances. It is frequently desirable also to have the two standard deviations expressed in algebraic form. Due to the logarithmic nature of the σ plots, a reasonable curve fit is obtained by the expressions below due to Martin [5],

$$\sigma_y = ax^b \quad \text{and} \quad \sigma_z = cx^d + f \tag{4-14}$$

Values for four of the stability-dependent constants are given in Table 4-3. Note that the constants are different when x is less than or greater than 1 km. The value of b is always 0.894, and x must be expressed in kilometers.

Other sources of σ_y and σ_z values are available in the literature. One such set of data is that prepared by the Brookhaven Laboratories, as seen in Stern [6]. Extensive data have also been collected from comprehensive field

Table 4-2 APPROXIMATE VALUES OF σ_y AND σ_z AS A FUNCTION OF DOWNWIND DISTANCE FOR VARIOUS STABILITY CLASSES IN METERS

DISTANCE	STABILITY CLASSES AND σ_y VALUES						STABILITY CLASSES AND σ_z VALUES					
(km)	A	B	C	D	E	F	A	B	C	D	E	F
0.1	27	19	13	8	6	4	14	11	7	5	4	2
0.2	50	36	23	15	11	8	29	20	14	8	6	4
0.4	94	67	44	29	21	14	72	40	26	15	11	7
0.7	155	112	74	48	36	24	215	73	43	24	17	11
1.0	215	155	105	68	51	34	455	110	61	32	21	14
2.0	390	295	200	130	96	64	1950	230	115	50	34	22
4.0		550	370	245	180	120		500	220	77	49	31
7.0		880	610	400	300	200		780	360	109	66	39
10.0		1190	840	550	420	275		1350	510	135	79	46
20.0		2150	1540	1000	760	500		2900	950	205	110	60

SOURCE: D. B. Turner, *Workbook of Atmospheric Dispersion Estimates*. Washington, D.C.: HEW, Rev., 1969.

Table 4-3 VALUES OF CONSTANTS TO BE USED IN EQUATION (4-14) AS A FUNCTION OF DOWNWIND DISTANCE AND STABILITY CONDITION

		$x \leqslant 1$ km			$x \geqslant 1$ km		
STABILITY	a	c	d	f	c	d	f
A	213	440.8	1.941	9.27	459.7	2.094	−9.6
B	156	106.6	1.149	3.3	108.2	1.098	2.0
C	104	61.0	0.911	0	61.0	0.911	0
D	68	33.2	0.725	−1.7	44.5	0.516	−13.0
E	50.5	22.8	0.678	−1.3	55.4	0.305	−34.0
F	34	14.35	0.740	−0.35	62.6	0.180	−48.6

SOURCE: D. O. Martin, *J. Air Pollu. Control Assoc.* **26**, no. 2 (1976): 145.

studies over a period of 20 years by the Tennessee Valley Authority. This information is based on dispersion from power plants over a varied range of unit sizes, stack heights, and meteorological conditions. Horizontal and vertical Gaussian standard deviations as a function of downwind distance are reported by TVA [2]. These curves are based on six different average values of the potential temperature gradient, ranging from neutral conditions to a strong inversion. Other σ-data appear in the literature [7, 8, 9, 10]. Although Turner's σ-data are those most frequently used, the availability of new field data indicates that a reevaluation of the σ-data for use in air pollution dispersion studies is in order.

Example 4-1

Sulfur dioxide is emitted at a rate of 160 g/s from a stack with an effective height of 60 m. The wind speed at stack height is 6 m/s, and the atmospheric stability class is D for the overcast day. Determine the ground-level concentration along the center line at a distance of 500 m from the stack, in micrograms per cubic meter.

SOLUTION

From Figures 4-6 and 4-7 the horizontal and vertical standard deviations, σ_y and σ_z, at 500 m for stability class D are 36 m and 18.5 m, respectively. Substitution of these values and other given data into Equation (4-13) yields, for $y = 0$,

$$
\begin{aligned}
C(500,0,0) &= \frac{160 \times 10^6}{\pi(6)(36)(18.5)} \exp\left[-0.5\left(\frac{60}{18.5}\right)^2\right] \\
&= 12.7 \times 10^3 (5.25 \times 10^{-3}) \\
&= 66.0\ \mu\text{g/m}^3 \text{ of } SO_2
\end{aligned}
$$

It is interesting to note that this value is just within the primary air quality standards of 80 μg/m^3 listed in Table 2-1. Equation (4-13) has been used since considerable reflection occurs at 500 m downwind.

Example 4-2

For the data given in Example 4-1, determine the concentration crosswind at 50 m from the center line for the downwind distance of 500 m.

SOLUTION

To account for the concentration in the crosswind direction at ground level, one must modify the above solution by the term $\exp[-0.5(y/\sigma_y)^2]$, found in Equation (4-13). Hence,

$$C(500, 50, 0) = 66.0 \exp\left[-0.5\left(\frac{50}{36}\right)^2\right]$$

$$= 66.0(0.38) = 23\ \mu g/m^3 \text{ of } SO_2$$

Thus at a crosswind distance which is 10 percent of the downwind distance, the estimated concentration has fallen off by nearly 60 percent.

Example 4-3

For the data of Example 4-1, determine sufficient values of C as a function of x on the ground-level center line so that the variation on either side of the maximum value is established.

SOLUTION

The general solution to this problem, in terms of Equation (4-13), is

$$C(x, 0, 0, 60) = \frac{160 \times 10^6}{\pi(6)\sigma_y\sigma_z} \exp\left[-0.5\left(\frac{60}{\sigma_z}\right)^2\right]$$

The table below summarizes the computations in a convenient manner, where the column headed "exp" represents the exponential term in the equation. Note that the preexponential factor decreases rapidly with increasing distance, owing to the steadily increasing values of σ_y and σ_z. The exponential factor, however, rapidly increases from an extremely small value toward a value of unity as x becomes large. Since these two terms are multiplied together to obtain C, there must be a maximum value of C at some distance x. In this case the maximum concentration occurs around 1.5 km. The data indicate that the concentration at ground level along the center line builds up rapidly as x increases, but the falloff in concentration is rather slow after the maximum point. This is fairly typical of the Gaussian-type solution for the atmospheric dispersion.

x (km)	σ_y	σ_z	$Q/\pi\sigma_y\sigma_z u$	$-\frac{1}{2}(H/\sigma_z)^2$	exp	C ($\mu g/m^3$)
0.5	36	18	13,090	5.55	0.0039	50
0.8	60	27	5,240	2.47	0.085	445
1.0	76	32	3,490	1.76	0.172	600
1.5	110	45	1,710	0.89	0.411	700
1.7	140	50	1,210	0.72	0.487	590
2.0	160	55	960	0.595	0.552	530
3.0	220	71	540	0.357	0.700	380
5.0	350	100	240	0.180	0.835	200
10.0	620	150	90	0.080	0.923	83

4-6 THE MAXIMUM GROUND-LEVEL IN-LINE CONCENTRATION

The effect of ground reflection, as noted in Section 4-4-B, is to increase the ground-level concentrations of gaseous pollutants as x increases, to a point well above the level expected without reflection. However, such an increase in C in the x-direction cannot continue indefinitely. Eventually diffusion outward (crosswind) in the y-direction and upward in the z-direction will diminish the concentration at ground level ($z = 0$) and along the center line ($y = 0$). Thus, as noted in Figure 4-5 and Example 4-3, the curve of C versus x has a maximum point before falling off toward zero at large x-values.

One method of determining the downwind distance for the maximum concentration, and the maximum concentration at that point, has been developed by Turner [4] in a graphical format. Figure 4-8 is based upon this work, which originally was developed on the basis of Equation (4-13). In Figure 4-8 the distance to the maximum concentration is plotted versus the maximum value of the parameter Cu/Q, with information on the stability class and the effective stack height appearing within the diagram. In the typical problem the known data are the stability class and the effective height. These data determine a particular point on the figure. From this point we read downward and to the left to ascertain C_{max} and x_{max}, respectively.

Rather than rely on reading values from Figure 4-8 to obtain the maximum concentration downwind, the data can be fit to a general equation and the solution found algebraically. A general equation developed by Ranchoux [11] is of the form

$$\left(\frac{Cu}{Q}\right)_{max} = \exp\left[a + b(\ln H) + c(\ln H)^2 + d(\ln H)^3\right] \qquad \text{(4-15)}$$

where H is in meters and Cu/Q is in m^{-2}. The values of the coefficients a,

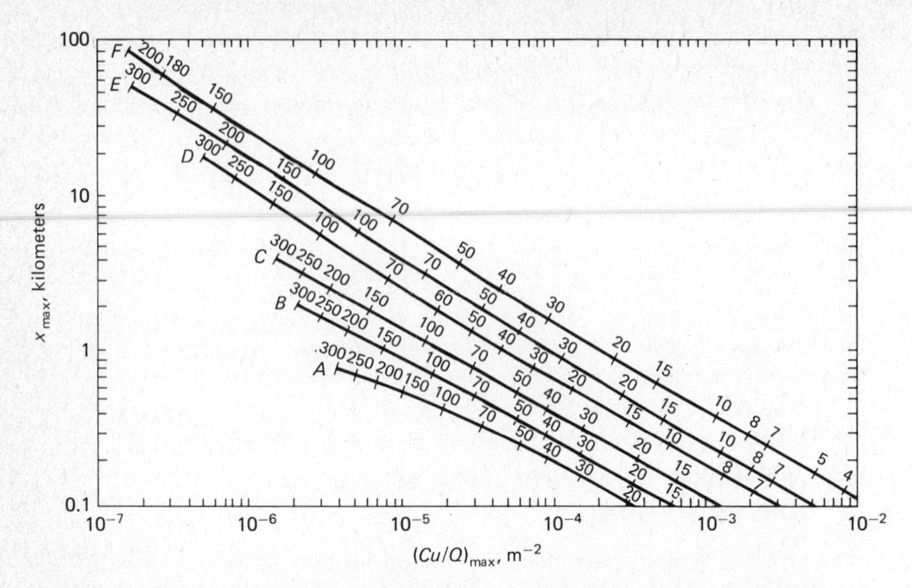

Figure 4-8 Distance of maximum downwind concentration and maximum downwind Cu/Q value as a function of stability class and effective height in meters. (SOURCE: D. B. Turner. *Workbook of Atmospheric Dispersion Estimates.* Washington, D.C.: HEW, 1969.)

b, *c*, and *d* for each stability class are shown in Table 4-4. The error between the equation and the actual curves is less than 2 percent for curves *A*, *B*, and *C*, and is less than 4.5 percent for the *D*, *E*, and *F* curves.

An alternate method of ascertaining the position and the value of the maximum concentration is based on a characteristic of the σ_y and σ_z charts. Under moderately unstable to near neutral conditions the ratio σ_y/σ_z is nearly independent of the distance x. If this ratio is taken to be constant and y is set equal to zero, then Equation (4-13) can be written so that C is solely a function of σ_z (which in turn is solely a function of x for a given stability class). Hence by the maximization technique of differential calculus one can obtain analytical information concerning the maximum concentration along

Table 4-4 VALUES OF CONSTANTS TO BE USED IN EQUATION (4-15) AS A FUNCTION OF STABILITY CLASSES

STABILITY	COEFFICIENTS			
CLASS	*a*	*b*	*c*	*d*
A	−1.0563	−2.7153	0.1261	0
B	−1.8060	−2.1912	0.0389	0
C	−1.9748	−1.9980	0	0
D	−2.5302	−1.5610	−0.0934	0
E	−1.4496	−2.5910	0.2181	−0.0343
F	−1.0488	−3.2252	0.4977	−0.0765

SOURCE: R. Ranchoux, *J. Air. Pollu. Control Assoc.* **26**, no. 11 (1976): 1089.

the center line. The position of maximum concentration can be found only implicitly by this method, not explicitly. The result of the differentiation of Equation (4-13) in this modified form results in the following expression:

$$\sigma_z = \frac{H}{(2)^{1/2}} = 0.707H$$

The value of σ_z is first determined from the best estimate of the corrected stack height, H. Then from a chart of σ_z versus x, for various stability conditions, the value of x can be read off, which gives the position of maximum C. The value of x determined by this method is only approximate, owing to the nature of the $\sigma_z - x$ chart.

If the condition $H^2 = 2\sigma_z^{\ 2}$ is substituted into Equation (4-13), and y is set equal to zero, then the maximum concentration downwind on the center line and at ground level is approximately given by

$$C_{\text{max, reflec}} = \frac{0.1171Q}{u\sigma_y\sigma_z} \tag{4-16}$$

Again, this expression gives better results when applied to unstable atmospheric conditions, for the reasons noted above. Example 4-4 illustrates both Turner's graphical format and the approximation equation developed above for the determination of the position and the value of the maximum concentration expected on the center line at ground level for an elevated point source.

Example 4-4

For the data of Example 4-1, determine the position downwind on the center line at ground level where the maximum concentration will occur, and determine the maximum value in micrograms per cubic meter.

SOLUTION

For an effective stack height of 60 m, the value of σ_z which leads to the position of maximum concentration is

$$\sigma_z = 0.707H = 0.707(60) = 42.4 \text{ m}$$

From Figure 4-7 we determine that the corresponding x-value is 1.55 km, which is the estimated point of maximum concentration. The value of C at this distance is given by Equation (4-16), namely,

$$C = \frac{0.117Q}{u\sigma_y\sigma_z} = \frac{0.117(160)(10^6)}{6(105)(42.4)} = 700\ \mu\text{g/m}^3$$

This value is in excellent agreement with the general results of Example 4-3.

The values of C_{max} and x_{max} may also be estimated through the use of Figure 4-8. For class D stability and an H-value of 60 m, it is found from the figure that

$$x_{max} \approx 1.5 \text{ km} \quad \text{and} \quad \left(\frac{Cu}{Q}\right)_{max} \approx 2.7 \times 10^{-5} \text{ m}^{-2}$$

Hence,

$$C_{max} = (2.7 \times 10^{-5} \text{ m}^{-2})\left[(160 \times 10^{6} \ \mu\text{g})/\text{s}\right](\text{s}/6 \text{ m}) = 720 \ \mu\text{g/m}^{3}$$

In this case the two methods are substantially in agreement. They do not always agree, however, since agreement depends upon the stability class involved. Note that the maximum value is nearly 10 times the value of 80 μg/m^3 set as the primary air quality standard (Table 2-1).

As a third method, Equation (4-15) may be used. In this case for class D stability the equation becomes

$$\left(\frac{Cu}{Q}\right)_{max} = \exp\left[-2.5302 - 1.5610(\ln 60) - 0.0934(\ln 60)^{2}\right]$$
$$= \exp(-10.487) = 2.8 \times 10^{-5} \text{ m}^{-2}$$

For the given values of u and Q we find that C_{max} is 740 μg/m^3, which is reasonably close to the two preceding answers.

4-7 CALCULATION OF THE EFFECTIVE STACK HEIGHT

Most of the analytical methods for predicting the concentrations of stack effluents involve the location of a virtual or equivalent origin, as shown in Figure 4-2. The elevation H of the virtual origin is obtained by adding a term Δh, the plume rise, to the actual height of the stack, h_s. There are numerous methods for calculating Δh, and these are discussed at some length by Stern [6]. Basically, three sets of parameters control the phenomenon of a gaseous plume injected into the atmosphere from a stack. These are stack characteristics, meteorological conditions, and the physical and chemical nature of the effluent. A large number of analytical expressions have been proposed to relate these factors to plume-rise predictions. It is not surprising that no one expression has proven superior for all stack geometries and atmospheric conditions.

In a fairly recent study, Carson and Moses [12] compared 711 observed plume-rise values with calculated values given by 11 different equations. Their conclusion was that much more study is needed before a completely satisfactory method can be developed for predicting plume-rise values. The majority of equations that predict plume rise contain a momentum term and a thermal buoyancy term. The first term accounts for the vertical momentum of the stack gas due to its own veloctiy, V_s. The second term takes into account, in some manner, the difference between the stack gas temperature, T_s, and the environmental temperature, T_a.

Carson and Moses concluded that the following equation gave the best agreement with all the observed data, regardless of stability condition.

$$\Delta h = -0.029\frac{V_s d}{u} + 2.62\frac{(Q_h)^{1/2}}{u} \tag{4-17}$$

where Δh is the plume rise in meters, V_s is the stack gas exit velocity, in meters per second, d is the stack exit diameter in meters, u is the speed at the stack exit in meters per second, and Q_h is the heat emission rate in kilojoules per second.

To further clarify Equation (4-17),

$$Q_h = \dot{m}c_p(T_s - T_a)$$

where $\dot{m}$ is the stack gas mass flow rate in kilograms per second ($\dot{m} = \pi d^2 V_s P/4RT_s$), c_p is the constant pressure specific heat of the stack gas, T_s is the stack gas temperature at the stack exit in degrees Kelvin, and T_a is the atmospheric air temperature at stack height in degrees Kelvin.

Thomas, Carpenter, and Colbaugh [13] compared the observed plume-rise data for large stacks of electric generation stations with values calculated employing 10 different equations. Several of the equations employed in reference 12 were included in the comparisons of Thomas et al. The Holland formula,

$$\Delta h = \frac{V_s d}{u}\left[1.5 + 2.68 \times 10^{-3} Pd\left(\frac{T_s - T_a}{T_s}\right)\right] \tag{4-18}$$

showed fairly good agreement with observations, with a slight tendency to underestimate the plume rise. The symbols and units for the Holland equation are the same as those listed above. In addition, the pressure P, must be expressed in millibars. This equation appears to be more accurate for tall stacks. The last term in the Holland formula can be replaced by $0.0096Q_h/V_s d$, if desired.

Moses and Kraimer [14] analyzed 17 equations for plume rise on the basis of 615 observations invovling 26 different stacks. Among those equations which gave reasonably good predictions were two proposed by Concawe [8]. The original Concawe formula, based on observations in Europe, is

$$\Delta h = 2.71\frac{Q_h^{\,1/2}}{u^{3/4}} \tag{4-19a}$$

When optimized by Thomas et al. [13] on the basis of TVA data, the equation took the form

$$\Delta h = 4.71\frac{Q_h^{\ 0.444}}{u^{0.694}} \tag{4-19b}$$

In addition, the three specific equations by Moses and Carson for unstable, neutral, and stable atmospheres were found to be reasonably accurate. The data for these three relations were used to develop the overall Moses and Carson equation given earlier as Equation (4-17). Specifically, these three equations are

$$\Delta h = 3.47\frac{V_s d}{u} + 5.15\frac{(Q_h)^{0.5}}{u} \quad \text{(unstable)} \tag{4-20a}$$

$$\Delta h = 0.35\frac{V_s d}{u} + 2.64\frac{(Q_h)^{0.5}}{u} \quad \text{(neutral)} \tag{4-20b}$$

$$\Delta h = -1.04\frac{V_s d}{u} + 2.24\frac{(Q_h)^{0.5}}{u} \quad \text{(stable)} \tag{4-20c}$$

Carpenter et al. [2] have made a comprehensive study of plume-rise models at coal-burning power plants for TVA. They found that a formula suggested by Briggs [15] is preferable for estimating plume rise at TVA power plants. A modification of the formula, based on empirical data, is given by Equation (4-21).

$$\Delta h = \frac{114C(F)^{1/3}}{u} \tag{4-21}$$

where $F = gV_s d^2(T_s - T_a)/4T_a$, $\text{m}^4\ \text{s}^{-3}$; g = gravitational acceleration, 9.8 m/s^2; $C = 1.58 - 41.4(\Delta\theta/\Delta z)$, dimensionless; $\Delta\theta/\Delta z$ = potential temperature gradient, °K/m.

The constant 114 in this equation has units of $\text{m}^{2/3}$. This equation is especially useful, because the potential temperature gradient factor in the dimensionless constant C permits adjusting for different stability conditions. The linear correlation between C and $\Delta\theta/\Delta z$ is based on data for $\Delta\theta/\Delta z$ which extend from $-0.001°$ to 0.013°K/m. The symbols and units in Equations (4-17) through (4-21) are the same as those noted earlier.

The preceding equations predict the effective plume rise above the top of a stack at some distance downwind where the plume has essentially reached its maximum height. This vertical displacement, Δh, is shown in Figure 4-2, as noted earlier. This figure shows, in addition, that the position of maximum rise may occur considerably downwind from the stack in the x-direction. As a result, the dispersion of pollutants from the plume at a position close to the stack occurs at a height which is overestimated by the effective stack height, H. Past methods for estimating the effective plume rise for a specific distance downwind from the stack have shown considerable ambiguity. Consequently, most models have necessarily been based on the total effective height. In 1972 the TVA [16] reported the results of a comprehensive study from which was developed a procedure for estimating the effective plume rise at a specific distance x from the stack. This required extensive evaluations of the relationships observed among stack height, plume rise, and atmospheric stability for single stacks at three coal-fired power plants.

The results for Δh as a function of x from this study are expressed as a function of three atmospheric stability conditions. For neutral atmospheric stability ($-0.17 < \Delta\theta/\Delta z < 0.16$, where $\Delta\theta/\Delta z$ is expressed in °K/100 m) and distances up to 3000 m,

$$\Delta h = 2.50x^{0.56}F^{1/3}u^{-1} \tag{4-22a}$$

For moderately stable atmospheric conditions ($0.16 < \Delta\theta/\Delta z < 0.70$) and distances up to 2800 m,

$$\Delta h = 3.75x^{0.49}F^{1/3}u^{-1} \tag{4-22b}$$

Under very stable atmospheric conditions ($0.70 < \Delta\theta/\Delta z < 1.87$),

$$\Delta h = 13.8x^{0.26}F^{1/3}u^{-1} \tag{4-22c}$$

For tall stacks, the use of equations which account for the variation of Δh with distance x probably will not significantly alter the concentration profiles predicted by dispersion models. However, such an accounting could have a pronounced effect on modeling of dispersion from relatively short stacks. Correct modeling is extremely important in this latter case, since effluents from short stacks are more likely to lead to ground-level pollutant concentrations which exceed desirable limits. In addition to these three specific equations, a general expression for the effective plume rise at 1824 m is also proposed [9]. At this distance the major part of the plume rise should already have occurred. For this selected distance,

$$\Delta h = \frac{173F^{1/3}}{u\exp(0.64\,\Delta\theta/\Delta z)} \tag{4-22d}$$

where $\Delta\theta/\Delta z$ is expressed in °K/100 m. Many other plume-rise equations are available in the literature; for a survey of these, see the references cited above. It is a matter of experience that plume-rise models necessarily change as the size of units and the stack height change. It is also evident that the type of atmospheric conditions is critical to the evaluation of Δh, especially when the plant size is altered.

Example 4-5

The heat-emission rate associated with a stack gas is 4800 kJ/s, the wind and stack gas speeds are 5 and 15 m/s, respectively, and the inside stack diameter at the top is 2 m. Estimate the plume rise by means of the Moses and Carson general equation, the Holland formula, and the Concawe formula.

SOLUTION

In summary, the basic equations to be used are:

$$\Delta h = -0.029\frac{V_s d}{u} + 2.62\frac{(Q_h)^{1/2}}{u} \qquad \text{(Carson and Moses)}$$

$$\Delta h = \frac{V_s d}{u}\left(1.5 + 0.0096\frac{Q_h}{V_s d}\right) \qquad \text{(Holland)}$$

$$\Delta h = 4.71\frac{(Q_h)^{0.444}}{(u)^{0.694}} \qquad \text{(Concawe)}$$

Substitution of the known data indicates that

$$\Delta h = -0.029\left[\frac{15(2)}{5}\right] + 2.62\left[\frac{(4800)^{1/2}}{5}\right] = -0.1 + 36.3$$

$$= 36.2 \text{ m} \qquad \text{(Carson and Moses)}$$

$$\Delta h = \frac{15(2)}{5}\left[1.5 + 0.0096\frac{4800}{15(2)}\right] = 6(1.5 + 1.54)$$

$$= 18.2 \text{ m} \qquad \text{(Holland)}$$

$$\Delta h = 4.71\left[\frac{(4800)^{0.44}}{(5)^{0.694}}\right] = 4.71\left(\frac{43.1}{3.09}\right)$$

$$= 66.3 \text{ m} \qquad \text{(Concawe)}$$

The spread in the answers determined above is not unusual.

Example 4-6

The wind and stack gas speeds are 3 and 6 m/s, respectively, and the stack diameter is 2 m. The atmospheric stability condition is neutral with a temperature of 300°K, and the stack gas temperature is 440°K. Estimate the plume rise in meters by the Briggs equation, Equation (4-21), and by the TVA model, Equation (4-22d).

SOLUTION

The equations to be used are

$$\Delta h = \frac{114C(F)^{1/3}}{u}$$

and

$$\Delta h = \frac{173(F)^{1/3}}{u} \exp(-0.64\,\Delta\theta/\Delta z)$$

where

$$F = \frac{gV_s d^2(T_s - T_a)}{4T_a}$$

For neutral stability $\Delta\theta/\Delta z = 0$. The value of F in this situation is

$$F = \frac{9.8(6)(2)^2(440 - 300)}{4(300)} = 27.4 \text{ m}^4 \text{ s}^{-3}$$

Since $C = 1.58$, Equation (4-21) shows that

$$\Delta h = \frac{114(1.58)(27.4)^{1/3}}{3} = 180 \text{ m}$$

For neutral atmospheric conditions the second expression for Δh reduces to

$$\Delta h = \frac{173(F)^{1/3}}{u}$$

$$= \frac{173(27.4)^{1/3}}{3} = 173 \text{ m}$$

The results from the two expressions are in good agreement.

4-8 SOME OTHER CONSIDERATIONS REGARDING GASEOUS DISPERSION

In preceding sections we developed specific equations for the dispersion of gaseous pollutants from ground and elevated sources and illustrated some general methods for solutions. In this section we consider some other ramifications of the dispersion problem.

4-8-A Determination of Required Stack Height

A typical air pollution problem for new industry is the following: The gaseous products from a new industrial process are to be dispersed into the atmosphere by means of a new stack. The emission rate Q from the process is known. Downwind from the plant at a given distance x lies someone else's property, at which position it is necessary that the concentration never exceed a stated value C. The expected general wind speed, u, is also known from meteorological data. The basic problem is to determine the required stack height, H.

The worst situation would occur when the maximum concentration downwind lies just at the distance x. Recall that the maximum concentration occurs roughly when $H = (2)^{1/2}\sigma_z$. If σ_z could be found, then the required H could be evaluated from this relation. The value of σ_z can be ascertained indirectly by rewriting Equation (4-16) in the form

$$\sigma_y \sigma_z = \frac{0.117Q}{C_{\max} u} \tag{4-23}$$

All the quantities on the right side are known; hence $\sigma_y\sigma_z$ is determined. Since σ_y and σ_z are a function of x (as well as stability conditions), the quantity $\sigma_y\sigma_z$ can be represented on a diagram similar to Figures 4-6 and 4-7. A $\sigma_y\sigma_z$ diagram is shown as Figure 4-9, where the stability classes A through F again appear. In our design problem, we now know $\sigma_y\sigma_z$ and the design

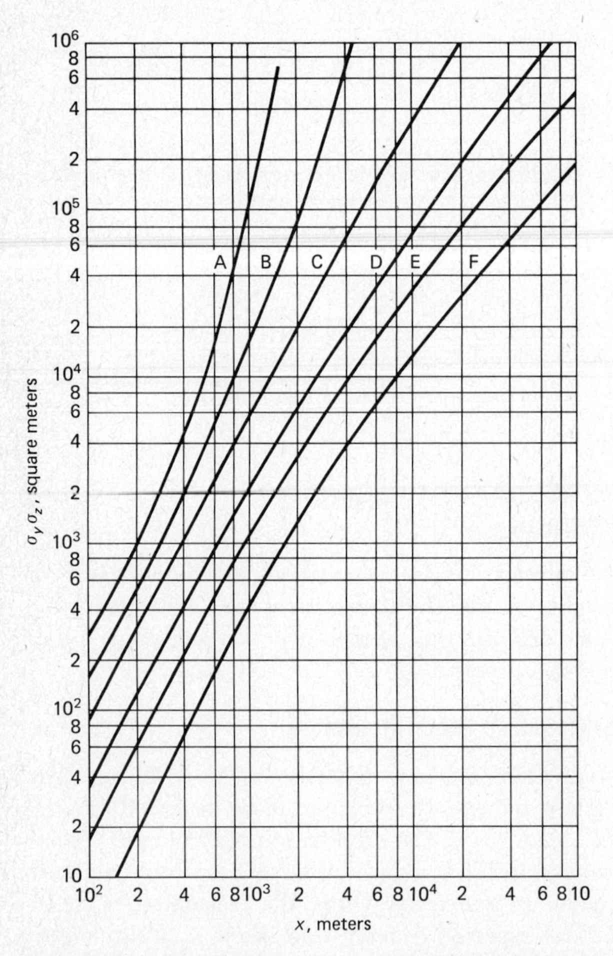

Figure 4-9 The product, $\sigma_y\sigma_z$, of the dispersion standard deviations as a function of downwind distance. (SOURCE: D. B. Turner. *Workbook of Atmospheric Dispersion Estimates.* Washington, D.C.: HEW, 1969.)

distance x_d. The location of these two values lies somewhere on Figure 4-9 and would usually lie between two stability classes. We now estimate the fractional distance the given point lies between the two stability class lines at position x_d. Moving to Figure 4-7 for σ_z versus x, we now measure off this same fractional distance between the same stability lines. Once this point is established on the σ_z chart, we move horizontally to the left and read off a value of σ_z. By substituting this quantity into the relation $H = (2)^{1/2}\sigma_z$ we then determined the stack height required to avoid exceeding a certain value of C at the design distance.

Note that the quantity H just determined is really the effective stack height, not the actual stack height. Hence the stack could be shorter than the height H. However, if H is used for the actual stack height, a safety factor will be built into the estimate. Since the dispersion equations are probably good only within a factor of 2 or 3, a safety factor on the stack height would be appropriate.

Example 4-7

Sulfur dioxide is emitted at a rate of 160 g/s into an atmosphere where the wind speed is expected to be approximately 6 m/s at stack height. It is desired that the ground-level concentration at the center line not exceed 200 $\mu g/m^3$ at a distance of 800 m. What effective stack height is required, in meters?

SOLUTION

On the basis of Equation (4-23) we find that

$$\sigma_y \sigma_z = \frac{0.117(160)(10^6)}{(200)(6)} = 1.56 \times 10^4$$

From Figure 4-9 we see that this value of $\sigma_y \sigma_z$ and a distance of 800 m represents a point which lies roughly 75 percent of the distance vertically between A and B stability classes. For the same spacing on Figure 4-7, the value of σ_z is approximately 115. The estimate of the required stack height then becomes

$$H = 1.414\sigma_z = 1.414(115) = 163 \text{ m}$$

An actual stack height of 150 to 160 m would provide a small margin of safety.

4-8-B Effect of Stack Height on Maximum Ground-Level Concentrations

We have seen how to determine the stack height required to avoid exceeding maximum permissible ground-level concentration at a certain distance downwind. We are also in a position to estimate qualitatively the direct effect of a change in effective stack height on the maximum concentration downwind on the center line at ground level. We again assume that Q and u have fixed values. If we now combine the relation $H = (2)^{1/2}\sigma_z$ with Equation (4-16), we find that

$$C_{\max} = \frac{0.117Q}{u\sigma_y\sigma_z}\left(\frac{2\sigma_z^{\ 2}}{H^2}\right) = \frac{2(0.117)Q}{u}\left(\frac{1}{H^2}\right)\left(\frac{\sigma_z}{\sigma_y}\right)$$

However, on the basis of Figures 4-6 and 4-7 for σ_y and σ_z, it can be determined for a given stability that σ_z/σ_y is essentially a constant over a range of x. As a result, we note that

$$C_{\max} \approx K\left(\frac{1}{H^2}\right)$$

where K is a constant. Thus the maximum concentration downwind is, roughly, inversely proportional to the square of the effective stack height. If the effective stack height is doubled, for example, the maximum concentration downwind at ground level on the center line should roughly be decreased by a factor of 4. Hence the effective stack height is a major variable in the control of ground-level concentrations.

4-8-C Concentration Estimates for Various Sampling Times

Most correlations of σ-data for atmospheric dispersion estimates, as typified by the work of Turner [6], lead to concentrations averaged over a 10-min time interval. If the sampling technique in use employs some other time interval, it will be necessary to correct the results predicted by a dispersion model. Information to date indicates that the effects of sampling time are exceedingly complex. If it is necessary to estimate concentrations from a single source for time intervals greater than a few minutes, the following equation gives approximate values for times less than 2 hr.

$$C_2 = C_1\left(\frac{t_1}{t_2}\right)^q \tag{4-24}$$

where C_2 is the desired concentration, C_1 is the concentration calculated by the dispersion equation, t_2 is the sampling time period in minutes, t_1 is 10 min, and q has a value between 0.17 and 0.20.

The value of the exponent q is substantiated by Nonhebel [17], who based his work on dispersion coefficients rather than sampling results. His base point was a 3-min sampling interval, also, rather than a 10-min period. Nevertheless, he suggested a concentration-time relationship which varied inversely to the 0.17 power. Recently Heno [18] has related the maximum concentration inversely to the sampling time to the 0.5 power.

4-8-D Effect of a Plume Inversion Trap

The presence of an elevated inversion can have a devastating effect on the ground-level concentration downind from a stack, since the inversion acts as a giant lid on the upward dispersion of pollutant gases. This situation is often modeled as a gas passing downwind between two reflecting surfaces—the ground and the bottom of the elevated inversion layer. In Section 4-4-B we pointed out that ground reflection can be modeled by a virtual image at distance $-H$ below the earth's surface. With the addition of an inversion layer, additional reflections are considered at heights L and $-L$, where L is the distance to the bottom of the inversion layer. An accounting of all of the stable layer and ground reflections can be made through a summation of terms. The end result is a center-line expression of the form

$$C = \frac{Q}{2\pi\sigma_y\sigma_z u}\sum\left\{\exp\left[\frac{-(z-H+2jL)^2}{2\sigma_z^{\ 2}}\right] + \exp\left[\frac{-(z+H+2jL)^2}{2\sigma_z^{\ 2}}\right]\right\}$$

where the summation is carried out from $j = -\infty$ to $+\infty$. This series usually converges rapidly, requiring only the first few terms, for example, values of j up to ± 2 or ± 3.

A good approximation to this equation may be made by assuming that the inversion layer has no effect on the vertical dispersion until a downwind distance x_L for which $\sigma_z = 0.47(L - H)$. If we know the height of the inversion layer, we can use this relation to estimate x_L through the use of Figure 4-7. The effect of reflections from the stable layer and the ground beyond the distance x_L is such that uniform vertical mixing has taken place by the downwind distance $2x_L$. Beyond $2x_L$ the appropriate equation is

$$C(>2x_L, y, z) = \frac{Q}{(2\pi)^{1/2}\sigma_y L u}\exp\left[-\frac{1}{2}\left(\frac{y}{\sigma_y}\right)^2\right] \tag{4-25}$$

Note that this expression contains only x and y as variables. For distances between x_L and $2x_L$ Turner [4] suggests that ground-level, center-line concentrations be read from a straight line drawn between the concentrations for points x_L and $2x_L$ on a log-log plot of ground-level, center-line concentration versus distance.

Example 4-8

Sulfur dioxide is emitted at a rate of 160 g/s from a stack with an effective height of 60 m. The wind speed at stack height is 6 m/s, and the atmospheric stability class is C. Estimate the distance at which reflection from the stable layer just begins to occur, in meters, for an inversion layer 150 m above ground level. Also find the concentration at a distance of $2x_L$.

SOLUTION

As an estimate, the value of x_L occurs when

$$\sigma_z = 0.47(L - H) = 0.47(150 - 60) = 42.3 \text{ m}$$

From Figure 4-7, for stability class C we find that reflection just begins to occur at a downwind distance of 680 m, approximately. Consequently, the value of $2x_L$ is 1360 m. From Figure 4-6, the value of σ_y is 140 m at this latter distance. For the center-line concentration at a distance of 1360 m Equation (4-25) yields

$$C(2x_L, 0, z) = \frac{Q}{(2\pi)^{1/2}\sigma_y L u} = \frac{160(10^6)}{2.51(140)(150)(6)} = 505\ \mu\text{g/m}^3$$

This value persists, of course, at all heights up to 150 m.

4-8-E Line Sources

In some situations, such as a series of industries located along a river or harbor, or heavy traffic along a straight stretch of highway, the pollution problem may be modeled as a continuous emitting infinite line source. When the wind direction is normal to the line of emission, the ground-level concentration downwind is given by

$$C(x, 0) = \frac{2q}{(2\pi)^{1/2}\sigma_z u} \exp\left[-\frac{1}{2}\left(\frac{H}{\sigma_z}\right)^2\right] \tag{4-26}$$

where q is the source strength per unit distance. For example, q might be expressed in terms of g/s·m. The horizontal standard deviation, σ_y, is absent from the equation, since crosswind diffusion from various portions of the emitted gases should be self-compensating. Note also that y does not appear in Equation (4-26) since the concentration should be uniform in the y-direction at a given x-distance. When the wind direction is not perpendicular to the line source, Turner [4] suggests that Equation (4-26) be divided by $(\sin \phi)$, where ϕ is the angle between the line source and the wind direction. This correction should not be used when ϕ is less than 45 degrees.

When the continuously emitting line source is reasonably short in length, we must account for the edge effects caused by the two ends of the source. These edge effects become more important, in the sense that they extend to greater crosswind distances, as the distance downwind from the source increases. If the line source is perpendicular to the wind direction, then it is convenient to define the x-axis in the direction of the wind and also passing through the sampling point downwind. The ends of the line source then are at two positions in the crosswind direction, y_1 and y_2, where y_1 is less than y_2. The concentration along the x-axis at ground level is then given by the expression

$$C(x,0,0) = \frac{2q}{(2\pi)^{1/2}\sigma_z u}\exp\left[-\frac{1}{2}\left(\frac{H}{\sigma_z}\right)^2\right]\int_{p_1}^{p_2}\frac{1}{(2\pi)^{1/2}}\exp(-0.5p^2)\,dp \tag{4-27}$$

where $p_1 = y_1/\sigma_y$ and $p_2 = y_2/\sigma_y$. Once the limits of integration are established, the value of the integral may be determined from standard statistical tables.

Example 4-9

Estimate the total hydrocarbon concentration at a point 300 m downwind from an expressway at 5:30 P.M. on an overcast day. The wind is perpendicular to the highway and has a speed of 4 m/s. The traffic density along the highway is 8000 vehicles per hour, and the average vehicle speed is 40 mi/hr. The average vehicle emission rate of hydrocarbons is 2×10^{-2} g/s.

SOLUTION

Assuming a reasonably straight section of highway, we will consider the pollutants emanating from a continuous, infinite line source. The emission rate per unit length, q, is determined from the product of the emission rate per vehicle times the number of vehicles per unit length. This latter quantity is found by dividing the rate of vehicle travel past a point by the average speed of the vehicle. Hence,

$$\frac{\text{vehicles}}{\text{m}} = \frac{8000(\text{vehicles/hr})}{40(\text{mi/hr})}\left(\frac{\text{mi}}{1600\text{ m}}\right) = 0.125$$

Therefore,

$$q = 0.125\text{ vehicle/m} \times 2 \times 10^{-2}\text{ g/vehicle(s)} = 2.5 \times 10^{-3}\text{ g/s·m}$$

For an overcast day the stability class is D. From Figure 4-7, at a downwind distance of 300 m the value of σ_z is 12 m. Recognizing that the exponential term in Equation (4-26) is unity for a ground-level source, we find that this equation yields

$$C(300,0,0) = \frac{2(2.5)(10^{-3})}{(2\pi)^{1/2}(12)(4)} = 42(10^{-6})\text{g/m}^3 = 42\ \mu\text{g/m}^3$$

This is the concentration estimate for a 10-min sampling period. The federal standard for hydrocarbons is 160 μg/m^3 averaged over a 3-hr sampling interval. Since the concentration usually decreases when averaged over a longer time period, the 42 $\mu g/m^3$ estimated above is well within the federal standard.

4-8-F Concentration Estimates due to Several Point Sources

Frequently a receptor or sampling point may lie downwind from two or more continuously emitting point sources, and these sources are not directly upwind. This type of dispersion problem is usually solved by the method of superposition. That is, the total concentration at the receptor point is found by summing the concentrations coming from the individual sources. Consequently, the problem becomes a matter of correctly identifying the geometry or orientation of the receptor with respect to the wind direction and each source location, separately. Then the standard equations developed earlier for an elevated or ground source are applied, in terms of the given wind speed and stability class of interest. An excellent review of techniques to be used in these situations appears in the literature [19].

APPENDIX—DEVELOPMENT OF THE GAUSSIAN-TYPE DISPERSION EQUATION

The following is one method for developing the Gaussian-type dispersion equation for a gas continuously released from a point source. On the basis of mass transfer theory, the mass rate of diffusion N_x of a gaseous species in the x-direction at some cross-sectional area A is given by the expression

$$N_x = -A\frac{\partial(D_x C)}{\partial x} \tag{a}$$

where N_x is the mass transfer per unit time; D_x is the mass diffusivity, area/time, in the x-direction; C is the concentration in mass per unit volume; and A is the cross-sectional area in the x-direction.

We now wish to apply this general relationship to the diffusion of a gas, which originates continuously at a point source, through a differential volume in space. Consider the situation represented in Figure 4-A. A gaseous pollutant, carried downwind in the x-direction with a wind speed u, enters a fluid element of size $dx\,dy\,dz$. The mass rate of diffusion into the fluid element (independent of the bulk motion of the gas stream) in terms of Equation (a) would be

$$N_x = -dy\,dz\frac{\partial(D_x C)}{\partial x}$$

The rate out of the differential volume, in the x-direction, is

$$N_{x+dx} = -dy\,dz\frac{\partial(D_x C)}{\partial x} + \frac{\partial}{\partial x}\left[\left(\frac{\partial D_x C}{\partial x}\right)dy\,dz\right]dx$$

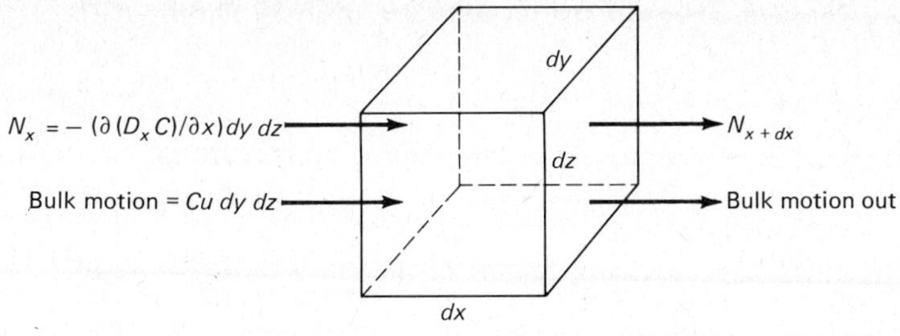

Figure 4-A Schematic for the development of Equation (e), which relates mass transfer due to diffusion and bulk motion in and out of a fluid element to the change in internal concentration.

The rate of change of the concentration within the differential volume, due to mass diffusion in and out in the x-direction, is the difference between N_{x+dx} and N_x. This quantity is

$$N_{x+dx} - N_x = \frac{\partial}{\partial x}\left(\frac{\partial(D_x C)}{\partial x}\right) dx\,dy\,dz \tag{b}$$

Similar expressions are valid for the y- and z-directions.

In addition, however, the movement of pollutant is also aided by the bulk motion of the fluid. The rate of influx of pollutant at position x through area $dy\,dz$ is found to be

$$\text{rate in (bulk motion)} = C(u)\,dy\,dz$$

and the rate out at position $(x + dx)$ is

$$\text{rate out (bulk motion)} = Cu\,dy\,dz + \frac{\partial}{\partial x}(Cu\,dy\,dz)\,dx$$

The net rate of change of concentration within the control volume $dx\,dy\,dz$, which results from the bulk motion of the fluid, then is

$$\text{net rate (bulk motion)} = -\frac{\partial}{\partial x}(Cu)\,dx\,dy\,dz \tag{c}$$

The net effect of mass diffusion and bulk motion is to change the amount of mass within the control volume. The rate of change of mass within the control volume symbolically is

$$\text{rate of change within } dx\,dy\,dz = \frac{\partial C}{\partial t}\,dx\,dy\,dz \tag{d}$$

Consequently, the general expression for the rate of change of mass within a differential volume $dx\,dy\,dz$, due to bulk transport as well as diffusional processes, is, from Equations (b), (c), and (d),

$$\frac{\partial C}{\partial t} = -\frac{\partial}{\partial x}(Cu) + \frac{\partial}{\partial x}\left(\frac{\partial(D_x C)}{\partial x}\right) + \frac{\partial}{\partial y}\left(\frac{\partial(D_y C)}{\partial y}\right) + \frac{\partial}{\partial z}\left(\frac{\partial(D_z C)}{\partial z}\right) \quad \text{(e)}$$

In developing Equation (e), the quantity $dx\,dy\,dz$ has been canceled throughout the equation.

Some basic idealizations with regard to the stack diffusion problem at this point are the following:

1. Mass transfer due to bulk motion in the x-direction far outshadows the contribution due to mass diffusion. That is, the second term on the right side of Equation (e) is far smaller than the first term and may be dropped from the equation.

2. We are primarily interested in the steady-state solution to the dispersion of pollutants in the atmosphere. Hence the quantity $\partial C/\partial t$ is zero.

3. Even though the wind speed does vary in the three coordinate directions, the variation is relatively small. Therefore it is appropriate to assume that the wind speed u is constant, since this leads to a simpler and more straightforward solution to the partial differential equation.

In the absence of sufficient information to the contrary, it is assumed that the mass diffusivities D_x, D_y, and D_z are constant. These four statements lead to the reduction of Equation (e) to the following form:

$$u\left(\frac{\partial C}{\partial x}\right) = D_y\left(\frac{\partial^2 C}{\partial y^2}\right) + D_z\left(\frac{\partial^2 C}{\partial z^2}\right) \quad \text{(f)}$$

The general solution to this second-order partial differential equation is

$$C = Kx^{-1}\exp\left\{-\left[\left(\frac{y^2}{D_y}\right) + \left(\frac{z^2}{D_z}\right)\right]\frac{u}{4x}\right\} \quad \text{(g)}$$

where K is an arbitrary constant whose value is determined by the boundary conditions on the specific atmospheric problem. One boundary condition that must be satisfied is that the rate of transfer of pollutant through any vertical plane downwind from the source is a constant in steady state, and this constant must equal the emission rate of the source, Q. That is, all pollutant emitted from the source must be accounted for somewhere downwind in the y-z plane. Hence it is assumed that no chemical reactions occur downwind to remove some of the pollutant, and no other removal mechanisms such as absorption or adsorption by other media are acknowledged. In a mathematical context this boundary condition is expressed by

$$Q = \iint uC\,dy\,dz \quad \text{(h)}$$

Generally the limits of integration on dy are minus to plus infinity. However the limits on dz depend on the physical situation of interest.

A. Point Source at Ground Level

For a point source at ground level the limits of integration on z are taken from 0 to ∞. With these limits and with substitution of Equation (g) into Equation (h), we find that

$$Q = \int_0^\infty \int_{-\infty}^\infty Kux^{-1} \exp\left[-\left(\frac{y^2}{D_y} + \frac{z^2}{D_z}\right)\frac{u}{4x}\right] dy\, dz$$

Now let $\bar{y} \equiv y/(D_y)^{1/2}$ and $\bar{z} \equiv z/(D_z)^{1/2}$. With this change in variables,

$$Q = Kux^{-1}(D_y)^{1/2}(D_z)^{1/2} \int_0^\infty \exp\left(-\frac{\bar{z}^2 u}{4x}\right) d\bar{z} \int_{-\infty}^\infty \exp\left(-\frac{\bar{y}^2 u}{4x}\right) d\bar{y}$$

However, from standard integral tables we find that

$$\int_0^\infty \exp(-a^2x^2)\, dx = \frac{(\pi)^{1/2}}{2a}$$

and the integral of the same quantity from minus to plus infinity is twice the given value, or $(\pi)^{1/2}/a$. Consequently,

$$\begin{aligned} Q &= Kux^{-1}(D_y)^{1/2}(D_z)^{1/2}\left(\frac{\pi x}{u}\right)^{1/2}\left[2\left(\frac{\pi x}{u}\right)^{1/2}\right] \\ &= 2\pi K(D_y)^{1/2}(D_z)^{1/2} \end{aligned}$$

or

$$K = \frac{Q}{2\pi(D_y D_z)^{1/2}} \tag{i}$$

This result is Equation (4-6) in Section 4-4-A, where it is used to formulate an expression for the concentration downwind from a ground-level point source.

B. Point Source at Elevation *H* Above the Ground Level

For a point source at an elevation H above the ground, the limits of integration on z in Equation (h) are taken to be from minus to plus infinity. The mathematical limit of minus infinity is physically meaningful in the following sense. Even if the ground were permeable to the diffusion of a pollutant gas, the Gaussian distribution is of such a nature that the majority of the pollutant would exist between the ground and the height H. Hence adding on to Q the integration from ground level to minus infinity leads to a reasonable small error, but makes the mathematics much more tractable.

The effect of this change in the lower limit of integration for z is to halve the value of K found previously for the point source at ground level.

That is, in the present case

$$K = \frac{Q}{4\pi(D_y D_z)^{1/2}} \tag{j}$$

This relationship for K is now substituted into Equation (4-5), and the quantities D_y and D_z are replaced by σ_y and σ_z in terms of Equation (4-9). As a result,

$$C = \frac{Qx^{-1}}{4\pi(D_y D_z)^{1/2}} \exp\left[-\left(\frac{y^2}{D_y} + \frac{z^2}{D_z}\right)\frac{u}{4x}\right]$$

$$C = \frac{Q}{2\pi u \sigma_y \sigma_z} \exp\left[-\frac{1}{2}\left(\frac{y^2}{\sigma_y{}^2} + \frac{z^2}{\sigma_z{}^2}\right)\right] \tag{k}$$

It is this formulation for a point source above ground level which leads to Equation (4-11) shown in Section 4-4-B.

QUESTIONS

1. What two physical phenomena account for the dispersal of pollutants in the atmosphere as a gas stream leaves a stack?
2. What are the deficiencies of the eddy diffusion model as represented by Equation (4-2)?
3. What is the format of a function $f(x)$ which is said to be normally distributed?
4. Why is a double-Gaussian format for a dispersion equation necessary in atmospheric studies?
5. What is the major difference in the development of a dispersion formulation for a ground source as opposed to an elevated source?
6. At what elevation is the wind velocity u evaluated for use in Equation (4-9)? Does the actual wind velocity vary with elevation across the plume? How would one take this factor into account?
7. It may be convenient to label the curves of Figures 4-6 and 4-7 according to atmospheric stability ranging from very unstable to very stable. Do this.
8. What data are usually required to determine the maximum ground-level concentration?
9. What three sets of parameters control the phenomenon of a gaseous plume injected from a stack into the atmosphere?
10. Most equations for predicting plume rise contain two terms which account for different physical reasons for the rise. What is the nature of these two terms?
11. Is it justifiable to use a constant value for the effective plume rise for a given situation, or should one, in reality, adjust for the distance downwind from the stack? Discuss.
12. What is the general effect of sampling time on the expected concentration at a given location?

13. What is the sampling time limitation when employing Turner's data for σ-values?
14. How does one modify the usual dispersion equation from a stack to account for the presence of an inversion trap?
15. In what manner does one express the source emission rate for a line source, such as cars along a highway?
16. By what general mathematical technique does one attack the problem of estimating ground-level concentrations resulting from several point sources?

PROBLEMS

4-1. Plot $f(x)$ versus x for a normally distributed (Gaussian) function for x values up to ± 5 (if necessary) and σ values of (a) 0.5, (b) 1.0, and (c) 2.0, for the case where $\mu = 0$. Show table of computations for each case.

4-2. Consider a normally distributed variable x. For values of σ of (a) 1.0 and (b) 2.0, determine the value of x for which the ratio given by $f(x)/f(x=0)$ is (1) 0.05, (2) 0.02, and (3) 0.01. Secondly, for the specified conditions, determine the value of x/σ in each case.

4-3. Consider a normally distributed variable x and σ values of (a) 1.0, and (b) 2.0. Determine the values of $f(x)/f(x=0)$ when the value of x/σ is (1) 2.5, (2) 3.0, and (3) 3.5.

4-4. Sulfur dioxide is being emitted at a rate of 0.90 kg/s from a stack with an effective height of 220 m. The average wind speed at stack height is 4.8 m/s, and the stability category is B. Determine the short-time period, downwind, center-line concentration in micrograms per cubic meter at ground-level distances from the stack of (a) 0.6, (b) 0.8, (c) 1.0, (d) 1.2, (e) 1.6, (f) 2.0, (g) 3.0, and (h) 4.0 km. Plot C versus the logarithm of the distance.

4-5. What is the expected short-time period ground-level concentration at (a) 150 m and (b) 250 m away from the downwind center line for the conditions of Problem 4-4 for cases (b) through (h)?

4-6. From the results of Problems 4-4 and 4-5, plot the isopleths (lines of constant concentration) for SO_2 concentrations of 50, 150, 250, 400, and 550 $\mu g/m^3$ on an x-y diagram. Plot x from 0 to 4 km and plot y from 0 to 400 m, full scale.

4-7. Consider the data of Problem 4-4. Estimate the distance downwind on the center line in kilometers at which the maximum concentration will occur at ground level, and estimate what that concentration will be in micrograms per cubic meter employing Figure 4-8.

4-8. Check the values of σ_y and σ_z that were used in Problem 4-4, using the basic equations and data of Martin.

4-9. Check the value of C_{max} determined in Problem 4-7, using the method discussed in the text where σ_y/σ_z is a constant. Also use this method to determine x_{max}, in kilometers.

4-10. Derive expressions for x_{max} and C_{max} for the situation where $\sigma_y = ax^p$ and $\sigma_z = bx^q$.

4-11. Evaluate C_{max} and x_{max} from the equations developed in Problem 4-10, using the σ_y and σ_z data of Martin, for the data of Problem 4-4.

4-12. Data from the Brookhaven National Laboratory for unstable conditions in the atmosphere indicate that $\sigma_y = 0.35(x)^{0.86}$ and $\sigma_z = 0.33(x)^{0.86}$. For these equations, determine x_{max} and C_{max} for the data of Problems 4-4 and 4-10.

4-13. Reconsider Problem 4-4. Change the stability class to C, and determine the downwind, center-line concentrations in micrograms per cubic meter at ground-level distances from the stack of (a) 1.2, (b) 1.6, (c) 2.0, (d) 2.5, (e) 3.0, (f) 5.0, (g) 10, and (h) 20 km. Plot C versus the logarithm of the distance.

4-14. What is the expected ground-level concentrations at (a) 300 m, and (b) 500 m crosswind for the conditions of Problem 4-13 for cases (b) through (h)?

4-15. From the results of Problems 4-13 and 4-14, plot the isopleths for SO_2 concentrations of 150, 200, 300, 400, and 450 $\mu g/m^3$ on an x-y diagram. Plot x from 0 to 10 km and plot y from 0 to 600 m, full scale.

4-16. Consider the data of Problem 4-13. Estimate the distance downwind on the center line in kilometers at which the maximum concentration will occur at ground level, and estimate what that concentration will be in micrograms per cubic meter employing Figure 4-8.

4-17. Check the values of σ_y and σ_z that were used in Problem 4-13, using the basic equations and data of Martin.

4-18. Check the value of C_{max} determined in Problem 4-16, using the method discussed in the text where σ_y/σ_z is a constant. Also use this method to determine x_{max}, in kilometers.

4-19. Evaluate C_{max} and x_{max} from the equations developed in Problem 4-10, using the σ_y and σ_z data of Martin, for the data in Problem 4-13.

4-20. Data from the Brookhaven National Laboratory for near neutral conditions in the atmosphere indicate that $\sigma_y = 0.32(x)^{0.78}$ and $\sigma_z = 0.22(x)^{0.78}$. For these data, determine x_{max} and C_{max} for the data of Problems 4-10 and 4-13.

4-21. It is convenient to know the distance y_p where the concentration has dropped to p percent of its value on the plume axis. Consider the horizontal spread in pollutant concentration at ground level. Making use of Equation (4-13), prove that the value of y_p at any distance x is found simply from the expression, $y_p = [2\sigma_y{}^2 \ln(100/p)]^{1/2}$.

4-22. Hydrogen sulfide is vented from a stack which has an effective height of 50 m. The wind speed is 2.5 m/s on an overcast night. For an emission rate of 0.06 g/s, (a) determine the maximum ground-level concentration on the plume center line downwind from the stack, and (b) plot the ground-level concentration as a function of y-distance from the center line at the x-location determined in part (a) for y values of 50, 100, 200, and 300 m.

4-23. The odor threshold for H_2S is 0.00047 ppm. Using the emission conditions given in Problem 4-22 in terms of x- and y-coordinates, estimate the region in which an average person could detect the hydrogen sulfide by smell.

4-24. The ground-level concentration of SO_2 downwind from a stack is to be limited to 80 $\mu g/m^3$. The wind speed is 4 m/s on a clear day, and the emission rate is 50 g/s. What is the minimum required effective stack height, in meters?

4-25. For the data given in Example 4-2, determine the crosswind distance in meters at which the concentration will be (a) 30 percent, (b) 20 percent, and (c) 10 percent of the center-line concentration.

4-26. The rate of emission of SO_2 from the stack of a power plant is 126.1 g/s. The effective height of the stack is 46 m. Calculate the SO_2 concentration in parts per million at a parking lot located 900 m downwind from the stack on a sunny October day when the wind velocity is 4 m/s. Use class *C* stability.

4-27. How many meters downwind from the stack does the maximum ground-level concentration of SO_2 occur for Problem 4-26? What is the value of the concentration in parts per million at that location?

4-28. You are located downwind from two oil-burning power plants. One is located 0.3 km NNE of your location and burns 1400 kg of 0.5 percent sulfur oil per hour. The second plant is located 0.5 km NNW of you and burns 1600 kg/hr of fuel oil containing 0.75 percent sulfur. Assume that both plant stacks have an effective height of 40 m. The wind is blowing from the north at 3.3 m/s. For a class *B* stability condition, what is the SO_2 concentration at your location at ground level, in micrograms per cubic meter? The wind speed is at the standard height of 10 m.

4-29. A plant is to be constructed which will emit 3.5 metric tons of hydrogen sulfide per day. One of the design criteria is that the concentration 1 km downwind from the stack must not exceed 120 $\mu g/m^3$, so that the odor threshold is not exceeded. For the purpose of estimation the plume rise is neglected initially. Estimate the required stack height, in meters, for wind speeds of (a) 4 m/s and (b) 8 m/s.

4-30. On an overcast day with a class *C* stability, the wind velocity at 10 m is 4 m/s. The emission rate of NO is 50 g/s from a stack having an effective height of 100 m. (a) Estimate the center-line, ground-level concentration 20 km downwind from the stack, in micrograms per cubic meter. (b) Estimate the ground-level concentration 20 km downwind and 900 m from the stack center line, in micrograms per cubic meter.

4-31. A power plant emits SO_2 on a day with class *C* stability when the wind speed at the top of the stack is 7 m/s. The effective stack height is 282 m. If the short-time ground-level concentration downwind is not to exceed (a) 1000 $\mu g/m^3$ and (b) 1300 $\mu g/m^3$, what is the maximum permissible SO_2 emission rate, in grams per second?

4-32. An existing power plant has been found to produce an SO_2 concentration of 20 $\mu g/m^3$ at a distance of 800 m directly downwind from the stack when the wind is 4 m/s from the north during a class *C* stability situation. At a later date another plant is built 200 m to the west of the original plant. It burns 4000 lb/hr of fuel oil which contains 0.5 percent sulfur. The second plant has an effective stack height of 60 m, and it has no SO_2 emission controls. For the same atmospheric conditions listed above, estimate the percentage of increase in SO_2 concentration at the downwind site due to the second plant.

4-33. The concentration of hydrogen sulfide, H_2S, is 55 ppb at a location 150 m downwind from an abandoned oil well. What is the rate of H_2S emission from the well if the winds are 2.7 m/s on a sunny June afternoon, in grams per second? Assume ground-level emission.

4-34. A fire burning at ground level is emitting nitric oxide at a rate of 3.6 g/s. The fire is assumed to be a point source with no effective rise of the plume. Determine the concentration of NO directly downwind at a distance of 2.5 km under the following atmospheric conditions: (a) overcast night, 6 m/s wind speed; (b) clear night, 3 m/s wind speed; and (c) partly cloudy afternoon, 4 m/s wind speed.

4-35. Sulfur dioxide is emitted at a rate of 0.9 kg/s into a class *B* stability atmosphere which has a potential temperature gradient of $-0.010°C/m$. The SO_2 concentration in the stack gas is 4.0 g/m^3, and the stack gas temperature and pressure are 175°C and 980 mbar, respectively. The wind speed is 4.8 m/s and its temperature is 18°C. The stack diameter at the top is 5.5 m, and it is assumed that other properties of the stack gas are the same as for air. Determine the plume rise above the stack in meters, using: (a) the Holland equation, (4-18); (b) the Carson-Moses equation, (4-20a); (c) the modified Briggs equation, (4-21); and (d) the modified Concawe equation, (4-19b).

4-36. Reconsider Problem 4-35, with the following changes. The stability class is *C* and the potential temperature gradient is 0.010°C/m. Also, for part (b) the correct equation is now (4-20b).

4-37. Prove that the last term of Equation (4-18) can be replaced by the quantity $0.0096 Q_h / V_s d$.

4-38. In Example 4-6 the effective plume rise is calculated for a neutral atmosphere by use of two empirical formulas. Use these formulas to compute the plume rise for the same conditions except that the atmosphere is (a) moderately stable with a potential temperature gradient of 0.003°K/m, and (b) stable with a potential temperature gradient of 0.008°K/m.

4-39. Using Equations (4-17), (4-18), (4-19b), and (4-20c), estimate the effective correction to stack height, Δh, in meters, for the following conditions: Q_h equal to 114,000 kJ/s; stack height of 250/m; gas exit velocity of 14.65 m/s; stack diameter of 9.13 m; wind velocity at stack exit of 7 m/s; atmospheric temperature of 280°K; gas exit temperature of 422°K; and atmospheric temperature gradient of $+0.534°C/100$ m.

4-40. Two coal-fired furnaces discharge into one stack, 100 m tall. Each furnace is fired with coal at the rate of 250 tons every 24 hr. The combustion air is supplied at the rate of 10 lb for each pound of coal. The gases exit from the stack with a velocity of 20 ft/s at 350°F. The atmospheric temperature at the stack outlet is 60°F. The wind velocity is 10 mi/hr at 10 m. Assume neutral atmosphere. Calculate the plume-rise value in meters employing the (a) Holland, (b) Concawe, and (c) Moses-Carson equations.

4-41. Gas with the composition essentially of air exits from a stack with a velocity of 10 m/s. The gas temperature is 200°C, and the atmospheric temperature at the top of the stack is 0°C. The stack diameter is 10 m. Determine the value of Δh for a neutral atmosphere by use of (a) Equation (4-17), (b) Equation (4-18), (c) Equation (4-19a), (d) Equation (4-20b), and (e) Equation (4-21). The wind speed at the top of the stack is 5 m/s.

4-42. The following data apply to a coal-burning steam power plant: stack height, 200 m; stack diameter, 9 m; coal-firing rate, 1.165×10^6 kg/day; air supply, 12 lb air/lb coal; stack gas temperature, 150°C; ambient air temperature at 350 m, 7°C; heating value of coal, 5250 kJ/kg; sulfur content of coal, 3.1 percent; ash content of coal, 8 percent; ash carried up the stack, 80 percent; atmospheric conditions, cloudy, daytime; wind speed 6 m/s at 10-m height; temperature gradient, neutral. Calculate the effective stack height using (a) Equation (4-17), (b) Equation (4-18), (c) Equation (4-19b), and (d) Equation (4-20b).

4-43. An estimate has been made of pollutant concentration at a given location downwind from a point source. This value is to be checked by actual sampling. What percentage of error, due to sampling time alone, might you expect in sample concentration if the sample time is (a) 2 min, (b) 20 min, and (c) 1 hr?

4-44. A long line of agricultural waste burning in a field may be considered an infinite line source. On a clear fall afternoon, the wind speed is 4.5 m/s. Determine the particulate concentration for the small airborne particles at 600 m downwind, if the source strength is 0.23 g/m·s.

4-45. The traffic density for an interstate highway is 10,000 vehicles/hr and the average vehicle speed is 80 km/hr. The wind speed perpendicular to the highway is 3 m/s. The average carbon monoxide emission per vehicle is 20 g/km. (a) For an overcast day, estimate the CO concentration 1 km downwind from the highway. (b) Does the concentration found in part (a) represent a health hazard as indicated by the primary air quality standard for carbon monoxide?

4-46. Show that Equation (k) in the appendix to this chapter is a solution to Equation (j).

References

1. W. P. Lowry and R. W. Boubel. *Meteorological Concepts in Air Sanitation*. Corvallis, Ore.: Oregon State University, 1967.
2. S. B. Carpenter et al. "Principal Plume Dispersion Models, TVA Power Plants." 63d Annual Meeting, Air Pollution Control Association, June, 1970.
3. K. W. Ragland and J. J. Peirce. "Boundary Layer Model for Air Pollutant Concentrations due to Highway Traffic." *J. Air Pollu. Control Assoc.* **25**, no. 1 (1975): 48–51.
4. D. B. Turner. *Workbook of Atmospheric Dispersion Estimates*. Washington, D.C.: HEW, 1969.
5. D. O. Martin, "The Change of Concentration Standard Deviation with Distance." *J. Air Pollu. Control Assoc.* **26**, no. 2 (1976): 145.
6. A. C. Stern, ed. *Air Pollution*. Vol. I, 2d ed. New York: Academic Press, 1968.
7. U.S. Weather Bureau. *Meteorology and Atomic Energy*. Report AECU-3066. Washington, D.C.: AEC, 1955.
8. N. E. Bowne. "Diffusion Rates." *J. Air Pollu. Control Assoc.* **24**, no. 9 (1974):832.
9. D. H. Stade, ed. *Meteorology and Atomic Energy*, TID-24190, Washington, D.C., 1968.
10. F. N. Frenkiel and R. E. Munn, eds., *Turbulent Diffusion in Environmental Pollution*. Advances in Geophysics Series, Vols. 18A and 18B. New York: Academic Press, 1974.
11. R. Ranchoux. "Determination of Maximum Ground Level Concentration." *J. Air Pollu. Control Assoc.* **26**, no. 11 (1976): 1089.
12. J. E. Carson and H. Moses. "The Validity of Several Plume Rise Formulas." *J. Air Pollu. Control Assoc.* **19**, no. 11 (1969): 862–866.
13. F. W. Thomas, S. G. Carpenter, and W. C. Colbaugh. "Plume Rise Estimates for Electric Generating Stations." *J. Air Pollu. Control Assoc.* **20**, no. 3 (1970): 170–177.

14. H. Moses and M. R. Kraimer. *J. Air Pollu. Control Assoc.* **22**, no. 8 (1972): 621.
15. G. A. Briggs. *Plume Rise*. AEC Critical Review Series, TID-25075, 1969.
16. T. L. Montgomery et al. "Results of Recent TVA Investigations of Plume Rise." *J. Air Pollu. Control Assoc.* **22**, no. 10 (1972): 779–784.
17. G. Nonhebel. *J. Inst. Fuel* **33** (1960): 479.
18. M. Heno. *Atmos. Environ.* **2** (March 1968): 149.
19. L. J. Shieh and P. K. Halpern. *Numerical Comparison of Various Model Representations for a Continuous Area Source*, G320-3293. Palo Alto, Calif.: IBM Data Processing Division, August, 1971.

Chapter 5
Particulate

5-1 INTRODUCTION

Of the total mass of air pollutants estimated for 1973, for example, approximately 9 percent was in the form of particulate. Of the 32,000,000 tons total, motor vehicles contributed 900,000 tons, industry 17,000,000 tons, electric power 4,100,000 tons, industrial boilers 6,900,000 tons, and refuse disposal 3,000,000 tons. Particulate matter from other sources includes ocean salt, volcanic ash, products of wind erosion, roadway dust, products of forest fires, and plant pollen and seed. Several terms are employed to classify airborne particulate. The definitions of these terms [1] are presented in Table 5-1.

Although particulate comprised only 9 percent of the total mass of man-made air pollutants for 1973, the potential hazard from this type of pollutant is much greater. Particulate presents a health hazard to the lungs; enhances chemical reactions in the atmosphere; reduces visibility; increases the possibility of precipitation, fog, and clouds; reduces solar radiation, with concomitant changes in environmental temperature and biological rates of plant growth; and soils materials extensively. The magnitude of the problem in each of the above areas is a function of the range of particle sizes in the local atmosphere, the particle concentration, and the chemical and physical composition of the particulate. Each of these factors needs examination.

In general, airborne particles range in size from 0.001 to 500 μm, with the bulk of the particulate mass in the atmosphere ranging from 0.1 to 10 μm. Particles below 0.1 μm in size display a behavior similar to that of

Table 5-1 DEFINITIONS OF TERMS THAT DESCRIBE AIRBORNE PARTICULATE

Particulate matter	Any material, except uncombined water, that exists in the solid or liquid state in the atmosphere or gas stream at standard condition
Aerosol	A dispersion of microscopic solid or liquid particles in gaseous media
Dust	Solid particles larger than colloidal size capable of temporary suspension in air
Fly ash	Finely divided particles of ash entrained in flue gas. Particles may contain unburned fuel
Fog	Visible aerosol
Fume	Particles formed by condensation, sublimation, or chemical reaction, predominantly smaller than 1 μm (tobacco smoke)
Mist	Dispersion of small liquid droplets of sufficient size to fall from the air
Particle	Discrete mass of solid or liquid matter
Smoke	Small gasborne particles resulting from combustion
Soot	An agglomeration of carbon particles

molecules and are characterized by large random motions caused by collisions with gas molecules. Particles larger than 1 μm but smaller than 20 μm tend to follow the motion of the gas in which they are borne. Particles larger than 20 μm have significant settling velocities; consequently, they are airborne for relatively short periods of time. The approximate settling velocities for particles having a density of 1 g/cm^3 are:

0.1 μm	4×10^{-5} cm/s
1 μm	4×10^{-3} cm/s
10 μm	0.3 cm/s
100 μm	30 cm/s

These values indicate why there is a significant difference in the airborne behavior of particulate matter. The center section of Figure 5-1 shows the range of particle sizes for various materials. Although the list is limited, note that it gives roughly a fourfold range of particles sizes, from 10^{-2} to 10^{2} μm. It is highly unlikely that any one type of collection equipment will be effective in removing particles over such a broad range. This observation is confirmed by the bottom section of Figure 5-1, which shows the spread of particle sizes for which various equipment is appropriate. However, these data are somewhat misleading. Although a given type of collector may remove particles over the entire range indicated, the efficiency of removal in many cases is a function of particle size. For example, a collector may remove large particles in a given spread with nearly 100 percent efficiency, but the collector's efficiency in removing the smaller particles may lie close to zero. We present typical collection efficiencies for specific collection systems later in this chapter. Other tables and charts similar to Figure 5-1 appear in the literature [2, 3, 4].

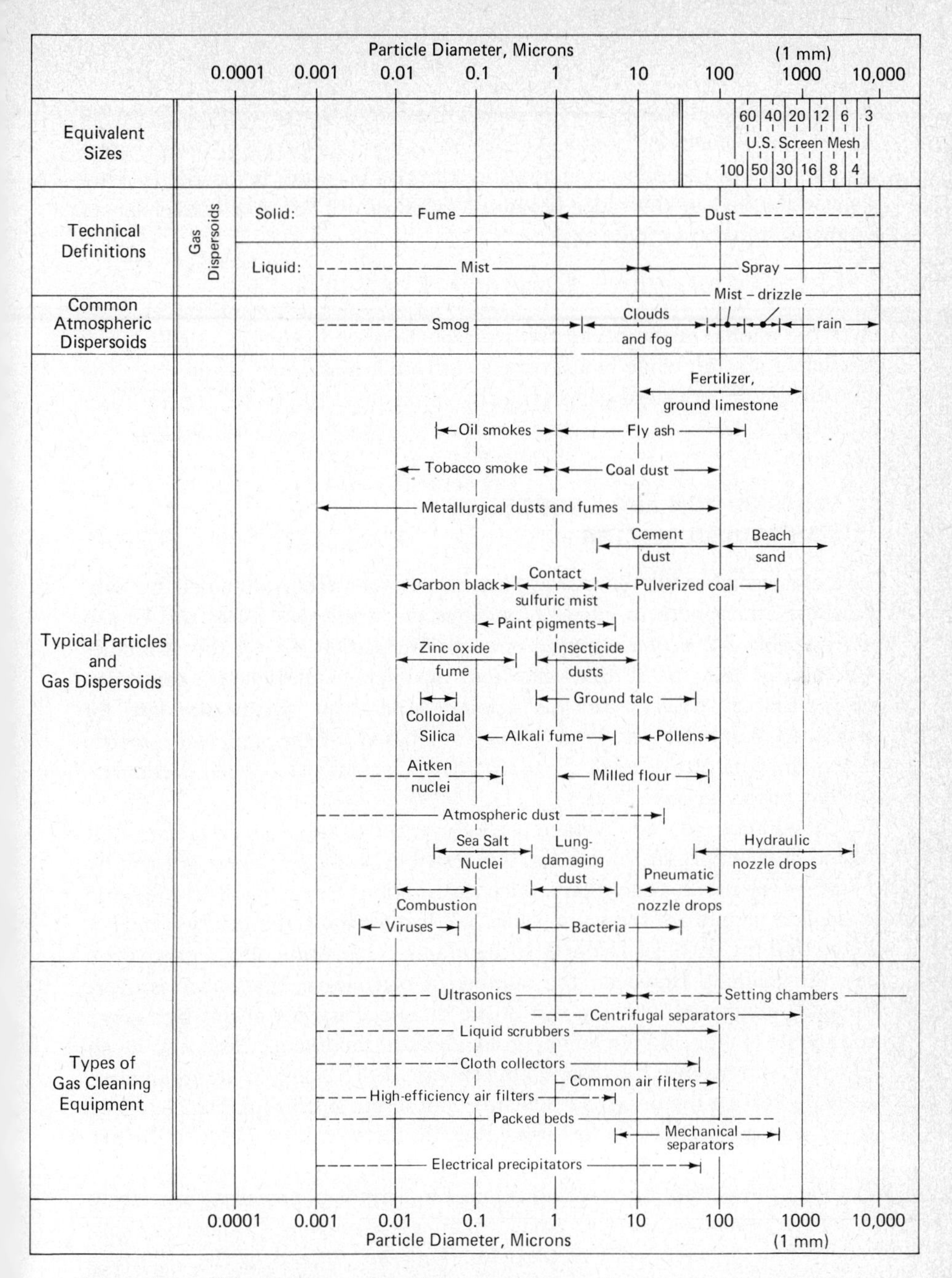

Figure 5-1 Characteristics of particles and particle dispersoids. (SOURCE: C. E. Lapple. *Stanford Research Institute Journal* **5**, 1961.)

Particulate concentration is usually expressed as the total mass of the particles in a given volume of gas. The basic units for particle concentration are micrograms per cubic meter, although units of grains per cubic foot are well established in the older literature (7000 gr = 1 lb). For conversion or comparison purposes, note that

$$1.0\ \mathrm{gr/ft^3} = 2.29\ \mathrm{g/m^3} = 2.29 \times 10^6\ \mu\mathrm{g/m^3}$$

Over the middle of the ocean the atmospheric dust loading is usually much less than 1 $\mu g/m^3$, while in a severe dust storm it may reach 10^9 $\mu g/m^3$. The dust loading in industrial gases typically varies from 10^4 to 10^8 $\mu g/m^3$ (0.01 to 100 gr/ft^3).

5-2 DISTRIBUTION AND SOURCES OF PARTICULAR MATTER

Data are available for the average particle concentration obtained by sampling the atmosphere in cities throughout the world as well as the United States. Table 5-2 is one example of such information [1]. Of the 401 cities sampled, 75 percent at that time had an average particle concentration greater that 80 $\mu g/m^3$. Yet the primary federal air quality standard for suspended particulate is set at 75 $\mu g/m^3$, based on the annual geometric mean. The data of this table show a positive correlation between particulate concentration and city size.

In addition to the average particle concentration in mass per unit volume, it is important to note the size distribution by particle count and by volume in the urban atmosphere. Such distributions for a typical atmospheric particulate sample are shown in Table 5-3. From data in the last two entries, we see that the particles in the 0- to 1-μm range constitute only 3 percent by mass (or volume). However, the number of particles in that range is overwhelming compared with the rest of the sample. Particles of this size range are capable of entering the lungs. From a health standpoint, it is not so much a question of lowering the overall atmospheric dust loading in an urban area, but of decreasing the heavy particulate count in the smaller size range.

In general, particles in the atmosphere in the size range below 1 μm are produced by condensation, while larger particles result from either comminution (pulverization) or combustion. Dry grinding processes are rarely efficient in producing particles smaller than a few microns. Combustion may produce four distinct types of particles. They are formed in the following ways:

1. Heat may vaporize materials which subsequently condense, yielding particles between 0.1 and 1 μm.
2. The chemical reactions of the combustion process may produce short-lived particles of unstable molecular clusters below about 0.1 μm.

TABLE 5-2 DISTRIBUTION OF SELECTED CITIES BY POPULATION CLASS AND PARTICLE CONCENTRATION, 1957–1967

POPULATION CLASS	AVERAGE PARTICLE CONCENTRATION ($\mu g/m^3$)										
	< 40	40–59	60–79	80–99	100–119	120–139	140–159	160–179	180–199	> 200	TOTAL
3 MILLION							1		1		2
1–3 MILLION							2	1			3
0.7–1 MILLION			1		2		4				7
400,000–700,000				4	5	6	1	1	1		18
100,000–400,000		3	7	30	24	17	12	3	2	1	99
50,000–100,000		2	20	28	16	12	6	5	1	3	93
25,000–50,000		5	24	12	12	10	2	1	2	3	71
10,000–25,000		7	18	19	9	5	2	3	1		64
10,000	1	5	7	15	11	2	1	2			44
TOTAL URBAN	1	22	77	108	79	52	31	16	8	7	401

SOURCE: NAPCA. *Air Quality Criteria for Particulate Matter*, AP-49. Washington, D.C.: HEW, 1969.

Table 5-3 PARTICLE DISTRIBUTION BY COUNT AND VOLUME PERCENT OF A TYPICAL ATMOSPHERIC SAMPLE AS A FUNCTION OF SIZE

SIZE RANGE (μm)	AVERAGE SIZE (μm)	PARTICLE COUNT[a]	VOLUME PERCENT[b]
10–30	20	1	27
5–10	7.5	112	53
3–5	4	167	12
1–3	2	555	5
$\frac{1}{2}$–1	0.75	4,215	2
0–$\frac{1}{2}$	0.25	56,900	1

[a] Count of other sizes relative to count of 20-μm size.
[b] Also mass percent if uniform specific gravity.

3. Mechanical processes may release ash or fuel particles 1 μm or larger.
4. If liquid fuel sprays are involved, a very fine ash may escape directly.
5. Partial combustion of fossil fuels may produce soot.

The stationary sources of particulate emissions may be divided into classes such as household and commercial, industrial, and power. Of the total particulate formed, roughly 85 to 90 percent come from power production sources [5], and the vast majority from power sources is due to the burning of bituminous and lignite coal. Fortunately, with the operation of electrostatic precipitators and other control devices, well over 90 percent of these potential emissions are ultimately removed before release to the atmosphere.

The major industrial sources of particulate pollution are presented in Table 5-4. Asphalt batching in the construction industry is another large potential source. And the giant food and feed industry generates particulates through such processes as soil preparation, insecticide spraying, grain milling and drying, and meat and fish processing.

To facilitate the estimation of industrial emission rates, the U.S. government has published several tables of emission factors based on the quantities of goods or materials processed. Table 5-5 is an extract from a much longer list of emission factors [5]. As an example of the specific sources of emission within a given general category, consider motor vehicles as listed in Table 5-5. Particulate matter emitted by gasoline-fueled vehicles consists of carbon, metallic ash, and hydrocarbon aerosols. Metal-based particles result from the combustion of fuels containing lead antiknock compounds. Carbon and unburned hydrocarbons are the result of incomplete combustion. The particulate matter discharged by diesel engines consists primarily of carbon and hydrocarbon aerosols resulting from incomplete combustion under conditions of severe engine loading. Both the spark-ignition and diesel engines will be treated in greater detail in Chapter 10.

Table 5-4 INDUSTRIAL PROCESS AND CONTROL SUMMARY

INDUSTRY OR PROCESS	SOURCE OF EMISSIONS	PARTICULATE MATTER	METHOD OF CONTROL
Iron and steel mills	Blast furnaces, steel-making furnaces, sintering machines	Iron oxide, dust, smoke	Cyclones, bag houses, electrostatic precipitator, wet collectors
Gray iron foundries	Cupolas, shake-out systems, core making	Iron oxide, smoke, oil dust, metal fumes	Scrubbers, dry centrifuge collectors
Nonferrous metallurgy	Smelters and furnaces	Smoke, metal fumes, oil, grease	Electrostatic precipitators, fabric filters
Petroleum refineries	Catalyst regenerators, sludge incinerators	Catalyst dust, ash from sludge	Cyclones, electrostatic precipitators, scrubbers, bag houses
Portland cement	Kilns, driers, material-handling systems	Alkali and process dusts	Fabric filters, ES precipitators, mechanical collectors
Kraft paper mills	Recovery furnaces, lime kilns, smelt tanks	Chemical dusts	Electrostatic precipitators, venturi scrubbers
Acid manufacture—phosphoric, sulfuric	Thermal processes, rock acidulating, grinding	Acid mist, dust	Electrostatic precipitators, mesh mist eliminators
Coke manufacture	Oven operation, quenching materials handling	Coal and coke dust, coal tars	Meticulous design, operation and maintenance
Glass and fiberglass	Furnaces, forming and curing, handling	Acid mist, alkaline oxides, dust, aerosols	Fabric filters, after-burners

SOURCE: W. Jost et al. Z. *Phys. Chem. N.F.* **45** (1965): 47.

Table 5-5 EMISSION FACTORS FOR SELECTED CATEGORIES OF UNCONTROLLED SOURCES OF PARTICULATES

EMISSION SOURCE	EMISSION FACTOR
Natural gas combustion	
Power plants	15 lb/million ft^3 of gas burned
Industrial boilers	18 lb/million ft^3 of gas burned
Domestic and commercial furnaces	19 lb/million ft^3 of gas burned
Distillate oil combustion	
Industrial and commercial furnaces	15 lb/thousand gal of oil burned
Domestic furnaces	8 lb/thousand gal of oil burned
Residual oil combustion	
Power plants	10 lb/thousand gal of oil burned
Industrial and commercial furnaces	23 lb/thousand gal of oil burned
Coal combustion	
Cyclone furnaces	(2 times ash %) lb/ton of coal burned
Other pulverized-coal furnaces	(13–17 times ash%) lb/ton of coal burned
Spreader stokers	(13 times ash %) lb/ton of coal burned
Other stokers	(2–5 times ash %) lb/ton of coal burned
Incineration	
Municipal, multiple chamber	17 lb/ton of refuse burned
Commercial, multiple chamber	3 lb/ton of refuse burned
Flue-fed incinerator	28 lb/ton of refuse burned
Domestic, gas-fired	15 lb/ton of refuse burned
Open burning of refuse	16 lb/ton of refuse burned
Motor vehicles	
Gasoline-powered engines	12 lb/thousand gal of gasoline burned
Diesel-powered engines	110 lb/thousand gal of fuel burned
Cement manufacturing	38 lb/barrel of cement produced
Kraft pulp mills	
Lime kiln	94 lb/ton of dried pulp produced
Recovery furnaces[a]	150 lb/ton of dried pulp produced
Steel manufacturing	
Open-hearth furnaces	1.5–20 lb/ton of steel produced
Electric arc furnaces	15 lb/ton of metal charged
Sulfuric acid manufacture	0.3–7.5 lb acid mist/ton acid produced

SOURCE: *Control Techniques for Particulate Air Pollutants*. Washington, D.C.: HEW, December 1968.

[a] With primary stack gas scrubber.

It has been wisely said that the prevention of air pollution from industrial sources starts within the factory or mill. Relying upon gas cleaning devices to reduce emissions and on tall discharge stacks to disperse and dilute offensive substances to tolerable ground-level concentrations is unnecessary when process and system control is effective in preventing the formation and discharge of air pollutants. The following generalized techniques are employed for controlling the source or reducing the concentration of air pollutants:

1. Gas cleaning
2. Source relocation

3. Fuel substitution
4. Process changes
5. Good operating practice
6. Source shutdown
7. Dispersion

Before suitable air pollution control methods are instituted, the answers must be found to the following pertinent questions:

1. Is the atmospheric contaminant in fact a necessary consequence of the operation?
2. Can the rate of generation of the contamination be reduced and can high bursts of release be avoided?
3. Does the process lend itself to control by local exhaust ventilation equipment such as hoods?

If consideration of a pollution problem leads to the conclusion that gas cleaning equipment is necessary to control particulate emission, numerous devices and techniques are available. It has been stated that an industry can remove any quantity of particulate if it is willing to pay for it in pressure drop, capital investment, and operating cost. The relationships between emission control or air quality index and cost or benefit are illustrated in Figure 5-2. The specific shapes of the cost and benefit curves in Figure 5-2(a) will vary with the type of air pollution problem. However, in general we might anticipate that initially the benefits increase rapidly for a relatively small increase in cost. Then, as greater control is desired, the cost increases disproportionately. While this type of plot is useful, Figure 5-2(b) frequently is considered more so. In Figure 5-2(b), the incremental cost or benefit per unit of air quality index is plotted against the air quality index. Mathematically, the incremental cost and benefit curves are simply plots of the slopes of the curves in Figure 5-2(a). Incremental curves give a better indication of the point at which an improved air quality index is justified. One such measure is the crossover point of the two curves in Figure 5-2(b). It may be noted that beyond this point, the incremental benefits are fast diminishing while the incremental costs are fast increasing. In any case, a compromise must be reached between the air quality that society desires and the associated cost burden that industry indirectly, and the consumer or society directly, can afford. It must be realized that society ultimately pays the cost for having clean air.

5-3 PARTICULATE COLLECTION EFFICIENCY

Several devices for cleaning gas streams are presented in the last column of Table 5-4. The general operating characteristics of these devices are summarized in Table 5-6. Note the column headed "Efficiency by Weight." This information is extremely important to have for any collection device, since it

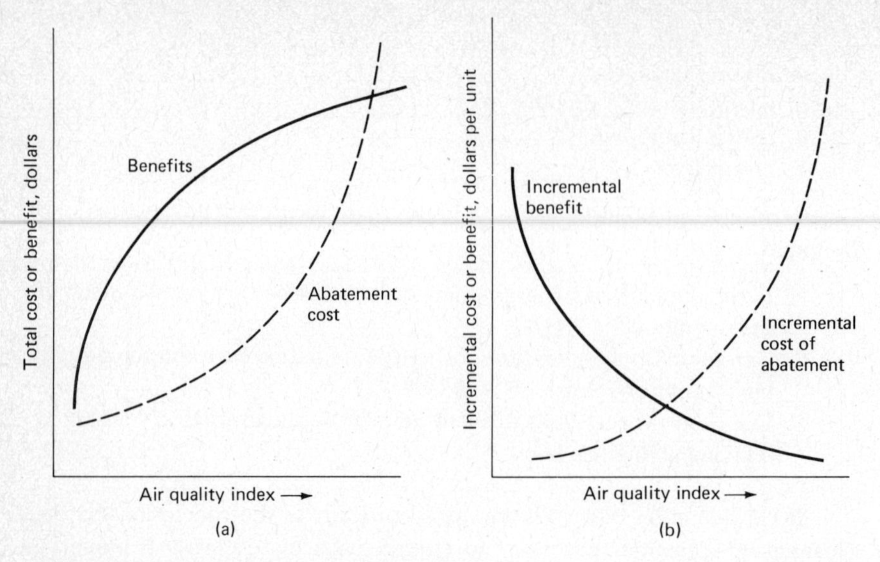

Figure 5-2 Economics of air pollution control.

indicates what fraction of the total particulate weight will be removed for the entire size range. The overall collection efficiency η_0 is a generalized parameter employed to indicate the performance level of a gas cleaning device. When the overall collection efficiency is expressed as a fraction, then η_0 is given by Equation (5-1), namely,

$$\eta_0 = \frac{C}{A} = \frac{C}{B+C} = \frac{A-B}{A} \tag{5-1}$$

where A is the entering loading or concentration, B is the leaving loading or concentration, and C is the amount caught or retained by the cleaning device. The values of A, B, and C must be expressed in the same units.

In general, the overall collection efficiency η_0 by weight is predicted from a knowledge of (1) the mass or weight distribution among the particles sizes of the dust of interest, and (2) the collection efficiency as a function of the particle diameter, d_p. A plot of the latter data is known as a fractional efficiency or grade efficiency curve. This is shown by the hypothetical curve in Figure 5-3, which is typical of certain collection equipment. It is a general characteristic of dust collectors that the fractional collection efficiency η_d increases with increasing particle size. Note, however, that the fractional efficiency varies drastically with particle size. There is a rapid rise in efficiency through a fairly small particle-size range, with the efficiency essentially 100 percent at the larger sizes. The shape of this curve varies for different types of equipment and for design variations and operating conditions within a given type of device. In addition, the fractional collection efficiency may be a function of the type of dust. This latter variable is due to the different physical characteristics of particulates, such as shape or density.

Table 5-6 MECHANICAL COLLECTION EQUIPMENT

COLLECTOR TYPE	SPACE REQUIREMENTS	VOLUME RANGE (ft^3/min)	EFFICIENCY BY WEIGHT	PRESSURE LOSS[a] (in. H_2O)	TEMPERATURE LIMITATIONS	POWER[b] (hp per 1000 ft^3/min gas)
Settling chambers	Large	Space available only limitation	Good above 50 μm	0.2–0.5	700–1000°F Limited only by materials of construction	0.04–0.12
Conventional cyclone	Large	Normal range up to 50,000 ft^3/min	Approx. 50% on 20 μm	1–3	700–1000°F Limited only by materials of construction	0.24–0.73
High-efficiency cyclone	Medium	Normal range up to 12,000 ft^3/min	Approx. 80% on 10 μm	3–5	700–1000°F Limited only by materials of construction	0.73–1.2
Multitube cyclones	Small	Normal range up to 100,000 ft^3/min	90% on $7\frac{1}{2}$ μm	4.5	700–1000°F	1.1
Dynamic precipitator	Small	17,000 ft^3/min	80% on 15 μm	No loss (true fan)	700°F	Power consumption will depend on selection point, mechanical efficiency in usual selection range, 40–50%
Impingement separator	Small	Space available only limitation	90% on 10 μm	1 – 5	700°F	0.24–1.20

SOURCE: American Industrial Hygiene Association. *Air Pollution Manual*, Part II, "Control Equipment, 1968 [6]."

[a] Pressure drop is based on standard conditions.

[b] Power consumption figured from hp = ft^3/min × t.p./6356 × M.E. (mechanical efficiency assumed to be 65 percent).

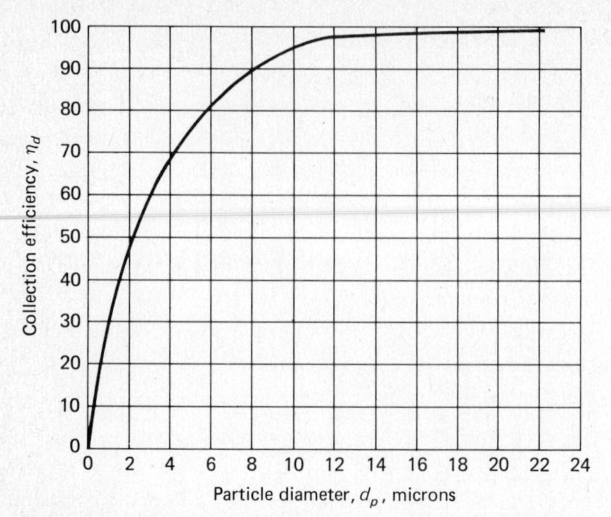

Figure 5-3 Hypothetical fractional collection efficiency curve as a function of particle diameter.

There are equations, based on suitable models of the process, which enable one to predict the fractional efficiency curves for various devices. Some of these models will be presented in the subsequent sections of this chapter. Nevertheless, in practice it is highly desirable to use data based on experimental measurement. Such information is frequently available from manufacturers of the equipment.

Before developing the characteristics of the number or mass (weight) distribution of a collection of particles, it is important to discuss the subject of particle diameter itself, since both the mass distribution and the fractional efficiency data are a function of particle diameter. In theory, if all actual particles were spherical, then the definition of particle diameter would be clear-cut. Most liquid droplets in industrial processes, for example, are essentially spherical. Thus in this case no ambiguity arises in the term droplet diameter. However, solid particles are usually not spherical, whatever their source in industrial or natural processes. Thus the measurement of a linear dimension which represents a diameter is not obvious. Furthermore, a diameter may be defined in terms of some other physical characteristic of a particle. For example, a surface area or a volumetric diameter could be considered. A particle diameter based on surface area, d_{SA}, is defined as the diameter of a sphere that will have the same surface area as the particle of interest. The volumetric diameter of an actual particle, d_V, is defined as the diameter of a sphere that will have the same volume as the particle in question. A diameter can also be defined in terms of a specific type of particle behavior. For example, two other diameters frequently mentioned are the Stokes' diameter and the aerodynamic equivalent diameter. The Stokes' diameter, d_S, is the diameter of a sphere with the same density of the nonspherical particle which falls freely in laminar flow at the same terminal speed as the nonspherical particle. The aerodynamic equivalent diameter, d_A, is defined similar to the Stokes' diameter, except the sphere is taken to have the density of 1 g/cm^3. Other types of diameters can be defined. As a result, there are a number of bases for the mass distribution and fractional efficiency data in terms of the specific type of diameter chosen for the analysis.

To discuss the number or mass distribution data more fully, it is helpful to introduce first some notation and nomenclature. Since samples of particulate contain a tremendous number of particles, we shall assume that the sizes (diameters) are continuously distributed (as opposed to discretely distributed). This means that the collection of particles can be analyzed in terms of a differential range of particle sizes, from d_p to $d_p + d(d_p)$.

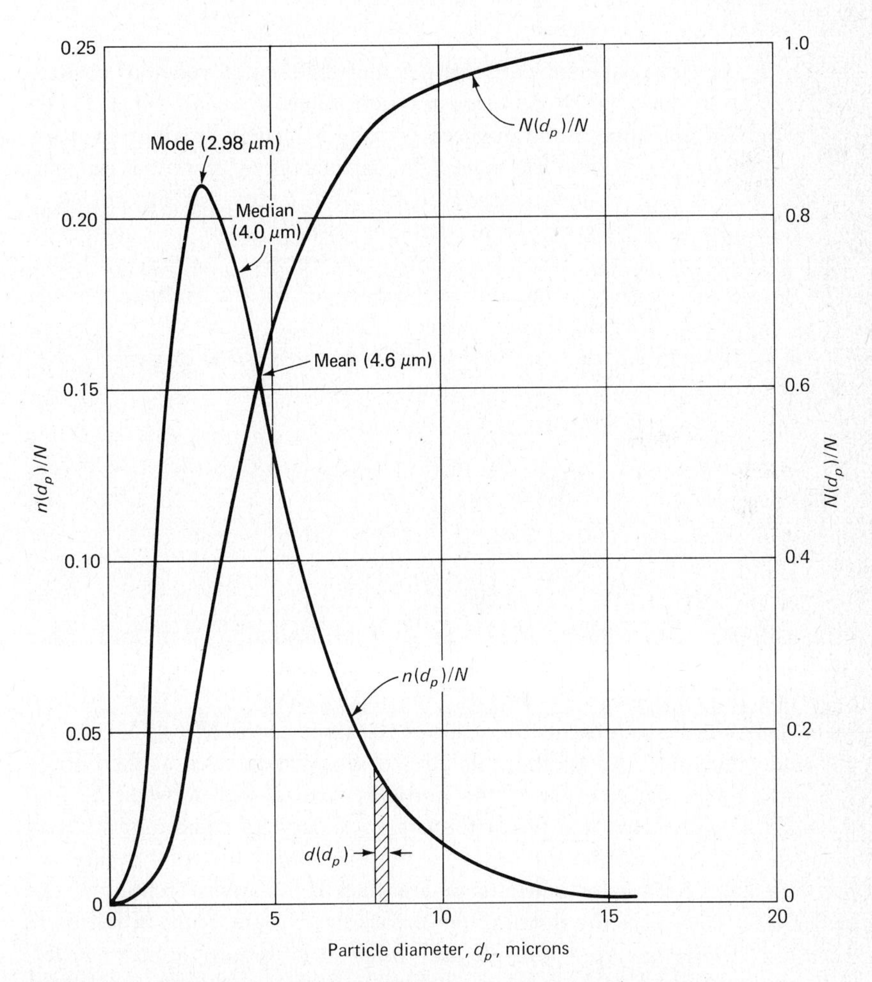

Figure 5-4 Hypothetical size distribution of a particulate sample.

Define $N(d_p)$ as the cumulative number distribution, that is, as the number of particles whose diameters are equal to or less than d_p, and N is the total number of particles of all sizes. A typical plot of $N(d_p)$ versus d_p is shown in Figure 5-4, where the curve has been normalized. That is, rather than plotting $N(d_p)$, the quantity $N(d_p)/N$ is plotted. As a result the ordinate runs from 0 to 1, rather than from 0 to N. The coordinate for d_p has been plotted in a linear fashion, and it is seen that nearly all the particles have a characteristic diameter less than 20 μm in this case. Although this distribution curve is of fundamental importance, the second curve shown in Figure 5-4 is equally important. It is determined in the following manner. Since $N(d_p)$ is a continuous function, it has a derivative $n(d_p)$ with respect to d_p which is

$$n(d_p) = \frac{dN(d_p)}{d(d_p)} \tag{5-2}$$

Thus $n(d_p)$ is the number of particles per unit differential range of particle size, that is, the number of particles per size range d_p to $d_p + d(d_p)$. The parameter $n(d_p)$ is quite useful in characterizing a particle distribution, since the quantity $n(d_p)/N$ is the fraction of the particles per differential range of particle size. A plot of $n(d_p)$ based on the $N(d_p)$ curve in Figure 5-4 is shown on the same figure. Note that the $n(d_p)$ curve has also been normalized in the figure. This plot of $n(d_p)/N$ versus d_p is very valuable, since this represents the particle distribution needed for the evaluation of the overall collection efficiency. Note, however, that the ordinate of the figure is plotted as the number fraction (or percent) per unit interval of particle size, and not number fraction or percent alone.

Obvious relations between $n(d_p)$ and $N(d_p)$ are

$$N(d_p) = \int_0^{d_p} n(d_p)\, d(d_p) \tag{5-3}$$

and

$$N = \int_0^{\infty} n(d_p)\, d(d_p) \tag{5-4}$$

[Although the upper limit of the integral in Equation (5-4) and other equations that follow is infinity for mathematical convenience, one recognizes that in practice there is a finite limit to the value of d_p.] Any point on the $n(d_p)/N$ curve represents the value of the slope of the $N(d_p)/N$ curve at the same value of d_p. In addition, the area of the vertical slice marked $d(d_p)$ is a measure of the fraction of the particles with a size between d_p and $d_p + d(d_p)$. The entire area under the $n(d_p)/N$ curve is equal to unity, and the area to the left of the slice at d_p is the fraction of the total number of particles with a diameter equal to or less than d_p. Various techniques are available to determine the distribution of particles by size. Some of these are discussed briefly in Appendix A, and other descriptions appear in the literature [7, 8].

The quantity $n(d_p)$ is used in Equation (5-4) to determine the number of particles N in a collection of particulate. As a more generalized problem, knowledge of $n(d_p)$ as a function of d_p is required to evaluate any property of the overall collection. Let Y be any property of interest, such as the volume or surface area of the particulate, and $Y(d_p)$ be the property value for a single particle. We assume $Y(d_p)$ to be continuously distributed. For all the particles in the range of size from d_p to $d_p + d(d_p)$ the value of Y is $Y(d_p)n(d_p)\,d(d_p)$. The total value of Y for the entire collection is found by integrating this expression over all particles sizes, that is,

$$Y = \int_0^\infty Y(d_p)n(d_p)\,d(d_p) \tag{5-5}$$

Equation (5-4) is just a special case of Equation (5-5), where $Y(d_p)$ is unity. We shall now use Equation (5-5) to develop a general expression for the overall collection efficiency by weight of any collector. If $m(d_p)$ is the mass of a particle of size between d_p and $d_p + d(d_p)$, then the total mass in this range of particle size is $m(d_p)n(d_p)\,d(d_p)$. The total mass in the collection of particles then is

$$m = \int_0^\infty m(d_p)n(d_p)\,d(d_p) \tag{5-6}$$

In addition, if the fractional collection efficiency in this range is η_d, then the mass collected in this size range is $\eta_d m(d_p)n(d_p)\,d(d_p)$. The total mass collected is the integral of this quantity from zero to infinity. When we divide the integral by the total mass, we obtain the overall collection efficiency based on mass, $\eta_{0,m}$. Thus

$$\eta_{0,m} = \frac{\int_0^\infty \eta_d m(d_p)n(d_p)\,d(d_p)}{\int_0^\infty m(d_p)n(d_p)\,d(d_p)} \tag{5-7}$$

This expression shows the direct need of expressions for η_d and $n(d_p)$ as a function of d_p. Once this information is available, Equation (5-7) can, in principle, be integrated and the overall collection efficiency determined for a particular dust and collector. Similar expressions for the overall efficiency can be developed based on number, surface area, and volume, for example. If the particle distribution is known for discrete intervals, Equation (5-7) becomes

$$\eta_{0,m} = \frac{\sum \eta_i m_i n_i}{\sum m_i n_i} \tag{5-8}$$

where the summation is taken over all discrete particle sizes. The quantity η_i is the average collection efficiency in the finite interval of particle size under consideration. The quantities m_i and n_i are the average mass of a particle and the total number of particles in the given size range, respectively. Equation (5-8) for the overall collection efficiency based on mass can also be written as

$$\eta_0 = \frac{\sum (\text{weight})_d \eta_d}{\sum (\text{weight})_d} = \sum (\text{weight fraction})(\eta_d) \qquad (5\text{-}9)$$

where $(\text{weight})_d$ is the weight of sample in the particle range around diameter d_p and η_d is the fractional collection efficiency in that size range.

Example 5-1

Assume the validity of the following size distribution and collection efficiency data.

PARTICLE RANGE, d_p (μm)	AVERAGE PARTICLE SIZE (μm)	WT. %	CUMULATIVE WT. %	COLLECTION EFFICIENCY
< 0.50	0.25	0.1	0.1	8
> 0.50– 1.5	1.0	0.4	0.5	30
> 1.5 – 2.5	2.0	9.5	10.0	47.5
> 2.5 – 3.5	3.0	20.0	30.0	60
> 3.5 – 4.5	4.0	20.0	50.0	68.5
> 4.5 – 5.5	5.0	15.0	65.0	75
> 5.5 – 6.5	6.0	11.0	76.0	81
> 6.5 – 7.5	7.0	8.0	84.0	86
> 7.5 – 8.5	8.0	5.5	89.5	89.5
> 8.5 –11.5	10.0	5.5	95.0	95
> 11.5 –16.5	14.0	4.0	99.0	98
> 16.5 –23.5	20.0	0.8	99.8	99
> 23.5		0.2	100.0	99 +

These are the data that were plotted previously on Figure 5-4. The collection efficiencies in the last column have been approximated from Figure 5-3. Determine the overall efficiency based on mass.

SOLUTION

Substituting the values from the third and fifth columns into Equation (5-9), we find that

$$\begin{aligned}\eta_{0,m} &= 0.001(8) + 0.004(30) + 0.095(47.5) + 0.2(6) + 0.2(68.5) \\ &\quad + 0.15(75) + 0.11(81) + 0.08(86) + 0.055(89.5) + 0.055(95) \\ &\quad + 0.04(98) + 0.008(99) + 0.002(100) \\ &= 72.44 \text{ percent}\end{aligned}$$

Thus for the specific particle-size distribution and collection efficiency, the particulate collector would remove about 72 percent by weight. To remove a larger percent of the total weight of the given particulate, one must select another collector which has larger fractional collection efficiencies at each of the given particle size range.

The numerical calculation of overall efficiency illustrates another point: two or more cleaning devices installed in series may be required to satisfy particulate emission standards. Such might be the case for coal-burning electric power plants where cyclone separators remove the larger particulate, followed by electrostatic precipitators which remove the very small particulate. The overall efficiency (expressed in percent) for series operation of two collectors is given by

$$\eta_0 = \eta_p + \frac{\eta_s(100 - \eta_p)}{100} \tag{5-10}$$

where η_p is the efficiency of the primary or first collector and η_s is the efficiency of the second collector based upon appropriate dust loading leaving the primary collector. Note that η_s is based on the size distribution entering the second collector, not on the original distribution entering the primary collector.

5-4 PARTICLE DISTRIBUTIONS

It is frequently useful to characterize an entire collection of particles by a single diameter. Typical diameters used for this purpose are the modal, median, and mean diameters. These quantities are shown in Figure 5-4. The modal diameter is defined as that diameter at which the greatest number of particles occur. This is found from the mathematical expression, $dn(d_p)/d(d_p) = 0$, and therefore is the maximum point on the $n(d_p)/N$ curve. Likewise, this is also the point of inflection on the $N(d_p)/N$ curve. Two median values for particle diameters which have proved useful in pollution studies are the number (count) median diameter and the mass median diameter. They are particularly useful when the particles are essentially spherical. The *number median* diameter d_{NM} is that diameter for which 50 percent of the particles are larger (smaller) than d_{NM} by count. Stated another way, the area under the $n(d_p)/N$ curve is the same on both sides of the vertical line representing the median value, with each area having a value of one-half. The *mass* (volume) *median* diameter d_{MM} is that diameter for which the mass of all particles larger than d_{MM} constitutes 50 percent of the total mass.

The third diameter shown on Figure 5-4 is the (arithmetic) mean diameter, d_{mean}. An arithmetic mean value, in general, is found by summing all values of the variable and then dividing the sum by the total number of samples. For a continuously distributed system, the mean diameter is given by

$$d_{\text{mean}} = \frac{1}{N}\int_0^\infty d_p n(d_p)\, d(d_p) \tag{5-11}$$

Mathematically, the centroid of the area under the $n(d_p)/N$ curve lies on the vertical line passing through d_{mean}. Like the median diameter, there are various values of the mean diameter, which depend on the type of diameter under consideration. Which value to be used normally depends upon the specific design situation under investigation. For a discretely distributed set of diameters, the mean diameter is found by direct summation of the number of particles $n_i(d_p)$ which have a diameter d_p. That is,

$$d_{\text{mean}} = \frac{1}{N} \sum n_i(d_p)\, d_p \tag{5-12}$$

where the summation is over all discrete particle sizes.

Thus the modal, median, and mean values of a particle distribution are three measures frequently used to indicate the general location of the distribution along the diameter coordinate of a plot similar to Figure 5-4. Once these values are established, it is also important to have some measure of the spread of the distribution. That is, how much of the area under the $n(d_p)/N$ curve is in the neighborhood of the mean value. One measure of the spread or dispersion is called the variance, σ^2. The positive square root of the variance is the standard deviation, σ. Mathematically speaking, the variance is given by

$$\sigma^2 = \frac{1}{N} \int_0^{\infty} (d_p - d_{\text{mean}})^2 n(d_p)\, d(d_p) \tag{5-13}$$

This quantity is also called the moment of inertia of the area under the $n(d_p)$ curve about the vertical line passing through d_{mean}.

The concepts of a mean diameter and the variance or standard deviation are quite useful in developing mathematical expressions for $n(d_p)/N$ as a function of d_p. Such functional relations could be used in Equation (5-7) to help evaluate the overall collection efficiency of a device, and thus are of considerable importance. In a number of industrial and natural phenomena the distribution of a random variable approaches that of a "normal" or "Gaussian" distribution. (See Chapter 4, Section 4-3.) If a collection of dust particles has a normal distribution, the frequency distribution curve would appear similar to that shown in Figure 5-5. The equation for the curve is given by

$$\frac{n(d_p)}{N} = \frac{1}{\sigma(2\pi)^{1/2}} \exp\left[\frac{-(d_p - d_m)^2}{2\sigma^2}\right] \tag{5-14}$$

where both the mean diameter d_m and the standard deviation σ appear. When a distribution is Gaussian, the modal, median, and mean diameters are identical. Although some collections of particulate have approximately a normal distribution of sizes, most do not. The more commonly found frequency distribution curve for particulate is shown by the $n(d_p)/N$ curve

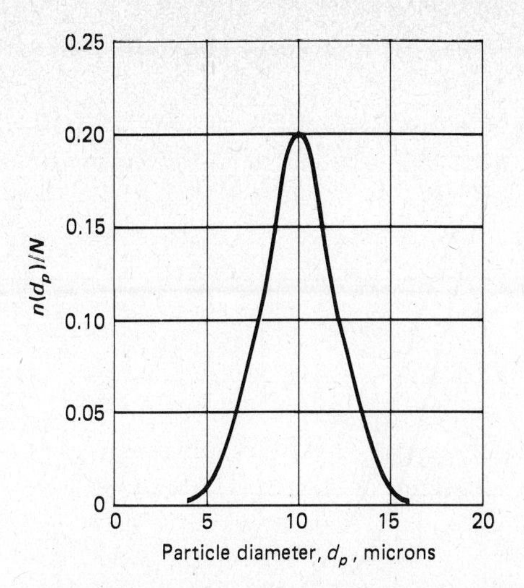

Figure 5-5 A normal distribution of particle sizes.

noted earlier in Figure 5-4. It is found by experience, however, that many particulate collections exhibit a normal distribution when the $n(d_p)/N$ data are plotted against $\ln d_p$ rather than d_p. Such collections are said to have a log-normal distribution.

When the $n(d_p)/N$ data of Figure 5-4 are replotted against the logarithm of the particle diameter, as shown in Figure 5-6, the shape of the curve more closely approximates that of a Gaussian distribution. Deviations from the Gaussian shape will occur when the experimental data are plotted, but the match is no worse than any other theoretical curve one might use. Now, let $w = \ln d_p$. Another way of presenting the data of Figure 5-6 is to plot $n(w)/N$ versus w. Note that this plot in Figure 5-6 also appears to have a normal distribution. The quantity $n(w)$ is defined as

$$n(w) = n(\ln d_p) = \frac{dN}{d(\ln d_p)}$$

That is, $n(w)$ is a measure of the number of particles per unit interval of size $d(\ln d_p)$. Recall that $n(d_p) = dN/d(d_p)$. The equation for the frequency distribution shown in Figure 5-6, in terms of w, is

$$\frac{n(w)}{N} = \frac{1}{\sigma_w(2\pi)^{1/2}} \exp\left[\frac{-(w - w_m)^2}{2\sigma_w^{\ 2}}\right] \tag{5-15}$$

where σ_w and w_m are the standard deviation and the mean value of w, respectively, of the $n(w)$ data shown in Figure 5-6. The relationship between

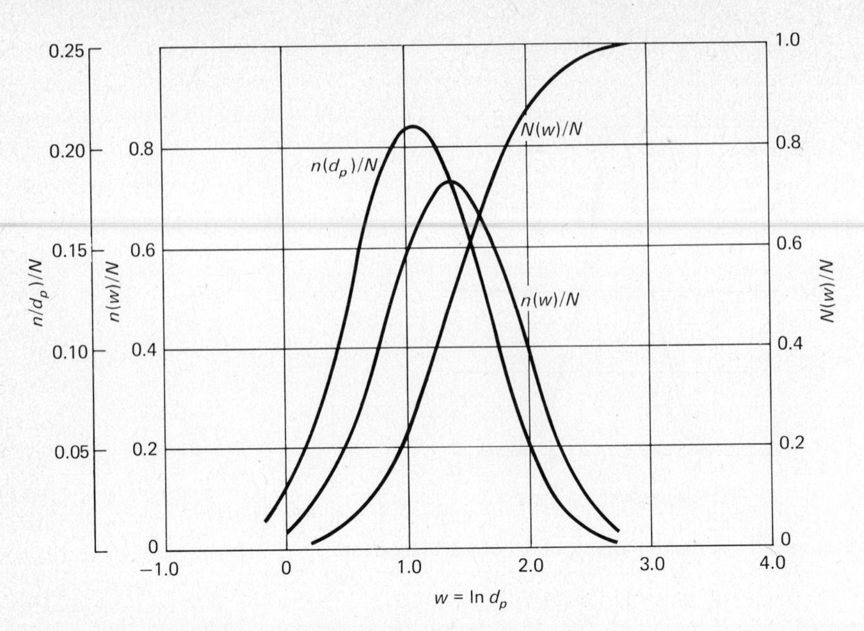

Figure 5-6 A log-normal distribution of particle sizes.

$n(d_p)$ and $n(w)$ is simply

$$\frac{n(d_p)}{N} = \frac{[n(w)/N]}{d_p} \tag{5-16}$$

In principle, Equation (5-15) could be used in conjunction with Equation (5-7) to ascertain the overall collection efficiency.

Of major importance is the fact that the distribution represented by Equation (5-15) will plot as a straight line on log-probability paper. Such a plot requires that the particle diameter be plotted on a logarithmic coordinate and the cumulative number percent be plotted on a probability coordinate. To determine the cumulative percent one uses the relation

$$\frac{N(w)}{N} = \frac{N(d_p)}{N} = \frac{1}{2} + \frac{1}{2}\text{erf}\left[\frac{\ln(d/d_g)}{2\ln\sigma_g}\right] \tag{5-17}$$

where $\sigma_g = \exp(\sigma_w)$ and $d_g = \exp(w_m)$. The diameter d_g is the geometric mean of the distribution, whereas d_m is the arithmetic mean diameter. The notation *erf* indicates the *error function*. The error function is tabulated in standard mathematical tables, and an abridged set of data appears in the appendix. A curve representing Equation (5-17) is also plotted on Figure 5-6. Figure 5-7 is the log-probability plot of the data used for Figure 5-6. The log-normal representation is useful since its linearity lends itself to easy interpolation of values, especially when only limited experimental data on the particle-size distribution are available. When read from the probability

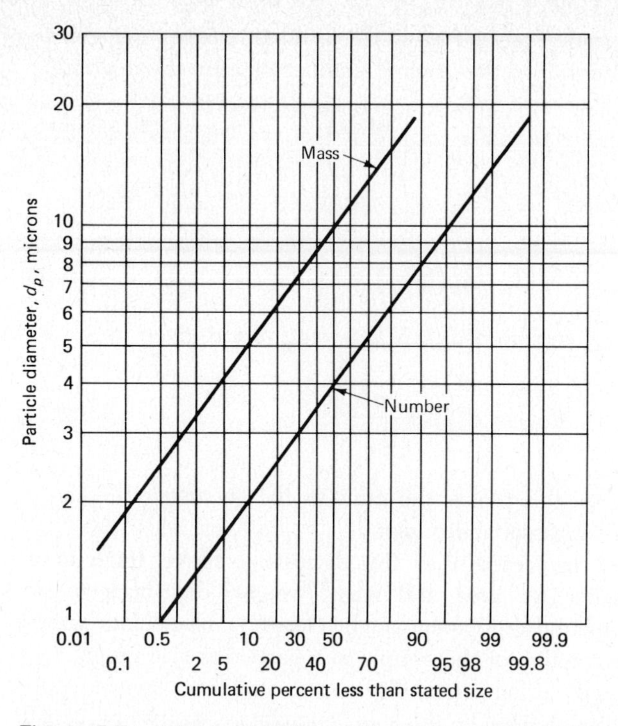

Figure 5-7 A log-probability plot of a hypothetical size distribution of a particulate sample.

plot, the value of d_p at the cumulative value of 50 percent is the geometric mean diameter d_g for that sample based on count. For the data of Figure 5-7, the geometric mean diameter is 4 μm. The slope of the line representing data on the log-probability plot is a measure of the geometric standard deviation, σ_g. It can be shown that the geometric standard deviation is found by

$$\sigma_g = \frac{d_{84.1}}{d_{50}} = \frac{d_{50}}{d_{15.9}} \tag{5-18}$$

where $d_{15.9}$, d_{50} and $d_{84.1}$ are the diameters on the cumulative curve where the cumulative percents are 15.9, 50, and 84.1, respectively. For the data of Figure 5-7, the value of σ_g is 1.72.

When only a limited amount of experimental data is available for a particulate sample, and the distribution is not normal, it is difficult to relate the particle diameters to the particle count in any simple manner. Experience shows, however, that many dust samples are log-normal. By fitting a straight line to the data as best as possible on a log-probability plot, the value of d_g and σ_g can be ascertained. (Note: it is quite typical that data at the low end and the high end of the diameter scale will deviate from a straight line drawn through the main body of the data.) It can be shown mathematically that these values of d_g and σ_g are related to the arithmetic mean diameter d_m and the standard deviation σ of the original number data by the relations

$$d_m = d_g \exp\left(\frac{\ln^2 \sigma_g}{2}\right) \tag{5-19}$$

and

$$\sigma = d_m\left[\exp\left(\ln^2 \sigma_g\right) - 1\right]^{1/2} \tag{5-20}$$

The modal diameter d_{mo} for the $n(d_p)/N$ versus d_p plot is given by

$$d_{mo} = d_g\left(1 + \frac{\sigma^2}{d_m{}^2}\right)^{-1} \tag{5-21}$$

The median diameter on this plot is the same as the geometric median d_g already found from the log-probability plot.

Finally, it should be noted that the discussion above for number distributions also applies for area and mass (weight) distributions. For example, some experimental equipment might classify a particulate sample according to the mass fraction of the sample in a particular size range. Such data can also be plotted on log-probability paper, and the data frequently can be fitted by a straight line. In this case cumulative weight percent is plotted, and d_g or d_{50} is the geometric mean diameter based on weight rather than on number. Generally d_g based on weight is larger than d_g based on number. In fact, they are related by the expression

$$\ln d_{g,\,\text{mass}} = \ln d_{g,\,\text{number}} + 3 \ln^2 \sigma_g \tag{5-22}$$

Both the number and mass distributions have the same geometric standard deviation σ_g. Therefore the lines representing the number and mass distributions on a log-probability plot have the same slope, and hence are parallel. The mass distribution is also shown on Figure 5-7, based on the same data as used in Figures 5-4, 5-6, and 5-7 for the number distribution. To make the conversion, spherical particle shape was assumed. The geometric mean diameter based on mass is roughly 9.7, rather than 4.0 based on number. The example below gives some of the particulars in the calculation of data for Figures 5-4, 5-6, and 5-7.

Example 5-2

A particulate sample is analyzed and found to have a number (count) distribution in various ranges of diameter as given in the first two columns of the table below. Determine values of $n(d_p)/N$ and $(mf)_i$ for the given ranges of size, and the values of the arithmetic mean diameter, the standard deviation, and the modal diameter.

SOLUTION

Column 3 data are obtained by summing the values in column 2. Then by plotting the data from columns 1 and 3 on log-probability paper (see Figure 5-7), it is estimated that $d_g = 4.0$ and $\sigma_g = 1.72$. On the basis of Equations (5-15) and (5-16) the corresponding values of $n(w)/N$ and $n(d_p)/N$ are calculated and tabulated in columns 4 and 5. These data are plotted also in Figure 5-6. The cumulative mass, plotted in Figure 5-7, is determined from the fact that, for spherical particles, the mass fraction is proportional to the number fraction in a given interval times the cube of the average diameter in that interval. Column 6 has data representing this proportionality, and column 7 has normalized these data to a percent basis. The cumulative weight percent, which is used in Figure 5-7, is not shown in the table. Finally, the arithmetic mean diameter, standard deviation, and modal diameter are calculated by Equations (5-19, (5-20), and (5-21).

$$d_m = 4.0\exp\left(\frac{0.294}{2}\right) = 4.63\ \mu\text{m}$$

$$\sigma = 4.63\left[\exp(0.294) - 1\right]^{1/2} = 2.71\ \mu\text{m}$$

$$d_{\text{mo}} = 4.0\left[1 + \left(\frac{2.71}{4.63}\right)^2\right]^{-1} = 2.98\ \mu\text{m}$$

d_p (μm)	$\Delta N(d_p)/N$	$N(d_p)/N$	$n(w)/N$	$n(d_p)/N$	$\sim (mf)_i$	$(mf)_i$
0–1	0.0053	0.0053	0.028	0.028	0	0
1–2	0.0957	0.101	0.325	0.1625	0.32	0.0014
2–3	0.197	0.298	0.639	0.213	3.76	0.0160
3–4	0.202	0.500	0.736	0.184	6.77	0.0289
4–5	0.160	0.660	0.676	0.135	14.58	0.0622
5–6	0.113	0.773	0.556	0.0927	18.80	0.0802
6–7	0.076	0.849	0.432	0.0617	20.87	0.0891
7–8	0.050	0.899	0.325	0.0406	21.09	0.0900
8–10	0.055	0.954	0.1765	0.0177	40.10	0.1711
10–14	0.0355	0.9895	0.0510	0.0036	61.34	0.2618
14–20	0.0095	0.999	0.0090	0.0005	46.67	0.1992
					234.3	

5-5 TERMINAL OR SETTLING VELOCITY

One basic method for removing particulates is simply by gravity settling. This technique is used both by nature and by the designers of industrial equipment. The important parameter that determines its usefulness is the terminal or settling velocity (speed) of a particle, V_t. This is defined as the constant downward speed that a particle attains in a direction parallel to the earth's gravity field as it overcomes the forces due to buoyancy and frictional drag. (The time to reach terminal speed is extremely short, and is usually neglected.) In terms of a force balance on a particle,

$$F_{\text{drag}} + F_{\text{buoyancy}} = F_{\text{gravity}}$$

The gravity force may be replaced by Newton's second law, namely, $F = ma = m_p g$. The buoyancy force is simply equal to the weight of the displaced fluid, which equals the volume of the fluid times its density and the local acceleration of gravity. The drag force is usually correlated to other physical variables through a drag coefficient, C_D. The overall force balance then becomes

$$\frac{\rho_g A V_t^{\,2} C_D}{2} + m_p\left(\frac{\rho_g}{\rho_p}\right)g = m_p g$$

where ρ_g is the fluid (atmospheric) density, ρ_p is the particulate (apparent) density, V_t is the terminal or settling velocity, m_p is the mass of particle, A is the frontal cross-sectional area, and g is the local acceleration of gravity. The general solution to this equation, in terms of V_t, is

$$V_t = \left[\frac{2m_p g(\rho_p - \rho_g)}{AC_D\rho_p\rho_g}\right]^{1/2} \tag{5-23}$$

The final evaluation of this expression requires some additional experimental evidence on C_D, as well as some appropriate approximations.

Some of the difficulties associated with Equation (5-23) are fairly apparent. The major difficulty is the determination of C_D. It is well established from fluid mechanics studies that the drag coefficient is strongly a function of the shape of the object. Particulate dust can take many shapes, not only as a result of its formation process but also through later agglomeration of particles in the exhaust dust or in the atmosphere. To simplify any initial study, it is appropriate to assume spherical particles. Any further improvement in the model will require additional experimental evidence. For spherical particles, Equation (5-23) becomes

$$V_t = \left[\frac{4gd_p(\rho_p - \rho_g)}{3\rho_g C_D}\right]^{1/2} \tag{5-24}$$

where d_p is the particle diameter. The drag coefficient for spherical particles is readily available from the literature, and it is usually plotted as a function of the dimensionless Reynolds number, Re. By definition, $\text{Re} = \rho_g V d/\mu$, where d is a characteristic length such as a diameter and μ is the fluid dynamic viscosity. Since the C_D-Re plot is an experimental correlation, exact analytical expressions relating the two variables are not possible except in restricted ranges of Re.

The region of Reynolds number between 10^{-4} and 0.5 is known as the Stokes or streamline flow regime. In this case experimental evidence for spheres indicates that

$$C_D = \frac{24}{\mathrm{Re}} \qquad \text{(streamline flow, spheres)} \tag{5-25}$$

Note also that in atmospheric air studies $\rho_p \gg \rho_g$. These two facts enable us to write Equation (5-24) for spherical particles in the form

$$V_t = \frac{g d_p{}^2 \rho_p}{18 \mu_g} \qquad \text{(streamline flow)} \tag{5-26}$$

This expression is known as Stokes' law. Generally speaking, this equation is quite accurate for spherical particles with diameters less than 50 μm and frequently is used with little error for particle sizes up to 100 μm. The lower limit of $\mathrm{Re} = 10^{-4}$ corresponds to a particle size of 5 μm, or somewhat less. This range of roughly 1 to 100 μm is an important size range for industrial dusts. For particle sizes less than about 5 μm, d_p approaches the mean free path of the gas molecules in the atmosphere. Particles now tend to slip past the gas molecules, and the settling velocity becomes greater than that predicted by Stokes' law. Below this lower limit it is necessary to apply Cunningham's correction to Equation (5-26). This correction factor, K_C, is given by [9],

$$K_C = 1 + \frac{2\lambda}{d_p}\left[1.257 + 0.400 \exp\left(\frac{-0.55 d_p}{\lambda}\right)\right] \tag{5-27}$$

where λ is the mean free path of the molecules in the gas phase. This quantity is given by

$$\lambda = \frac{\mu}{0.499 \rho u_m} \tag{5-28}$$

where u_m is the mean molecular speed, μ is the gas viscosity, and ρ is the gas density. From the kinetic theory of gases u_m is given by

$$u_m = \left[\frac{8 R_u T}{\pi M}\right]^{1/2} \tag{5-29}$$

where M is the molar mass or molecular weight of the gas. At 25°C and 1 atm the value of u_m is 467 m/s (1530 ft/s) and λ is 0.067 μm. At this same pressure and for spherical particles with a diameter greater than 1 μm, the correction, K_C, is reasonably approximated by

$$K_C = 1 + \frac{9.73 \times 10^{-3} T^{1/2}}{d_p} \tag{5-30}$$

where d_p is expressed in micrometers and T in degrees Kelvin. On the basis of Equation (5-27), the Cunningham factor increases rapidly as the particle size decreases and approaches the mean free path. When d_p is 10 μm, the correction is less than 2 percent. However, for a 1 μm-diameter particle the value of V_t will be over 15 percent greater than the Stokes' law value. The relationship of V_t at these small sizes is

$$V_t = K_C V_{t,\text{ Stokes}} \tag{5-31}$$

and it is known as the Stokes-Cunningham law.

In the transition region between laminar and turbulent flow around a spherical particle the standard drag coefficient curve can be modeled by the expression

$$C_D = 0.22 + \frac{24}{\text{Re}}\left[1 + 0.15(\text{Re})^{0.6}\right] \tag{5-32}$$

This equation for flow around spheres is valid for Reynolds numbers from 1 to 500. When the Reynolds number is greater than 0.5 (or the particle size appreciably greater than 50 μm), it is convenient to use the experimental C_D-Re data directly to evaluate V_t versus d_p at various particle densities, and then present the results graphically. This is done in Figure 5-8, which covers the particle-size range from 0.1 to 3000 μm on a log-log plot. The terminal or settling velocity varies between 10^{-3} to 10^3 cm/s. For particle sizes above 30 μm (or for Reynolds numbers greater than roughly 0.1) four curves are shown, for particle densities of 1.0, 1.5, 2.0, and 3.0 g/cm^3. (For these curves, the left and upper axes must be used.) The single line to the right represents the Stokes and Stokes-Cunningham regions of flow. Note that the correlating line is essentially straight between V_t values of 5×10^{-3} and 1 cm/s, as predicted by Stokes' law for a log-log plot. Below 5×10^{-3} cm/s the line curves as a result of the Cunningham correction. This single line is valid only for spherical particles with an apparent density of 1 g/cm^3. From Equation (5-26) we note that V_t is proportional to ρ. Thus the terminal velocity in the Stokes region for other particle densities may be found by the relation

$$V_t(\text{at } \rho_p) = V_t\left(\text{at } \rho_p \text{ of } 1 \text{ g/cm}^3\right) \times \rho_p \tag{5-33}$$

where the first term on the right is found directly from Figure 5-8. It should be noted that Figure 5-8 is valid only for spherical particles settling in air at room temperature and pressure.

Example 5-3

To illustrate the evaluation of terminal velocities, assume a particle with unit density (1 g/cm^3) and a diameter of (a) 10, (b) 100, and (c) 1000 μm. Estimate the terminal velocity in room air in centimeters per second. Also estimate the value for the 10-μm particle when its density is 2 g/cm^3.

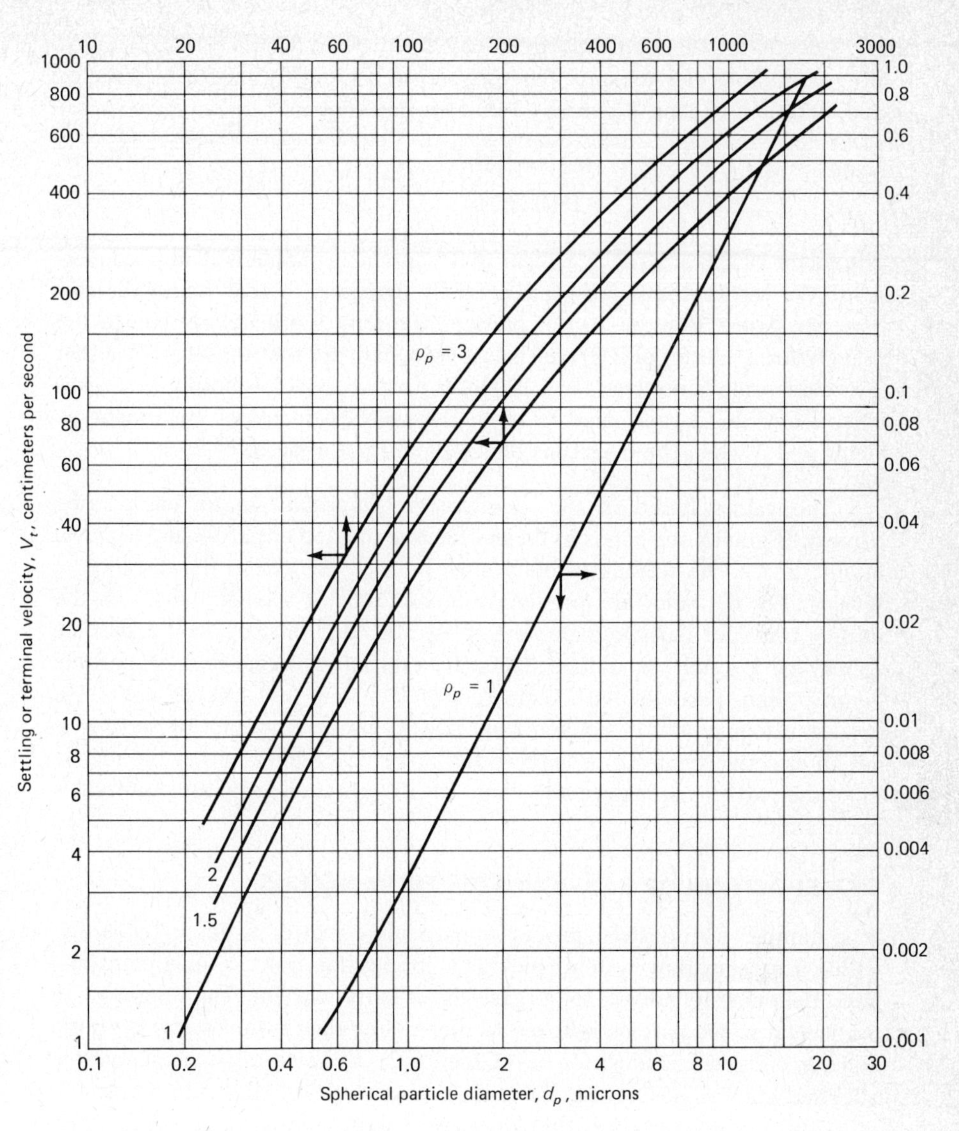

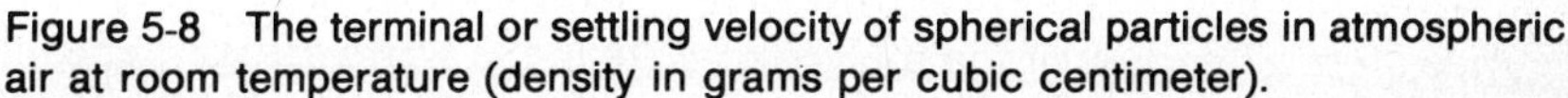

Figure 5-8 The terminal or settling velocity of spherical particles in atmospheric air at room temperature (density in grams per cubic centimeter).

SOLUTION

The terminal velocities can be read directly from Figure 5-8 in the required units. From this figure, as an estimate,

$$
\begin{aligned}
d_p &= 10\ \mu\text{m}, & V_t &= 0.3\ \text{cm/s} \\
&= 100\ \mu\text{m}, & &= 25.5\ \text{cm/s} \\
&= 1000\ \mu\text{m}, & &= 400\ \text{cm/s}
\end{aligned}
$$

The velocity for the 10-μm particle may also be found from Stokes' law. At room conditions the dynamic viscosity of air is roughly 0.045 lb/ft·hr or 0.067 kg/m·hr. Using Equation (5-26), we find that

$$V_t = \frac{gd_p{}^2\rho_p}{18\mu_g} = \frac{9.8(10)^2(1)}{18(0.067)}(3.6\times10^{-4}) = 0.29 \text{ cm/s}$$

The factor of 3.6×10^{-4} is a term that contains a number of unit conversions. We see that this answer essentially agrees with that found directly from the figure. For the 10-μm particle, the flow is streamline. Hence the velocity for a density of 2 g/cm^3 is just twice that for unit density. Note that this relationship is not true for a 1000-μm particle, which is not in the Stokes region. For this larger sized particle the velocity increases by roughly a factor of 1.5 when the density is doubled, and not a factor of 2.

The values found in this example are important in an engineering context. If gravity settling is to be an effective method of particulate removal in any device, the settling velocity must be high relative to the distance of settling. For the velocities previously found, and for a height of 10 m, the settling times for particle diameters of 10, 100, and 1000 μm and a particle density of 1 g/cm^3 would be 3300, 40, and 2.5 s. Regardless of particle density, then, particles with diameters of less than 100 μm do not settle quickly in equipment of conventional size. Consequently, gravity settling is not an effective removal mechanism for particulates unless either the majority of the particles are reasonably large ($>$ 100 μm) or a large time interval is permissible.

5-6 DEPOSITION OF PARTICULATES FROM STACKS

In a number of industrial processes particulate matter is generated either within a gaseous flow system (such as a combustion process) or a plant and then deliberately added to a gaseous exhaust system. The quantity of particulate which may be released to the environment is limited by law for a number of specific industries (see Chapter 2). In these cases, some type of effective control device usually must be inserted into the flow stream to maintain emission levels within prescribed limits. Even with such control devices, a small portion of the total particulate in the waste gas will probably be emitted into the atmosphere. In uncontrolled industries, of course, the mass of solid matter released can be extremely large. In order to plan emission standards it is important to be able to predict the deposition rate of particulate at the ground level at various distances from the source.

There are several approaches to predicting the particulate deposition rate that results from stack emissions. Probably the most conventional of these is a modification of the Sutton equation used for gaseous emissions from stacks (see Chapter 4). This method assumes a Gaussian-type distribution of the pollutant in the y- and z-coordinate directions. The mass diffusion coefficients in these two coordinate directions are correlated in terms of standard deviations σ_y and σ_z, which in turn are functions of the spatial coordinates as well as the atmospheric stability. It was shown in Chapter 4 that an appropriate expression for gaseous dispersion from a stack of equivalent height H and without reflection is

$$C(x, y, z, H) = \frac{Q}{2\pi u \sigma_y \sigma_z} \exp\left\{ -\frac{1}{2}\left[\left(\frac{y}{\sigma_y}\right)^2 + \frac{(z-H)^2}{\sigma_z{}^2}\right]\right\} \tag{4-11}$$

where u is the average wind speed (usually taken at the top of the stack) and Q is the emission rate of the gaseous pollutant, mass per unit time. As we mentioned before, the expression "without reflection" is quite important, since for particulate emissions the ground acts as a sink for the emitted particles. For gaseous emissions the ground acts as a reflective surface in most cases, and this reflection enhances the gas phase concentration of the pollutant.

In addition to the nonreflective correction, one other major modification of the Gaussian-type dispersion equation must be made. Gaseous pollutant species are unaffected by gravity force. However, the motion of solid particles is strongly affected by gravity as well as by the fluid forces associated with atmospheric movement. Consequently, the dispersion equation must reflect these forces. In essence, the effect of gravity on particulate dispersion is to make the center line for emission appear to slope downward as the pollutant stream passes farther downstream. In comparison to the fairly horizontal plume center line for gaseous emissions, the particulate plume appears to tilt downward.

As a consequence of gravity, the H quantity in the Gaussian dispersion equation must be corrected for the general settling of the particulate matter. The free-fall distance after leaving the stack of a particle with a terminal speed of V_t is simply $V_t t$, where t is the time for the main pollutant stream to reach a distance x downwind. The time t is also given by x/u. Hence the free-fall distance by which H must be corrected is $V_t x/u$. This is equivalent to shifting the center line of the Gaussian distribution in the z-direction downward by the amount $V_t x/u$. By accounting for the gravitational effect in this manner, the particulate concentration in general is given by

$$C = \frac{Q_p}{2\pi u \sigma_y \sigma_z} \exp\left[-\frac{1}{2}\left\{\left(\frac{y}{\sigma_y}\right)^2 + \left[\frac{z-(H-V_t x/u)}{\sigma_z}\right]^2\right\}\right] \tag{5-34}$$

Of particular interest is the ground-level concentration along the center line. By setting $y = 0$ and $z = 0$, we find that

$$C(x, 0, 0, H) = \frac{Q_p}{2\pi u \sigma_y \sigma_z} \exp\left\{ -\frac{1}{2}\left[\frac{H-(V_t x/u)}{\sigma_z}\right]^2\right\} \tag{5-35}$$

where Q_p is the particulate emission rate. Typically, Q_p may be expressed in such units as grams per second, with σ_y and σ_z in meters and u in meters per second. Q_p refers to one specific particle size with a terminal speed V_t.

Rather than evaluate the concentration C of particulate at a given location, as given by Equation (5-35), it is more appropriate to express the results of particulate dispersion in terms of the mass deposited per unit time and unit area, w. The relation between C and w is indicated below.

$$w = \frac{\text{mass rate of transport}}{\text{area}} = \frac{(\text{volume rate})(\text{concentration})}{\text{area}}$$
$$= (\text{velocity})(\text{concentration}) = V_t C$$

Consequently, the deposition rate of particulate at ground level along the center line of the stack is estimated by

$$w = \frac{Q_p V_t}{2\pi u \sigma_y \sigma_z} \exp\left\{ -\frac{1}{2}\left[\frac{H - (xV_t/u)}{\sigma_z} \right]^2 \right\} \tag{5-36}$$

where typically Q_p is in grams per second, V_t and u are in meters per second, σ_y and σ_z are in meters, and w is in $\text{g/m}^2 \cdot \text{s}$.

Since the terminal speed V_t appears in Equation (5-36), it is apparent that the value of w is the deposition rate of a particle of a given density and mean diameter. To obtain a true indication of the total deposition rate from a stack, the total emission rate Q_p must be weighted according to the mass fractions of particles within various size classes.

Another formulation for predicting dust deposition from stacks has been developed by Bosanquet et al. [10]; this formulation has been checked experimentally against data for power plants in England. Along the axis of the plume, the deposition rate based on Bosanquet et al.'s findings is given by

$$w = \frac{16.4 Q_p}{H^2} F\left(\frac{V_t}{u}, \frac{x}{H} \right) \tag{5-37}$$

where Q_p is expressed in milligrams per second, H in meters, and w in $\text{mg/m}^2 \cdot \text{hr}$. The value of the function F is obtained from a graph with the dimensionless parameter x/H as the abscissa and the dimensionless parameter V_t/u as a series of parametric lines. This graph appears as Figure 5-9. The value of the function F varies roughly from 0.5 to 50, with the upper range occurring for large values of V_t and small values of x.

If uniform particles are being emitted from a stack of equivalent height H into a wind of known speed u, then Equation (5-37) enables us to calculate the axial rate of deposition directly. More generally, the dust is composed of particles that range in size up to several hundred microns. In this case it is suggested that a mean value of F be estimated for use in

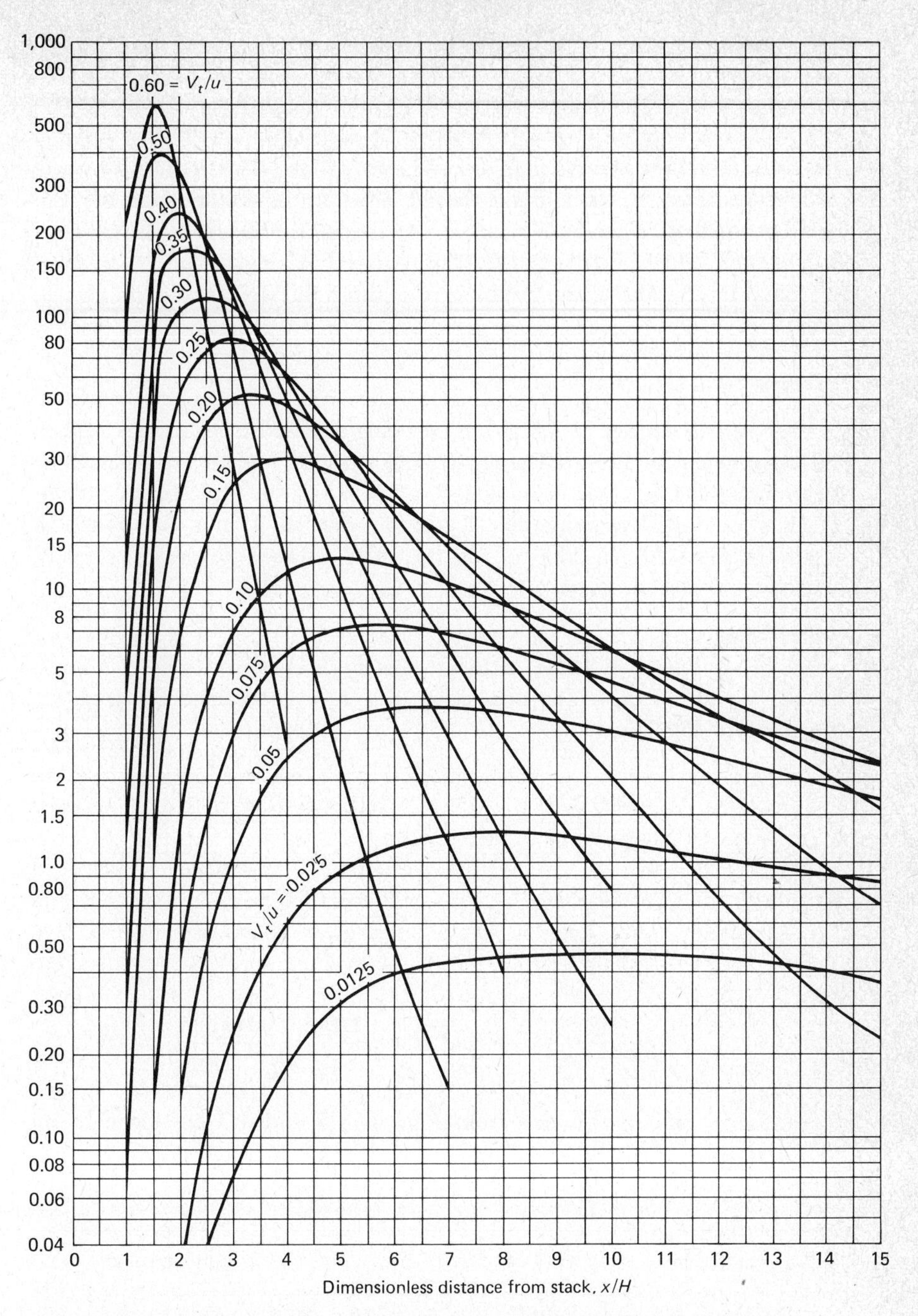

Figure 5-9 Variation of rate of deposition of particulate with distance from stacks for various particle sizes. (SOURCE: C. H. Bosanquet, W. F. Carey, and E. M. Halton. "Dust Deposition from Chimney Stacks." *Proc. Inst. Mech. Engineers* **162** (1950): 355.)

Equation (5-37). This is done by dividing the overall range of particle sizes into various groups or grades. It might be appropriate to divide, for example, the overall particle size range into six or eight groups. The fraction by weight of the particles in each group would need to be determined by some standard method. Then an F_i value would be determined for each size group for selected values of u, x, and H. The required mean value of F then would be found by the sum of the products of F_i times the weight fraction for each group.

Example 5-4

A dust with a density of 1.5 g/cm^3 is being released from a stack with an effective height of 120 m. The particulate emission rate for particles 40 μm in diameter is 4 g/s. The wind speed is 3 m/s and the atmospheric stability is class D. (a) Determine the deposition rate in g/m^2·s for downwind distances from 200 to 5000 m. (b) At what distance downwind does the maximum deposition occur?

SOLUTION

(a) From Figure 5-8 the settling velocity is approximately 7.3 cm/s. Hence $V_t = 0.073$ m/s. The deposition in g/m^2·s is obtained by employing Equation (5-36). Thus

$$w = \frac{4(0.073)}{(2\pi)(3)\sigma_y\sigma_z}\exp\left\{-\frac{1}{2}\left[\frac{120-(0.073x/3)}{\sigma_z}\right]^2\right\}$$

$$= \frac{1.55\times10^{-2}}{\sigma_y\sigma_z}\exp\left[-\frac{(120-0.0243x)^2}{2\sigma_z^{\ 2}}\right]$$

Let

$$H = \frac{1.55\times10^{-2}}{\sigma_y\sigma_z} \text{ and } G = \frac{(120-0.0243x)^2}{2\sigma_z^{\ 2}}$$

Values of σ_y and σ_z obtained from Figures 4-6 and 4-7 are given in the tabulation.

x (m)	σ_y (m)	σ_z (m)	H	G	e^{-G}	w (μg/m$^2\cdot$s)
200	18	8.5	1.01×10^{-4}	91.7	1.50×10^{-40}	1.52×10^{-38}
500	40	19	2.04×10^{-5}	16.1	1.01×10^{-7}	2.06×10^{-6}
1000	75	31	5.67×10^{-6}	4.77	8.52×10^{-3}	4.83×10^{-2}
1500	110	40	3.52×10^{-6}	2.18	1.13×10^{-1}	3.98×10^{-1}
2000	160	55	1.76×10^{-6}	0.843	4.31×10^{-1}	7.59×10^{-1}
3000	210	70	1.05×10^{-6}	0.226	7.97×10^{-1}	8.37×10^{-1}
4000	290	84	6.36×10^{-7}	0.037	9.64×10^{-1}	6.13×10^{-1}
5000	350	100	4.43×10^{-7}	0	1.0	4.43×10^{-1}

(b) From the table it can be seen that the maximum deposition rate occurs at a distance of approximately 3000 m downwind from the stack.

Example 5-5

If the average wind direction and wind speed are constant for a day, what is the maximum daily deposition rate for the conditions of Example 5-4?

SOLUTION

For the purpose of evaluation, we assume that the maximum deposition does occur at 3000 m downwind. On this basis,

$$\text{daily deposition} = \frac{8.37 \times 10^{-1}\,\text{g}}{\text{m}^2\cdot\text{s}}\left(\frac{8.64 \times 10^{4}\,\text{s}}{\text{day}}\right) = 72{,}400\ \text{g/m}^2$$

This is a significant quantity and would result in an easily visible layer on a smooth surface.

5-7 HOOD AND DUCT DESIGN

In general, most air pollution control equipment is more efficient when handling higher concentrations of contaminants, all else being equal. Therefore the gas handling system should be designed to concentrate contaminants in the smallest possible volume of air. This is important since, exclusive of the blower, the cost of control equipment is based principally upon the volume of gas to be handled and not on the quantity of particulate to be removed. The reduction of air polluting emissions by process and system control is an important adjunct to gas cleaning technology.

In designing local exhaust hoods an attempt is made to create a controlled air velocity that will prevent the escape of contaminants from the controlled area to the general environment. The air velocity that will just overcome the dispersive motions of the contaminant, plus a suitable safety factor, is termed the "control velocity." The control velocity is adjusted to obtain the least air flow rate that gives satisfactory control results for minimum gas volume and maximum contaminant loading. Optimum control velocities depend upon the size and shape of the hood, its position relative to the points of emission, and the nature and quantity of the air contaminants. The rate of exhaust ventilation must exceed the rate of air displacement created by the process being controlled or otherwise induced around it.

The velocity patterns for air entering a circular duct representing a simple hood are not radial lines but are modified as illustrated in Figure 5-10. The lines of constant velocity are given in percentage of the duct velocity. An equation giving the approximate velocity, developed by Dalle Valle [11], is

$$\frac{y_x}{100 - y_x} = \frac{0.1A}{x^2} \tag{5-38}$$

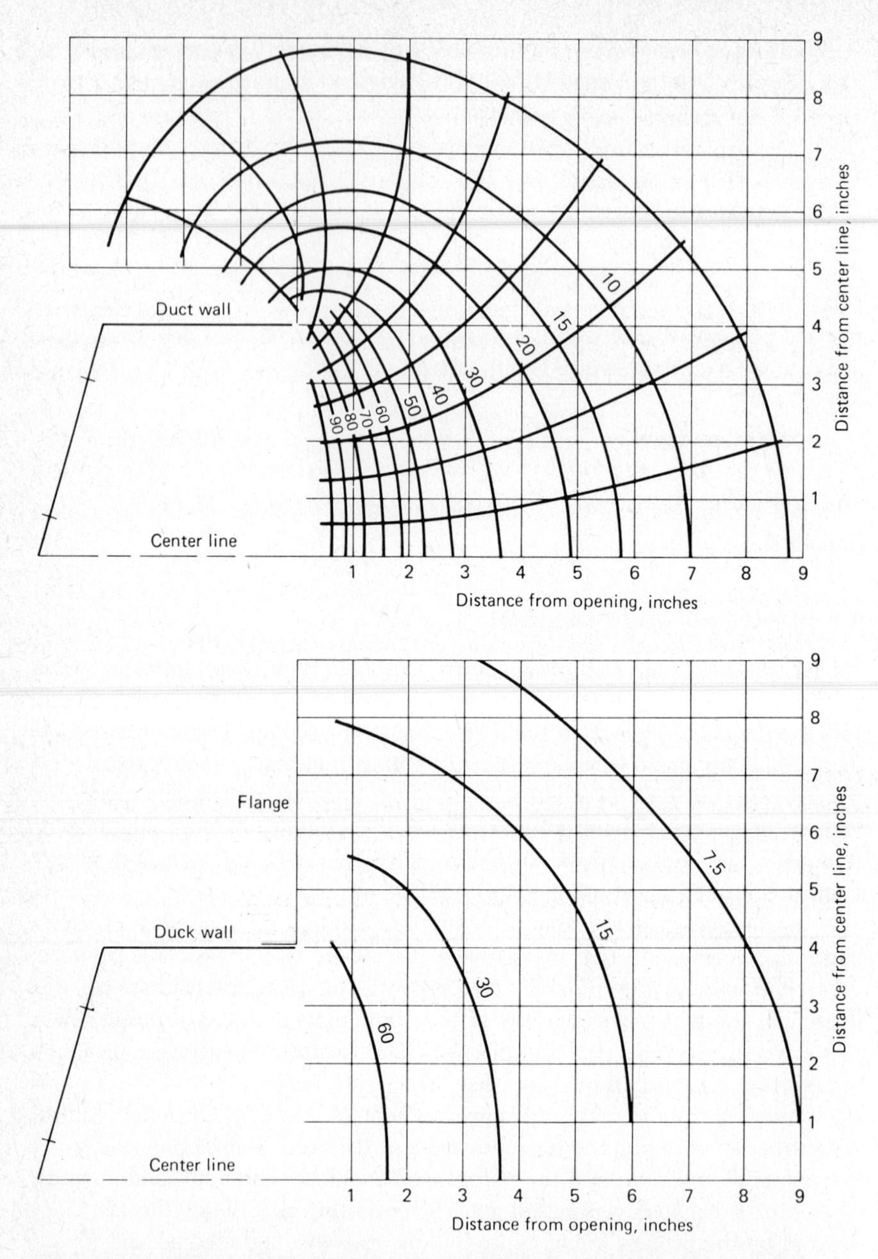

Figure 5-10 Actual flow contours and streamlines for flow into circular openings. Contours are expressed as percentage of opening velocity. (SOURCE: NAPCA. *Air Pollution Engineering Manual*, No. 999-AP-40. Washington, D.C.: HEW, 1967.)

where y_x is a percentage of duct velocity at a point x on the duct axis; x is the distance outward along the axis from the opening, in feet; and A is the area of the opening, in square feet.

For square, round, and rectangular openings (3 : 1 length to width ratio), Hemeon [12] recommends the following equation when the flow into the hood is unrestricted.

$$Q_h = V_x(10x^2 + A_h) \tag{5-39a}$$

where Q_h is the volume of air entering the hood, in cubic feet per minute; V_x is the air velocity at point x, in feet per minute; x is the distance to any point measured along the duct axis; and A_h is the area of the hood face, in square feet.

When rectangular hoods are bounded on one side by a plane surface such as the floor, the hood is considered to be twice its actual size by the addition of a mirror image. Equation (5-39a) becomes

$$Q_h = V_x\left(\frac{10_x{}^2 + 2A_h}{2}\right) \tag{5-39b}$$

Air contaminants may be released into the environment with considerable velocity at their point of origin. Since the mass is normally small, the momentum is soon lost and the particles are easily captured. The general position of the particle when its original velocity has been reduced to approximately zero is termed the *null point*, as illustrated in Figure 5-11. If an adequate velocity toward a hood is provided at the most distance null point from the hood, most of the air contaminants will be captured. What constitutes an adequate control velocity toward the hood depends upon drafts in the area. Experience has shown that a velocity of less than 100 ft/min (30 m/min) at a null point can seldom be tolerated without a loss in the hood's effectiveness.

Only the basic concepts of hood design have been presented here. An extensive treatment of the subject may be found in reference 2. The fans, blowers, and compressors required to move the gases through the ventilating system are discussed in the literature [2, 13].

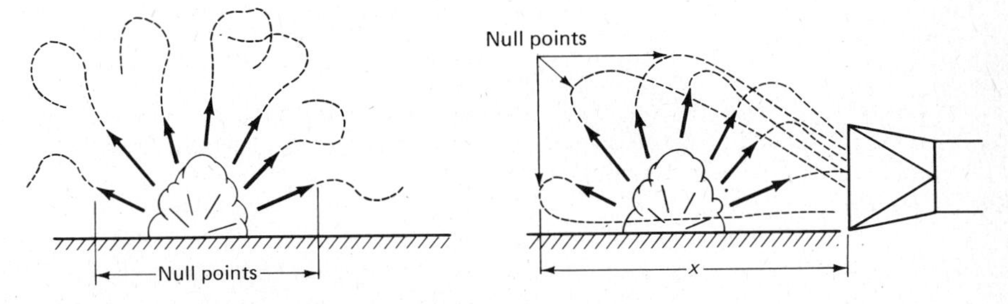

Figure 5-11 Illustration of null points.

5-8 PARTICULATE COLLECTION MECHANISMS

The six available mechanisms for collecting particles may be classified as gravitational settling, centrifugal impaction, inertial impaction, direct interception, diffusion, and the electrostatic effects. One or more of these mechanisms are responsible for the removal of particulate in any of the industrial collection devices discussed in the following sections. The phenomena involving gravitational, centrifugal, and electrostatic forces are well known to engineers and scientists. However, it may be appropriate at this point to distinguish the subtle differences among inertial impaction, direct interception, and diffusion. Figure 5-12 illustrates simple models of the three mechanisms. Particles are carried along at approximately the same velocity as the main gas stream. Owing to its extreme lightness, the gas moves in streamlines around any object in its path. However, solid particulate with a much heavier mass resists changes in motion. The larger the particle, the less its tendency to change direction. Inertial impaction, shown in Figure 5-12(a), is associated with the relatively larger particles which travel on a collision

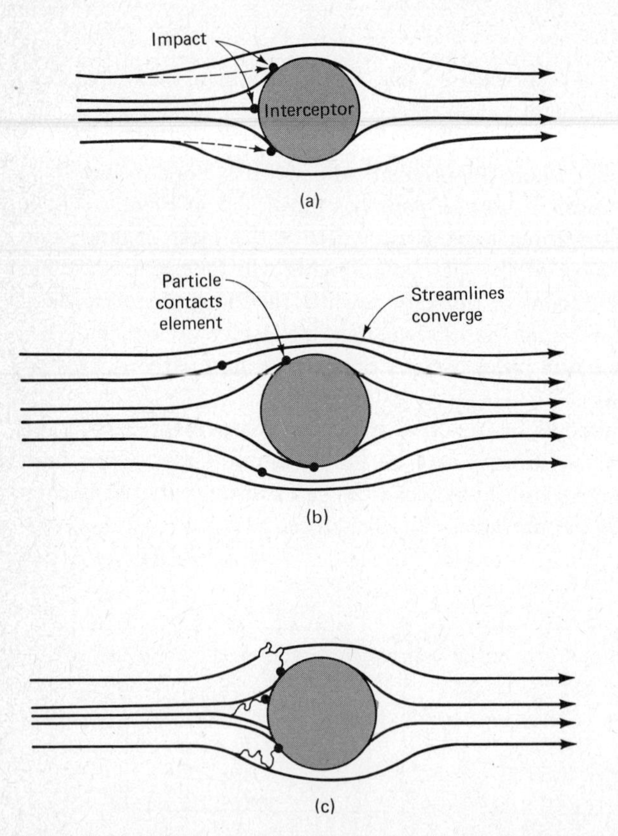

Figure 5-12 Three mechanisms for mechanical removal of particulates: (a) inertial impaction; (b) direct interception; (c) diffusion.

course with the interceptor. Inertia keeps them on this path, even though the gas and the smaller particles tend to diverge and pass around the interceptor.

In direct interception, illustrated in Figure 5-12(b), some of the smaller particles, even though they tend to follow the streamlines, may contact the interceptor at the point of closest approach. This occurs because the streamlines tend to converge as the gas passes around the element, and the particle radius is greater than the distance between the streamline and the element. Finally, collection by diffusion is shown in Figure 5-12(c). In this case, very small particles (usually less than 1 μm) impinge upon the collector as a result of random molecular (Brownian) motion or diffusion. It may be necessary to coat the inceptor with some substance that improves the adhesion, or the dust particles may bounce off upon impact and return to the gas stream. Coating also reduces the likelihood that particles which hit the target later will knock off already collected particles.

Agglomeration is another mechanism to improve collection efficiency. With this technique, the average particle size is increased, so that collectors downstream will have a better chance of removing the pollutant. Particles sometimes agglomerate simply by mutual intermolecular attraction, but often external inducement by electrostatic or ultrasonic devices may be required to produce the necessary agglomeration. One or more of these mechanisms are responsible for the removal of particulate in any of the industrial collection devices discussed in the following sections. More extensive treatments of gas cleaning devices for particulates are presented in the literature [2, 5, 6, 13].

5-9 PARTICULATE CONTROL EQUIPMENT

A number of factors must be determined before a proper choice of collection equipment can be made. Among the most important data required are the following: the physical and chemical properties of the particulate; the range of the volumetric flow rate of the gas stream; the range of expected particulate concentrations (dust loadings); the temperature and pressure of the flow stream; the humidity; the nature of the gas phase (such as corrosive and solubility characteristics); and the required condition of the treated effluent. The last piece of information may be the most important, for it indicates the collection efficiency that must be met, either by a single piece of equipment or several operating in series. In many cases, the above considerations restrict the engineer to one or two basic types of equipment.

Following are the five basic classes of particulate collection equipment:

1. Gravity settling chambers
2. Cyclone (centrifugal) separators
3. Wet collectors
4. Fabric filters
5. Electrostatic precipitators

Within a class, devices may bear different descriptive names based on their individual operating and construction differences. Detailed descriptions of the following devices are readily available in the literature as well as from equipment manufacturers.

5-9-A Gravity Settling Chamber

Gravitational force may be employed to remove particulate in settling chambers when the settling velocity is greater than about 25 ft/min (13 cm/s). In general, this applies to particles larger than 50 μm if the particle density is low, down to 10 μm if the material is reasonably dense. Smaller particles than this would require excessive horizontal flow distances, which would lead to excessive chamber volumes. One possible configuration of a gravity settling chamber is shown in Figure 5-13. For a gravity chamber to be effective in preventing reentrainment of the settled particles, the gas velocity must be uniform and relatively low—certainly less than 10 ft/s (300 cm/s) and preferably less than 1 ft/s (30 cm/s).

Theoretically, the minimum particle size that can be removed with 100 percent efficiency can be ascertained in the following manner. With reference to Figure 5-13, the time required for a particle of size d_p to fall a distance H, and hence escape from the gas stream, must be equal to or less than the time required to move horizontally a distance L. For the minimum particle size that can be removed 100 percent, these two times are equal. Hence for uniform duct flow with no macroscopic mixing,

$$t = \frac{H}{V_t} = \frac{L}{V} \tag{5-40}$$

where it is assumed that all particles pass through the chamber with the gas velocity V. To introduce the particle diameter d_p, it is now necessary to relate d_p to the settling speed V_t. Such relationships were presented in Section 5-5 for various flow regimes. A general solution for spherical par-

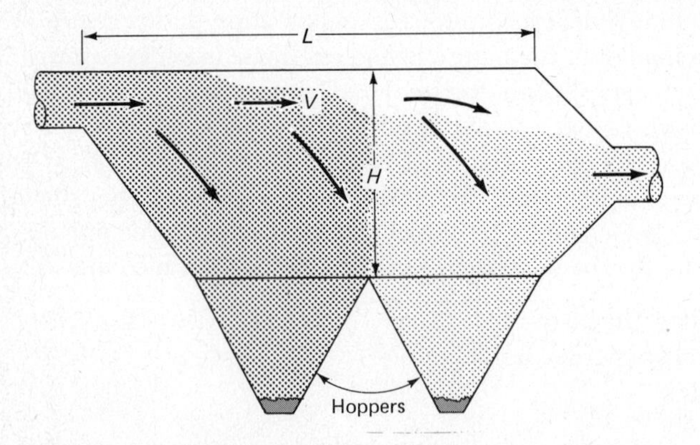

Figure 5-13 A gravity settling chamber.

ticles at room temperature is provided by Figure 5-8. Thus,

$$V_t = f(d_p) = \frac{VH}{L} = \frac{Q}{LW} \tag{5-41}$$

where Q is the volume flow rate and W is the width of the chamber. From a knowledge of V, H, L, Q, and W one calculates V_t. This latter value is then used in conjunction with Figure 5-8 (or the more general relations for V_t) to find the particle size with this terminal speed. This value of d_p should be a rough estimate of the minimum particle size retained with 100 percent collection efficiency.

As a first approximation in estimating the minimum particle size with 100 percent separation, it is frequently acceptable to employ Stokes' law, as given by Equation (5-26). (Recall that Stokes' law normally is not valid when d_p is more than about 50 μm.) Substitution of Equation (5-26) into Equation (5-41) leads to

$$d_{p,\min} = \left(\frac{18\mu HV}{gL\rho_p}\right)^{1/2} \tag{5-42}$$

In this expression the effect of the gas density, ρ_g, has been neglected. As usual, a consistent set of units must be used.

From Equation (5-42) we see that the minimum particle size removed with 100 percent efficiency is made smaller by geometrically reducing the value of H/L. Rather than a single low and long chamber, a reasonably high, short chamber can be used by installing horizontal baffle plates across the chamber, spaced several inches apart. Commercially available, such a device is really a number of gravity settling chambers in parallel. Although more efficient, it is more costly to construct and more difficult to clean. For particle sizes below that given by Equation (5-41) or (5-42), the fractional collection efficiency for a given particle size d_p can be estimated for uniform duct flow by the relation

$$\eta_d = \frac{V_t L}{HV}(100) = \frac{V_t LW}{Q}(100) \tag{5-43}$$

where η_d is in percent. For settling chambers with horizontal baffles the equation corresponding to Equation (5-41) for the minimum particle size with 100 percent collection would be

$$V_t = \frac{VH}{nL} = \frac{Q}{nLW} \tag{5-44}$$

where n is the number of flow channels, or the number of horizontal baffles plus the bottom surface. In this case the corresponding fractional efficiency equation is

$$\eta_d = \frac{nV_t L}{HV}(100) = \frac{nV_t LW}{Q}(100) \tag{5-45}$$

Turbulence within the settling chamber causes deviations in particle settling speed and direction of flow from the uniform duct flow model with no macroscopic mixing. Reentrainment of dust already on the collection surface should also be considered. As an engineering approximation, it has been found that reducing the terminal particle velocity by a factor of 2 leads to better agreement between theory and practice. Thus the last two quantities in Equation (5-41) should be multiplied by 2 if a more conservative estimate is desired. Likewise, the factor 18 in Equation (5-42) may be replaced by a value of 36 if desirable. Finally, if a conservative approach is used to evaluate the minimum particle size with a 100 percent collection efficiency, then Equation (5-43) for the fractional collection efficiency should also be corrected by a factor of 2 in the denominator. The same comments also are valid when evaluating the performance of a settling chamber with horizontal baffles. Also note that since industrial gas flow rates are frequently not constant, the overall efficiency of a settling chamber increases at low loads and decreases at overloads.

Example 5-6

Estimate the value of $d_{p,\min}$ for 7.0-m-long settling chamber with a 1.20-m height and a gas velocity of 30 cm/s. The air temperature is 80°F and ρ_p is 2.50 g/cm^3.

SOLUTION

The air viscosity at this temperature is 0.067 kg/m·hr. Employing a factor of 36 in Equation (5-42), we find that

$$d_{p,\min} = \left[\frac{36(0.067)(1.2)(30)}{9.8(7)(2.5)(36{,}000)} \right]^{1/2} = 3.75 \times 10^{-3}\ \text{cm} = 37.5 \mu\text{m}$$

The factor 36,000 in the denominator is a composite conversion factor to make the units self-consistent. The minimum diameter found here for 100 percent removal efficiency is reasonably low, since we have chosen a fairly large particle density. For a density of 1 g/cm^3, the value of d_p would be 60 μm.

The preceding discussion introduces an empirical correction factor of 2 into equations developed for plug flow in order to account for turbulent effects in real settling chambers. Another approach is to model the flow as "well-mixed" or "well-stirred" with a laminar layer next to the collection tray. Particles of all sizes continually settle downward into the laminar layer. Any particle entering this layer is assumed to be captured and does not return to the turbulent or well-mixed flow region. The basic assumption associated with a well-mixed flow region is that a uniform particle distribution exists in the turbulent region for all particle sizes, regardless of how far downstream in the chamber. In order to determine the collection efficiency for any given particle size d_p we need to find a relation for the number of particles N_p of size d_p left in the gas stream at any position x in the settling chamber. The physical situation is illustrated in Figure 5-14 for a differential length dx of the chamber. A well-mixed turbulent zone fills most of the overall height H of the channel, while a laminar sublayer of thickness y^* exists along the tray at the bottom. The overall length and width of the settling chamber are L and W, respectively. The fraction of the total number of particles of size d_p which reaches the laminar layer and therefore is removed from the main flow stream while flowing a distance dx is dN_p/N_p. The time required for the main flow to travel a distance dx is simply dx/V. In this same time interval the particles of size d_p are falling downward with a terminal speed V_t. The maximum distance y in the vertical direction that a particle can traverse and still reach the laminar sublayer is tV_t. Therefore, $y/V_t = dx/V$. But for a well-mixed system, the ratio y/H is also the fraction of the particles which have reached the laminar layer and thus have been removed from the main gas stream. Consequently,

$$\frac{dN_p}{N_p} = \frac{y}{H} = -\frac{V_t\,dx}{VH} \tag{5-46}$$

The negative sign has been introduced to account for the fact that the number of particles is decreasing with increasing x. Integration of Equation (5-46) leads to

$$\ln N_p = -\frac{V_t x}{VH} + \ln C$$

There are two boundary conditions on this equation, namely, (1) at $x = 0$, $N_p = N_{p,0}$ and (2) at $x = L$, $N_p = N_{p,L}$. Substitution of these limits into the preceding equation gives

$$N_{p,L} = N_{p,0}\exp\left(-\frac{V_t L}{VH}\right) \tag{5-47}$$

where the subscript p is a reminder that this equation applies to a particular

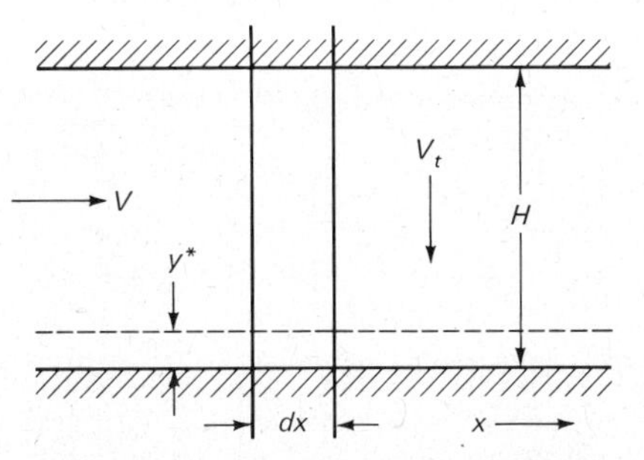

Figure 5-14 Schematic for the separation of particulate in a gravity collector for turbulent flow.

particle size d_p. The fact that V_t is a function of d_p emphasizes this point, also. Finally, the fractional collection efficiency η_d for particle sizes between d_p and $d_p + d(d_p)$ is given by the relation

$$\eta_d = 1 - \frac{N_{p,L}}{N_{p,0}} = 1 - \exp\left(-\frac{V_t L}{VH}\right) \tag{5-48a}$$

$$= 1 - \exp\left(-\frac{LV_t W}{Q}\right) \tag{5-48b}$$

where Q is the volume flow rate. A comparison of results obtained when Equations (5-43) and (5-48) are used is presented in Example 5-7.

Example 5-7

Determine the length of a simple gravity collector required to obtain an efficiency of 90 percent when collecting particles 50 μm in diameter and having a density of 2.0 g/cm^3. The bulk gas velocity is 0.5 m/s and the chamber is 3 m high.

SOLUTION

Assume the gas is air at standard temperature and pressure. From Figure 5-8 the settling velocity is found to be 15 cm/s or 0.15 m/s. Substitution of the appropriate values into Equation (5-43) gives

$$L = \frac{\eta HV}{50V_t} = \frac{90(3)(0.5)}{50(0.15)} = 18 \text{ m}$$

As a comparison, Equation (5-48) yields

$$0.90 = 1 - \exp\left(\frac{-V_t L}{VH}\right)$$

or

$$-\frac{V_t L}{VH} = \ln 0.1 = -2.303$$

$$L = 2.303\frac{VH}{V_t} = \frac{2.303(0.5)(3)}{0.15} = 23.3 \text{ m}$$

The answers obtained by the two methods are in reasonable agreement. In this particular case the answers indicate that perhaps the designer should use a number of trays within the 3 m height, in order to reduce the effective height and therefore reduce the required length L. For example, installing two additional trays, so that the effective height is 1 m, will reduce the estimated required length to around 6 to 8 m.

The basic characteristics of gravity settling chambers include: (1) very low energy cost, (2) low maintenance cost, (3) low installed cost, (4) excellent reliability, (5) very large physical size, and (6) low to very low collection efficiency. This last item is particularly true when the dust loading contains substantial amounts of fine to medium-size particles. As a result, settling chambers are used primarily to provide economical precleaning of coarse particles from a gas stream. Precleaning is especially helpful when either very high dust loadings or extremely coarse particles might damage a downstream collector in series with the settling chamber. Although chambers can be manufactured from almost any material, they are infrequently used because of the large space requirements.

5-9-B Cyclone Separators

Cyclone separators are gas cleaning devices that employ a centrifugal force generated by a spinning gas stream to separate the particulate matter (solid or liquid) from the carrier gas. The separator unit may be a single large chamber, a number of small tubular chambers in parallel or series, or a dynamic unit similar to a blower. Units in parallel provide increased volumetric capacity, while units in series provide increased removal efficiency. Two major classes of cyclone separators are the vane-axial and the involute types. The only difference between the two is the method of introducing the gas into the cylindrical shell in order to impart sufficient spinning motion. In the simple dry cyclone separator depicted in Figure 5.15(a), the circular motion is attained by a tangential gas inlet. The rectangular involute inlet passage has its inner wall tangent to the cylinder, and the inlet is designed to blend gradually with the cylinder over a 180-degree involute. Figure 5-15(b), shows a vane-axial cyclone. In this case, the cyclonic motion is imparted to the axially descending dirty gas by a ring of vanes. In either case, the operation depends upon the tendency (inertia) of particles to move in a straight line when the direction of the gas stream is changed. The centrifugal force due to a high rate of spin flings the dust particles to the outer walls of the cylinder and cone. The particles then slide down the walls and into the storage hopper. The gradually cleaned gas reverses its downward spiral and forms a smaller ascending spiral. A vortex finder tube extending downward into the cylinder aids in directing the inner vortex out of the device.

The cyclone separator is usually employed for removing particles 10 μm in size and larger. However, conventional cyclones seldom remove particles with an efficiency greater than 90 percent unless the particle size is 25 μm or larger. High-efficiency cyclones, which are effective with particle sizes down to 5 μm, are available. Regardless of the design, the fractional removal efficiency of any cyclone drops rapidly beyond a certain particle size. This is illustrated by Figure 5-16, which shows typical curves for several types of equipment. A high-volume design sacrifices efficiency for high rates of

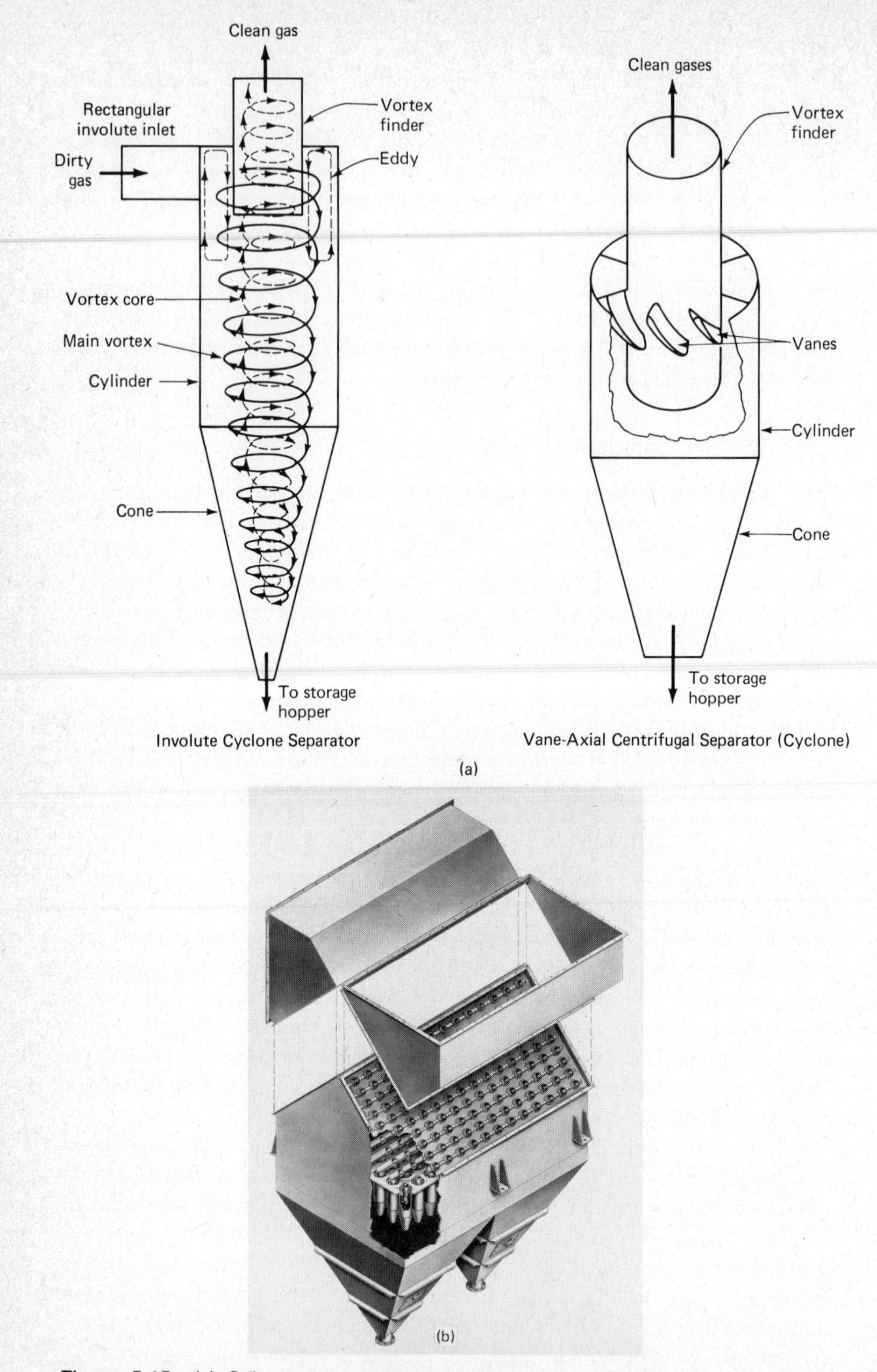

Figure 5-15 (a) Schematics of cyclone separators. (b) Multitube cyclone separator. (Courtesy of Research-Cottrell, Inc., Bound Brook, N.J.)

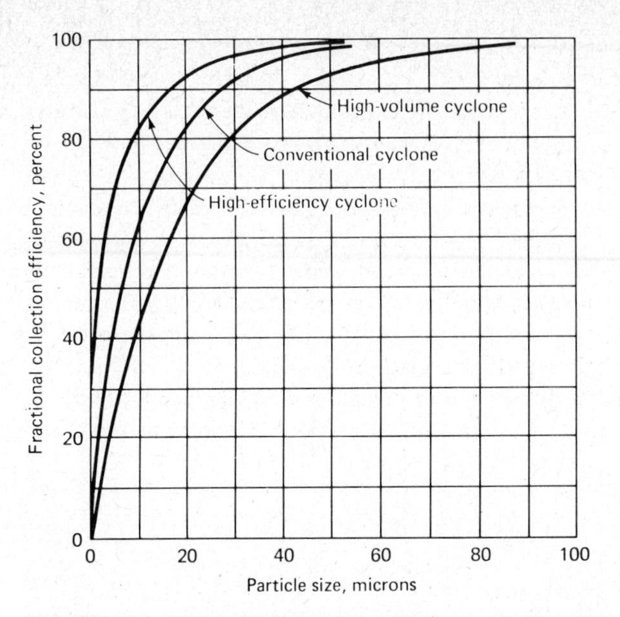

Figure 5-16 Fractional collection efficiency as a function of particle size for several types of cyclones.

collection. It might be used as a precleaner to remove the larger particles before the gas passes through another piece of collection equipment. Representative overall cyclone efficiencies are presented in Table 5-7. A given cyclone may be in more than one class, depending upon the particle size being collected and mode of operation.

The major variables that affect the fractional collection efficiency of a cyclone can be ascertained from the simple model presented below. As the dust-laden gas enters the cyclone, it spins through N_e revolutions in the main outer vortex (see Figure 5-15) before entering the inner vortex and passing upward toward the exit of the cyclone. As an approximation, the value of N_e is given by

$$N_e = \frac{1}{H}\left[L_1 + \left(\frac{L_2}{2}\right)\right] \tag{5-49}$$

Table 5-7 OVERALL CYCLONE COLLECTION EFFICIENCY

PARTICLE SIZE (μm)	CONVENTIONAL CYCLONE	HIGH-EFFICIENCY CYCLONE
<5	<50	50–80
5–20	50–80	80–95
15–50	80–95	95–99
>40	95–99	95–99

SOURCE: A. C. Stern, P. D. Bush, and K. J. Kaplan. *Cyclone Dust Collectors*. New York: American Petroleum Institute, 1955 [14].

where L_1 is the height of the main upper cylinder, L_2 is the height of the lower cone, and H is the height of the rectangular inlet through which the dirty gas enters. We shall assume that the particulate is uniformly distributed at the entrance to the cyclone. For dust of any given diameter d_p to be collected with 100 percent efficiency, all the particulate of that size which enters at the inner radius R_i of the cyclone must be thrown by centrifugal force to the outer wall at radius R_o before the N_e revolutions are completed. All particulate of size d_p which enters at some intermediate radius R (between R_i and R_o) obviously will also be collected with 100 percent efficiency. For any size d_p which will not be collected completely in N_e revolutions, only those particles that lie within the distance $R_o - R^*$ will be collected completely. R^* is the minimum radius for which particles of size d_p will reach the outer wall during the N_e revolutions. Hence the fractional collection efficiency in general may be expressed as

$$\eta_d = \frac{R_o - R^*}{R_o - R_i} \tag{5-50}$$

The quantity $(R_o - R_i)$ is also the width W of the rectangular inlet to the cyclone. By developing an expression for $R_o - R^*$, we can obtain an equation for the fractional collection efficiency.

The distance $R_o - R^*$ is directly related to the product of the particle velocity normal to the gas flow and the time the gas is in the outer vortex. The normal (radial) velocity is found by equating centrifugal and drag forces in the radial direction, and neglecting the effect of gravity. As a model, Stokes flow will be assumed for the particle movement. For a spherical particle the force balance results in the equation,

$$3\pi d_p \mu V_n = \rho_p\left(\frac{\pi d_p^{\,3}}{6}\right)\left(\frac{V_t^{\,2}}{R}\right)$$

or

$$V_n = \frac{\rho_p d_p^{\,2} V_t^{\,2}}{18\mu R} \tag{5-51}$$

where V_n and V_t are the normal (radial) and tangential velocities of the particle. In practice V_t is a function of R. As a simplifying assumption, V_t will be taken as the inlet gas velocity V_g and R as an average radius between R_o and R_i. In addition, if V_n is essentially constant for a particle as it moves outward, then $V_n = (R_o - R^*)/\Delta t$, as discussed above. Therefore,

$$R_o - R^* = \frac{\rho_p d_p^{\,2} V_g^{\,2} \Delta t}{18\mu R}$$

But the time that a particle remains in the outer vortex is also given by

$$\Delta t = \frac{2\pi R N_e}{V_g}$$

where N_e again is the number of effective turns or revolutions. Eliminating Δt from these two expressions, we find that

$$R_o - R^* = \frac{\pi N_e \rho_p d_p{}^2 V_g}{9\mu} \tag{5-52}$$

Substitution of this expression into Equation (5-50) for the fractional collection efficiency yields

$$\eta_d = \frac{\pi N_e \rho_p d_p{}^2 V_g}{9\mu W} = \frac{\pi N_e \rho_p d_p{}^2 Q}{9\mu H W^2} \tag{5-53}$$

where Q is the volume flow rate.

Equation (5-53) establishes the variables that influence the collection efficiency. This equation has one main deficiency, however. For a given geometry and flow situation, η_d varies proportionally to $d_p{}^2$, and there is a finite value of d_p beyond which the efficiency is always 100 percent. Experimental evidence dictates, however, that the fractional efficiency increases somewhat exponentially with increasing d_p. That is, there is no sharp cutoff in the efficiency curve. To circumvent this problem, several authors [13] have developed equations relating the size of the particle collected with an efficiency of 50 percent to other parameters of interest. Lapple [15] has developed an empirical expression for the collection efficiency based on Equation (5-53) and the particle *cut size*, $d_{p,50}$. If we set η_d in the preceding equation equal to 0.5 and solve for d_p, then

$$d_{p,50} = \left[\frac{9\mu W}{2\pi N_e V_g \rho_p} \right]^{1/2} = \left[\frac{9\mu W^2 H}{2\pi N_e \rho_p Q} \right]^{1/2} \tag{5-54}$$

where all the terms have been defined previously, and $d_{p,50}$ is the particle size collected with 50 percent efficiency. The units in this equation must be internally consistent. Lapple then correlated data from cyclones of similar proportions and presented the results in the generalized curve presented as Figure 5-17. This curve correlates better for cyclones of "standard proportions". Table 5-8 lists the proportions of a standard cyclone in terms of the outer diameter, D_o. The significance of the curve in Figure 5-17 is that the fractional efficiency curve approaches the 100 percent value asymptotically, rather than having the abrupt cutoff predicted by the simple theory developed earlier. Although the correlation by Lapple is based on experimental data, some curves from manufacturers of cyclones show a slightly lower efficiency than Lapple's work for $d_p/d_{p,50}$ values greater than unity.

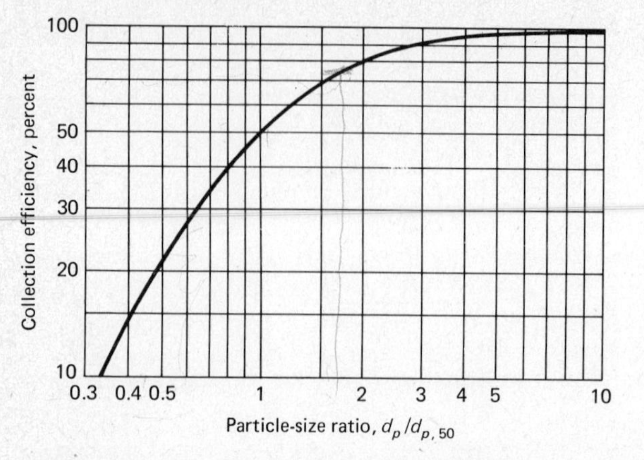

Figure 5-17 Cyclone efficiency versus particle-size ratio. (SOURCE: C. E. Lapple. "Processes Use Many Collection Types." *Chem. Eng.* **58** (May 1951): 145.)

Example 5-8

A cyclone is designed with an inlet width of 12.0 cm and four effective turns. The inlet gas velocity is to be 15.0 m/s, and the particle density is 1.70 g/cm^3. Estimate the particle size that will be collected with 50 percent efficiency, if the gas is air and its temperature is 350°K.

SOLUTION

The particle size with a fractional collection efficiency of 50 percent can be estimated through the use of Equation (5-54). The gas density is assumed to be negligible when compared with the particle density. From Appendix B, the viscosity of air at 350°K is found to be 0.0748 kg/m·hr. Substitution of the proper values into Equation (5-16) yields

$$d_{p,50} = \left[\frac{9(12.0\ \text{cm})(0.0748\ \text{kg/m}\cdot\text{hr})}{2\pi(4)(15.0\ \text{m/s})(1.70\ \text{g/cm}^3)} \times \frac{10^3\ \text{g/kg}}{(10^4\ \text{cm}^2/\text{m}^2)(3600\ \text{s/hr})} \right]^{1/2}$$

$$= (3.50 \times 10^{-7}\ \text{cm}^2)^{1/2} = 5.92 \times 10^{-4}\ \text{cm} = 5.92\ \mu\text{m}$$

This value is fairly typical of a cyclone, and indicates that only particles with diameters greater than roughly 10 μm will be effectively collected in this case.

Table 5-8 STANDARD CYCLONE PROPORTIONS

Length of cylinder	$L_1 = 2D_o$
Length of cone	$L_2 = 2D_o$
Height of entrance	$H = D_o/2$
Width of entrance	$W = D_o/4$
Diameter of exit cylinder	$D_e = D_o/2$
Diameter of dust exit	$D_d = D_o/4$

In review, the collection efficiency of a cyclone is found to increase with an increase in the right-hand term found in Equation (5-55), namely,

$$\eta_{col} \propto \frac{\text{centrifugal force}}{\text{drag force}} \propto \frac{V_p \rho_p d_p^{\ 2}}{R\mu_g} \tag{5-55}$$

As a first approximation, then, the collection efficiency is found to increase with an increase in inlet or tangential velocity, particle density, and particle diameter; it is found to decrease with an increase in carrier gas viscosity and cyclone diameter. (Recall that gas viscosity increases with increase in temperature.) Equation (5-55) shows very clearly why small particles are not captured efficiently, since the efficiency is related to the square of the particle diameter. In addition, it is found experimentally that efficiency also increases with an increase in cyclone body length (or number of vortex revolutions) and the amount of dust loading. Smoothness of the inner cyclone wall is another important parameter, since turbulence must be avoided in the vicinity of the wall. Finally, the above relation indicates that small-diameter cyclones are more efficient than large-diameter cyclones, but this is true only for a collection of cyclones whose ratios of physical dimensions are fixed. By changing certain dimensions, a large-diameter cyclone can be made more efficient than one of small diameter. However, the increased efficiency will mean an increase in overall size for the same flow rate or an increase in pressure drop, or both.

This tradeoff is extremely important, since it illustrates a basic dichotomy or contradiction in cyclone design. A compromise must be made between collection efficiency and pressure loss. Pressure drop, of course, is directly related to energy expenditure. In general, higher collection efficiencies are related to higher pressure losses. This is easily seen when empirical pressure loss equations are compared to an efficiency relation such as Equation (5-55). One such relation [6] is

$$\Delta P = \frac{KQ^2 P \rho_g}{T} \tag{5-56}$$

where ΔP is the pressure drop, in inches of water; Q is the volume flow rate of gas, in cubic feet per minute; P and T are the pressure and temperature of the gas, in atmospheres and degrees Rankine, respectively; ρ_g is the gas density, in pounds per cubic foot; and K is the empirical design factor, as given in Table 5-9.

Table 5-9 PRESSURE DROP PARAMETER *K* VERSUS CYCLONE DIAMETER

Cyclone diameter (in.)	29	16	8.1	4.4
K	10^{-4}	10^{-3}	10^{-2}	10^{-1}

The factor K versus cyclone diameter is a straight line on log-log paper. If the carrier gas is air, Equation (5-56) can be simplified to the form

$$\Delta P = \frac{39.7KQ^2P^2}{T^2} \tag{5-57}$$

where the units are the same as above. Since the gas velocity is related directly to Q, the pressure drop increases proportionately to the square of the velocity, as might be expected. If the particle velocity initially is taken as the same as the gas velocity, then Equation (5-55) indicates that the collection efficiency is proportional to the gas velocity directly. Hence both collection efficiency and pressure loss are tied directly to the velocity or volume rate through a given device. For simple cyclones, the pressure drop ranges from 0.5 to 2 in. water, while high-efficiency cyclones may experience losses of 2 to 6 in. water (1 in. water = 2.5 mbar).

Cyclone sizes such as 6, 8, 9, and 10 in. (15 to 25 cm) are quite common. Typical inlet velocities are 50 to 60 ft/s (15 to 20 m/s), with volumetric rates of 500 to 1000 ft^3/min (15 to 30 m^3/min) per tube. (However, single involute cyclones with capacities varying from 30 to 30,000 ft^3/min are available.) The absence of moving parts gives the cyclone high reliability with simplicity, especially with standardized tubes. It is the least expensive of all high-efficiency-type particulate collectors. The vane-axial type of collector has disadvantages which include vane erosion (particularly with abrasive dusts) and possible plugging of passages between vanes. A major disadvantage of involute cyclones is that the size and cost increase as the required efficiency increases. Thus a vane-axial cyclone is usually a first choice if its operating characteristics match the requirements. Other refinements are possible within the two general classes of vane-axial and involute cyclones.

It is often necessary to estimate the performance of cyclone collectors when operating at off-design conditions. Performance curves for specific collectors should be used when available. However, the following approximate relationships are available for estimation purposes [13]. For variable flow rate,

$$\frac{100 - \eta_a}{100 - \eta_b} = \left(\frac{Q_b}{Q_a}\right)^{0.5} \tag{5-58}$$

For constant gas flow rate, efficiency and gas viscosity are related by

$$\frac{100 - \eta_a}{100 - \eta_b} = \left(\frac{\mu_a}{\mu_b}\right)^{0.5} \tag{5-59}$$

For variations in gas density,

$$\frac{100-\eta_a}{100-\eta_b}=\left(\frac{\rho_p-\rho_{gb}}{\rho_p-\rho_{ga}}\right)^{0.5} \tag{5-60}$$

For moderate changes in gas particle loadings,

$$\frac{100-\eta_a}{100-\eta_b}=\left(\frac{C_{bi}}{C_{ai}}\right)^{0.183} \tag{5-61}$$

where, for operating conditions a and b, η is the collection efficiency, in weight percent; Q is the volume flow rate; μ is the gas viscosity; ρ_p is the particle density; ρ_g is the gas density; and C is the particulate concentration, in mass per unit volume.

The last type of cyclone separator we consider here is the dynamic precipitator. Again, collection of particulate is based on the use of centrifugal forces, this time generated by a rotating impeller. Rotation of the centrifugal turbinelike impeller blades sucks in the dirty gas axially and turns it 90 degrees so that it flows outward radially, as shown in Figure 5-18. For the same capacity, a device of this type can develop a force upward to seven times as high as a conventional cyclone. The dust particles are thrown outward and collected in a secondary air circuit at the outer periphery of the device. Since the device acts as a centrifugal fan or prime mover as well, there is no pressure drop associated with the device. However, the power input will be higher than with a pure centrifugal fan (no particulate collection) of the same capacity. Especially useful in the collection of fine

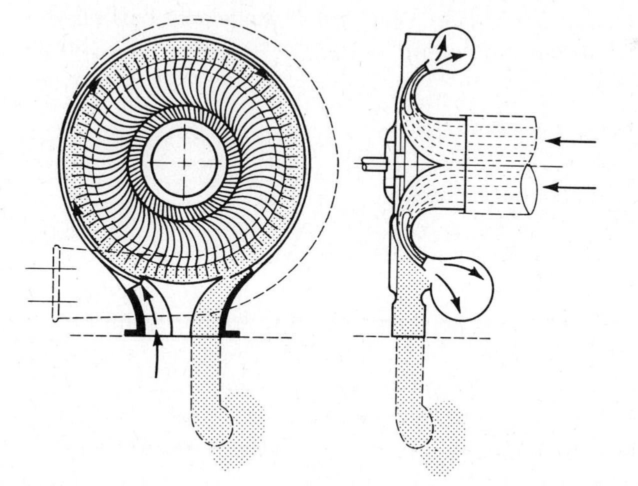

Figure 5-18 Sketch of a dynamic precipitator. (SOURCE: *Air Pollution Manual.* Part II—"Control Equipment." Detroit, Mich.: American Industrial Hygiene Assoc., 1968.)

particles, this type of device is widely used in the food, pharmaceutical, and wood-working industries.

The costs of purchasing, installing, and operating cyclones for dust removal depend upon the volume throughput, among other factors. During 1979 the fabrication cost for a standard design in the 100,000 ft^3/min range was typically \$0.25 to \$2.00/actual ft^3/min, with installation adding an additional 100 percent to the cost. For small units in the 1000 to 5000 ft^3/min size, initial costs may run double that of the larger size quoted above. Prices also vary according to whether light or heavy duty is anticipated, and whether special operating conditions are present. Additional information on cyclone collectors and descriptions of dynamic precipitators may be found in references 2, 5, 6, 13, 14, and 16.

5-9-C Wet Collectors

In a wet collector a liquid, usually water, is used to capture particulate dust or to increase the size of aerosols. In either case the resulting increased size facilitates the removal of the contaminant from the gas stream. Fine particulates, both liquid and solid, ranging from 0.1 to 20 μm, can be effectively removed from a gas stream by wet collectors. The actual mechanism used in the removal equipment may be any of those discussed in Section 5-8. One of the primary aims of the device, however, should be adequate dispersion of the liquid phase in order to achieve good contact between the particulate (or aerosol) phase and the liquid phase. Many different arrangements of basic equipment are sold industrially. To simplify the discussion, we restrict ourselves primarily to three major types of wet collectors, namely:

1. Spray chamber scrubbers (with or without impingement baffles).
2. Cyclonic scrubbers (wet cyclones).
3. Venturi scrubbers.

Packed towers could be added to the list, but they are used primarily for gas absorption. In some cases they are used for the dual purpose of removing both gaseous and particulate contaminants.

Wet collectors have certain disadvantages not found with dry equipment. One major problem is how to handle and dispose of the wet sludge which is an inherent product of the process. However, in some applications the sludge may be easier to manage than dry dust. If the equipment is installed in the natural environment, the question of freezing in cold weather must be considered. The presence of water also has a tendency to increase the corrosiveness of materials. Finally, to achieve high collection efficiencies for fine particles requires good dispersal of the liquid phase, and this in turn requires relatively high power input.

Although one or more collection mechanisms may be used within the range of wet collector equipment, the major requirement for any such device is the initiation of impingement or interception of a particle with a droplet.

As a result, it is important to determine what particle, droplet, and fluid properties influence such impingement. As a simplified approach to the problem, consider the following model. A particle approaches a droplet, as shown in Figure 5-12(a), and undergoes inertial impaction. At some distance upstream from the droplet the particle leaves the streamline of the gas flow and proceeds toward the droplet. The particle is now under two main forces: its own inertial force and the drag force due to the surrounding gas. (Forces such as gravitational, electric, magnetic, and thermal are neglected.) As a result of these two forces, the particle eventually will come to a stop, relative to the droplet. If the stopping distance x_S is greater than the original distance from the point where it left the streamline to the droplet, impaction will result. We define an impaction number, N_I, as the dimensionless ratio of the stopping distance, x_S, to the droplet diameter, d_D. That is,

$$N_I \equiv \frac{x_S}{d_D} \tag{5-62}$$

The efficiency of collisions between particles and droplets, and hence removal of particles from the gas stream, is found to be related to the impaction number.

An explicit expression for the stopping distance can be derived if we assume the validity of Stokes' law for the motion of the particle. In terms of a force balance on a particle, we find that

$$F_{\text{inertial}} + F_{\text{drag}} = 0$$

or

$$m_p \frac{dV_p}{dt} + 3\pi V_p \mu_g d_p = 0$$

where V_p is the relative velocity of particles with respect to the liquid droplets. If we assume that the particles are spherical with a density ρ_p, and note that $dV_p/dt = (dV_p/dx)(dx/dt) = V_p\, dV_p/dx$, then

$$\frac{\pi d_p{}^3 \rho_p}{6} V_p \frac{dV_p}{dx} + 3\pi V_p \mu_g d_p = 0$$

Upon canceling common terms and rearranging, we find that

$$-\int_{V_{p,o}}^{0} \frac{d_p{}^2 \rho_p}{18\mu_g} dV_p = \int_0^{x_s} dx$$

The left-hand term is integrated from $V_{p,o}$, the initial particle velocity relative to the gas stream, to 0, while the right side is integrated from 0 to the stopping distance, x_S. The result of integration is

$$x_S = \frac{V_{p,o} d_p{}^2 \rho_p}{18\mu_g} \tag{5-63}$$

In most cases, $V_{p,o}$ also represents the velocity of the gas stream relative to the liquid droplets, initially.

Substitution of Equation (5-63) into Equation (5-62) shows that the impaction number, N_I, is given approximately by

$$N_I = \frac{V_{p,o} d_p{}^2 \rho_p K_C}{18\mu_g d_D} = \frac{d_p{}^2 \rho_p K_C (u_p - u_D)}{18\mu_g d_D} \tag{5-64}$$

where u_D is the droplet velocity and u_p is the particle velocity in the direction of flow. The Cunningham correction factor, K_C, has been added to the equation for N_I. Introduced in Section 5-5, it is applicable for particle diameters less than roughly 5 μm. [It should be carefully noted that some investigators define the impaction number in Equation (5-62) as the ratio of the stopping distance to the radius of the droplet. In such cases, the constant 18 in the denominator of Equation (5-64) is replaced by a value of 9. This alternate definition of N_I will change analytical and graphical presentations in the literature. Hence some care must be exercised when employing such data.] The impaction number is dimensionless; hence appropriate units must be chosen for the terms on the right side. For a particle of given size and density, the impaction number is directly proportional to the relative velocity, $V_{p,o}$, and inversely proportional to the diameter of the liquid droplet. Thus a large relative velocity between the two phases (solid and liquid) and a finely dispersed liquid phase are highly desirable in order to achieve a relatively large impaction number. Generally, on the basis of the physical interpretation of the impaction number, we anticipate that the larger the value of N_I, the higher the inertial or impaction collection efficiency. This is confirmed by both theoretical and experimental [17] studies. Theoretically, the collection efficiency can be evaluated for both potential and viscous flow

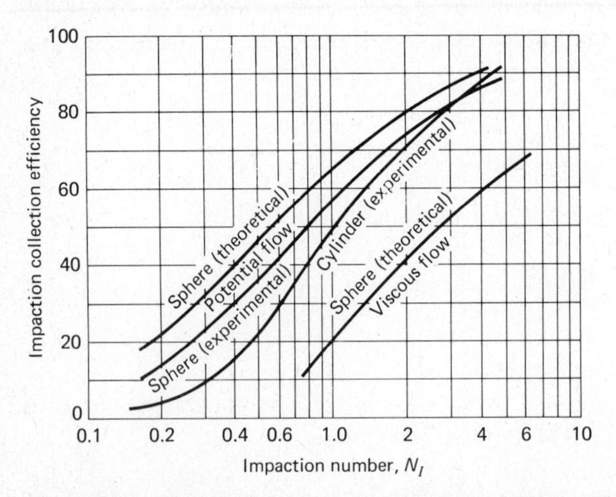

Figure 5-19 Theoretical and experimental impaction collection efficiencies for spheres and cylinders.

in terms of the impaction number. Theoretical and experimental curves for spheres and cylinders are shown in Figure 5-19.

Whether the collection mechanism is due to inertial forces as in a venturi scrubber or gravitational forces as in a spray chamber, the collection efficiencies presented in Figure 5-19 are not representative of actual performance. The reason is that the relative velocity in the impaction number is not a constant over a range of droplet sizes for a given particle size. In a high-velocity system such as a venturi scrubber, the smaller liquid droplets are more easily accelerated to the carrier gas velocity than are large droplets. Thus the relative velocity between a dust particle and a liquid collection particle varies with droplet size. The same type of phenomenon also occurs in spray chambers. In this latter case, the gas velocity does not affect the droplet speed, but gravitational forces do. The terminal or settling velocity is higher for larger droplets (see Section 5-5), and hence the relative velocity between the particulate carried in the gas stream and the falling liquid droplets is greater for large droplets. The overall effect in either case is that the collection efficiency eventually falls off with decreasing liquid droplet size. This effect, coupled with the general predictions of Figure 5-19, leads to the important result that there is an optimum droplet size for maximum target collection efficiency for a given dust size. This point is illustrated by Figure 5-20 for particles having a density of 2 g/cm^3 in a gravitational spray tower. Note the rapid decrease in efficiency when the droplet size falls below 100 μm. Also note the general increase in impaction efficiency with increase in particle size.

SPRAY CHAMBER SCRUBBERS

One of the simplest devices for wet collection of particulate is the circular or rectangular spray tower, as shown in Figure 5-21. The polluted gas flows upward and the particles collide with liquid droplets produced by suitable

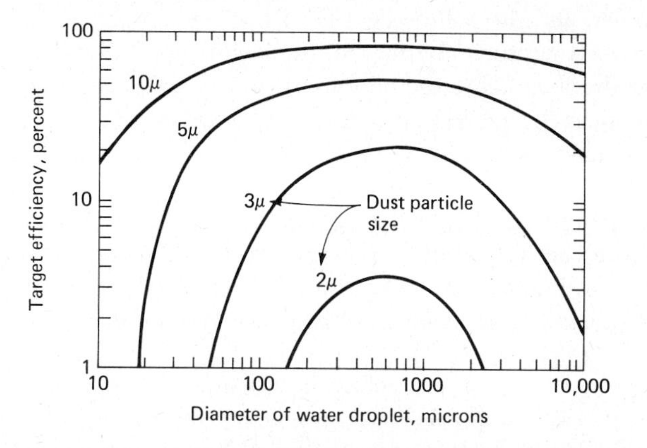

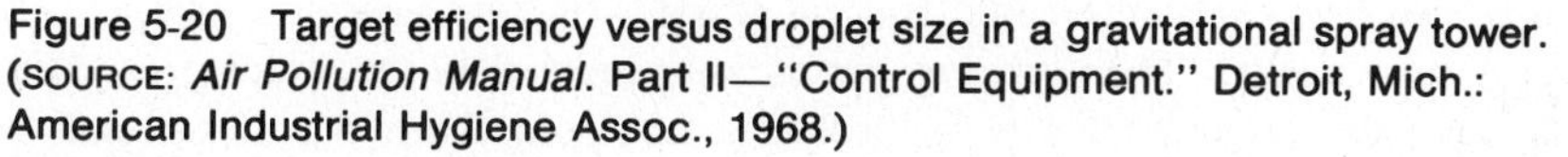

Figure 5-20 Target efficiency versus droplet size in a gravitational spray tower. (SOURCE: *Air Pollution Manual.* Part II—"Control Equipment." Detroit, Mich.: American Industrial Hygiene Assoc., 1968.)

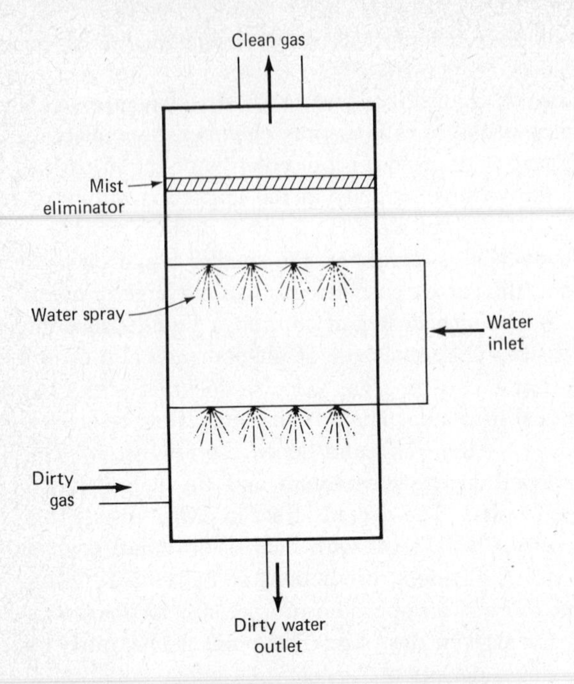

Figure 5-21 **Sketch of a spray tower scrubber.**

nozzles located across the flow passage. If the gas flow rate is relatively slow, the contaminated liquid droplets will settle by gravity to the bottom of the tower. A mist eliminator is usually placed at the top of the tower to remove both excess clean water droplets and dirty droplets which are very small and thus are carried upward by the gas flow. Rather than the counterflow type described above, an alternate design is a crossflow type. Water is sprayed from the top of a chamber, and the polluted gas flows horizontally across the chamber. The particulate is captured by inertial impingement or diffusion, and the resulting larger droplets fall to the bottom by gravity settling. Again, some type of demister must be placed after the spray section in order to remove droplets which have not yet settled to the water surface at the bottom. A typical technique is to place vertical baffles after the spray section. Improvement in the collection efficiency of the crossflow scrubber can be achieved by placing additional vertical baffles in the spray section as well. This device, shown in Figure 5-22, is known as a wetted impingement baffle scrubber. A vertical-flow spray tower arrangement of the impingement plate scrubber is also available.

The water rate through a chamber scrubber is in the range of 2 to 10 gal/min for every 1000 ft^3/min of gas flow. Makeup water must be added to replenish that evaporated into the gas stream. In most cases the water used must be recycled. Thus settling ponds are required in the plant area, and, because the recycled water is probably not completely clean, special or

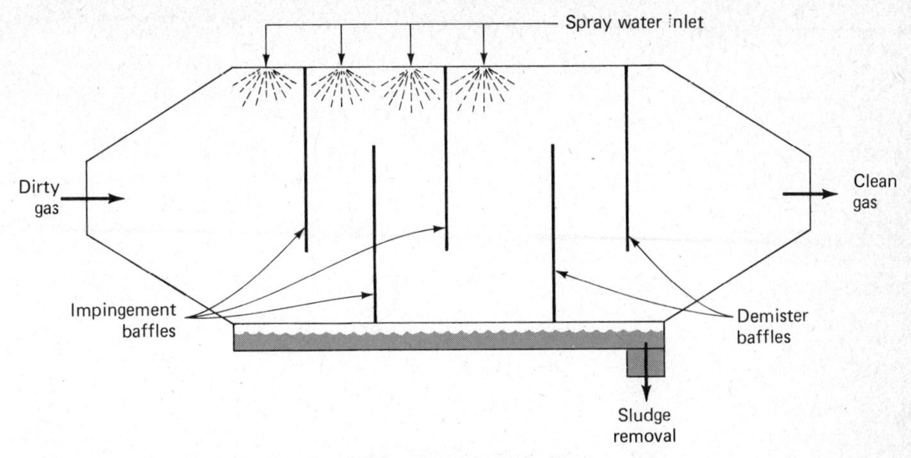

Figure 5-22 Sketch of a wetted impingement baffle scrubber.

coarse nozzles must be used to prevent plugging and eroding of the spray nozzles. The pressure drop is quite small in this type of equipment, usually on the order of 1 to 2 in. water. The collection efficiency is quite acceptable for particle sizes over 10 μm. The effectiveness of a conventional spray tower ranges from 94 percent for 5-μm particles to 99 percent for 25-μm particles. The use of baffles (or packed beds or trays) to increase the contact time between gas and liquid yields efficiencies of 97 percent for 5-μm particles and nearly 100 percent for 10-μm particles. High efficiencies for particle sizes down to 1 μm can be attained by using high-pressure fog sprays. However, the power requirements go up accordingly, and recycled water may not be adequate to prevent plugging of the nozzles.

CYCLONIC SCRUBBERS (WET CYCLONES)

The simplest type of cyclonic scrubber is achieved by inserting banks of nozzles in ring fashion inside a conventional dry cyclone. The spray acts on the particles in the outer vortex, and the dust-loaded liquid particles are thrown outward against the wet inner wall of the cyclone. The dust-laden solution flows down the walls of the cyclone to the bottom, where it is removed. The water spray may also be located in the cyclone inlet. A mist eliminator usually is required at the outlet. Another version of a cyclonic scrubber, shown in Figure 5-23, has the dirty gas introduced into the lower portion of the vertical cylinder. Water is introduced through an axially located multiple nozzle, which throws the water radially outward across the spiraling gas flow. If there is enough height above the spray region, the upper section of the cylinder can act as a demister, obviating the need for further demister equipment at the outlet.

The water circulation rate in wet cyclones runs from 1 to 8 gal/1000 ft^3 of treated gas. The draft loss or pressure drop typically runs between 1 to 4

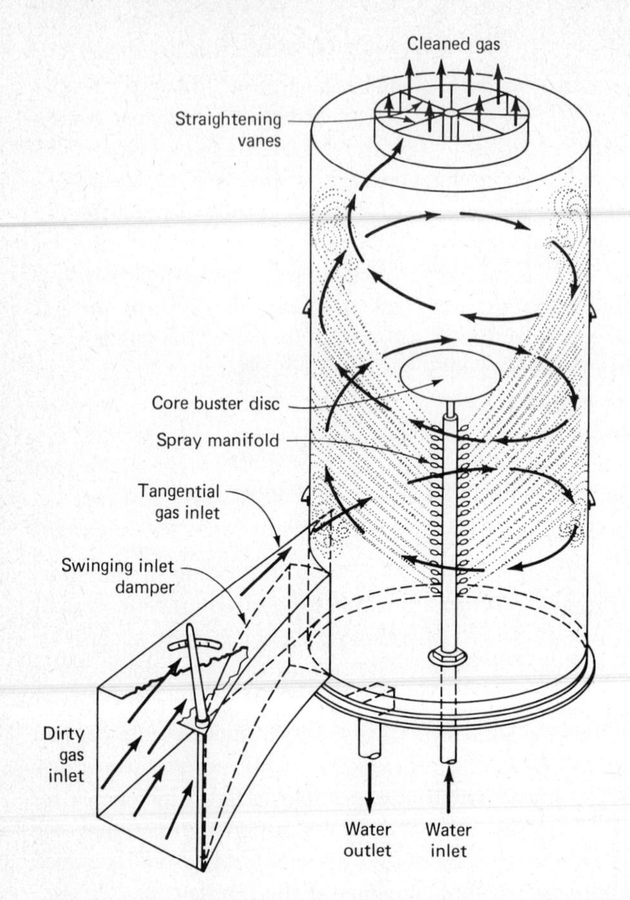

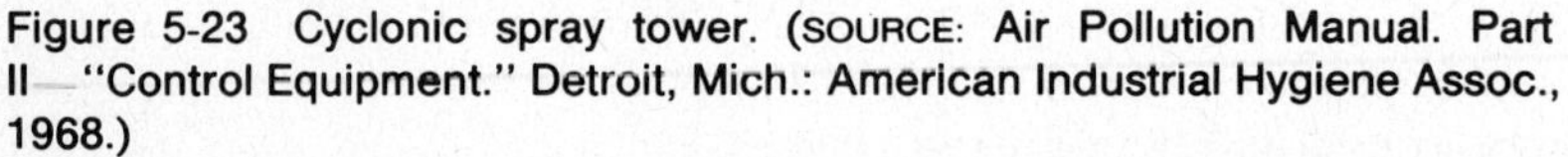
Figure 5-23 Cyclonic spray tower. (SOURCE: Air Pollution Manual. Part II—"Control Equipment." Detroit, Mich.: American Industrial Hygiene Assoc., 1968.)

in. water. This latter range depends upon the internal arrangement of the equipment. In general, wet cyclones have a collection efficiency of 100 percent for droplets of 100 μm and over, around 99 percent for droplets from 50 to 100 μm, and from 90 to 98 percent for droplets between 5 and 50 μm. The use of cyclones in series with a venturi scrubber as a collection device is discussed in the following subsection.

VENTURI SCRUBBERS

A venturi is a rectangular or circular flow channel which converges to a narrow throat section and then diverges back to its original cross-sectional area. In the converging section flow work associated with the fluid is converted into kinetic energy, with a concomitant decrease in static pressure and rise in velocity. The latter reaches values of 160 to 600 ft/s (50 to 180 m/s) in the throat section. The area ratio between the inlet and the throat typically is 4 : 1 in a venturi scrubber. The angle of divergence is around 5 to 7 degrees in order to achieve good static pressure recovery. The scrubbing action occurs through the introduction of water either in the throat region (recommended) or at the beginning of the convergent section. Figure 5-24 shows a vertical downward rectangular venturi scrubber with throat injection. A bank of nozzles on either side of the throat injects water into the high-velocity gas stream. The scrubbing liquid can also be injected into the gas stream through slots, or in some cases weirs, located on either side of the venturi throat. The high-velocity gas atomizes the liquid injected into the gas stream. Good atomization is essential if sufficient targets for inertial impaction are to be available. It is normally assumed that the fine particulate enters the venturi throat with a velocity equal to that of the gas stream. The droplets of the scrubbing liquid, on the other hand, are assumed to have no axial velocity initially, and to be accelerated through the venturi by the aerodynamic drag of the gas stream. The collection of the fine particles by the liquid droplets is accomplished by inertial impact during the time the droplets are being accelerated. As the droplet velocity approaches that of the gas, and the relative velocity between particulate and droplet approaches zero, the probability of inertial impaction downstream from the throat decreases rapidly. Since the impaction collection efficiency increases with an increase in relative velocity, high inlet gas velocities are essential.

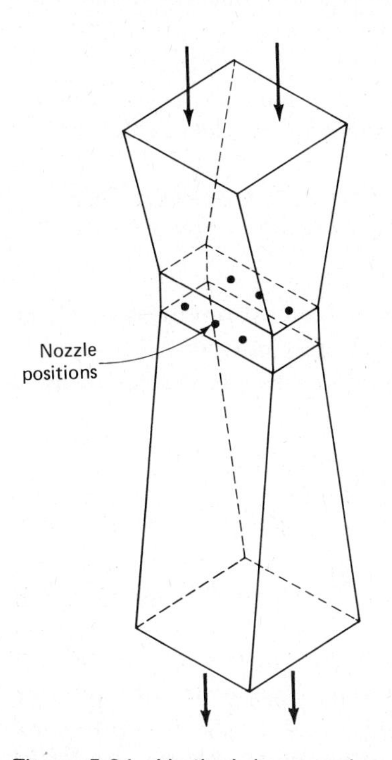

Figure 5-24 Vertical downward venturi scrubber with throat injection.

It has been shown [16] that the acceleration of liquid droplets is a function of their diameter. If we use as a criterion the distance x necessary to accelerate a droplet to 90 percent of the gas velocity, then, as an approximation,

$$\frac{x_1}{x_2} = \frac{2d_1}{d_2} \tag{5-65}$$

where 1 and 2 represent two different sizes and $d_1 > d_2$. Consider a 100-μm droplet injected normally into a 100-ft/s gas stream. It is found that the droplet would reach 90 percent of the gas velocity in roughly 16 in. From the above expression, it can be determined that a 50-μm droplet will reach the same condition in 4 in. A major inference can be drawn from these data: if a high relative velocity between particulate and liquid droplet is a basic requirement for high impaction efficiencies, then most of the impaction removal must occur within the first few inches of the divergent section. The added length is necessary for adequate pressure recovery, but it has little influence on the overall target efficiency.

Since the energy of the high-velocity gas stream in a venturi scrubber is used to increase liquid atomization and to accelerate the liquid droplets, it is not surprising that the pressure drop in the venturi is large in comparison to that in other dry and wet collectors. One method of estimating the pressure drop in a venturi scrubber, developed by Calvert et al. [18], is based on a model that all the energy loss of the gas stream is employed to accelerate the liquid droplets to the gas velocity in the venturi throat. If a force balance is made on an element of length dx in the direction of flow in the venturi throat, one finds that

$$dP = -\rho_L u_G \left(\frac{Q_L}{Q_G} \right) du_D$$

where u is a velocity, Q is a volume flow rate, L and G represent the liquid and gas, respectively, and D represents the droplet. Integrating from $x = 0$ (the point of liquid injection) where the droplet velocity in the x-direction is assumed to be zero, we find that

$$\Delta P = -\rho_L u_G \left(\frac{Q_L}{Q_G} \right) u_{D2} \tag{5-66}$$

where u_{D2} is the droplet velocity downstream at distance x_2. To determine the pressure drop for a given throat length, a relationship between u_{D2} and x_2 must be found.

Among the assumptions made in order to model the flow in the venturi throat are: (1) the gas velocity u_G is constant and equal to that at the throat, (2) the flow is one-dimensional, incompressible, and adiabatic, (3) the liquid fraction of the flow is small at any cross-section, (4) droplet evaporation is negligible, so that d_D is constant, and (5) the pressure forces around the droplet are symmetrical, and hence ignored. On this basis a force balance is made on a droplet, that is, the sum of the inertial and drag forces on the droplet are set equal to zero. The result is

$$m\frac{du_D}{dt} = -\frac{1}{2}\rho_G(u_G - u_D)^2 A_D C_D$$

where A_D is the droplet projected cross-sectional area perpendicular to the flow and C_D is the drag coefficient. For a spherical droplet this becomes

$$\frac{du_D}{dt} = \frac{3\rho_G C_D}{4\rho_L d_D}(u_G - u_D)^2 \tag{5-67}$$

To convert this to a distance basis, note that $du_D/dt = u_D(du_D/dx)$. Hence

$$\frac{du_D}{dx} = \frac{3\rho_G C_D}{4\rho_L d_D u_D}(u_G - u_D)^2 \tag{5-68}$$

To solve this equation we must first express C_D as a function of u_G. The value of C_D in terms of its value at the injection point, C_{Do}, is given by the Hollands and Goel expression [19],

$$C_D = C_{Do}\left(\frac{u_g}{u_g - u_D}\right)^{0.5} \tag{5-69}$$

where C_{Do} is given by Equation (5-32). The above equation is adequate in the Reynolds number range from 10 to 500. Substitution of Equation (5-69) into Equation (5-68), and subsequent rearrangement, yields

$$\frac{u_D du_D}{(u_G - u_D)^{1.5}} = \frac{3\rho_G C_{Do}}{4\rho_L d_D}(u_G)^{0.5}\,dx \tag{5-70}$$

The limits of integration are: (1) $u_D = 0$ at $x = 0$ and (2) $u_D = u_{DL}$ at $x = L$, the total length of the throat. Integration, using standard integral tables, yields

$$u_{DL} = 2u_G\left[1 - m^2 + (m^4 - m^2)^{0.5}\right] \tag{5-71}$$

where

$$m = \frac{3\rho_G C_{Do} L}{16\rho_L d_D} + 1 \tag{5-72}$$

Returning to Equation (5-66) for the pressure drop in the venturi throat, we may now write in dimensionless form,

$$\frac{P_1 - P_2}{\frac{1}{2}\rho_G {u_G}^2} = 4\left(\frac{\rho_L}{\rho_G}\right)\left(\frac{Q_L}{Q_G}\right)\left[1 - m^2 + (m^4 - m^2)^{0.5}\right] \tag{5-73}$$

where $\frac{1}{2}\rho_G u_G{}^2$ is "one velocity head." If the venturi throat is long enough, the u_{DL} will approach u_G. In this case the quantity in brackets in Equation (5-73) approaches 0.50 in value. In this limiting case,

$$\Delta P = P_2 - P_1 = -\rho_L u_G{}^2 \left(\frac{Q_L}{Q_G} \right) \tag{5-74}$$

If it is desired to express the pressure drop in an equivalent head of water, then the above equation becomes

$$\Delta P = -1.02 \times 10^{-3} u_G{}^2 \left(\frac{Q_L}{Q_G} \right) \tag{5-75}$$

where ΔP is in centimeters of water, u_G is in centimeters per second, and Q_L and Q_G are expressed in the same set of units.

Based upon a correlation of experimental data obtained from many different venturi scrubbers, Hesketh [20] developed the following equation for the pressure drop across a venturi scrubber:

$$\Delta P = \frac{V_{g,t}{}^2 \rho_g (A)^{0.133}}{507} (0.56 + 0.125L + 2.3 \times 10^{-3} L^2) \tag{5-76}$$

where ΔP is the pressure drop across the venturi, in inches of water; $V_{g,t}$ is the gas velocity at the throat, in feet per second; ρ_g is the gas density downstream from the venturi throat, in pounds per cubic feet; A is the cross-sectional area of the venturi throat, in square feet; and L is the liquid-to-gas ratio, in gal/1000 actual ft^3. Equations (5-75) and (5-76) are two of the more widely accepted expressions for the pressure drop in venturi scrubbers.

Although venturi scrubbers have enjoyed wide application in removing particulate from gas streams, reliable design equations for the collection efficiency have been lacking. Calvert et al. [18] summarize a development for the particle penetration based upon an analysis which takes into account droplet size, the inertial impaction parameter, droplet concentration across the venturi throat, and the continuously changing relative velocity between particle and droplet. Penetration (Pt) is defined as

$$\text{Pt} = \text{penetration} = 1 - \text{efficiency} \tag{5-77}$$

It is a convenient parameter when the collection efficiency is a number approaching 100 percent. The resulting equation, due to Calvert et al., after introducing several simplifying assumptions, is

$$\text{Pt} = \exp\left[-\frac{6.1 \times 10^{-11} \rho_L \rho_p K_C d_p{}^2 f^2 \Delta P}{\mu_g{}^2} \right] \tag{5-78}$$

where ΔP is the pressure drop across the venturi, in centimeters of water; μ_g

is the gas viscosity, in kg/m·s; ρ_L is the liquid density, in grams per cubic centimeter; ρ_p is the particle density, in grams per cubic centimeter; and d_p is the particle diameter, in micrometers. In this expression, f is an experimental coefficient which commonly varies from 0.1 to 0.4.

In a more recent paper by Calvert, et al. [21] another expression for the penetration has been developed, based on work in earlier papers [22, 23]. A material balance on the particulate over a differential scrubber volume leads to the following differential equation for the change in concentration dC with length

$$-\frac{dC}{C} = \frac{1.5(u_G - u_D)}{u_G d_D}\left(\frac{Q_L}{Q_G}\right)\eta\, dx \tag{5-79}$$

where η is the single droplet target efficiency. This latter quantity may be given the following physical interpretation. Consider a stream-tube of diameter D which contains a single droplet of size d_D (where $D > d_D$) and a uniformily dispersed group of particles of diameter d_p. Even though the droplet is accelerating, a certain fraction of the particles will be captured by the droplet before it reaches the gas or particle velocity, u_G. This fraction is defined as the single droplet target efficiency, η. From empirical considerations this is given by

$$\eta = \left(\frac{2N_I}{2N_I + 0.7}\right)^2 \tag{5-80}$$

where N_I again is the inertial impaction number defined by Equation (5-64). The third required equation is Equation (5-71) for the droplet velocity. These three equations can be nondimensionalized by using the inlet gas velocity u_{Go} as the characteristic velocity, the quantity $2d_D\rho_L/3C_{Do}\rho_G$ as the characteristic length, and the inlet particulate concentration C_i as the characteristic particle concentration. The resulting differential equation (see Reference [21]), is integrated, and the result is a fairly complex equation in terms of $(\ln \text{Pt})/B = \ln(C_L/C_i)/B$, where $B = Q_L\rho_L/Q_G\rho_G C_{Do}$. Rather than show the equation, a plot of $(\ln \text{Pt})/B$ versus the dimensionless throat length $L^* = 3C_{Do}\rho_G L/2d_D\rho_L$ is shown in Figure 5-25. The plot demonstrates that the longer the throat length L, the higher the scrubber collision efficiency. This dependency is greater for larger values of N_I, that is, larger particles. In general, the plot shows that L^* values of 2–3 are sufficient. Larger L^* values would show little increase in scrubber performance.

From an examination of experimental data, Hesketh [20] concluded that the venturi scrubber is essentially 100 percent efficient in removing particles larger than 5 μm and therefore studied the penetration for fine particles less than 5 μm in diameter. On the basis of this study he concluded that the overall collection efficiency of particles less than 5 μm in diameter, expressed as a penetration value, is approximately related to the pressure drop

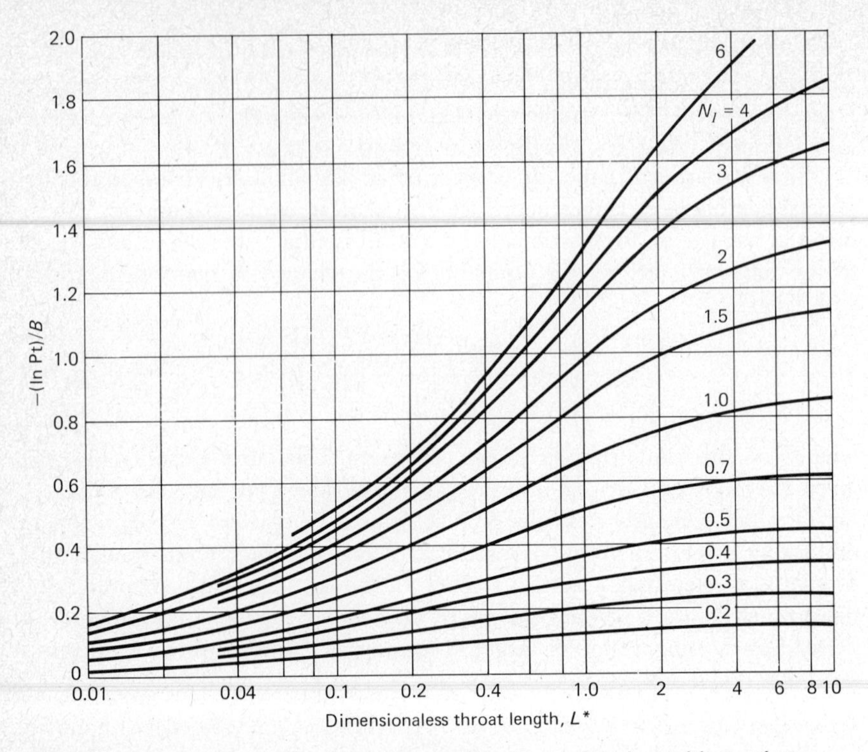

Figure 5-25 Plot for determining penetration in a venturi scrubber, where $B = Q_L \rho_L / Q_G \rho_G C_{Do}$ and $L^* = 3C_{Do}\rho_G L / 2d_D \rho_L$.

across the venturi by the equation

$$\mathrm{Pt} = \frac{C_o}{C_i} = 3.47(\Delta P)^{-1.43} \tag{5-81}$$

where C_i is the weight concentration of particles less than 5 μm in diameter into the venturi scrubber; C_o is the weight concentration of particles less than 5 μm in diameter out of the venturi scrubber; and ΔP is the pressure drop, in inches of water.

Hesketh also presents equations similar to Equation (5-78) for the collection of charged particles and the penetration when wetting agents are used in the scrubber.

Example 5-9

Water is introduced into the throat of a venturi scrubber at a rate of 1.0 l/m^3 of gas flow. The air velocity is 400 ft/s, its density is 0.072 lb/ft^3, and its temperature is 170°F. The throat area is 125 in.^{2}, the parameter f has a value of 0.25, and the particulate density is 1.50 g/cm^3. For a particle size of 1.0 μm, determine (a) the pressure drop by means of Equations (5-75) and (5-76), and (b) the penetration by means of Equations (5-78) and (5-81).

SOLUTION

(a) To use Equation (5-75) the gas velocity must be in centimeters per second. A velocity of 400 ft/s is equivalent to 12,200 cm/s. Hence,

$$\Delta P = 1.03 \times 10^{-6}(12{,}200)^2(1.0) = 153.3 \text{ cm water} = 60.4 \text{ in. water}$$

Now, using Equation (5-76) attributed to Hesketh, and noting that L is 7.48 gal/1000 ft^3, we find that

$$P = \frac{(400)^2(0.072)(0.868)^{0.133}}{507}\left[0.56 + 0.125(7.48) + 2.3 \times 10^{-3}(7.48)^2\right]$$
$$= 22.3(1.62) = 36.2 \text{ in. } H_2O$$

(b) In order to use Equation (5-78) we need the value of the Cunningham factor, plus the gas viscosity. Using Equation (5-30) for K_C,

$$K_C = 1 + \frac{0.182}{d_p} = 1.182$$

From Appendix B, the viscosity of air at 170°F is 0.0748 kg/m·hr or 2.08×10^{-5} kg/m·s. Thus, on the basis of Equation (5-78),

$$\text{Pt} = \exp - \frac{6.1 \times 10^{-11}(1)(1.5)(1.182)(1)^2(0.25)^2(153.3)}{(2.08 \times 10^{-5})^2}$$
$$= \exp(-2.36) = 0.095$$

Finally, the penetration can also be obtained from Equation (5-81) by employing the pressure drop value determined from Equation (5-76). Thus,

$$\text{Pt} = 3.47(36.2)^{-1.43} = 0.0205$$

One notes from Example 5-9 that there is a considerable difference in the answers given by the models of Calvert and Hesketh. However, the major point to be gained from Example 5-9 is that large pressure losses are necessary if small values of the penetration are to be achieved for small particles.

The water circulation rate to a venturi scrubber varies from 2 to 12 gal/1000 ft^3. But more significant for this device, as compared with other dry and wet collectors, is the pressure loss. This varies from 3 to 100 in. water, depending upon the removal efficiency desired. The collection efficiency is directly related to the energy expenditure, and hence to the pressure loss. The main advantage of the venturi scrubber, however, is that by taking a large draft loss a very high efficiency can be attained even for very small particles. The efficiency may reach 99 percent in the submicron range, and 99.5 percent for 5-μm particles. One relationship between collection efficiency and particle size for venturi scrubbers is shown in

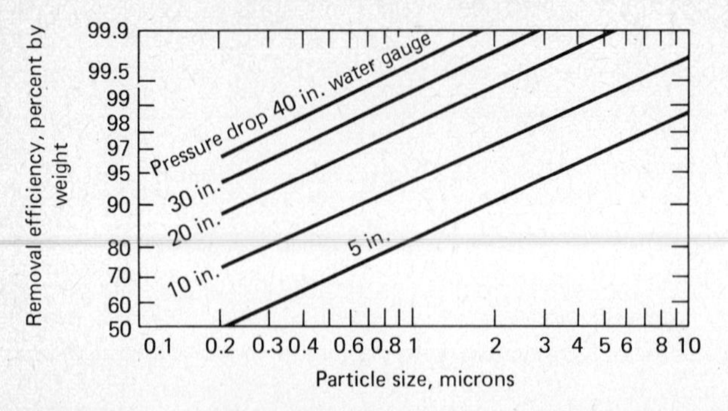

Figure 5-26 Relationship between fractional collection efficiency and particle size in venturi scrubbers. (SOURCE: *Air Pollution Manual.* Part II—"Control Equipment." Detroit, Mich.: American Industrial Hygiene Assoc., 1968.)

Figure 5-26. For particles in the submicron range, it is apparent from the figure that high removal efficiencies require relatively high pressure drops. Data presented elsewhere [24] show efficiencies somewhat lower than those presented in Figure 5-26, for the same pressure loss, but the trends are equivalent.

In order for a collection device to attain these efficiencies, the dust-laden droplets must be removed from the gas stream after passing through the venturi scrubber. A common method of achieving this is to pass the gas stream through a cyclone separator in series with the venturi passage. Another technique is to pass the gas from the venturi through a fluidized packed bed. Modifications of venturi design other than the one shown in Figure 5-24 are available; for example, it can be used with a jet injector, with a water supply through overflow weirs, in parallel with other venturis, and in the configuration known as the flooded-disk venturi.

Although we have restricted the discussion to three basic types of wet collectors (and modifications permitted by impingement baffles), other types are available. The packed-bed scrubber is a towerlike device, which contains any of a variety of packings, including spheres, raschig rings, berl saddles, and so on (see Figure 6-8). Water flows downward through the packing, counterflow to the gas stream. The large wetted surface area enhances the chance that the dust will be removed from the gas before it leaves the packed section of the tower. Subclasses of the packed-bed scrubber include the fixed-bed, fluidized-bed, and flooded-bed scrubbers. The submerged-orifice and the centrifugal-fan types of scrubbers are two other important classes, as well as a general class of mechanical scrubbers.

Wet scrubbers range in size from 500 to 120,000 ft^3/min and cost, as of 1979, from $0.35 to $5.00 per ft^3/min of cleaning capacity. Installation costs are up to 100 percent more and operating costs range from $0.75 to $2.50 per ft^3/min/yr. Maintenance can be high when corrosive materials are being collected. More detailed information on wet scrubbers is given in references 2, 5, 6, 13, and 25.

5-9-D Fabric Filters

Filtration is one of the oldest and most widely used methods of separating particulate from a carrier gas. A filter generally is any porous structure composed of granular or fibrous material which tends to retain the particulate as the carrier gas passes through the voids of the filter. The filter is constructed of any material compatible with the carrier gas and particulate, and may be arranged in deep beds, mats, or fabric. Mat and deep bed filters have large void space amounting to from 97 to 99 percent of the total volume. They are used for very light dust loads and are designed to remain in service for long periods. In general, they are cleaned in place periodically at relatively short intervals. Fabric filters are usually formed into cylindrical tubes and hung in multiple rows to provide large surface areas for gas passage. Fabric filters have efficiencies of 99 percent or better when collecting 0.5-μm particles and can remove substantial quantities of 0.01-μm particles. Typical dust loadings handled are from 0.1 to 10 gr/ft^3 of gas (0.23 to 23 g/m^3).

Woven fabrics usually have air-space-to-cloth area ratios of 1 : 1 to 5 : 1; therefore, some phenomenon other than simple sieve action is responsible for initial particle separation. Of the six general mechanisms of particulate collection discussed in Section 5-8, the important filtering mechanisms are inertial impaction, direct interception, and diffusion. Electrostatic attraction may play a role with certain types of dusts. Particles larger than 1 μm are collected mainly by impaction and direct interception, whereas particles from 0.001 to 1 μm are removed mainly by diffusion and electrostatic attraction. The relative magnitudes of the dust collected by the different mechanisms cannot be calculated at the present time, and additional research is needed in this area.

One of the disadvantages of fabric filters is the necessity of relatively frequent cleaning in order to avoid unreasonable pressure drops. As a result, the basic design of industrial filters is usually predicated on a geometry which lends itself to relative ease of cleaning. However, another geometric consideration must be a large surface area per volume flow rate of polluted gas. The inverse of this ratio is known as the air-to-cloth ratio, or the filtering ratio, and optimum values range from 1 to 8 ft/min (0.5 to 4 cm/s). This is also the superficial gas velocity. One basic method of meeting these criteria is a collection system based on the bag house, shown in Figure 5-27. Fabric cylinders (supported internally in some manner) ranging from 5 to 14 in. in diameter and up to 40 ft long are arranged in vertical rows. Many individual bags must be employed in one bag house when large gas volumes must be cleaned. One large bag house capable of removing 6 tons of dust and fumes per hour from a large automotive foundry consists of 16 compartments having a total of 4000 Dacron bags which are 7.64 in. in diameter and $22\frac{1}{2}$ ft long. This leads to one inherent disadvantage of bag house design, namely, the overall size of the equipment is large in comparison to other competitive types of removal devices.

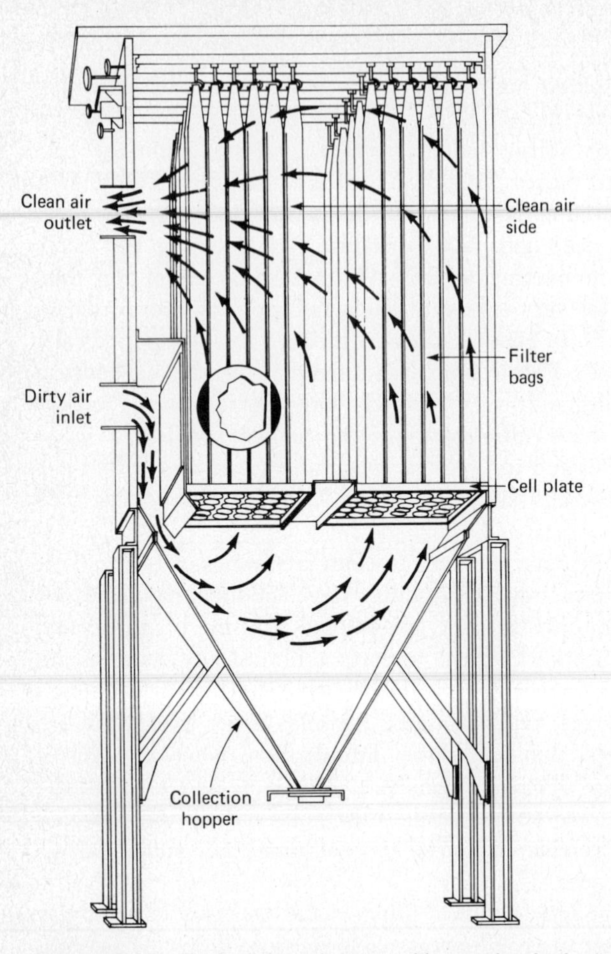

Figure 5-27 Typical bag house with mechanical shaking. (Courtesy of Wheelabrator Frye, Inc., Mishawaka, Ind.)

Generally the dirty gas enters the bag at the bottom and passes through the fabric, while the particulate is deposited on the inside of the bag. Although there is a wide variety of bag house designs, they may generally be classified by the method of cleaning and by whether the operation is periodic (or intermittent) and continuously automatic. Periodic operation requires shutdown of portions of the bag house at regular intervals for cleaning. Continuous automatic operation is required where periodic shutdown is not desirable or practical. (Intermittent operation requires shutting the entire process down in order to clean the filter cloth. The dirty gas bypasses the equipment during the cleaning cycle. This type of design is of limited use, since very few effluents could be discharged to the atmosphere without cleaning, even on a temporary basis.) Cleaning is accomplished in a variety of ways, including mechanical vibration or shaking, pulse jets, and reverse air flow. A bag house with mechanical shaking is shown in Figure 5-27. This type of cleaning frequently provides the lowest cost per volume flow rate. The maximum filtering velocity for shaker bag houses with woven fabrics is in the range of 2 to 3.5 ft/min (0.6 to 1.0 m/min).

A system with reverse air flow is shown in Figure 5-28. The particulate is collected on the inside of the bag, similar to the mechanical shaking method. At the proper time the flow of polluted air is cut off from the compartment, and cleaning air flows through the bag in the opposite direction. Economics of reverse-flow cleaning dictates the use of bags which are 20 ft (6–7 m) or taller. Bag diameters typically are in the 8 to 12 in. (20–30 cm) range, and recommended filtering velocities are around 6 to 12 ft/min (2 to 4 m/min). Figure 5-29 illustrates the pulse-jet system. Particulate is collected on the outside of the bags. At an appropriate time a short pulse of compressed air is directed downward through a venturi at the top of a bag. The pulse passes quickly down the bag, knocking large dust layers from the bag. These large chunks of dust settle by gravity to the bottom of the bag house. Since the time of cleaning a bag is very short, and only a fraction of the bags are cleaned at one time, continuous flow is maintained through the bag house. Typical filtering velocities in a pulse-jet baghouse are in the range of 5 to 15 ft/min (2 to 5 m/min), dependent upon the application. Bag heights in this latter case are usually less than 15 ft (5 m), since the bottom of extremely long bags frequently are not cleaned too well. Further descriptions of these systems appear in the literature [2, 6, 25].

A variety of fabrics is available. Among those in common use are wool, cotton, nylon, glass fiber, polyesters, and aromatic polyamides. Choice of fabric depends upon chemical composition, temperature, and moisture content of the gas, as well as the physical and chemical composition of the particulate. The chemical composition of both gas stream and particulate has

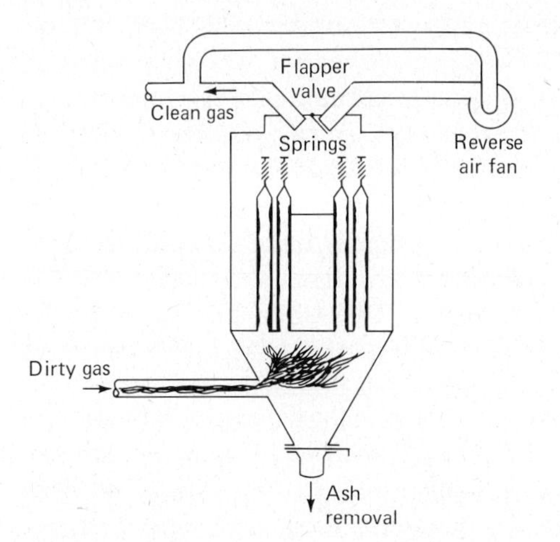

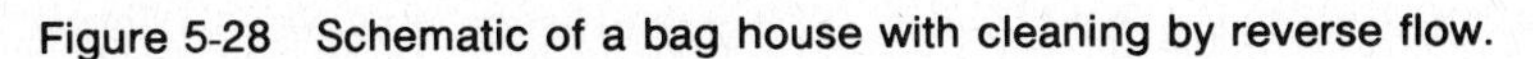
Figure 5-28 Schematic of a bag house with cleaning by reverse flow.

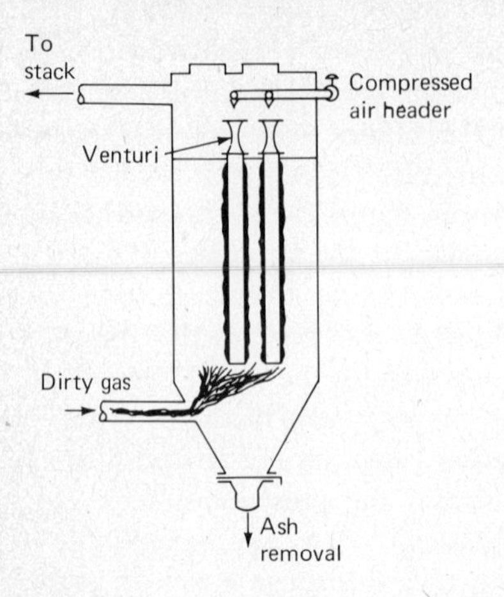

Figure 5-29 Schematic of a bag house with cleaning by pulse jet.

an effect on the rate at which the fabric wears out. While a cotton or wool fabric would be acceptable for gas up to a temperature of 175° to 200°F (350°K), a glass fiber is more suitable when the gas temperature approaches 500° to 550°F (about 550°K). This latter range of temperature is about the current upper limit, although research continues on fiber development with emphasis on increasing the limit beyond 550°F. The advantage of higher operating temperatures is the concomitant reduction in thermal control requirements. That is, less precooling of a hot gas stream before it enters the fabric filter equipment is required. Moreover, excessive gas cooling is not usually permissible, because bag houses commonly will not work for a gas which is cooled below the dew point. However, the savings brought about through less gas cooling is offset somewhat by the increase in gas volume passing through the filters as a result of increased temperature.

The collection theory for inertial impaction on filter beds is much more complex than the simple theory developed earlier for wet collectors (see Section 5-9-C). In this latter case it is appropriate to model the impaction as occurring between two spherical objects with a certain relative velocity. In the case of fabric filtration, the initial model might be one of a moving sphere approaching a stationary cylinder. However, once an initial layer of dust adheres to the fibers, the geometry becomes quite complicated. The retained dust as well as the fabric fibers act as targets for the deposition of additional particulate. Obviously the targets are continually changing spatially with time. Nevertheless, a reasonable starting place for a filtration model is to consider the impaction of spheres on cylinders.

Consider a dust-laden gas approaching a cylindrical fiber of diameter d_f. If we project the cross-sectional area of the fiber (per unit length of fiber, for example) upstream in a direction parallel to the flow, only those particles in the undisturbed flow upstream which lie within the projected area can ever impact directly on the cylinder. Consequently, an impaction or target efficiency in this case is defined as the fraction of particles of a given size which pass through the projected area upstream and subsequently strike the collecting surface of the cylinder. This efficiency is not 100 percent for the simple reason that some of the particles are carried around the cylinder along the streamlines of the gas flow, which diverge around the cylinder [see Figure 5-12(a)]. Note that the efficiency is for a given size and thus may be expected to vary considerably with size because of the dependency of inertia on mass. In a manner similar to the theory developed earlier for wet collectors, the target efficiency is found to be a function of the impaction or separation number N_I given by Equation (5-64), namely:

$$N_I = \frac{V_p d_p^{\,2} \rho_p K_C}{18 \mu_g d_f} \tag{5-82}$$

where the droplet diameter d_D has been replaced in this case by the fiber diameter d_f. It is assumed that Stokes' law is valid for the particle motion. One experimental correlation of the impaction or target efficiency versus the impaction number is shown in Figure 5-19 for cylindrical targets. For a given particle size and density and gas stream properties this correlation indicates that the highest target efficiency will be achieved when the filter is composed of small-diameter fibers.

Another important consideration is the pressure drop across fabric filters. As the filter "cake" accumulates on the supporting fabric, the removal efficiency would be expected to increase. At the same time, however, the resistance to flow also increases. For a relatively clean filter cloth, the pressure drop is around $\frac{1}{2}$ in. water (1.2 mbar) and the efficiency of removal is low. For a draft loss of 2 to 3 in. water, after sufficient cake buildup, the efficiency reaches 99 percent. When the pressure drop approaches 5 to 6 in. water, it usually is necessary to clean the filter by some method in order to reduce the pressure drop to a more reasonable value. Otherwise the energy expenditure becomes excessive. The overall pressure drop is the sum of that due to the fabric plus that contributed by the dust layer. The flow through the fabric and the deposited dust layer is assumed to be viscous. Hence the pressure drop for both the dust layer and the cleaned filter may be represented by Darcy's equation, which states that, in general,

$$\frac{\Delta P}{x} = \frac{V \mu_g}{K} \tag{5-83}$$

where K is the dust or filter permeability, V is the superficial gas velocity, and x is the particulate bed depth or filter thickness.

Note that V in the Darcy equation is the superficial velocity through the bed, that is, the volumetric flow rate approaching the filter divided by the cloth area. Equation (5-83) is really a definition of permeability, K, since all of the other terms in the equation are readily determined parameters. K is a fairly difficult quantity to predict without direct measurement, since it is a function of properties of the deposited dust such as porosity, specific surface area, pore size distribution, and particle size distribution, among others. The dimension of the term permeability is length squared.

The overall pressure drop, ΔP_o, is the sum of the pressure drop across the filter cloth and across the deposited particulate. Hence it may be written, in terms of the Darcy equation, as

$$\Delta P_o = \Delta P_f + \Delta P_p = \frac{x_f \mu_g V}{K_f} + \frac{x_p \mu_g V}{K_p} \tag{5-84}$$

where the subscripts f and p refer to the cleaned fabric and the particulate or dust layer, respectively. The filter cloth pressure drop, ΔP_f, should essentially be a constant for a given cloth material and dust. Consequently, the overall pressure drop depends primarily on the variation of ΔP_p as the dust layer builds up on the cleaned cloth. For given operating conditions (gas viscosity and superficial velocity), ΔP_p is primarily a function of the dust permeability K_p and the dust-layer thickness, x_p. The dust-layer thickness, in turn, is a direct function of the time of operation, t.

An expression for the variation of ΔP_p, the pressure drop across the dust layer, with time can be developed from first principles. This is accomplished by noting that the mass of dust collected in time t (after startup) equals the volume flow rate of gas times the time interval times the mass of dust in the gas stream per unit volume, L_d. However, the mass collected also equals the density of the particulate, ρ_c, on the filter surface times the volume of dust collected in time t. Equating these two quantities, we can write the following expression:

$$\text{mass collected} = (VA)(t)(L_d) = \rho_c(Ax_p) \tag{5-85}$$

where A is the cross-sectional area of the filter, V is the gas velocity normal to the filter, and L_d is known as the dust loading. Equation (5-85) shows that $x_p = VL_d t/\rho_c$. Hence the pressure drop across the fresh dust deposit becomes

$$\Delta P_p = \frac{x_p \mu_g V}{K_p} = \frac{VL_d t}{\rho_c}\left(\frac{\mu_g V}{K_p}\right) = \frac{V^2 L_d t \mu_g}{K_p \rho_c} \tag{5-86}$$

Since μ_g, ρ_c, and K_p are unique values of a given gas stream and particulate, we lump them together into a resistance parameter R_p, such that by definition

$$R_p = \frac{\mu_g}{\rho_c K_p} \tag{5-87}$$

As a result, Equation (5-86) can be simplified to the form

$$\Delta P_p = R_p V^2 L_d t \tag{5-88}$$

L_d and V are assumed not to vary during the filter cycle in this development. For given gas characteristics and filter cake permeability, it is seen that the pressure drop across the deposited particulate varies linearly with dust loading in the gas stream and with time. It also varies as the square of the superficial gas velocity, on the basis of this theory. It has been indicated elsewhere [26] that the pressure drop varies more like an exponential function of the filtering velocity, in practice, than like a squared function.

Example 5-10

Air at 170°F passes through a fabric filter for a period of 5.40 hr, after which the total pressure drop is measured as 4.74 in. water. The filter cake density is 1.28 g/cm^3, and the residual pressure drop across the cleaned filter before the test is 0.55 in. water. The air velocity is maintained at 4.20 ft/min during the test, and the initial dust loading is 14.0 gr/ft^3. Estimate the permeability, K_p, of the dust layer in units of square feet.

SOLUTION

Equation (5-86) will be used as the basis for the calculation, since it contains time as a variable. Rewriting this equation, we find that

$$K_p = \frac{V^2 L_d \mu_g t}{\rho_c \Delta P_p}$$

The viscosity of air at 170°F is 0.0503 lb/ft·hr from Appendix B, and from the same source we note that 1 in. water pressure difference equals 0.0361 lbf/in.^{2}. The pressure drop across the deposited dust is (4.74–0.55) in. water, which then equals 0.151 lbf/in.^{2} Employing the proper values, we find that

$$K_p = \frac{(4.20\ \text{ft/min})^2(14.0\ \text{gr/ft}^3)(0.0503\ \text{lb/ft·hr})(5.40\ \text{hr})}{(1.28\ \text{g/cm}^3)(0.151\ \text{lbf/in.}^2)(32.2\ \text{lbm·ft/lbf·s}^2)}$$

$$= 10.78\frac{(\text{gr})(\text{cm}^3)(\text{in.}^2)(\text{s}^2)}{(\text{min}^2)(\text{ft}^3)(\text{g})} = 4.76 \times 10^{-11}\ \text{ft}^2$$

where the factor 32.2 has been introduced to relate lbm and lbf by Newton's second law, and conversion factors found in Appendix B have been used to convert to the final answer. It is usually found that the permeability of dusts is an extremely small number when expressed in units of feet squared.

The ratio of the particulate pressure drop, ΔP_p, to the superficial gas velocity, V, is called the filter drag, S. It is frequently expressed in units of in. water/ft/min, although other units would be equally appropriate, such as N·min/m^3. It is convenient to derive an expression for S in terms of the mass of collected dust at time t per unit area normal to the direction of gas flow,

W. First, Equation (5-88) may be written in a form for the filter drag, namely,

$$S = \frac{\Delta P_p}{V} = R_p V L_d t \tag{5-89}$$

In addition, in conjunction with Equation (5-85), the value of W defined above may be expressed as

$$W = \frac{\text{mass collected}}{A} = \frac{VAtL_d}{A} = VL_d t \tag{5-90}$$

By combining these two equations, we see that the filter drag is given simply by

$$S = R_p W \tag{5-91}$$

The units of W might be grains per square foot or grams per square meter. For given conditions of gas flow and particle characteristics, Equation (5-91) indicates that the filter drag should vary linearly with W. In actuality, the trend appears more like the curve in Figure 5-30(a). In the initial nonlinear portion of the curve, a new filter cake is being formed on a cleaned filter cloth. Hence the surface of the dust layer is fairly irregular for an initial time period, and the resistance to flow increases rapidly until nonhomogeneous regions in the dust layer are filled. Once a fairly uniform bed is established, the filter drag then varies linearly with areal density, W.

Once the cleaning cycles begin, the position of the filter drag curve is normally quite different from that shown in Figure 5-30(a). Referral to Figure 5-30(b) shows several possibilities. Curve (1) is the situation when the maximum possible cleaning occurs. Since some dust remains on the fabric filter after cleaning, a residual drag, S_R, exists. Curve (2) represents a highly efficient cleaning of a woven fabric, where the residual loading on the fabric may be 5 to 10 percent of the terminal loading. The value of W after cleaning in this case might be around 20 to 50 g/m^2 (0.07 to 0.15 oz/ft^2). On woven fabrics the value of W_T at the end of the cycle may reach 750 to 1000 g/m^2. Curve (3) in Figure 5-30(b) represents an average cleaning range for mechanical shaking, where a sizable fraction of the dust has remained on the filter surface. The actual position of the performance curve depends upon the type of fabric used as well as the method of removal. From experimental measurements of this type, the permeability of the deposited dust may be determined. Values of the residual drag S_R typically vary from 0.2 to 0.8 in. water/ft/min (100 to 500 $N\cdot\text{min/m}^3$) for a cleaned fabric with a minimum of retained dust.

Since the dust collected on a filter cloth acts as the filtering medium, the removal of the dust layer during cleaning does affect the efficiency of collection. This may be noted in Figure 5-31, where the fractional efficiency is plotted versus particle size for a typical fabric filter in a cleaned state and

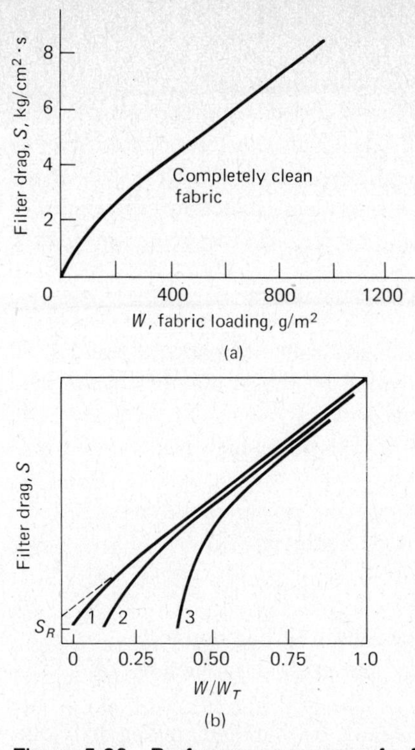

Figure 5-30 Performance curves of a fabric filter in terms of the parameters S and W.

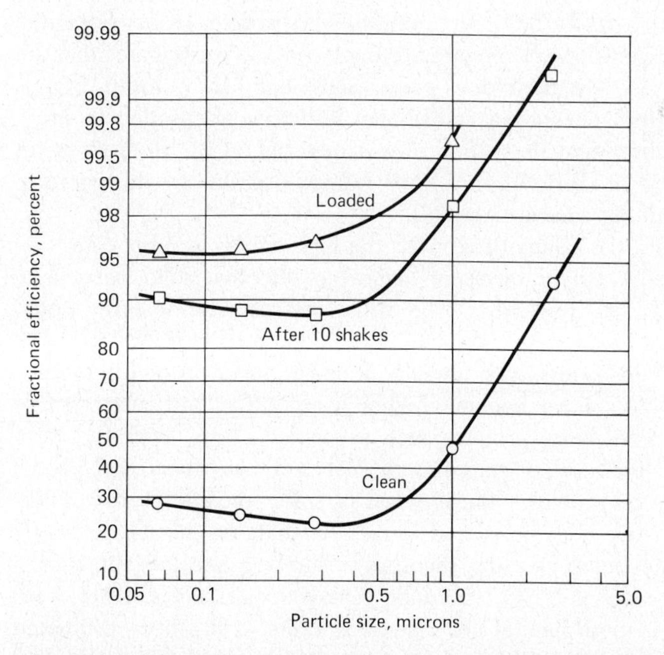

Figure 5-31 Fractional efficiency curves of a fabric filter. (Courtesy of the Torit Corporation.)

loaded state. In the cleaned state the efficiency dipped below 90 percent for sizes between 0.1 and 0.5 μm, but it remained above 98 percent for particles above 1 μm. As the loaded state was approached, the fractional efficiency was above 95 percent at all sizes, and was above 99.6 percent for sizes above 1 μm. In general, fabric filters will have fractional collection efficiencies of 90 percent and above after a few cleaning cycles, and the overall efficiency will usually exceed 99 percent.

The advantages of fabric filters include:

1. High collection efficiency over a broad range of particle sizes.
2. Extreme flexibility in design provided by the availability of various cleaning methods and filter media.
3. Volumetric capacities in a single installation which may range from 100 to 5 million ft^3/min.
4. Reasonable operating pressure drops and power requirements.
5. Ability to handle a diversity of solid materials.

Among acknowledged disadvantages are:

1. Space factors may prohibit consideration of bag houses.
2. Possibility of explosion if sparks are present in vicinity of a bag house.
3. Hydroscopic materials usually cannot be handled, owing to cloth cleaning problems.

This last item severely limits the nature of a gas stream that can be cleaned through the use of a fabric filter. In addition, wet particles can agglomerate on a filter cloth when the equipment is trying to handle waste gases that are at temperatures close to their dew point. Sufficient heat transfer from a waste gas as it approaches a bag house may lead to condensation of water vapor, and the subsequent deposition of wet material on the filter. Another major disadvantage of fabric filters is their limited usefulness with regard to gas streams containing caustic materials.

The fabric filter is competitive with the high-energy scrubber and the electrostatic precipitator in terms of achieving effective solid particulate control, especially with fine dusts. The initial cost of a fabric filter would probably be somewhat higher than that of a venturi scrubber, but the operating cost would be less, with the overall cost in favor of the filter. Since the late 1970s filters have become highly competitive with electrostatic precipitators for very large projects such as power plants, with the filter holding the edge for smaller scale design [27]. For new fossil fuel plants, future technology may involve the removal of sulfur dioxide by dry scrubbing. (Dry scrubbing is a process designed so that the scrubbing liquid droplets evaporate before leaving the spray chamber. See Section 7-4-H.) Then both particulate and SO_2 could be removed by a fabric filter, with some reductions in both capital and operating costs. It has been estimated [28] that a spray drier-fabric filter dual system for dust and SO_2 might cost

\$80/kW based on 1978 dollars. In comparison, a combination of an electrostatic precipitator with a wet scrubbing (lime-limestone) system might run around \$165/kW. When cloth filters are used to collect fly ash from coal-fired power plants, the air-to-cloth velocity is kept low (around 2 ft/min or 1 m/min) and reverse flow cleaning is used.

The size of the bag house (number of bags) determines to a large extent the cost of the filtering system. Initial costs, as of 1979, are roughly \$7.00 per ft^3/min capacity with installation costs about 0.75 to 1.3 times equipment costs. Maintenance costs are high, ranging from 10 to 25 percent of the initial cost per year, as the bags must be replaced. The proceedings of a comprehensive symposium on the use of fabric filters for the control of submicron particles appears in the literature [29]. A handbook available from the federal government [30] covers fairly extensively the cloth characteristics, performance, design details, and economics of fabric filters.

5-9-E Electrostatic Precipitators

Particulate and aerosol collection by electrostatic precipitation is based on the mutual attraction between particles of one electrical charge and a collecting electrode of opposite polarity. This technique has been an important part of industrial cleaning methods since pioneering work by F. G. Cottrell in 1910. Its advantages are a capacity to handle large gas volumes, high collection efficiencies even for submicron-size particles, low energy consumption and draft loss, and ability to operate with relatively high-temperature gases. Electrostatic precipitators have been built for volumetric rates from 100 to 4,000,000 ft^3/min, and they are used to remove particles from 0.05 to 200 μm. For most applications, the collection efficiency runs from 80 to 99 percent. With the advent of stricter air pollution codes, efficiencies in the range of 98 to 99 percent have become quite common, and in some cases are in the 99.5 to 99.9 percent range. The pressure drop is generally quite small, extending from 0.1 to 0.5 in. water (0.25 to 1.25 mbar). Gas temperatures up to 1200°F (920°K) and pressures up to 150 psi (10 bars) can be accommodated. Desirable characteristics such as these account for the wide use of electrostatic precipitators by industry, especially in the electrical power generating field. In addition, it is important to note that the energy expended in separating particulate from a waste gas stream by means of an electrostatic precipitator acts solely on the particulate, and not on the gas stream. This is unique for collection equipment in the air pollution field, since other devices based on different principles of separation require that energy be expended on the entire gas stream in order to accomplish the desired effect. Some typical data from industrial applications of the electrostatic precipitator are presented in Table 5.10.

Several basic geometries are used in the design of electrostatic precipitators. One of these is a tube type, in which the electrodes consist of wires suspended axially in a tube. A very high dc voltage is applied between the

Table 5-10 SOME TYPICAL DATA ON ELECTROSTATIC PRECIPITATOR APPLICATIONS

INDUSTRY (APPLICATION)	GAS FLOW (10^3 acfm)	TEMPERATURE °F(°C)	DUST CONCENTRATION, gr/ft^3(g/m^3)	TYPICAL EFF, %
Electric power (fly ash)	50–750	270–600 (130–320)	0.40–5 (0.9–12)	98–99.6
Portland cement (kiln dust)	50–1000	30–750 (150–400)	0.50–15 (1.1 – 35)	85–99 +
Steel (open hearth)	30–75	100–150 (40–65)	0.02–0.5 (0.04–1.1)	95–99
Pulp and paper (kraft mill fume)	50–200	275–350 (135–180)	0.50–2 (1.1–4.5)	90–95
Petroleum (catalyst recovery)	50–150	350–550 (180–290)	0.10–25 (0.2–60)	99–99.9

SOURCE: *Air Pollution Engineering Manual*, AP-40, 2nd ed., EPA, May, 1973.

wire and the tube, and the dirty gas flows down the tube and across the electric field established between the electrodes. The schematic of Figure 5-32 shows the design typical of a full-sized commercial unit. Although only two vertical plates are shown, a large number of parallel plates normally would be used, depending on the number of parallel passages necessary to accommodate the required volume flow rate. The flow usually is horizontal, and the passageways are fairly narrow—on the order of 8 to 10 in. (20 to 25 cm). The overall height of a plate may be as great as 30 to 40 ft, with a length of 25 to 30 ft. The high-voltage wires must be hung precisely on the center lines of each gas passage. In most cases, the wires are charged at 20 to 100 kV below ground potential, with 40 to 50 kV being fairly typical. (If the air leaving the precipitator goes directly to an inhabitable region, the center electrode is charged positively to avoid excessive ozone formation.) The gas velocity is generally held within the range of 1 to 20 ft/s. Gas flow at 8 ft/s through a passageway 9 in. wide and 30 ft high would have a volume rate of 10,800 ft^3/min. Higher overall flow rates could be accommodated by placing a number of 9-in.-wide sections in parallel.

The very high voltage differential between electrodes causes electrons to pass at the high rate from the center wire into the passing gas stream. The electrons in turn attach themselves to the gas molecules passing through the device and form negative ions. The existence of the heavily ionized gas in the vicinity of the wire is made evident by a visible blue corona effect. The corona power input may range from 0.05 to 0.5 kW per 1000 ft^3/min of gas flow. Under the influence of the large electrostatic force present, the negative ions migrate toward the grounded outer plates, while positive ions return to the center wire (which is negative relative to the grounded plates). Thus the first physical step in the collection mechanism is one of gas ionization. The second step is the charging of the dust particles in the gas stream. This charging results from the collision of negatively charged gas ions

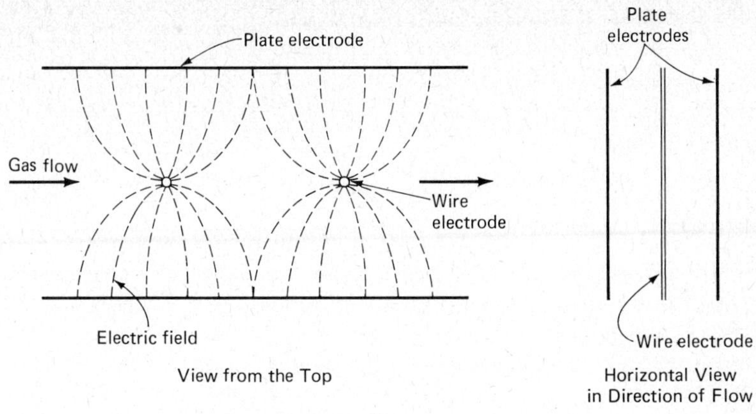

Figure 5-32 (a) Schematic of a plate and wire electrostatic precipitator. (b) Assembled electrostatic precipitator. (Courtesy of Research-Cottrell, Inc., Bound Brook, N.J.)

with the particulate. There is no difficulty in charging the dust particles, since the ions are present in abundance and their size is at least several orders of magnitude smaller than even submicron-size dust. Theoretically, the limited charge q given to a spherical particle of diameter d_p greater than approximately 1 μm is

$$q = p\pi\epsilon_o E_c d_p^{\ 2} \qquad \text{(5-92)}$$

where $p = 3D/(D+2)$, D is the dielectric constant for the particulate, and E_c is the strength of the charging field. Since the dielectric constant for many types of dusts falls between 2 and 8, the factor p in Equation (5-92) typically lies between 1.50 and 2.40.

The next step in the collection mechanism is migration of the charged dust particles to the plate electrodes, where dust collection occurs. The speed at which the migration takes place is known as the migration or drift velocity, w. It depends upon the electrical force on the charged particle as well as the drag force developed as the particle attempts to move perpendicular to the main gas flow toward the collecting electrode. The electrostatic force is proportional to the charge on the particle and the precipitating or collecting field strength, E_p. Employing Equation (5-92), we find that the electrostatic force F_e is given by

$$F_e = qE_p = p\pi\epsilon_o E_c E_p d_p^{\ 2}$$

The drag force on a particle which is in the Stokes' flow region is represented by

$$F_d = \frac{3\pi\mu_g d_p w}{K_C} \qquad \text{(5-93)}$$

where K_C again is the Cunningham correction factor which should be applied for particles with a diameter less than roughly 10 μm. [K_C may be evaluated by Equation (5-27).] Upon equating the electrical and drag forces, the drift velocity for spherical particles in the Stokes' flow region (d_p of roughly 1 to 100 μm in air) is given by

$$w = \frac{p\epsilon_o E_c E_p d_p}{3\mu_g} K_C \qquad \text{(5-94)}$$

where w is in meters per second, μ_g in kg/s·m, d_p in meters, E in volts per meter, and the permittivity $\epsilon_o = 8.854 \times 10^{-12}$ coulombs/volt-meter (C/V·m). As a result,

$$w = \frac{2.95 \times 10^{-12} p E_c E_p d_p}{\mu_g} K_C \qquad \text{(5-95)}$$

when employing the units noted above. When μ_g is expressed in kg/m·hr and d_p in microns, then the constant in Equation (5-95) becomes 1.06×10^{-14}.

The viscosity of air at room conditions (25°C and 1 bar) is 0.067 kg/m·hr. This theoretical equation shows that the migration velocity is directly proportional to the particle diameter and the square of the field strength (if E_c and E_p are equal) and inversely proportional to the gas viscosity. While essentially independent of pressure, the viscosity of air increases by roughly 50 percent between 300° and 500°K (80° to 460°F). Hence the drift velocity is sensitive to changes in temperature. In practice, the actual drift velocity may deviate considerably from the theoretical value predicted by Equation (5-95). The major reason is that actual precipitators cannot be modeled by pure electrostatic mass transfer normal to a laminar gas flow. Various degrees of turbulent diffusion will be present, along with inertial effects. A phenomenon known as electric wind also is thought to contribute to the migration. The criterion of laminar flow for Stokes' law is violated when $\mathrm{Re} > 1$, roughly. This condition may occur under various combinations of large particle size, high gas densities due to elevated pressures, and high electric fields. Nevertheless, the particle migration velocity is a fundamental quantity in all theories which attempt to explain and predict electrostatic precipitator performance. If possible, its value in design consideration should be based on observations of equipment working under conditions similar to those for which the design is intended. For particles smaller than 5 μm the theoretical migration (drift) velocity is usually less than 0.3 m/s. For fly ash the migration velocity typically lies between 0.01 to 0.2 m/s (0.03 to 0.6 ft/s) for the small particles of interest in electrostatic particulate collection devices.

The third physical mechanism for electrostatic particulate collection is the actual deposition of the charged particles on the electrode, with subsequent buildup of a dust layer. Adhesive, cohesive, and electrical forces must be sufficient to prevent reentrainment of the particles in the gas stream. One property of the dust layer which is extremely important in precipitator operation is the dust electrical resistivity. Owing to the widely varying nature of industrial dusts, the resistivity may vary from 10^{-3} to 10^{14} ohm·cm. When the resistivity is less than 10^4 ohm·cm, there is a rapid movement of charge from the deposited dust to the collector plate. Thus insufficient electrostatic charge remains on the collected dust particles to hold them together. Reentrainment back into the gas stream frequently results, and collection efficiency suffers. Carbon black, an industrial product of importance, is an example of a low-resistivity dust. On the other end of the scale, resistivities greater than about 10^{10} ohm·cm are a major source of poor performance in precipitators. First, a sizable fraction of the total voltage drop between electrodes occurs across a high-resistivity dust layer, as a result of the electrical insulating effect. Hence only a portion of the total corona power is available to ionize and drive the charged particles to the collection electrode. A second problem due to high resistivity is known as back corona or back ionization. This effect occurs when the voltage drop across the layer exceeds the dielectric strength of the layer. Air trapped in the collected dust layer becomes ionized as a result of the large potential drop across the layer.

Any positive ions formed will tend to migrate away from the collector plate and neutralize the ionized particles approaching the plate. This decreases the amount of particulate deposited. Also, ionization can lead to sparking in the dust layer, which can blow collected dust back into the gas stream. These effects reduce the collection efficiency of a precipitator.

Electrostatic precipitation is most effective in collecting dust in the resistivity range of 10^4 to 10^{10} ohm·cm. Since many industrial dusts do not fall into this range, it frequently is necessary to change operating conditions in order to enhance collection effciencies. Two gas properties that have a sizable influence on dust resistivity are temperature and humidity. The effect of these two parameters is shown in Figure 5-33(a), which is representative of cement kiln dust. The dome shape of the temperature-resistivity curve at any given moisture content is fairly typical of industrial particulate. Note that the peak of the curves shifts to lower temperatures as the humidity is lowered. For reasonably dry dusts, the resistivity peak generally is around 300°F (420°K), and the curves in Figure 5-33(a) substantiate this value. The shape of the curves is due to a change in the mechanism of conduction through the bulk layer of particles as the temperature is altered. Below 300°F the predominant mechanism is *surface conduction*. For this type of conduction the electric charges are carried in a surface film adsorbed on the particulate. As the temperature is increased above 300°F, the phenomenon of adsorption becomes less effective. The other mechanism is *volume* or *intrinsic conduction*. Volume conduction involves passage of electric charge through the particles. Such passage obviously depends upon the temperature and composition of the particles. For most materials the relationship between electrical resistivity ρ_e and absolute temperature T due to intrinsic conduction is given by an Arrhenius-type equation,

$$\rho_e = A \exp\left(\frac{-E}{kT}\right) \tag{5-96}$$

where A is a constant, E is the electron activation energy (a negative value), and k is Boltzmann's constant. The left side of the curves in Figure 5-33 is more difficult to match with a theoretical expression, since the conduction occurs through an adsorbed film. The influence of even small quantities of moisture (humidity) in the gas stream is shown in Figure 5-33(b), which is based on a typical fly ash. The water vapor content in a flue gas from hydrocarbon combustion is in the range of 5 to 10 percent in most cases.

Another approach to achieving electrical resistivities in the desired range is the addition of conditioning agents to the gas stream. This is usually done in the lower temperature range, in order to enhance surface conduction. The addition of small amounts of SO_3 or NH_3, which act as electrolytes when adsorbed on the deposited dust particles, sharply reduces resistivity. These two substances, and compounds closely related to them, are the only conditioning agents which are economically and technically feasible in commercial practice [34]. Figure 5-34(a) shows the conditioning effect of low

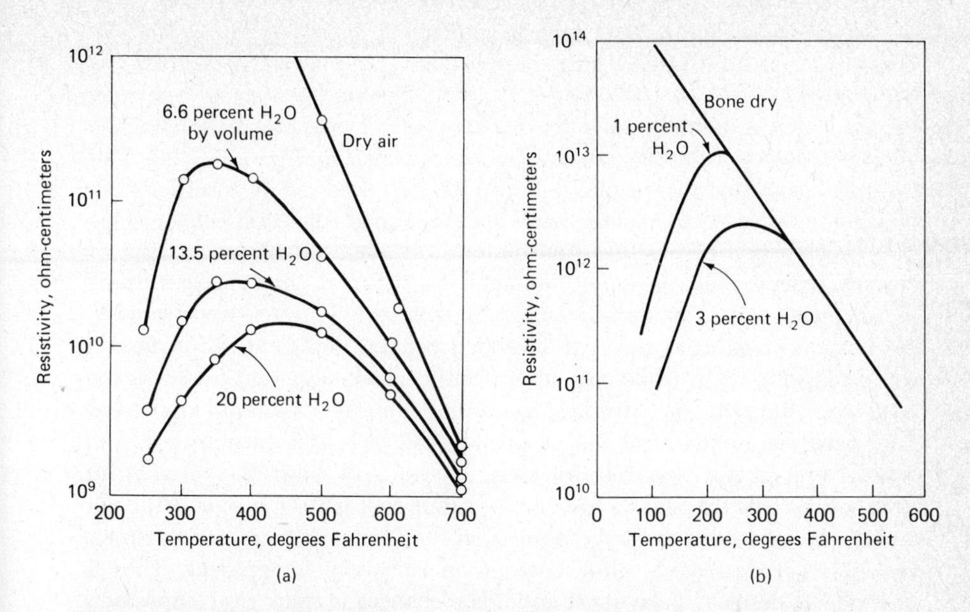

Figure 5-33 Effect of temperature and moisture content on the electrical resistivity of dusts. (a) Moisture conditioning of cement kiln dust; (b) effect of gas humidity in increasing the conductivity of a typical fly ash. (SOURCE: H. J. White. "Resistivity Problems in Electrostatic Precipitation." *J. Air Pollu. Control Assoc.* **24** (April 1974): 314.)

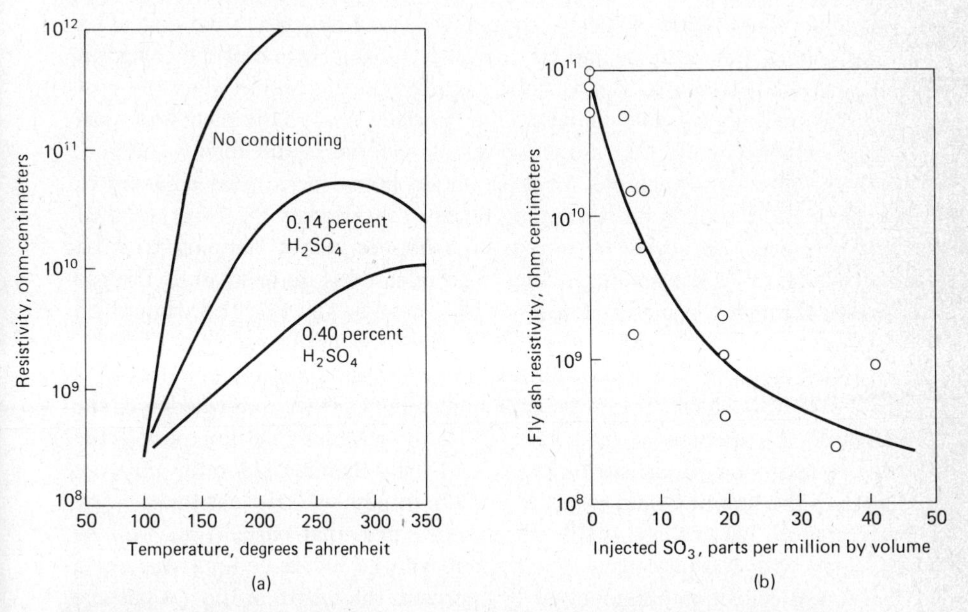

Figure 5-34 Effect of conditioning of fly ash by (a) H_2SO_4 fume and (b) SO_3 injection into the flue gas. (SOURCE: H. J. White. "Resistivity Problems in Electrostatic Precipitation." *J. Air Pollu. Control Assoc.* **24** (April 1974): 314.

concentrations of sulfuric acid on the resistivity of fly ash. For this particular fly ash a concentration of 0.40 percent H_2SO_4 fume at a typical stack gas temperature of 300°F (420°K) lowers the resistivity to the 10^{10} ohm·cm region, which is desirable for effective precipitator operation. The effect of the direct injection of SO_3 into a flue gas is shown in Figure 5-34(b). This curve indicates that the injection of only 10 to 20 ppm so SO_3 is sufficient to reduce the resistivity to a value that will permit good collection efficiencies.

The addition of SO_3 has some interesting implications in the light of current federal SO_2 emission standards. In the past, many coal-burning power plants and other installations considered the SO_2 produced from the combustion of sulfur in the coal an added benefit. Sufficient sulfur dioxide was oxidized to the trioxide and subsequently adsorbed to help condition the dust layer. But with the advent of low-sulfur fuels, necessitated by emission standards, the resistivity of the particulate changes sufficiently to alter the original precipitator removal efficiency. Figure 5-35 shows the variation in resistivity of coal fly ash as a function of sulfur content of the coal. Although more data are needed to fix the position of the curves with greater certainty, the effect of decreased sulfur content on resistivity is apparent. Thus a precipitator designer today must anticipate changes in stack gas composition predicated on changing federal standards. The use of SO_3 to enhance precipitator performance with low-sulfur fuels does not increase SO_x emissions, since the SO_3 is adsorbed and removed with the collected particulate.

The deposited particulate matter is usually removed from the collection electrodes by rapping or vibration. Both solid and liquid particles can be collected. When liquid particles are involved, they migrate to the collection plates, where they coalesce and fall by gravity to the bottom of the collector. Rapping of electrodes to remove solid particles usually occurs when the dust layer has built up a $\frac{1}{8}$- to $\frac{1}{4}$-in. layer. The particles next to the plate lose most of their charge, while the latest deposit of material is still highly charged. With the relatively weak electrical attraction between particles adjacent to the plate and the plate itself, rapping tends to shear away the collected dust in large pieces. These fall by gravity to a hopper below. Rapping must be controlled so that the amount of fine dust released is quite small. If the gas velocity through the plate passages is low enough, any fine dust should be recharged and redeposited before leaving the precipitator.

The length of the precipitator passage required for the removal of a particular size of particle can be estimated roughly from a knowledge of the drift velocity, discussed earlier. After allowing for a charging time period, the time required for a particle to migrate to the collection electrode must be less than the time it would take the particle to pass with the gas through the precipitator. When these times are exactly equal, that particle size will be collected with 100 percent efficiency, but with no tolerance for error. For a theoretical collection efficiency of 100 percent, the length of the gas passage required is given by

$$L = \frac{sV_g}{w} \tag{5-97}$$

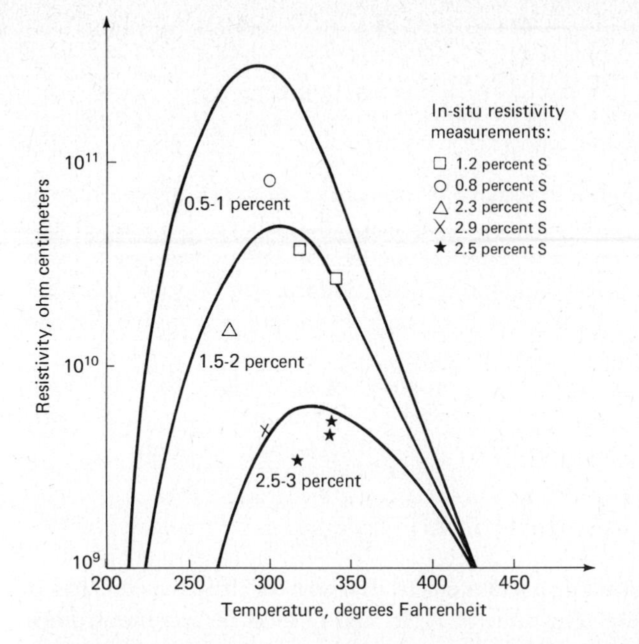

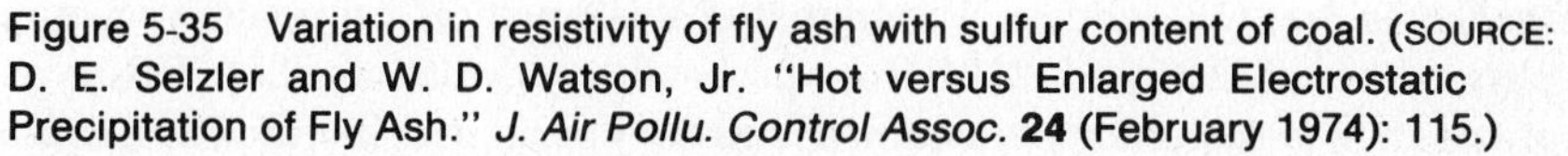

Figure 5-35 Variation in resistivity of fly ash with sulfur content of coal. (SOURCE: D. E. Selzler and W. D. Watson, Jr. "Hot versus Enlarged Electrostatic Precipitation of Fly Ash." *J. Air Pollu. Control Assoc.* **24** (February 1974): 115.)

where L is the length of the collecting electrode, s is the distance between charging and collecting electrodes, and V_g is the gas velocity in the flow passage. A consistent set of units must be used in Equation (5-97). An additional length of upstream flow passage must be added to account for the time required for charging.

Example 5-11

Consider a plate-type collector with an overall spacing of 23 cm (9 in.) and an applied voltage of 50 kV. The mean gas velocity through the collector is 1.5 m/s. Estimate the collecting plate length required for a collection efficiency of 100 percent for 0.5-μm particles at 420°K.

SOLUTION

Assuming a value of 2 for p, we first use Equation (5-95) to evaluate the migration or drift velocity, w.

$$w = \frac{1.1 \times 10^{-14} p E^2 d_p}{\mu_g} = \frac{1.1 \times 10^{-14}(2)(50{,}000/0.115)^2(0.5)}{0.0863}$$
$$= 0.024 \text{ m/s}$$

On the basis of Equation (5-97), the theoretical length of the collecting electrode is

$$L = \frac{sV_g}{w} = \frac{(0.23/2)(1.5)}{0.024} = 7.2 \text{ m}(23.6 \text{ ft})$$

The actual length required to remove 99 percent or more of a given size may vary widely from the theoretical value calculated above, because of the assumptions imposed on the theoretical model. Deviations from such basic assumptions as a uniform precipitating field, uniform gas velocity, spherical particles, uniform electrical properties, and so on, are responsible for the differences between theory and practice.

Several equations have been proposed which relate the collection efficiency of an electrostatic precipitator to various operating parameters. One of the expressions frequently quoted is the Deutsch equation,

$$\eta = 1 - \exp\left(\frac{-Aw}{Q}\right) \tag{5-98}$$

where A is the area of the collection electrodes, w is the drift velocity, and Q is the volume flow rate. The units of A, w, and Q must be consistent, since the factor Aw/Q is unitless. The format of this equation can be derived from physical considerations [32] or from probability calculations [33]. This type of equation is useful for estimation purposes. For an overall efficiency, some sort of average migration velocity must be used, since w generally is the drift velocity for a given size, as denoted by Equation (5-94). Because of the exponential term, a 100 percent collection efficiency is not possible, since A, w, and Q are always finite. Since w varies directly with particle size, the fractional efficiency is exponentially related to the particle diameter. In the preceding example, a plate-type collector had a spacing of 23 cm and a gas flow of 1.5 m/s. The general equation for the drift velocity from that example is

$$w = 0.048 d_p$$

where d_p is in microns and w in meters per second. If we choose, in addition, a plate height of 10 m and a length of 8 m, then Equation (5-98) reduces to

$$\eta = 1 - \exp(-2.23 d_p)$$

Figure 5-36 is a plot of this equation for particle sizes up to 3 μm, beyond which size the collection efficiency is essentially 100 percent. This fractional efficiency curve is misleading, in that it predicts very low efficiencies at small particle sizes. Measured fractional efficiencies for many electrostatic precipitators show an increase in the submicron range. If the efficiency is around 90 to 95 percent at 1 μm, it may rise to 99 percent or more for a 0.1-μm particle size. Measurements [34] indicate that in many instances the minimum collection efficiency occurs in the 0.1- to 0.5-μm size range. The

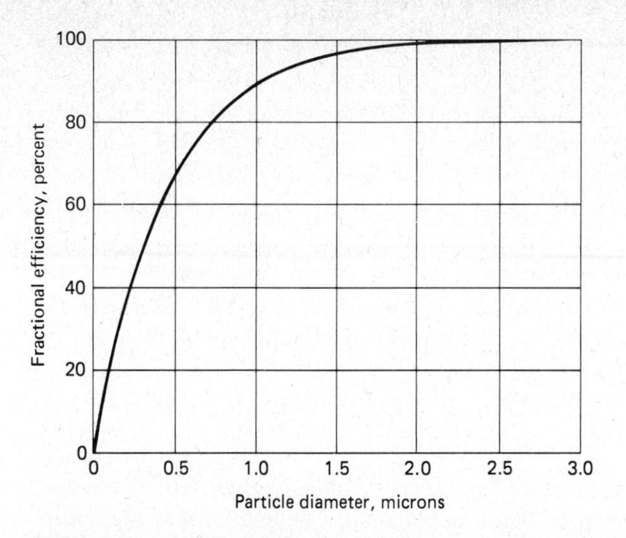

Figure 5-36 A typical fractional efficiency curve for electrostatic precipitators based on Equation (5-98).

increase in collection efficiency at lower particle diameters suggests that an electrostatic precipitation unit is a promising device for the removal of very fine particulate.

The Deutsch equation should not be used to predict the behavior of an industrial dust directly, since the collection efficiency cannot, as yet, be determined from fundamentals [35]. If its format is used, it is best to consider w as a performance factor. Then w should be ascertained from pilot studies under a variety of conditions. Nevertheless, Equation (5-98) is useful for estimating the effect of operating at off-design conditions. For example, let us assume that a particle size of 2 μm is representative of the dust in the preceding example. That is, the average drift velocity is

$$w = 0.048d_p = 0.048(2) = 0.096 \text{ m/s}$$

and the overall collection efficiency is

$$\eta = 1 - \exp[-2.23(2)] = 0.988(98.8 \text{ percent})$$

Now, as a result of a change in process requirements, the gas flow through the precipitator is increased by 20 percent. The overall collection efficiency becomes

$$\eta = 1 - \exp\left[\frac{-2.23(2)}{1.2}\right] = 0.976(97.6 \text{ percent})$$

Thus a 20 percent overload in flow rate decreases the efficiency by 1.2 percent. However, this small decrease in collection efficiency increases the mass of particulate not removed by the precipitator by a factor of 2. As a general rule, when exceedingly high efficiencies are required a small change in removal efficiency due to operating changes leads to a drastic change in the particulate emissions. One of the disadvantages of electrostatic precipitators is that they are not easily adaptable to varying loads and operating conditions. Other disadvantages are high initial cost and large space requirements. Also, a precleaner may be required if very high dust loadings are expected.

In the above paragraph an average drift velocity was selected for the dust, in order to estimate performance at off-design conditions. This average velocity is also known as the "effective" drift or migration velocity, and is frequently used in design calculations. Based on experimental data, examples of typical effective migration velocities are presented in Table 5-11 for some common industries.

The Deutsch equation, even when fitted with empirical migration velocity data, does not fit actual data on ultrahigh efficient electrostatic precipitators now required in many applications. Thus there are numerous attempts to modify it so that reasonable predictions can be made for new equipment. It has been pointed out [36] that an equation based on concepts from waste water disposal methods is useful in predicting overall efficiencies of ESP equipment, especially in the high efficiency range. The equation, based on the work of Hazen, has the format

$$\eta = 1 - \left(1 + \frac{wA}{nQ}\right)^{-n} \tag{5-99}$$

where n is an adjustable constant which varies from 1 to infinity. For electrostatic preciptators the values of n which fit most data run from 3 to 5, with an overall range of 2 to 8 possible. When n is infinite, the above equation reduces to the Deutsch equation. As noted earlier, experience has shown that effective migration velocities are in the range of 4 to 20 cm/s (0.1 to 0.7 ft/s). Another possible empirical equation also containing two coefficients might be

$$\eta = 1 - \exp\left(\frac{-wA}{Q}\right)^{x}$$

Table 5-11 TYPICAL EFFECTIVE MIGRATION VELOCITIES OF INDUSTRIAL DUSTS

APPLICATION	w (ft/s)	w (cm/s)
Utility fly ash	0.13–0.67	4–20
Sulfuric acid mist	0.19–0.25	6–8
Cement (dry)	0.19–0.23	6–7
Blast furnace	0.20–0.46	6–14
Catalyst dust	0.25	7.5

which is a modification of the original Deutsch model. In this case x is usually set at 0.5. In either case, w is the effective drift velocity applicable to a particular industry and for a given range of precipitator geometries and operating conditions.

Example 5-12

An electrostatic precipitator with a specific collection area (A/Q) of 300 $ft^2/1000$ acfm is found to have an actual overall collection efficiency of 97.0 percent. If the value of A/Q is increased to 400 $ft^2/1000$ acfm, estimate the anticipated overall collection efficiency on the basis of a constant effective migration velocity and using (a) the Deutsch equation and (b) the Hazen-type equation with an n value of 4.

SOLUTION

(a) By substitution of the given data into Equation (5-98) we obtain an effective migration velocity for the Deutsch equation. Thus

$$0.97 = 1 - \exp(-0.300w)$$
$$w = 11.69 \text{ ft/min } (5.9 \text{ cm/s})$$

Use of this effective w value with A/Q value of 0.400 yields

$$\eta = 1 - \exp - 11.69(0.400) = 0.991$$

(b) Substitution of the given data into the Hazen-type efficiency equation yields a somewhat different effective migration velocity.

$$0.97 = 1 - \left(1 + \frac{0.30w}{4}\right)^{-4}$$
$$w = 18.71 \text{ ft/min } (9.5 \text{ cm/s})$$

Solving Equation (5-99) now for the new efficiency, we find that

$$\eta = 1 - \left[1 + \frac{18.71(0.40)}{4}\right]^{-4} = 0.985$$

Thus if the precipitator is upgraded by increasing the specific collection area, the Hazen equation predicts a somewhat smaller increase in the overall collection efficiency than that given by the Deutsch equation.

The total installed cost of electrostatic precipitators ranged from \$5.00 to \$9.00 per actual ft^3/min in 1979. The operating cost is quite low, due in part to the very low draft (pressure) loss in such units, which is around 0.5 in. water gauge (1.25 mbar). Excellent discussions of electrostatic precipitators are found in references 32 and 37. The latter reference also contains an extensive bibliography through the late 1960s. Reference 38 gives a detailed report on a 1974 symposium on fine-particle electrostatic precipitator technology.

One interesting device for particle removal from gas streams combines an electrostatic precipitator with a wet scrubber. Such a device is frequently termed an electrostatic (droplet) spray scrubber, or a wet electrostatic precipitator. On the basis of theoretical calculations, this combination should result in substantially improved collection efficiency for small particles (0.5- to 5-μm diameter range). A number of experimental investigations have confirmed these findings since Penney [39] proposed the original theory in 1944. The voltages used in the various studies have varied from 5,000 to 30,000 V. The data from a study reported in 1975 [40] on a small unit (140 ft^3/min) operating on dioctyl phthalate aerosol are shown in Figure 5-37. Note that the collection efficiency increased from 35 percent for uncharged 0.3-μm diameter particles to around 87 percent for charged particles of the same size. In the same study a 1000-ft^3/min unit was operated off a magnesium sulfite recovery boiler of a pulp mill. At high recovery efficiencies the indicated power requirement was about 0.5 hp/1000 actual ft^3/min (350 W/1000 actual ft^3/min), which included gas and water pressure drop as well as the electrostatic charging of the aerosol and water droplets. Overall collection efficiencies greater than 98 percent were reported for this same unit operating on hot (about 350°F) gases from a coal-fired boiler. Outlet dust loading of around 0.003 gr/scf were listed for this latter test. In the study reported above, the aerosol in the gas stream and water droplets in

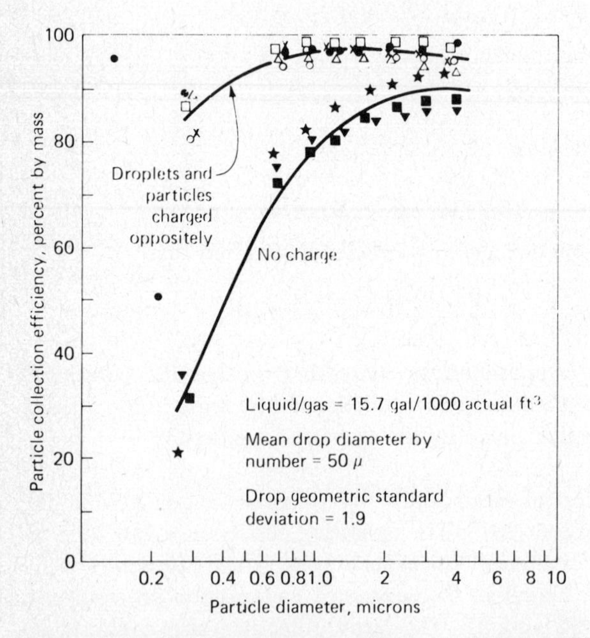

Figure 5-37 Particle collection efficiency of electrostatic spray droplet scrubber as a function of particle size. (SOURCE: M. J. Pilat. "Collection of Aerosol Particles by Electrostatic Droplet Spray Scrubbers." *J. Air Pollu. Control Assoc.* **25**, no. 2 (1975): 176.)

the spray were charged to opposite polarity, thus enhancing the force of attraction between particles and droplets.

In addition to the charged droplet scrubber described above, a wetfilm electrostatic precipitator is also commercially available. In this latter system a flow of finely divided liquid droplets is charged electrically and the droplets are attracted to the collecting plates. The precipitated water droplets create an even water film on the entire collection surface. The charged solid particulate is captured by the moving film of water. The slurry formed on the plates flows by gravity into a trough for disposal. An added advantage of this system is that gaseous pollutants can also be absorbed by the liquid spray, before the droplets are attracted to the collector plates. It may be necessary to add chemicals in the spray liquid in this latter case to enhance absorption of the gaseous pollutants. In both the scrubber systems described, the dry disposal problems associated with conventional electrostatic precipitators are replaced by wet disposal problems.

Increased interest in the wet precipitator has arisen because of the acknowledged shortcomings of other high-efficiency particulate collectors. The venturi scrubber, for example, is characterized by an extremely high energy consumption, which makes the venturi less attractive in an era of increasing energy costs. The dry electrostatic precipitator has dust resistivity limitations which restrict its usefulness in collecting certain classes of dusts. The fabric filter's usefulness is limited by the physical and chemical nature of the particulate. In addition, the filter material itself may have restricted ranges of chemical resistance and temperature. The wet electrostatic precipitator overcomes many of the limitations of these other devices. The nature of the particles is of considerably less importance, and temperature has little effect on performance. Because the collector plate is continually washed, dust resistivity of the collected layer is no longer a variable, and the resistivity of the water film is very low. In addition, gaseous pollutants are also removed by the washing process; the only real limitation is the solubility of the gaseous component in the washing liquid. With these characteristics, along with its relatively low total power consumption, the wet electrostatic precipitator offers certain economic and technical advantages over more traditional equipment in certain applications. This is especially true when a significant portion of the particulate is in the submicron range.

5-10 COMPARISON OF PARTICULATE CONTROL EQUIPMENT

The specific particulate control equipment a plant should choose in order to meet emission standards depends on many operational variables. A major variable is total volume flow rate to be handled. If the equipment under consideration does not have the capacity to remove particulate from the maximum or top output of the plant, then meeting standards will be impossible. The maximum collection efficiency rating should be a major consideration, since operation at 10 percent above design conditions is expected in many industrial processes. Collection efficiencies of many types of equipment are affected by operation above or below design capacities. The physical and chemical characteristics of the particulate influence equipment selection, also. For example, the degree of coarseness or fineness of the particles may preclude certain choices. The dust loading of the gas stream (in grains per cubic feet or grams per cubic meter) is another critical factor. For heavy dust loadings (for example, 100 gr/ft^3 or 230 g/m^3), an inexpensive precleaner may be necessary before the gas stream enters more expensive and more efficient collectors. The temperature range and the possibility of sudden temperature increases beyond the limit of the equipment must be considered. Equally important, the maintenance requirements of various types of equipment must be examined.

Listed below are four basic types or particulate control equipment and some of the situations under which each is most effective. Bear in mind that the list is general, and any number of exceptional situations may exist.

1. Cyclones are typically used when
   a. the dust is coarse;
   b. concentrations are fairly high ($> 1\ gr/ft^3$);
   c. classification is desired;
   d. very high efficiency is not required.
2. Wet scrubbers are typically used when
   a. fine particles need to be removed at a relatively high efficiency;
   b. cooling may be desirable and moisture is not objectionable;
   c. gases are combustible;
   d. gaseous as well as particulate pollutants need to be removed.
3. Fabric filters are typically used when
   a. very high efficiencies are required;
   b. valuable material is to be collected dry;
   c. the gas is always above its dew point;
   d. volumes are reasonably low;
   e. temperatures are relatively low.
4. Electrostatic precipitators are typically used when
   a. very high efficiencies are required for removing fine dust;
   b. very large volumes of gas are to be handled;
   c. valuable material needs to be recovered.

With a list of operational requirements and a knowledge of the capabilities of various general classes of collection equipment, it is possible to make a preliminary screening which will lead to a limited number of choices in terms of present technology.

A detailed comparison of the capabilities of various collection equipment is not a productive exercise. Capabilities change with the growth of technology, and up-to-date information is readily available from equipment manufacturers. However, there are certain important characteristics that are

Table 5-12 SEVERAL OPERATING CHARACTERISTICS OF PARTICULATE COLLECTORS

GENERAL CLASS	SPECIFIC TYPE	TYPICAL CAPACITY	PRESSURE LOSS (IN. WATER)[a]	POWER REQUIRED ($W/ft^3/min$)[b]
Mechanical collectors	Settling chamber	15–25 ft^3/min per ft^3 of casing volume	0.2 – 0.5 (0.5–1.3)	0.03–0.10 (1–4)
	Baffle	100–3500 ft^3/min per ft^2 of inlet area	0.5 (1.3)	
	High-efficiency cyclones	2500–3500 ft^3/min per ft^2 of inlet area	3–5 (7.5–12.5)	0.5–1.0 (15–35)
Fabric filters	Automatic	1–6 ft^3/min per ft^2 of fabric area	4–6 (10–15)	1.0–1.3 (35–45)
Wet scrubbers	Impingement baffle	400–600 ft^3/min per ft^2 of baffle area	2–5 (5–13)	0.2–1.0 (7–35)
	Packed tower	500–700 ft^3/min per ft^2 of bed cross-sectional area	6–8 (15–20)	
	Venturi	6000–30,000 ft^3/min per ft^2 of throat area	10–50 (25–125)	4–12 (140–425)
Electrostatic precipitators	Dry, single-field	2–8 ft^3/min per ft^2 of electrode collection area	0.2–0.5 (0.5–1.3)	0.4–1.0 (15–35)
	Wet (charged-drop scrubber)	5–15 ft^3/min per ft^2 of electrode collection area	0.5–0.7 (1.3–1.8)	0.3–0.5 (10–15)

[a] Values in parentheses under "pressure loss" are in mbars.
[b] Values in parentheses under "power required" are in $W/m^3/min$.

worth comparing at this point. Table 5-12 lists the capacity and the draft loss to be expected from four general classes of equipment. The excessive pressure drop in the venturi scrubber is noted again. Another factor of prime importance is the collection efficiency as a function of particle diameter. Figure 5-38 shows typical curves for four classes of equipment, including two types of cyclones. It is apparent from this figure that collection efficiencies vary widely in a given particle-size range. These curves are presented to establish trends. The specific location of an efficiency curve depends upon the characteristics of the particulate as well as the equipment design from a given manufacturer. Table 5-13 also indicates the variation in collection efficiency with particle size. This table substantiates the general curves shown in Figure 5-38.

In the final analysis, the selection of control equipment must be based also on capital investment, operating costs, and the cost of capital in terms of interest, taxes, and amortization. Equipment manufacturers and construction companies can provide information as to current initial costs for cubic feet per minute of gas flow, as well as costs for installation, operating, and maintenance. McKenna et al. have recently presented a study comparing performance and cost between fabric filters and alternative particulate control techniques [40].

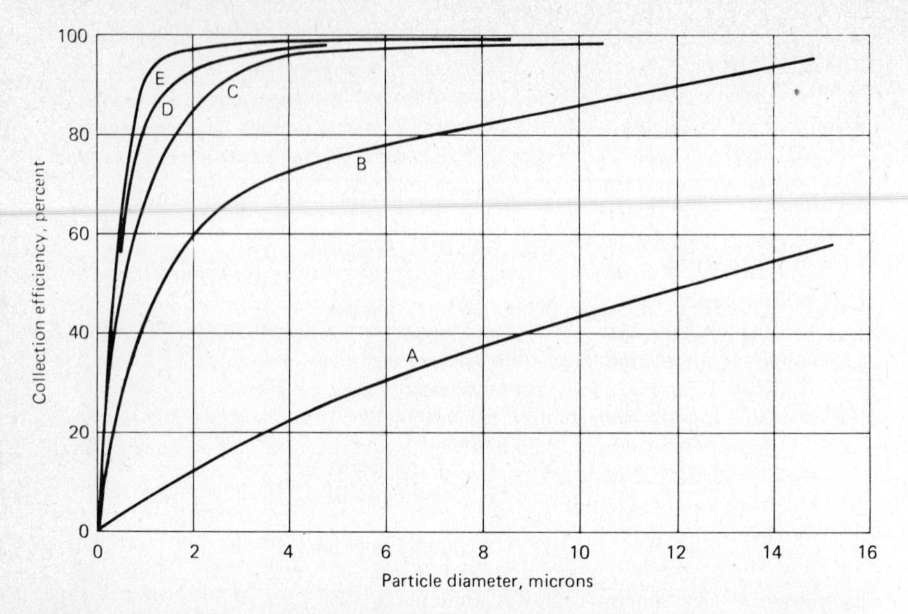

Figure 5-38 Typical collection efficiency curves for various types of dust collectors: *A*, high-throughput cyclone; *B*, high-efficiency cyclone; *C*, spray tower; *D*, dry electrostatic precipitator; *E*, venturi scrubber.

TABLE 5-13 TYPICAL FRACTIONAL COLLECTION EFFICIENCIES OF PARTICULATE CONTROL EQUIPMENT

	EFFICIENCY AT GIVEN SIZE		
	5 μm	2 μm	1 μm
Medium-efficiency cyclone	30	15	10
High-efficiency cyclone	75	50	30
Electrostatic precipitator	99	95	85
Fabric filter	99.8	99.5	99
Spray tower	95	85	70
Venturi scrubber	99.7	99	97

QUESTIONS

1. In micrograms per cubic meter, what range of particle concentration is typical for the atmosphere in an urban area with a population greater than 100,000 persons and less than 100,000 persons in the 1960s in the United States? What is the federal ambient air quality standard for particulate in micrograms per cubic meter?
2. In general, for a typical urban atmospheric sample, how do the particle count and the mass percent vary with particle size?
3. In general, how does the fractional collection efficiency curve vary for most particulate collectors?
4. For a given dust loading initially and two identical collectors in series, why is the overall collection efficiency of the second collector usually less than that of the first collector?

5. In general, why is the terminal or settling velocity an important parameter in the design of certain particulate collection equipment?
6. What is the physical basis for needing the Cunningham correction to Stokes' law? At roughly what particle size does this correction become important?
7. What are the two major modifications of the Gaussian-type dispersion equation for gaseous emissions in order to apply this form of equation to particulate dispersion?
8. What is meant by the null point in hood design?
9. What six mechanisms are responsible for particulate collection under varying circumstances?
10. List five basic classes of particulate collection equipment and describe which physical mechanisms of collection are prevalent in each case.
11. What is the basic use of gravity settling chambers?
12. Describe how the collection efficiency of a cyclone is affected by changes in the inlet velocity, particle density, particle diameter, and cyclone diameter.
13. Describe the major types of wet scrubbers.
14. What is a major disadvantage of a wet collector as opposed to a dry collector?
15. What is a major disadvantage of a venturi scrubber?
16. How does the collection efficiency of a wet scrubber vary with the relative velocity and the liquid droplet size?
17. What is meant by the air-to-cloth ratio for a fabric filter, commonly expressed in feet per minute or centimeters per second?
18. What is meant by the drift velocity for an electrostatic precipitator? Why is it an important design parameter?
19. Why is electrical resistivity of the particulate in a waste gas stream an important parameter in the operation of an electrostatic precipitator operation?
20. Why is the removal of SO_x from stack gases by using low-sulfur fuels sometimes a detriment to electrostatic precipitator operation?
21. In general, how do changes in the area of the collecting electrodes, in the gas flow rate, and in the drift velocity affect electrostatic precipitator collection efficiency?

PROBLEMS

5-1. A power plant burns residual oil having value of 140,000 Btu/gal. Based on the emission factor given in Table 5-5, is a particulate collector required to meet the 1970 federal emission standard for new sources?

5-2. An industrial furnace burns residual oil with a heating value of 135,000 Btu/gal. (a) Based on the emission factors given in Table 5-5, is a particulate collector required to meet the 1970 federal emission standard for new sources for particulates? (b) Estimate the required efficiency of the collector.

5-3. Coal with an ash content of 7 percent is burned in a pulverized coal-fired furnace. The heating value is 10,800 Btu/lb. Estimate the collector efficiency required to meet the 1980 federal emission standards for particulates from new sources. (See Tables 2-6 and 5-5.)

5-4. A coal having an ash content of 5 percent is burned in an underfed stoker. The heating value of the coal is 11,000 Btu/lb. What is the collection efficiency required to meet the 1970 federal standards for particulates from new sources (based on an average emission factor)?

5-5. What is the required collection efficiency for the coal-fired unit in Problem 5-4 if the coal is burned on a spreader stoker?

5-6. What is the collection efficiency required for a sulfuric acid plant to meet the 1970 federal emission standards for particulates from sulfuric acid plants? Give two answers, using the best and worst conditions listed in Table 5-5 as a guide.

5-7. The dust loading of a gas from a lime kiln is 3.25 gr/scf. The design loading before exhaust is 0.11 gr/scf. (a) What is the collection efficiency required to obtain the desired exit dust loading if one collector is to be used? (b) What is the efficiency required if two collectors are used in series, each with the same efficiency? (c) What is the efficiency required for the first collector of two in series, if the efficiency of the second unit is only 75 percent that of the first unit?

5-8. The dust loading desired before exhausting gases from a lime kiln is 0.05 gr/actual ft^3. The dust loading of the gas from the kiln is 3.60 gr/actual ft^3. (a) What is the required collector efficiency? (b) What is the dust loading of a gas leaving a collector having an efficiency of 90 percent when the dust loading of the gas to the collector is 8 gr/actual ft^3?

5-9. What is the maximum inlet dust loading that can be fed to a collector having an efficiency of 92 percent if the outlet dust loading from the collector is to be maintained below 0.1 gr/actual ft^3 (0.23 g/m^3)?

5-10. Reconsider Problem 5-7. Change in inlet dust loading to 4.1 g/m^3 and the exhaust loading to 0.065 g/m^3, and then determine the three required answers.

5-11. Two particulate collectors are in series. The fractional efficiency for size d_p in the upstream device is 80 percent, and for the downstream device the efficiency is 60 percent. Find the overall removal efficiency for size d_p.

5-12. What would be the dust loading in grains per cubic foot (or grams per cubic meter) for the combustion gas leaving an industrial boiler fired with natural gas if the volume ratio of flue gas produced to natural gas burned is 13.5 for a process? (See Table 5-5.)

5-13. A spreader stoker is used to burn coal containing 8 percent ash. The flue gas from burning 1 lb of coal is 170 ft^3. What is the maximum dust loading of the combustion gas, in grains per cubic foot? (See Table 5-5.)

5-14. The ambient air quality standard for particulate is 75 $\mu g/m^3$. (a) Does the dust loading of the flue gas in Problem 5-13 exceed this value? (b) What collection efficiency would be required to bring the flue gas to within 100 times the ambient standard?

5-15. What is the required minimum efficiency of an electrostatic precipitator used as the second collector, when it is preceded by a cyclone separator having an efficiency of (a) 90 percent and (b) 80 percent, if the desired overall efficiency is 98 percent?

5-16. A 1000-MW pulverized coal-fired steam power plant of 40 percent thermal efficiency uses coal with an ash content of 12 percent and a heating value of 26,700 kJ/kg. Assume that 50 percent of the ash goes up the stack as particulate in the flue gas. The particulate emission is controlled by an electrostatic precipitator which has the following removal efficiencies and weight distribution in the given size ranges.

Particle size, μm	0–5	5–10	10–20	20–40	>40
Efficiency, %	70	92.5	96	99	100
Wt., %	14	17	21	23	25

Determine the amount of fly ash emitted, in kilograms per second, with the flue gas.

5-17. Reconsider Problem 5-16. Change the efficiency to 38 percent and the ash content to 14 percent, and assume that 75 percent of the ash goes up the stack. In addition, change the weight percent in the five particle size ranges to the following: 12, 16, 22, 27, and 23 percent, respectively. Determine the fly ash emitted in kilograms per second.

5-18. Compute the specific surface of a sphere of (a) 1.0 μm and (b) 10 μm in diameter if its density is 1400 kg/m^3.

5-19. Compute the specific surface in square meters per kilogram of a group of spheres with a density of 1200 kg/m^3 if the diameter of the spheres is (a) 1.0 μm and (b) 10 μm.

5-20. Determine the number of spherical particles of (a) 1.0 μm and (b) 10.0 μm contained in 1 kg of material with a density of 1400 kg/m^3.

5-21. Determine the number of spherical particles contained in 1 kg of material with a density of 1200 kg/m^3 if the particle diameter is (a) 1.0 μm and (b) 10 μm.

5-22. A particle size distribution for a collection of spherical particles is given by, $n(d_p) = 3 \times 10^3 (d_p - 0.01)^{1/2}$, where d_p is in microns. There are no particles below 0.01 μm or above 100 μm. Determine (a) the total number of particles, (b) the specific surface in square meters per kilogram, and (c) the specific volume in cubic meters per kilogram. The particle density is 1.40 g/cm^3.

5-23. A collection of particles with a density of 1400 kg/m^3 has a number size distribution given by $n(d_p) = 6 \times 10^6 (d_p - 1)$ for particle diameters between 1 and 49 μm, and is zero for all other particle sizes. In these relations d_p is in microns and $n(d_p)$ in reciprocal microns. Determine (a) the total number of particles, (b) the mass of the particles in kilograms, and (c) the specific surface in square meters per kilogram, if the particles are spheres.

5-24. Consider the particle distribution given in Problem 5-23. For particle diameters between 15 and 40 μm the fractional collection efficiency in percent is given by $\eta_d = 4d_p - 60$, where d_p is in microns. Above 40 μm the efficiency is 100 percent. Calculate the overall collection efficiency based on number.

5-25. Employing the data of Problems 5-23 and 5-24, determine the overall collection efficiency based on mass.

5-26. A collection of spherical particles with a density of 1200 kg/m^3 has a number distribution given by $n(d_p) = 8 \times 10^6 (d_p - 1)$ for particle diameters between 1 and 30 μm, and is zero for all other particle sizes. In these relations d_p is in microns and $n(d_p)$ is in reciprocal microns. Determine (a) the total number of particles, (b) the mass of the particles in kilograms, and (c) the specific surface in square meters per kilogram.

5-27. Consider the particle distribution in Problem 5-26. For particle diameters between 5 and 30 μm the fractional collection efficiency in percent is given by $\eta_d = 4d_p - 20$, where d_p is in microns. Below 5 μm the efficiency is zero, and above 30 μm the efficiency is 100 percent. Calculate the overall collection efficiency based on number.

5-28. Employing the data of Problems 5-26 and 5-27, determine the overall collection efficiency based on mass.

5-29. A catalyst powder has the following weight distribution in the given particle size range: 10–20 μm, 10.0 percent; 20–30 μm, 42.0 percent; 30–40 μm, 33.0 percent; 40–50 μm, 9.0 percent; 50–80 μm, 5.9 percent. From a log-probability plot estimate (a) the geometric mean diameter, (b) the standard deviation of that mean diameter, (c) the arithmetic mean diameter and the arithmetic standard deviation, all based on mass. Then, (d) determine the same four items as above on a number or count basis. Finally, (e) make a plot of $n(d_p)/N$ versus d_p.

5-30. Reconsider the weight distribution data given in Problem 5-29. The fractional collection efficiencies for the five particle size ranges are given below for three types of collectors, in percent: device A, 40, 64, 72, 80, and 92; device B, 60, 73, 85, 90, and 97; and device C, 70, 85, 92, 96, and 99. Determine the overall collection efficiency based on mass for (a) device A, (b) device B, and (c) device C.

5-31. A particulate sample in a gas stream has the following average particle-size distribution by mass: 2 μm, 2.0 percent; 5 μm, 6.8 percent; 7 μm, 7.4 percent; 10 μm, 12.8 percent; 16 μm, 15.2 percent; 25 μm, 13.0 percent; 40 μm, 15.0 percent; 65 μm, 12.0 percent; 100 μm, 8.0 percent; 200 μm, 5.8 percent; 340 μm, 1.5 percent. (a) What is the particle size for which 50 percent of the total mass is less than that size and (b) what is the standard deviation, σ, for the distribution, if it is log-normally distributed?

5-32. A hypothetical sample has a number (count) distribution in various ranges of diameter as shown below.

d_p (μm)	$\Delta N(d_p)/N$	d_p (μm)	$\Delta N(d_p)/N$
0–0.4	0.059	3.0–4.0	0.095
0.4–0.6	0.055	4.0–6.0	0.100
0.6–0.8	0.056	6.0–8.0	0.080
0.8–1.0	0.055	8.0–10.0	0.050
1.0–2.0	0.195	10.0–20.0	0.055
2.0–3.0	0.165	20.0–40.0	0.030

Determine values for (a) d_g, (b) σ_g, (c) $n(w)/N$, (d) $n(d_p)/N$, (e) mf_i, (f) d_m, (g) σ_m, and (h) the modal diameter. Then prepare plots of $n(w)/N$ versus w, and $n(d_p)/N$ versus d_p.

5-33. A particulate sample in a gas stream has the following characteristics:

Particle size, μm	2	3	5	7	10	25
Cum. wt., %	2	10	40	70	92	99.9

(a) Determine if the distribution is log-normal. (b) Determine the geometric mean diameter and the standard deviation. (c) Using Figure 5-3 as representative efficiency data for a collector, determine the overall collection efficiency, in percent.

5-34. A particulate sample in a gas stream has the following particle size and weight distribution:

Average particle size, μm	1	2	5	10	15	40	60
Cumulative wt., %	2	14	50	80	92	99.5	99.9

Answer the questions found in Problem 5-33.

5-35. Consider a particle with an apparent density of 1.7 g/cm^3 falling through ambient air at 80°F (27°C). (a) Determine the terminal speed in feet per minute and centimeters per second for a 10-μm diameter particle. (b) What is the ratio of vertical distance to horizontal distance the particle travels if the horizontal air speed is 33 ft/s? (c) Determine the terminal speed in feet per minute and centimeters per second for a 1200-μm diameter particle. (d) Repeat part (b) for the 1200-μm particle.

5-36. Determine the terminal or settling speeds for spherical particles falling through atmospheric air. Particle densities of (a) 1.0, (b) 2.0, and (c) 3.0 g/cm^3 should be considered, with particle diameters of 10, 100, and 1000 μm.

0.6 cm/sec B

5-37. Determine the Cunningham correction factor for a 0.60-μm particle by evaluating (a) the mean molecular speed $\bar{u}$ for air in meters per second, (b) the mean free path λ in meters, at 1 atm and 300°K, and (c) K_C from an equation requiring d_p and λ as input data.

5-38. Determine the Cunningham correction factor for a 0.50-μm particle by calculating (a) the mean molecular speed $\bar{u}$ for air in meters per second, (b) the mean free path λ at 1 atm and 300°K, in meters, and (c) K_C from an equation requiring d_p and λ as input data.

5-39. A spherical particle with a 0.60-μm diameter and a density of 1.5 g/cm^3 falls under the influence of gravity in air at 1 atm and 300°K. (a) Determine the terminal speed in microns per second. (b) Determine the time required to reach 95 percent and 99 percent of its terminal speed, in microseconds.

5-40. Repeat Problem 5-39 for a particle with a diameter of 6.0 μm.

5-41. A spherical particle with a 0.50-μm diameter and a density of 2.0 g/cm^3 falls under the influence of gravity in air at 1 atm and 300°K. (a) Determine the terminal velocity in microns per second. (b) Determine the time required to reach 95 percent and 99 percent of its terminal speed, in microseconds.

5-42. Repeat Problem 5-41 for a particle with a diameter of 5 μm.

5-43. A plant with no fly-ash removal equipment burns 300 tons of coal per day. Using the emission rate given in Table 5-5 for a coal containing 9 percent ash and fired on a spreader stoker, determine the deposition rate for 50-μm particles having a density of 1.5 g/cm^3 at distances of 1 and 3.5 km downwind from the stack. The effective stack height is 100 m and the mean wind speed at the top of the stack is 4 m/s. The stability class is C. Use the Gaussian-type dispersion equation.

5-44. Determine the distance downwind from a stack having an effective height of 300 m where 20-μm fly-ash particles will theoretically have all settled out. The following data apply: specific gravity of the particles is 2.0, the wind speed at stack height is 10 m/s, and the day is clear. Use the Gaussian-type dispersion equation.

5-45. The power plant of Problem 5-45 burns 6000 tons of coal per day in a pulverized coal furnace, with the coal containing 7 percent ash. If a stability class *C-D* exists, what is the deposition rate of fly ash of 20-μm diameter at a position 10 km downwind from the stack? Assume no fly-ash control equipment and use a Gaussian-type dispersion equation.

5-46. Plot the deposition rate in $\mu g/m^3 \cdot s$ as a function of distance for the power plant of Problem 5-45 for distances of 3, 5, 6, 7, 10, and 12 km.

5-47. A dust with a density of 2.6 g/cm^3 is being released from a stack with an effective height of 100 m. For a wind speed of 3 m/s and a distance 500 m downwind on the center line at ground level, estimate the deposition rate in $g/m^2 \cdot s$ for particle diameters of (a) 10 μm, (b) 50 μm, (c) 25 μm, and (d) 100 μm. The particulate emission rate is 100 g/s, the atmospheric stability class is *D*, and the temperature is 25°C. Use a Gaussian-type dispersion equation.

5-48. Work Problem 5-47 by the Bosanquet dispersion method, employing the data of Figure 5-9.

5-49. A cement plant releases particles of 30 μm or less in diameter at a rate of 400 kg/hr from a 85-m stack. The wind speed is 2.5 m/s on a day with a stability class of *C*. What suspended particulate concentration, in micrograms per cubic meter, do you estimate would be measured at an air sampling station located at ground level and 1250 m directly downwind for 30-μm particles? Assume the particles are spherical with a particle density of 1.5 g/cm^3, and use a Gaussian-type dispersion.

5-50. Work Problem 5-49 by the Bosanquet dispersion method, employing the data of Figure 5-9.

5-51. A dust with a density of 1.0 g/cm^3 is being released from a stack with an effective height of 100 m. The particulate emission rate is 200 g/s, the wind speed is 3 m/s, the stability class is *E*, and the atmospheric temperature is 25°C. Estimate (1) the deposition rates in $g/m^2 \cdot s$ for downwind distances from 200 m up to 7000 m, if necessary, and (2) the distance in meters at which the maximum concentration occurs. Use particulate diameters of (a) 5 μm, (b) 25 μm, (c) 75 μm, and (d) 100 μm. Employ a Gaussian-type dispersion.

5-52. Work Problem 5-51 by the Bosanquet dispersion method, employing the data of Figure 5-9.

5-53. On a particular evening with a stability class of *E* two similar heavy industries are in line with each other with respect to the wind direction. Both industries are emitting particles which average 55 μm in diameter, and the density of the particles is 2.7 g/cm^3. The industry upwind has a stack with an effective height of 55 m, while the industry downwind has a stack with an effective height of 35 m; the two stacks are 100 m apart. The wind speed during the night is 1.8 m/s. A residential area begins 160 m downwind from the second industry. Which industry is responsible for the greater dust deposition at the residential boundary? Estimate the ratio of the dust deposition rates, greatest to least, using a Gaussian-type dispersion equation.

5-54. A particulate with a terminal settling speed of 3.0 cm/s is emitted from a stack with an effective height of 100 m at a rate of 100 g/s. The average wind speed is 4.0 m/s, and at a distance x downwind from the center line of the stack the dust concentration of this particulate size is 40 $\mu g/m^3$. If the values of σ_y and σ_z are 600 m and 150 m, respectively, estimate the value of x in meters.

5-55. A gravity settling collector is to be used to remove particulate having a diameter of 100 μm and a density of 1.5 g/cm^3. What is the maximum gas velocity that can be used if the chamber is 10 m long and 2 m high, and the desired collection efficiency is to be not less than 90 percent?

5-56. What is the smallest diameter particle in microns having a density of 2.0 g/cm^3 which can be collected with an efficiency of 85 percent in a gravity separator which is 12 m long and 3 m high? The gas velocity is 0.78 m/s.

5-57. A settling chamber is to be used to collect particles with a density of 1200 kg/m^3 from an air stream at 1.013 bars and 350°K. The volume flow rate is 2.0 m^3/s. The overall chamber is 2.5 m high, 3 m wide, and has 30 trays including the bottom surface. Assuming laminar flow, (a) what is the value of Re for the gas flow, (b) how long a chamber is required for 100 percent removal efficiency of 60-μm diameter particles, and (c) what is the fractional collection efficiency of particle sizes of 50, 40, 30, 20 and 10 μm, for the chamber length found in part (b)? (d) Plot on a fractional efficiency curve for this device on rectangular graph paper.

5-58. Reconsider the data of Problem 5-57, except that the volume flow rate is increased to 20 m^3/s. Assuming turbulent flow, determine (a) what length of chamber is required for 98 percent removal of 60-μm particles, (b) what length is required for 99.5 percent removal of 60-μm particles, and (c) what is the fractional collection efficiencies for particles sizes of 50, 40, 30, 20, and 10 μm, if the length determined in part (a) is used? (d) Plot a fractional efficiency curve for this device.

5-59. A settling chamber is to be used to collect particles of 50-μm diameter and 1400 kg/m^3 density from a stream of air at 1 atm and 300°K with a volume flow rate of 2.5 m^3/s. The overall chamber is 2 m high, 4 m wide, and will have 25 trays including the bottom surface. Assuming laminar flow, (a) how long is the chamber for 100 percent removal efficiency? (b) What is the value of Re for the gas flow? (c) What is the collection efficiency for particle sizes of 40, 30, 20, and 10 μm, using the length L found in part (a)? (d) Plot on a rectangular Cartesian plot, with efficiency as the ordinate and particle size as the abscissa, a fractional efficiency curve for this device.

5-60. Reconsider the data of Problem 5-59, except that the volume flow rate is increased to 10 m^3/s. Assume turbulent flow. (a) How long is a chamber for 98 percent removal efficiency of 50-μm particles, in meters? (b) How long is it for 99.5 percent removal of 50-μm particles? (c) What is the collection efficiency for 40-μm particles if the length determined in part (a) is used? Repeat calculations for 30-, 20-, and 10-μm particles. (d) Plot a fractional efficiency curve for this device.

5-61. A gravity settling chamber is used to collect 70-μm particles having a specific gravity of 1.5. If the particles are carried in standard air, what is the maximum gas velocity that can be used in a collector 3 m high and 5 m long, neglecting recirculation effects within the chamber? Assume laminar flow.

5-62. A particulate with a density of 1.6 g/cm^3 enters at 0.2 ft/s into a 20-ft-long settling chamber which is 10 ft high. The gas stream is atmospheric air. Determine the minimum-diameter particle, in microns, which will completely settle out under laminar flow conditions.

5-63. Determine the diameter of the smallest particle which can be captured with 100 percent efficiency in a settling chamber 25 ft long, 7 ft wide, and 2.5 ft high. The air stream is moving with a velocity of 7 ft/s and the particle density is 1.9 g/cm^3. The air temperature is 75°F. Account for turbulence within the chamber.

5-64. It is desired that all particles with a size of 50 μm and larger be removed with 100 percent efficiency in a settling chamber which has a height of 2.0 ft. The air moves at 5 ft/s and its temperature is 80°F. The particulate density is 2.2 g/cm^3. What minimum length of chamber should be provided, in feet, if turbulence within the chamber is to be taken into account?

5-65. For the chamber described in Problem 5-63, two horizontal baffle trays are placed across the chamber so that the chamber is divided vertically into three equal sections. Determine the diameter of the smallest particle that can be captured with 100 percent efficiency, accounting for turbulence within the chamber.

5-66. A particulate sample in an air stream has the following weight-percent distribution: 5 μm, 10 percent; 10 μm, 20 percent; 50 μm, 40 percent; 100 μm, 20 percent; 200 μm, 10 percent. A cyclone separator is employed on the basis of the following data: inlet width, 12 in.; effective turns, 5; inlet gas velocity, 20 ft/s; particle density, 1.6 g/cm^3; gas density, 0.074 lb/ft^3; gas viscosity, 0.045 lb/ft·hr. Using Figure 5-17, estimate the percentage of the total weight that would be removed in the cyclone separator.

5-67. Work Problem 5-66 on the basis of the conventional cyclone fractional efficiency curve shown in Figure 5-16, rather than the curve shown in Figure 5-17.

5-68. Work Problem 5-66 as stated, except let the number of effective turns in the cyclone be 4 instead of 5.

5-69. Estimate the overall efficiency of the cyclone separator in Problem 5-68 when (a) the volume of the gas flow is doubled, (b) the gas viscosity is 0.09 lb/ft·hr, (c) the gas density is tripled, and (d) the dirt loading is increased 40 percent. Consider that all other conditions remain the same with the exception of the one specified.

5-70. A particulate sample in an air stream is represented by the following weight-percent distribution: 5 μm, 10 percent; 10 μm, 20 percent; 40 μm, 40 percent; 70 μm, 20 percent; 100 μm, 10 percent. A cyclone separator is employed on the basis of the following data: inlet width, 12 in.; effective turns, 4; inlet gas velocity, 20 ft/s; particle density, 1.6 g/cm^3; gas density, 0.074 lb/ft^3; gas viscosity, 0.045 lb/ft·hr. (a) Estimate the percentage of the total weight removed in the cyclone. (b) If the volume flow rate is doubled, all other parameters remaining the same, estimate the percent total weight removed.

5-71. A cyclone is designed with a 5-in. inlet width of 6 effective turns. In one particular test the inlet air velocity is 35 ft/s and the particle density is 2.2 g/cm^3. Using Figure 5-17, estimate a sufficient number of fractional efficiencies for particle sizes from 2 to 50 μm in order to plot an efficiency curve for this particle-size range.

5-72. A cyclone with a 12.0-cm inlet width and 5 effective turns undergoes a test with an air inlet velocity of 14 m/s and a particle density of 1.8 g/cm^3. Using Figure 5-17, estimate a sufficient number of fractional efficiencies for particle sizes from 2 to 50 μm in order to plot an efficiency curve for this particle-size range.

5-73. A cyclone is designed with a 6-in. inlet width and 5 effective turns. (a) When the device is operating with an inlet velocity of 40 ft/s and a particle density of 2.4 g/cm^3, estimate from Figure 5-17 what particle size, in microns, will be collected with 80 percent efficiency from an air stream. (b) What will the particle size be if the air velocity is reduced to 20 ft/s?

5-74. A cyclone of standard proportions with an outer radius of 50 cm handles 2.0 m^3/s of dirty air at 300°K and 1 atm. The density of the particulate is 1400 kg/m^3. Determine (a) the gas velocity at the entrance of the cyclone in meters per second, (b) the particle size in microns collected with 50 percent efficiency based on Equation (5-54), if the number of effective turns is 5.0, and (c) the particle size in microns collected with 10, 30, 70, and 100 percent efficiency for this same model. Finally, (d) plot on rectangular coordinates the fractional efficiency versus the particle diameter for the above model. Then also plot on the same figure a curve based on Lapple's work, using the correlation shown in Figure 5-17.

5-75. A cyclone of standard proportions with an outer radius of 60 cm handles 2.5 m^3/s of dirty air at 300°K and 1.01 bars. The density of the particulate is 1200 kg/m^3. (a) Determine the gas velocity at the entrance of the cyclone in meters per second. (b) Determine the particle size, in microns, collected with 50 percent efficiency based on Equation (5-54), if the number of effective turns for the cyclone is 4.8. (c) Determine the particle size collected with 10, 30, 70, and 100 percent efficiency for this same model. (d) Plot on rectangular coordinates the fractional efficiency versus the particle diameter, for diameters from 3 to 25 μm, for the above model. Then plot on the same figure a curve based on Lapple's work, using the correlation shown in Figure 5-17.

5-76. A standard cyclone with an entrance area of 60 $in.^2$ and a diameter of 16 in. operates with air entering at 80 ft/s at 1.1 atm and 240°F. The particle density is 2.4 g/cm^3 and the number of effective turns is 5. (a) Estimate the pressure drop across the cyclone in inches of water. (b) Repeat part (a) for a cyclone diameter of 8.1 in.

5-77. A cyclone of standard proportions with an outer diameter of 29 in. operates with air entering at 75 ft/s at 1.05 atm and 140°F. The particulate density is 1.80 g/cm^3. (a) Estimate the pressure drop across the cyclone in inches of water. (b) Repeat part (a) for a cyclone diameter of 16 in.

5-78. A cyclone has a pressure drop of 0.10 $lb_f/in.^2$ and the gas enters the inlet duct which is 2.5 ft high and 1.0 ft wide with a velocity of 40 ft/s. Calculate the fan power requirement in kilowatts.

5-79. To bring the cyclone fractional efficiency equation based on Equation (5-53) more in line with experimental data, it might be modified into an exponential format of the type, $\eta_d = 1 - \exp[-\pi N_e \rho_p d_p{}^2 V_g / 9\mu W)^n]$, where n is an adjustable constant. In terms of the cut size, $d_{p,50}$, (a) show that this equation may be written as $\eta_d = 1 - \exp[-0.707(d_p/d_{p,50})^{2n}]$. (b) Determine and plot the fractional efficiency given by this equation for an n value of 0.6 and $d_p/d_{p,50}$ values of 0.5, 1.0, 2.0, and 4.0. Then plot on the same graph the Lapple curve from Figure 5-17.

5-80. Estimate the stopping distance x_s and the impaction number N_I for a droplet diameter of 1000 μm in a 60 m/s air stream at 1 atm and 25°C. The particles in the air stream have a density of 1400 kg/m^3. Consider dust particle diameters of (a) 1.0, (b) 2.0, and (c) 3.0 μm. Finally, on the basis of Figure 5-19 of experimental data for sphere, estimate the impaction collection efficiency.

5-81. Estimate the stopping distance x_s and the impaction (separation) number N_I for a 3-μm particle with a density of 1200 kg/m^3 in a 30 m/s air stream at 1 bar and 27°C. Consider droplet sizes of (a) 250, (b) 500, and (c) 1000 μm.

5-82. On the basis of Figure 5-19 in the text, what is the fractional collection efficiency for cases (a), (b), and (c) of Problem 5-81 based on experimental data for spheres?

5-83. Estimate the stopping distance x_s and the impaction number N_I for a 2.0-μm particle with a density of 1400 kg/m^3 in a 50 m/s air stream at 1 atm and 27°C. Consider droplet sizes of (a) 200, (b) 500, and (c) 1000 μm.

5-84. On the basis of Figure 5-19 in the text, what is the impaction collection efficiency for cases (a), (b), and (c) of Problem 5-83 based on experimental data for spheres?

5-85. Calculate the effective target efficiency for an inertial impactor where the targets are 120-μm water droplets moving at an average speed of 7 cm/s and the dust being collected has a diameter of (a) 5, (b) 10, and (c) 15 μm and a density of 2.4 g/cm^3. Assume that the dust moves with the same speed as the air that transports it, which is 35 cm/s, and that the dust and droplets are moving in opposite directions. Use Figure 5-9 for data.

5-86. Air at 80°F loaded with particulate with a density of 2 g/cm^3 enters the bottom of a spray chamber and moves upward at 5 ft/min. A spray of water at the top of the chamber releases droplets which fall under gravity against the gas flow. Consider droplet sizes of 40, 100, 200, 400, 1000, and 2000 μm. Estimate the gravitational collection efficiency for particle sizes of (a) 4, (b) 7, and (c) 10 μm for the six droplet sizes, employing the experimental data for spheres shown in Figure 5-19. Settling velocities for the droplets may be estimated from Figure 5-9. Check your results roughly against the curves in Figure 5-20.

5-87. An approximate expression for the target efficiency of dust particles impinging upon water droplets is given in the form $\eta = \exp - [0.018(M)^{0.5+R}/R - 0.6R^2]$, where M is the square root of the impaction number, N_I, and $R = d_p/d_D$. For dust particles of a density of 2 g/cm^3 carried in an air stream of 75°F, and an initial relative dust-water droplet velocity of 100 ft/s, determine the target efficiency for water-droplet sizes of 10, 20, 50, 100, 200, 500, and 1000 μm for a dust-particle size of (a) 10 μm, (b) 5 μm, and (c) 20 μm.

5-88. With reference to the data of Example 5-9, determine the penetration for (a) a particle size of 0.9 μm, a particle density of 1.8 g/cm^3, and an L value of 1.4, (b) a particle size of 0.8 μm, a particle density of 1.7 g/cm^3, and an f value of 0.20, and (c) a particle size of 1.0 μm, a particle density of 1.4 g/cm^3, an air velocity of 450 ft/s, and an f value of 0.20. In each case all other data remain the same.

5-89. Water enters a venturi scrubber at a rate of 1.2 l/m^3 of gas flow. The air temperature is 80°F, its density is 0.075 lb/ft^3, and its velocity is 380 ft/s. The throat area is 150 in.^{2} and the parameter f may be taken to be 0.22. If the particulate density is 1.7 g/cm^3 and the average particle size is 1.2 μm, estimate the pressure loss by means of Equations (5-75) and (5-76).

5-90. Using the Calvert model for the variation of droplet velocity with distance in a venturi throat, determine (a) the drag coefficiency for the droplet at the throat entrance. Then determine the ratio of particle velocity to gas velocity for distances from the injection point of velocity to gas velocity for distances from the injection point of (b) 0.05 m, (c) 0.1 m, and (d) 0.5 m. Then (e) plot this velocity ratio versus x (in meters) and show the general shape of the overall curve. The water droplet size is 90 μm, the gas stream is air at 300°K and 1 bar, and the inlet droplet Reynolds number is 260.

5-91. Using the data of Problem 5-90, determine (a) the gas velocity for the given inlet Reynolds number. Then determine the pressure drop across the venturi section, in centimeters of water, using (b) the simplified Calvert equation, and (c) the equation proposed by Hesketh. The cross-sectional area of the venturi throat is 0.010 m^2, and the ratio of gas to liquid volume flow rate is 600/1, in the same set of units.

5-92. On the basis of the data and results of Problem 5-91, determine the penetration given by Calvert's equation for particulate diameters of (a) 0.3 μm, (b) 0.5 μm, (c) 1.0 μm, (d) 1.2 μm, and (e) 1.5 μm. The value of f is 0.3, and the particulate density is 1600 kg/m^3. Plot the fractional efficiency versus particle diameter.

5-93. Repeat the calculation made in Problem 5-92, except use the newer model by Calvert in conjunction with Figure 5-25. Apply the Cunningham correction factor, and use a value of 0.5 m for the length of the venturi throat.

5-94. Using the Calvert model for the variation of droplet velocity with distance in a venturi throat, determine (a) the drag coefficient for the droplet at the throat entrance. Then determine the ratio of droplet velocity to gas velocity for distances from injection of (b) 0.05 m, (c) 0.1 m, and (d) 0.5 m. Then (e) plot this velocity ratio versus x and show the general shape of the overall curve. The water droplet size is 100 μm, the gas stream is air at 300°K and 1 atm, and the inlet droplet Reynolds number is 350.

5-95. Using the data of Problem 5-94, determine (a) the gas velocity for the given inlet Reynolds number. Then determine the pressure drop across the venturi section, in centimeters of water, using (b) the simplified Calvert equation, and (c) the equation proposed by Hesketh. The cross-sectional area of the throat is 0.0125 m^2, and the dimensionless ratio of gas to liquid volume flow rates is 1000/1.

5-96. On the basis of the data and results of Problem 5-95, determine the penetration given by Calvert's equation for particulate diameters of (a) 0.3 μm, (b) 0.5 μm, (c) 0.8 μm, and (d) 1.3 μm. The value of f is 0.32, and the particulate density is 1800 kg/m^3. Plot the fractional efficiency versus particle diameter.

5-97. Repeat the calculation made in Problem 5-96, except use the newer model by Calvert in conjunction with Figure 5-25. Apply the Cunningham correction factor, and use a value of 0.5 m for the length of the venturi throat.

5-98. Air at 150°F with a dust loading of 5 gr/ft^3 passes through a fabric filter with a superficial velocity of 3 ft/min (air-to-cloth ratio). The dust permeability is 2.50×10^{-11} ft^2, and the filter cake density averages 1.4 g/cm^3. (a) If the residual pressure drop across the cleaned cloth is 0.40 in. water, estimate the time required for the total pressure drop to reach 3.5 in. water, in hours, on the basis of Equation (5-86). (b) What is the value of the filter drag, S, in in. water/ft/min?

5-99. A test is run on a cleaned filter cloth to determine the dust permeability, K_p. The pressure drop across the cleaned cloth is 2.52 mbar. An air stream at 27°C with a velocity of 1.80 m/min is used, and the filter cake density is 1.20 g/cm^3. A record kept of the overall pressure drop versus the mass of deposited particulate is as follows:

ΔP_o, mbars	6.12	6.66	7.74	9.00	9.90	10.62	11.52
M_p, kg	0.002	0.004	0.010	0.020	0.028	0.034	0.042

Determine the dust permeability, in square meters, if the filter cloth area is 100.0 cm^2.

5-100. Air at 100°C with a dust loading of 20.0 g/m^3 passes through a fabric filter with an air-to-cloth ratio (or superficial velocity) of 1.00 m/min. The average filter cake density is 1.25 g/cm^3, and the residual pressure drop across the cleaned cloth before the test begins is 1.00 mbar. On the basis of Equation (5-86), estimate (a) the total pressure drop across the filter after 4 hr, in millibars, if the dust permeability K_p is known to be 1.40×10^{-11} m^2, and (b) the filter drag, S, in mbar/m/min.

5-101. Air at 200°F with a dust loading of 10 gr/ft^3 passes through a fabric filter. After a period of 7 hr the total pressure drop across the filter is measured as 3.60 in. water. The filter cake density is 1.10 g/cm^3 and the residual pressure drop across the cleaned filter is 0.45 in. water. If the air velocity is maintained at 4 ft/min during the test, estimate the dust permeability, K_p, in units of (a) lb·in./in. water·min^2 and (b) square feet, on the basis of Equation (5-86).

5-102. Air at 127°C with a dust loading of 10.0 g/m^3 passes through a fabric filter with an air-to-cloth ratio (or superficial velocity) of 2.0 m/min. The particulate cake density on the filter averages 0.90 g/cm^3 and the residual pressure drop across the cleaned filter is 1.20 mbar. (a) After a period of 6 hr the total pressure drop is 8.00 mbar. Estimate the dust permeability, K_p, in square meters. (b) If the dust permeability is 5.00×10^{-12} m^2, estimate the overall pressure drop in millibars after 4.0 hr. (c) If the dust permeability is 7.70×10^{-12} m^2, estimate the time required for the total pressure drop to reach 7.50 mbar. Use Equation (5-86) for the estimates.

5-103. A cloth filter has R_f and R_p resistance values of 24,000 kg/m^2·s and 56,000 s^{-1}, respectively. The filter area is 1600 m^2 and the volume flow rate of air is 12 m^3/s with a dust loading L_d of 8 g/m^3. Determine (a) the pressure drop at startup in newtons per square meter and millibars, (b) the mass-area concentration W after 5 hr of operation, in kilograms per square meter, and (c) the pressure drop after 5 hr in newtons per square meter and in millibars.

5-104. A filter with 6000 m^2 of surface area is used to clean 50 m^3/s of air which has a dust loading of 4 g/m^3. The values of R_f and R_p are 22,000 kg/m^{2}s and 40,000 s^{-1}, respectively. If it is desirable for the overall pressure drop not to exceed 18 mbars, determine the maximum allowable cleaning period, in hours.

5-105. A test is carried out to determine the values of R_f and R_p. The following data were measured: ΔP (after cleaning) = 4 mbars, mass collected = 44 kg, ΔP (after test run) = 21 mbars, filter area = 36 m^2, and flow rate = 0.48 m^3/s. Determine (a) R_f in kg/m^2·s, and (b) R_p in s^{-1}.

5-106. A test is run on a cleaned fabric filter cloth to determine the dust permeability, K_p, in square meters and the dust resistance coefficient, R_p, in s^{-1}. An air stream at 27°C and with a filtering velocity of 0.80 m/min is used, and the filter cake density is 1200 kg/m^3. The following is a record of the overall pressure drop versus the mass-area concentration w_d on the filter:

ΔP_o, N/m^2	360	440	580	710	780	900	980
w_d, kg/m^2	0.20	0.60	1.58	2.62	3.11	3.97	4.50

The pressure drop across the cleaned cloth is found to be 300 N/m^2. Estimate the values of (a) K_p and (b) R_p in the specified units.

5-107. During a test on a cleaned bag filter the filter drag due to dust at the end of the test period is 1000 N·min/m^3, and the resistance R_p of the dust built up on the cloth is found to be 30,000 s^{-1}. The dust loading in the dirty air stream is 5.0 g/m^3. If the pressure drop increases by 10 mbar during the test, estimate the time of the test period in hours.

5-108. A cloth filter has R_f and R_p values of 4000 $lb_m/ft^2 \cdot s$ and 50,000 s^{-1}, respectively. The filter area is 70,000 ft^2 and the volume flow rate of air is 1530 ft^3/s with a dust loading of 4.4 gr/ft^3. Determine (a) the pressure drop at startup in psi and inches of water, (b) the mass-area concentration W after 4 hr of operation in lb_m/ft^2, and (c) the pressure drop after 4 hr, in psi and inches of water.

5-109. A filter with 5000 m^2 of face area is used to clean 40 m^3/s of air which has a dust loading of 3 g/m^3. The values of R_f and R_p are 25,000 kg/$m^2 \cdot s$ and 35,000 s^{-1}, respectively. If it is desirable for the overall pressure drop not to exceed 20 mbar, determine the maximum allowable cleaning period, in hours.

5-110. A test is carried out to determine the values of R_f and R_p. The following data were measured: ΔP (after cleaning) = 3 mbar, ΔP (after test run) = 16 mbar, mass collected = 40 kg, filter area = 32 m^2, and the flow rate = 0.40 m^3/s. Determine (a) R_f in kg/$m^2 \cdot s$ and (b) R_p in s^{-1}.

5-111. A test is run on a cleaned fabric filter cloth to determine the dust permeability, K_p, in square meters and the dust resistance coefficient, R_p, in s^{-1}. An air stream at 27°C and a filtering velocity of 0.60 m/min is used, and filter cake density is 1100 kg/m^3. The following is a record of the overall pressure drop versus the mass-area concentration, W, on the filter:

ΔP_o, N/m^2	320	400	480	580	720	880	1040
W, kg/m^2	0.20	0.52	0.90	1.45	2.40	3.02	3.82

The pressure drop across the cleaned cloth is found to be 200 N/m^2. Estimate the values of (a) K_p and (b) R_p in the specified units.

5-112. Calculate the drift velocity in meters per second at 25°C in air for particles with a diameter of (a) 0.5 μm, (b) 1 μm, (c) 5 μm, (d) 10 μm, and (e) 50 μm. The charging and collecting voltages are 50,000 and 40,000, respectively, and the anode-cathode spacing is 10.0 cm. Assume a dielectric factor p of 2.50, and apply the Cunningham correction.

5-113. Carry out the calculations required in Problem 5-112, except that the charging and collecting voltages are 45,000 and 35,000, respectively, the anode-cathode spacing is 12.0 cm, and the dielectric factor p is 2.20. Again apply the Cunningham correction factor.

5-114. An electrostatic precipitator has collecting plates 3 m tall and 1 m long in the direction of flow. The spacing between the charging electrode and the collecting plate is 7.56 cm. The device is to be used to collect particles having a dielectric constant of 4 and an effective diameter of 4 μm. The carrier gas (air at 25°C) has a velocity of 1.2 m/s. Calculate the collection efficiency for a charging voltage of (a) 10 kV, (b) 20 kV, (c) 30 kV, (d) 40 kV, and (e) 50 kV. Omit the Cunningham correction.

5-115. Determine the collection efficiency for the conditions given in Problem 5-114 when the charging voltage is 20 kV and the length of the collecting plate in the direction of flow is (a) 2 m, (b) 3 m, and (c) 4 m. Compare your results with the original value for a length of 1 m.

5-116. Determine the collection efficiency for the precipitator of Problem 5-114, part (b), if the air temperature is changed to 230°C, all other data remaining the same. Compare your results with the original value for 25°C operation.

5-117. Determine the collection efficiency for the conditions given in Problem 5-114 when the charging voltage is 30 kV and the length of the collecting plate in the direction of flow is (a) 2 m, (b) 3 m, and (c) 4 m. Compare your results with the original value for a 1-m length.

5-118. Determine the collection efficiency for the precipitator of Problem 5-114, part (c), if the air temperature is changed to 230°C, all other data remaining the same. Compare your results with the original value when the temperature is 25°C.

5-119. For an electrostatic precipitator of given geometry and operating conditions, it is found that the collection efficiency for 10-μm particles is 99.80 percent. On the basis of a Deutsch-type efficiency equation, estimate the collection efficiency for (a) 5-μm, (b) 1-μm, and (c) 0.2-μm particles.

5-120. For an electrostatic precipitator of given geometry and operating conditions, the collection efficiency for 2-μm particles is 99.90 percent. On the basis of a Deutsch-type equation for efficiencies, estimate the collection efficiency for (a) 1-μm, (b) 0.5-μm, and (c) 0.1-μm particles.

5-121. For a fixed drift velocity and volume flow rate, what percent change in electrode collection area is required to increase the collection efficiency from (a) 90 to 99 percent, and (b) 95 to 99.5 percent, on the basis of the Deutsch equation?

5-122. Reconsider Problem 5-119 on the basis of a Hazen-type equation for parts (a), (b), and (c), with an n value of 4.0.

5-123. Reconsider Problem 5-120 on the basis of a Hazen-type equation for parts (a), (b), and (c), with an n value of 5.0.

5-124. A test on an ESP indicates that a mean or effective migration velocity for the entire range of particle sizes is 0.50 ft/s. The length of the device is 20 ft and the width of the passage between electrodes is 0.40 ft. What is the maximum gas velocity permitted, in feet per second, through the ESP in order for the overall collection efficiency to reach 99.2 percent based on a Deutsch-type equation?

5-125. Reconsider Problem 5-121 on the basis of a Hazen-type equation for parts (a) and (b), if n is 4.0.

5-126. An electrostatic precipitator with a specific collection area of 350 ft^2/1000 acfm is found to have an actual overall collection efficiency of 97.7 percent.

If the value of A/Q is increased to (a) 450 $ft^2/1000$ acfm and (b) 500 $ft^2/1000$ acfm, estimate the new overall collection efficiency using the Deutsch equation. Assume that the migration velocity remains constant as A/Q is altered.

5-127. The overall efficiency of collection of an ESP is found to be 98.4 percent. It is desired to decrease the penetration to one-fourth its initial value, without drastically changing the physical setup. One way of achieving this is to change the charging and collecting field strengths. If both initially are equal and are changed the same amount, what is the required percent change in the field strengths to reach the desired collection efficiency or penetration based on (a) the Deutsch equation and (b) a Hazen-type equation with an n value of 4?

5-128. An electrostatic precipitator is employed to remove glass particles having a mean diameter of 4 μm from air flowing through the collector at 4 ft/s. The precipitator is 8 ft tall and 6 ft wide with electrodes having a spacing of 6 in. and a length of 2 ft in the direction of flow. The charging voltage is 20,000 V, and the dielectric constant averages 4.75. Estimate (a) the efficiency of the precipitator, (b) the efficiency if the voltage is increased to 35,000 V, and (c) the efficiency if the length of the plates is increased to 3 ft, for parts (a) and (b).

5-129. Assume that the installation cost of a large electrostatic precipitator is \$1.00 per ft^3/min for an overall collection efficiency of 97.0 percent. Estimate the percentage of increase in cost per ft^3/min if it is desired to have an efficiency of 99.7 percent, with all other general operating variables remaining the same.

5-130. Selzler and Watson [27] have proposed the following empirical Deutsch-type equation for the collection efficiency for an electrostatic precipitator:

$$\eta_{col} = 100\left\{1 - \exp\left[-K\left(\frac{A}{V}\right)^{1.4}\left(\frac{KW}{V}\right)^{0.6}\left(\frac{S}{AH}\right)^{0.22}\right]\right\}$$

where A is the collecting plate area, in 1000 ft^2; V is the volumetric flow rate in 1000 acfm; KW is the power input to the discharge electrode in kilowatts; S/AH is the sulfur-to-ash ratio, by weight, for the coal used; and K is 116 for pulverized coal or 90 if a cyclone furnace is used or if the precipitator is preceded by a mechanical collector. For the data listed below, predict the collection efficiency for the three cases. The first and third cases use pulverized coal, while the second case uses a mechanical collector.

CASE	SIZE (MW)	V (1000 acfm)	S/AH (BY WT.)	A (ft^2)	POWER (kW)
(a)	200	810	2.5/12.4	80,000	6000
(b)	500	1720	0.9/11.7	240,000	1700
(c)	800	2480	1.1/9.3	355,000	2360

5-131. In Problem 5-130 Selzler and Watson's empirical equation for the collection efficiency of an electrostatic precipitator is given. Consider the following operational changes and their effect on the collection efficiency.

(a) A precipitator with an efficiency of 97.7 percent operates on a coal with a sulfur-to-ash ratio of 1.1 : 12.3. To help meet SO_2 emission standards, a coal is substituted with a sulfur-to-ash ratio of 0.7 : 10.9. If all other operating variables are the same, estimate the new collection efficiency. Was the change a wise one?

(b) A precipitator with an efficiency of 98.8 percent operates with a collector area of 80,000 ft^2 and a volume flow rate of 150,000 actual ft^3/min. What is the collection efficiency if a 10 percent overload in flow rate passes through the collector? What is the percent increase in dust released to the atmosphere? Estimate the collection efficiency for a 10 percent underload also, and the percent change in dust released.

(c) Consider the data in part (a). What percent change in corona power is necessary if, for the new sulfur-to-ash ratio, it is desired to restore the collection efficiency to its original value?

References

1. NAPCA. *Air Quality Criteria for Particulate Matter*, AP-49, Washington, D.C.: HEW, 1969.
2. NAPCA. *Air Pollution Engineering Manual*, No. 999-AP-40. Washington, D.C.: HEW, 1967.
3. *Air Conservation*. Washington, D.C.: American Association for the Advancement of Science, 1965.
4. A. C. Stern, *Air Pollution*, Vol. 1. New York: Academic Press, 1968.
5. NAPCA. *Control Techniques for Particulate Air Pollutants*, AP-51. Washington, D.C.: HEW, 1969.
6. *Air Pollution Manual*. Part II—"Control Equipment." Detroit, Mich.: American Industrial Hygiene Association, 1968.
7. W. L. Faith and A. A. Atkisson, Jr. *Air Pollution*. 2nd ed. New York: Wiley, 1972.
8. ASME. *Determining the Properties of Fine Particulate Matter*, ASME Power Test Code PTC 28-1965. New York, 1965.
9. W. Strauss. *Industrial Gas Cleaning*. London: Pergamon Press, 1966.
10. C. H. Bosanquet, W. F. Carey, and E. M. Halton. "Dust Deposition from Chimney Stacks." *Proc. Inst. Mech. Engineers* **162** (1950): 355.
11. J. M. Dalle Valle. *Exhaust Hoods*. 2nd ed. New York: Industrial Press, 1963.
12. W. C. L. Hemeon. *Plant and Process Ventilation*. 2nd ed. New York: Industrial Press, 1963.
13. H. L. Green and W. R. Lane, *Particulate Clouds: Dusts, Smokes, and Mists*, 2nd ed. Princeton, N.J.: Van Nostrand, 1964.
14. A. C. Stern, P. D. Bush, and K. J. Kaplan. *Cyclone Dust Collectors*. New York: American Petroleum Institute, 1955.
15. C. E. Lapple. "Processes Use Many Collection Types." *Chem. Engr.* **58** (May 1951): 145.
16. R. D. Ingebo. *Drag Coefficients for Droplets and Solid Spheres in Clouds Accelerating in Airstreams*, NACA Tech. Note 3762 (1965).
17. H. W. Walton and A. Woolcock. *Int. J. Air Pollu.* **3** (1960): 129.
18. S. Calvert et al. *Wet Scrubber System Study*: Vol. I, *Scrubber Handbook*, PB 213-016, 1972.
19. K. G. T. Hollands and K. C. Goel. "A General Method for Predicting Pressure Loss in Venturi Scrubbers." *Ind. Eng. Chem. Fundam.* **14** (1975): 16–22.

20. H. E. Hesketh. "Fine Particle Collection Efficiency Related to Pressure Drop, Scrubbant and Particle Properties, and Contact Mechanism." *J. Air Pollu. Control Assoc.* **24**, no. 10 (1974): 939–942.
21. S. Yung, S. Calvert, H. F. Barbarika, and L. E. Sparks. "Venturi Scrubber Performance Model." *Environ. Sci. Tech.* **12**, no. 4 (1978): 456–459.
22. S. Yung, H. F. Barbarika, and S. Calvert. "Pressure Loss in Venturi Scrubbers." *J. Air Pollu. Control Assoc.* **27**, no. 4 (1977): 348–351.
23. S. Calvert. "Source Control by Liquid Scrubbers." in *Air Pollution*, A. Stern, ed., Chap. 46. New York: Academic Press, 1968.
24. E. I. Shaheen. *Environmental Pollution*. Mahomet, Ill.: Engineering Technology, Inc., 1974.
25. R. D. Ross, ed. *Air Pollution and Industry*. New York: Van Nostrand Reinhold, 1972.
26. F. R. Culhane. "Fabric Filters Abate Air Emissions." *Environ. Sci. Tech.* **8** (February 1974): 127.
27. "Future Bright for Fabric Filters." *Environ. Sci. Tech.* **8** (June 1974): 508.
28. E. R. Frederick. "Fabric Filtration for Fly Ash Control." *J. Air Pollu. Control Assoc.* **29**, no. 1 (1979): 81–85.
29. "Proceedings of a Symposium on Fine Particle Fabric Filtration." *J. Air Pollu. Control Assoc.* **24**, no. 12 (1974): 1140–1197.
30. *Fabric Filter Systems Study, Handbook of Fabric Filter Technology*, Vol. 1. PB 200-648, APTD-0690, National Technical Information Service, December, 1970.
31. H. J. White. "Resistivity Problems in Electrostatic Precipitation." *J. Air Pollu. Control Assoc.* **24** (April 1974): 314.
32. H. J. White. *Industrial Electrostatic Precipitation*. Reading, Mass.: Addison-Wesley, 1973.
33. H. J. White. *Ind. Engr. Chem.* **47** (1955): 932.
34. J. D. McCain, J. P. Gooch, and W. B. Smith. "Results of Field Measurements of Industrial Particulate Sources and Electrostatic Precipitator Performance." *J. Air Pollu. Control Assoc.* **25**, no. 2 (1975): 117–121.
35. G. Penney. "Some Problems in the Application of the Deutsch Equation to Industrial Electrostatic Precipitation." *J. Air Pollu. Control. Assoc.* **19** (1969): 596.
36. J. O. Ledbetter. "Electrostatic Precipitator Efficiency Calculations." *J. Air Pollu. Control Assoc.* **28**, no. 12 (1978): 1228–1229.
37. W. Strauss, ed. *Air Pollution Control, Part 1*. New York: Wiley, 1971.
38. "Proceedings of a Symposium on Fine Particle Electrostatic Precipitation." *J. Air Pollu. Control Assoc.* **25**, no. 2 (1975): 98–189.
39. G. W. Penney. "Electrical Liquid Spray Dust Precipitator," U.S. Patent No. 2,357,354 (1944).
40. J. D. McKenna, J. C. Mycock, and W. O. Lipscomb. "Performance and Cost Comparisons between Fabric Filters and Alternate Particulate Control Techniques." *J. Air Pollu. Control Assoc.* **24**, no. 12 (1974): 1144–1150.
41. P. L. Feldman. "Effects of Particle Size Distribution on the Performance of Electrostatic Precipitators." Research-Cottrell, Inc., Presented at APCA Annual Meeting, Boston, 1975.
42. J. R. Bush, P. L. Feldman, and M. Robinson. "Development of a High Temperature, High Pressure Electrostatic Precipitator," EPA-600/7-77-132, November, 1977.
43. M. Crawford. *Air Pollution Control Theory*. New York: McGraw-Hill, 1976.

Chapter 6
General Control of Gases and Vapors

6-1 INTRODUCTION

By far the major portion of the recognized air pollutants are gases such as carbon monoxide, the oxides of nitrogen, the oxides of sulfur, and unburned hydrocarbons. Data from Table 1-8 for 1975 indicate that on a mass basis approximately 93 percent of the air pollutants are gases, with carbon monoxide contributing 52 percent. Combustion of fossil fuels is the prime source of gaseous pollutants. Other gases and vapors emitted during numerous common industrial processes may cause the local concentrations of the pollutants to reach objectionable levels. Specific information on those substances may be found in references 1 and 2 and other publications of the Enviromental Protection Agency (EPA) released periodically.

Generally, the concentrations of gaseous pollutants in gas mixtures are relatively low. The reduction of these concentrations to desirable levels can be accomplished by several methods. First, the pollutant species may be adsorbed on the surface of selective solid adsorbers. Second, the pollutants may be absorbed by liquid solvents. Third, the pollutant may be oxidized by means of catalytic or direct flame incineration to another chemical form which is not a pollutant. Fourth, the concentration of the pollutant may be reduced by restricting the quantity of the pollutant formed in the original chemical process. For the successful application of the third and fourth methods of pollution control, the reaction kinetics and chemical equilibrium evaluation of the chemical species involved must be studied. Consequently, the basic principles of chemical kinetics and equilibrium chemistry are presented in this chapter, in order to form a basis for the subject matter presented in this and later chapters on the control methods for carbon monoxide, sulfur dioxide, the oxides of nitrogen, and unburned hydrocarbons.

Because of the importance of sulfur dioxide, and the oxides of nitrogen, these topics will be discussed in separate chapters. Unburned hydrocarbons are emitted by motor vehicles, stationary combustion processes, and other industrial processes such as petroleum refining and the drying of industrial finishes. We discuss the control of hydrocarbon and carbon monoxide emissions from motor vehicles in Chapter 10. Here we deal with the basic principles of chemical kinetics and give some specific information on the control of carbon monoxide and hydrocarbons from stationary sources. We begin with a discussion of two methods of general importance in the control of many gaseous pollutants in waste gases, namely, adsorption and absorption.

6-2 ADSORPTION

Adsorption is a separation process based on the ability of certain solids to remove gaseous (or liquid) components preferentially from a flow stream. The pollutant gas or vapor molecules present in a waste stream collect on the surface of the solid material. The solid adsorbing medium is frequently termed the *adsorbent*, while the gas or vapor adsorbed is called the *adsorbate*. In addition to dehumidifying air and other gases, adsorption is useful in removing objectionable odors and pollutants from industrial gases as well as recovering valuable solvent vapors from air and other gases. Adsorption is a particularly useful technique when:

1. The pollutant gas is noncombustible or difficult to burn.
2. The pollutant is sufficiently valuable to warrant recovery.
3. The pollutant is in a very dilute concentration in the exhaust system.

Adsorption also makes it economically possible to purify gases containing only small amounts of pollutants that are difficult, if not impossible, to clean by other means.

The actual adsorption process is classified as either *physical adsorption* or *chemisorption*. In physical adsorption, the gas molecules adhere to the surface of the solid adsorbent as a result of intermolecular attractive forces (van der Waals forces) between them. The adsorption process is exothermic. The heat liberated, which depends upon the magnitude of the attractive force, is on the order of the enthalpy (heat) of condensation of the vapor. This generally is in the range of 2 to 20 kJ/g·mole. The advantage of physical adsorption is that the process is reversible. By lowering the pressure of the adsorbate in the gas stream or by raising the temperature, the adsorbed gas is readily desorbed without a change in chemical composition. The temperature effect is most commonly used. This reversibility feature is extremely important if recovery of the adsorbed gas or recovery of the adsorbent for reuse is economically desirable. The amount of gas physically adsorbed decreases rapidly with increasing temperature and is quite small when the temperature is above the critical temperature of the adsorbed component. Physical adsorption is usually directly proportional to the amount of solid surface available. However, this buildup is not restricted to a monomolecular layer; a number of layers of molecules can build up on the surface. Another desirable characteristic of physical adsorption is that the rate generally is quite rapid.

Chemisorption results from a chemical interaction between the adsorbate and the adsorbing medium. The bonding force associated with this type of adsorption is much stronger than that for physical adsorption. Consequently, the heat liberated during chemisorption is much larger than that liberated during physical adsorption, for example, on the order of an enthalpy (heat) of reaction. This may be in the range of 20 to 400 kJ/g·mole. Because of this high heat of adsorption, the energy associated with chemisorbed molecules is significantly different from that associated with molecules in the gas stream. Hence the energy (or activation) required for the chemisorbed molecules to react with another molecular species may be considerably less than the energy required when the two species react directly in the gas phase. This lower energy requirement is one basis of explanation of the catalytic effect of solid surfaces in enhancing the rate of some chemical reactions. Another important difference exists. In addition to being a much more highly exothermic process, chemisorption also frequently is irreversible. On desorption the chemical nature of the original adsorbate will have undergone a change. If either regeneration of the adsorbent or recovery of the adsorbate is desirable, the adsorbing medium must be chosen so that a physical adsorption process controls. On the other hand, it is the chemisorption process which is responsible for the catalytic effect. Catalysis is an extremely important process in a number of air pollution control systems. When the rate of chemisorption varies with temperature, the adsorption is called activated chemisorption. When chemisorption occurs very rapidly, the adsorption process is termed nonactivated chemisorption. Finally, another distinguishing feature of chemisorption is that only a monomolecular layer of adsorbate appears on the adsorbing medium. This effect is due to the extremely short distances over which the valence forces holding the adsorbate to the adsorbent are effective.

The adsorbent itself must have some fairly obvious properties. Particle diameter may range from $\frac{1}{2}$ in. to as small as 200 μm or less. A large surface area per unit weight is essential. For gas adsorption it is not the outside apparent surface area which is important, but rather the surface provided by the internal pores of the solid. When the effective pore diameter is just a few times larger than the molecular diameter of the gas of interest, then an enormous surface area will be available for adsorption. As an example, charcoal may contain an effective surface as large as 10^5 to 10^6 m^2/kg (10^6 to 10^7 ft^2/lb). In addition to a sufficient surface area, a certain chemical specificity is frequently required of the adsorbant. This is a function of the chemical nature of the solid medium. It is well known that charcoal (activated carbon) has an affinity for hydrocarbons, while silica gel adsorbs water readily. A physical property of concern in fixed-bed operation is the air-flow resistance. The pressure drop through the bed must be held to a tolerable level. If a fluidized bed is to be used, the particles of adsorbent must not be easily carried away by the flow stream. In order to withstand size reduction or crushing, the solid must also meet certain standards of strength and hardness.

The activated carbon referred to earlier is made by the carbonization of coal, wood, fruit pits, and coconut shells. It is made active by treatment with hot air or steam, and is available in pelleted or granular form. Charcoal of this type is particularly useful for the recovery of solvent vapors. Silica gel is a granular product made from the gel precipitated by sulfuric acid treatment of a sodium silicate solution. It is used primarily to dehydrate air and other gases, but it is limited to temperatures below 250°C (500°F). Both activated carbon and silica gel may be desorbed. Another dehydrating agent is activated alumina, which is an aluminum oxide. It is reactivated by heating to 175° to 325°C (350° to 600°F). Fuller's earths are natural clays used in the petroleum industry as well as with vegetable and animal oils. The adsorbed organic material can be removed by washing and burning.

There is considerable interest in the use of synthetic zeolites for the removal of pollutants from gas streams. The use of zeolites for water-softening purposes is well established. The synthetic zeolites now in use are crystalline metal aluminosilicates, usually called molecular sieves [3]. Their great advantage is their selectivity or specificity, which is attained by tailor-making their crystalline structure so that they will adsorb only certain molecules. The pores on any particular type of molecular sieve are uniform in molecular dimensions. Depending upon the size of these pores, gaseous molecules will be readily adsorbed, slowly adsorbed, or completely excluded. In addition to separating molecules by size, sieves can also be made to separate molecules of different dipole moment and hydrocarbons which differ in their degree of unsaturation. Molecular sieves can be regenerated by heating or elution (solvent extraction).

Molecular sieves have recently been developed [4] for the control of SO_2, NO_x, and Hg emissions. In these cases the pore diameter of the metal aluminosilicates ranges from 3 to 10 Ångstroms (10^{-10} m or Å). Specific applications are to the exhaust gases from sulfuric and nitric acid plants, for example. In these two cases a special benefit accrues, since there is no secondary waste problem. When the sieves are exhausted (saturated), they are regenerated by a heated purge gas. However, the purge gas is recycled back into the main plant stream, which results in an increase in productivity of the acid plant. For a 200 ton/day sulfuric acid plant the SO_2 recovery was greater than 99 percent, and the emissions were one-tenth that of the federal standard for new plants. For a 55 ton/day nitric acid plant, the emissions were also one-tenth of the federal standard and the increase in nitric acid production was 2.5 percent. This approach is not inexpensive, and hence should be used only where a recoverable product exists. Using molecular sieves to deal with the SO_2 problem of the public utilities or the NO_x problem from fossil-fueled steam generating plants would probably not be economically feasible.

The effect of pore size on separation selectivity is illustrated by some data on molecular sieves compiled by the Union Carbide Corporation. When the nominal diameter is 0.003 μm (30×10^{-8} cm), the molecules adsorbed typically might be H_2O and NH_3. By increasing the size to 0.004 μm, larger molecules, such as CO_2, SO_2, H_2S, C_2H_4, C_2H_6, and C_2H_5OH, are also adsorbed. A further increase to 0.005 μm allows the additional adsorption of normal paraffins to the exclusion of branched and cyclic hydrocarbons. Thus a fairly strong specificity can be built into molecular sieves, which enhances the engineer's control over the adsorption phenomena.

The general requirements that must be met in the design or selection of suitable adsorption equipment include: (1) provision for sufficient dwell time, (2) pretreatment of the gas stream to remove nonadsorbable matter that would impair the operation of the adsorption bed, (3) pretreatment to remove high concentrations of competing gases by other more effective processes to prevent overburdening the adsorption system, (4) good distribution of flow through the bed, and (5) provision for renewing or regenerating the adsorbent bed after it has reached saturation. In general, one has a choice between a regenerative type and a nonregenerative (throwaway) type, and between batch versus continuous operation. Generally, when the pollutant concentration in the gas stream entering the adsorber is less than 1 or 2 ppm, the adsorbent and adsorbed material are discarded. In this case, single-pass nonregenerative replacement adsorbent elements are used. However, discarding the elements without removal of the adsorbate is strictly safe only if the adsorbate is relatively nonvolatile. When the pollutant concentration exceeds several parts per million, one-pass or two-pass regenerative-type multiple-chamber systems are employed. When continuous operation is required, one chamber regenerates while the other adsorbs, as shown in Figure 6-1. In general, gas velocities at the face of the bed are from 20 to 100 ft/min with dwell times within the bed from 0.6 to 6 s. References 5, 6, and 7 provide much additional information on adsorbers.

Among the numerous theories on adsorption phenomena, none at present explains a majority of the experimental observations [8, 9]. Several of these observations are worth noting. First of all, a great deal of the experimental adsorption data pertain to equilibrium data, which resemble the equilibrium solubility data of gases in liquids. That is, the amount of gas adsorbed per unit of adsorbent at equilibrium is measured against the partial

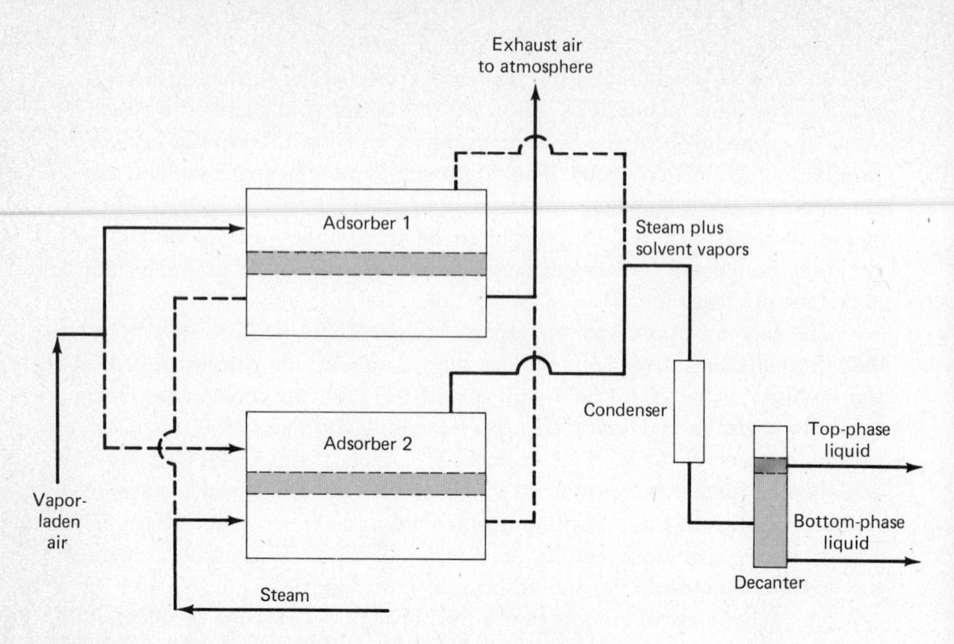

Figure 6-1 Sketch of a two-unit fixed-bed adsorber.

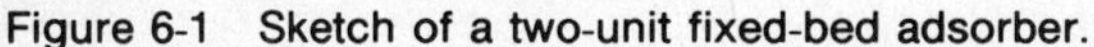

pressure of the adsorbate maintained in the gas phase. Figure 6-2 shows the data for two gases, acetone and benzene, adsorbed on activated carbon. As would be expected, increasing the pressure of the adsorbate in the gas stream causes more of it to be adsorbed at a given temperature. An increase in the temperature of operation for a given adsorbate and partial pressure in the gas stream decreases the concentration of adsorbed gas. As a consequence, it is usually desirable to operate an adsorption bed at as low a temperature as possible. It is also generally true that adsorption improves with an increase in the molar mass of the adsorbate, although the degree of unsaturation is also important. Figure 6-2 indicates that for the same partial pressure in the gas stream and the same temperature, considerably more benzene than acetone will be adsorbed at equilibrium.

The data of Figure 6-2 are examples of equilibrium *adsorption isotherms* for a specific combination of adsorbent and adsorbate. In general, an adsorption isotherm relates the volume or mass adsorbed to the partial pressure or concentration of the adsorbate in the main gas stream at a given temperature. As noted in Figure 6-2, the equilibrium concentration adsorbed can be quite sensitive to temperature. The adsorption isotherms do not need to have the shape of the curves in Figure 6-2, however. Although experimental data reveal a number of possibilities, Figure 6-3 illustrates three representative isotherms. The curve of Figure 6-3(a) is convex upward throughout. This type of behavior is quite favorable to the uptake of the adsorbate, since the quantity adsorbed remains essentially the same as the partial pressure (or concentration) of the adsorbate decreases. In Figure 6-3(b) the equilibrium

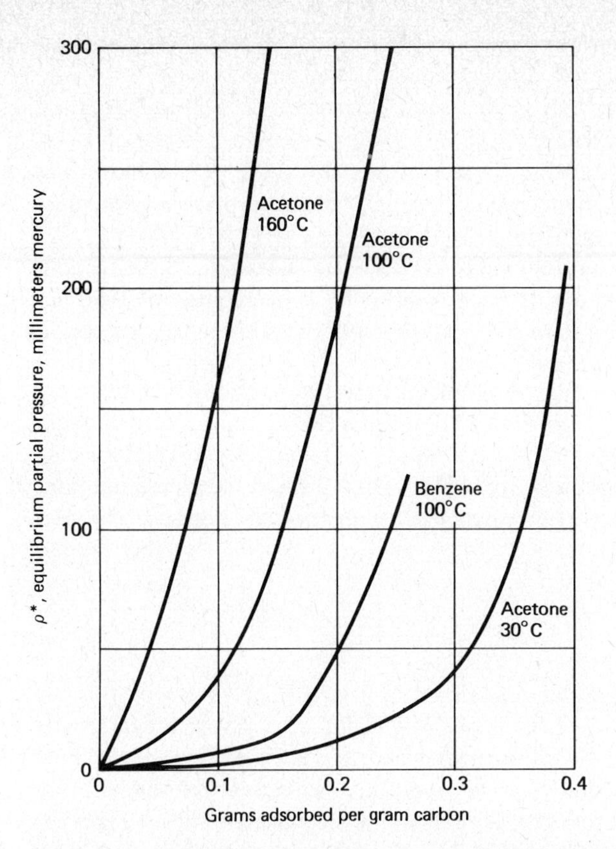

Figure 6-2 Equilibrium adsorption on an activated carbon.

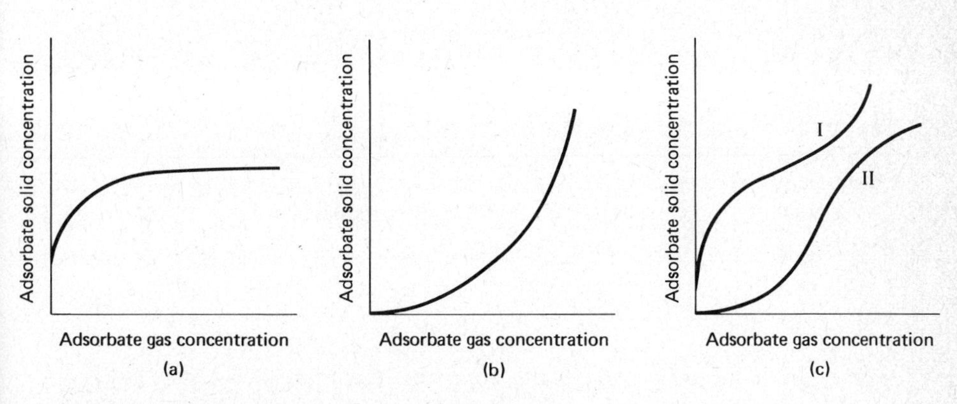

Figure 6-3 Representative adsorption isotherms.

curve is concave upward throughout. This is an unfavorable situation for adsorption, since at low gas concentrations the quantity adsorbed will be quite low. As shown in Figure 6-3(c), some adsorption isotherms contain an inflection point in the curve.

The equilibrium adsorption isotherm is equally valid for the adsorption process and the desorption or elution process, if the adsorption is reversible. Consequently, it can be argued on the basis of the curves in Figure 6-3 that isotherms which are favorable to adsorption are not favorable to desorption. In the regeneration of an adsorbent this difficulty is overcome by desorbing at a different temperature, thus altering the characteristics of the desorption process to a more favorable condition.

In the literature a number of equations are proposed to fit analytically the various experimental isotherms. In physical adsorption the approximation equation of Brunauer, Emmett, and Teller is frequently used. The development [10] of this equation is based on the rates of condensation and evaporation from various layers of molecules on the solid surface. The end result is in the form

$$\frac{V}{V_m} = \frac{cP}{(P_o - P)[1 + (c-1)(P/P_o)]} \tag{6-1}$$

which is known as the Brunauer-Emmett-Teller (B.E.T.) equation. V is the volume the amount of adsorbed gas would fill at a given pressure and temperature, and V_m is the volume adsorbed if a layer one molecule thick fills the surface. P_o is the vapor pressure of the adsorbate at the temperature of the system, P is the actual partial pressure of the adsorbate in the main gas stream, and c is a parameter of the particular adsorption process.

The value of V_m and c are found from experimental data on the adsorption system of interest. First, it is convenient to rearrange the above equation to the form

$$\frac{1}{cV_m} + \frac{c-1}{cV_m}\frac{P}{P_o} = \frac{P}{V(P_o - P)} \tag{6-2}$$

On this basis, a plot is made of $(P/V)/(P_o - P)$ versus P/P_o, with the best possible straight line drawn through the data points. Equation (6-2) indicates that its intercept is $1/cV_m$ and its slope is $(c-1)/cV_m$. From these two values we can determine V_m and c, which in turn establishes the equilibrium relationship between V and P through Equation (6-1). It is easily shown that the volume V_m corresponding to a filled monomolecular layer of adsorbed gas, in terms of a B.E.T. plot, is simply

$$V_m = \frac{1}{I + s} \tag{6-3}$$

where I is the intercept and s is the slope from the linear plot. Figure 6-4 is a plot of some experimental data for the adsorption of CO and CO_2 on silica

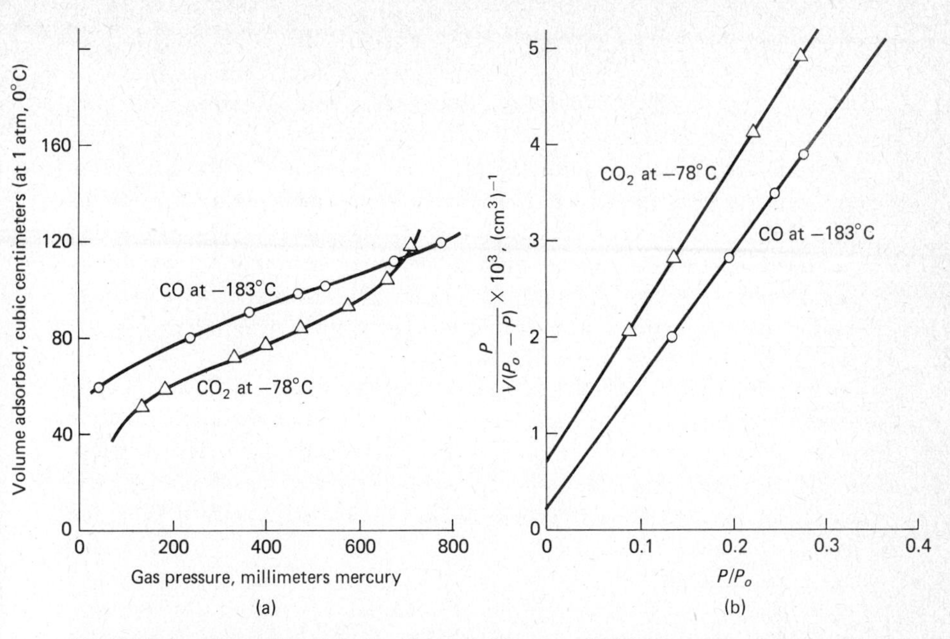

Figure 6-4 Adsorption data for CO and CO_2 on a 0.606-g sample of silica gel. (a) Adsorption isotherm. (b) Plot of B.E.T. equation. (DATA SOURCE: *J. Am. Chem. Soc.* **59** (1937): 2682.)

gel. Figure 6-4(a) shows the adsorption isotherm, while Figure 6-4(b) illustrates a B.E.T. plot of the data. The linearity of these data on the B.E.T. plot is apparent. Research indicates that frequently the B.E.T. plot is not linear for P/P_o values of less than 0.05 and greater than 0.35. In some cases the B.E.T. plot of data leads to an S-shaped isotherm. Other techniques then must be used to evaluate V_m. The data of Figures 6-2 and 6-4 are a measure of the capacity of the adsorbent at equilibrium at a given temperature and gas stream concentration and represent upper limits to real processes.

6-3 THE ADSORPTION WAVE

Consider the situation where a polluted gas stream is passed through a stationary (fixed) bed of adsorbent. The gas stream will be taken to be a binary mixture composed of a dilute pollutant gas of initial concentration C_o in an inert carrier gas such as air. As the time of operation of the adsorption process increases, the amount of adsorbed gas on the bed increases. However, this adsorption does not occur uniformly throughout the bed. During the initial period of operation the layer of adsorbent nearest to the inlet effectively absorbs most of the solute (adsorbate) in the carrier gas. Once past this initial layer of adsorbent, the carrier gas is relatively free of pollutant. The remaining portion of the bed also is essentially free of the pollutant. This situation is illustrated in Figure 6-5(a), where the relative

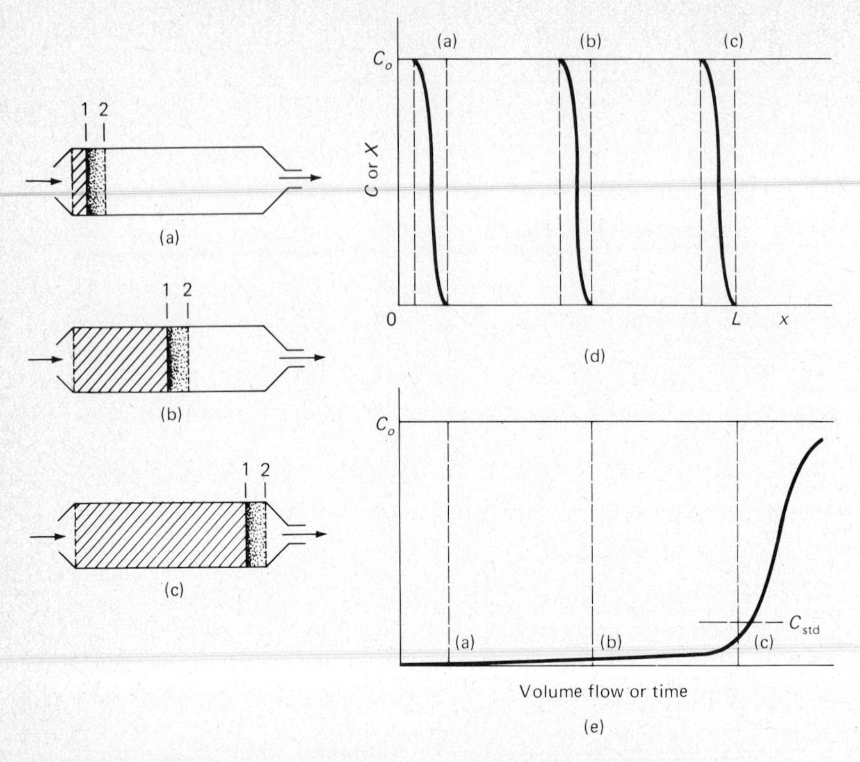

Figure 6-5 Passage of an adsorption wave through a fixed bed.

density of dots between positions 1 and 2 indicates the relative concentration of pollutant adsorbed on the bed. The region from the inlet to position 1, indicated by the cross-hatched area, is essentially saturated with pollutant. The region between positions 1 and 2 is known as the *adsorption zone*. In this zone the pollutant concentration of the adsorbent goes from an essentially saturated state at position 1 to a nearly pollutant-free state at position 2. Likewise, the concentration of pollutant in the carrier gas changes from a value C_o at position 1 to a value near zero at position 2. In part (d) of Figure 6-5 the value of either C (the mass of pollutant per volume of polluted gas) or X (the mass of pollutant per mass of adsorbent) is plotted versus position in the bed at various times since startup. Note the rapid drop in pollutant concentration from the beginning to the end of the absorption zone.

As more polluted gas enters the adsorption bed, more of the initial layers of the bed become saturated. Thus the adsorption zone moves through the bed toward the outlet. At some later time, illustrated by Figure 6-5(b), the bed is saturated through roughly the first half of its length. Figure 6-5(e) is a plot of pollutant concentration in the gas phase at the exit of the bed versus either the volume of carrier gas passed through the bed or the time of operation. When the adsorption zone is at either (a) or (b), the value of C at the exit is finite, but nearly zero.

At still some later time, or for the passage of a still larger amount of polluted gas through the bed, the leading edge of the adsorption zone reaches the end of the fixed bed of length L. See Figure 6-5(c). Since the concentration profile through the adsorption zone is frequently S-shaped, the adsorption process appears as a "wave" passing through the bed. Hence the adsorption phenomenon in a stationary bed is frequently referred to as an *adsorption wave*. As this wave reaches the last layer of adsorbent, the concentration of pollutant in the exit gas stream begins to rise rapidly, as shown in Figure 6-5(e). This condition of a rapid change in C at the exit is known as the *breakthrough point*, or *break point*. Arbitrarily, this might be the condition when the first measurable pollutant concentration is detected at the exit. From an air pollution control viewpoint, the value of C after breakthrough cannot be allowed to exceed the emission standard, if the gas stream is vented directly from the adsorption bed to the environment. It is probably not advisable to have the leading edge of the adsorption zone approach the last layer of adsorbent, since variations in flow rates, inlet concentration, temperature, or other factors may cause the outlet pollutant concentration to fluctuate and exceed that permitted by regulation.

The shape of the adsorption wave is extremely important. The concentration profile within the zone could be extremely steep or quite flat. In the limit, for an infinite rate of adsorption, the adsorption wave curve would be vertical. Hence instant breakthrough would occur at some moment, which would be a disadvantage from a control standpoint. However, an advantage would be that the entire bed would be saturated, so that a larger volume of polluted gas could pass through a given size of bed before shutdown would be necessary. In practice, it is highly desirable that breakthrough not occur until bed saturation reaches 75 or 80 percent. The other extreme would be the situation where the S-shaped curve were very flat. In this case breakthrough would come relatively early in terms of total gas flow through the bed, and much of the bed would not be saturated at the time of breakthrough. In some cases saturation may be as low as 15 percent under this circumstance. Thus longer beds would be required to keep the time of operation of a bed at a reasonable length. This leads to other disadvantages, such as a larger pressure drop across the bed. Hence some intermediate shape of the adsorption wave is more appropriate. In order to determine the time of breakthrough, for a given concentration profile within the adsorption zone, one must determine the velocity of the adsorption zone through the bed. The factors that determine the shape and velocity of the adsorption wave are discussed in the next section. It is also apparent that a uniform distribution of adsorbent throughout the fixed bed is imperative. If any type of channeling results from improper distribution of the solid material, the time for breakthrough may be significantly shortened.

6-4 TRANSIENT ANALYSIS OF AN ADSORPTION WAVE

The physical interpretation of an adsorption wave was presented in the preceding section. We now wish to describe this phenomenon mathematically, so that the shape of the concentration profile and the velocity of the wave can be determined under specified conditions. In Figure 6-6, the general features of an adsorption wave are shown. The gas-phase concentration C of the pollutant (expressed as mass of pollutant per volume of gas) or the solid-phase concentration X (expressed as mass of pollutant per mass of adsorbent) is plotted versus position x along the bed of total length L. Concentration C varies from zero to C_o, the inlet pollutant concentration. It is assumed that the bed is saturated at the trailing edge of the wave, that is, the gas and solid phases are essentially in equilibrium at any position behind the adsorption zone. Thus the concentration X varies between zero and X_{sat}. Therefore at position 1 the concentrations are C_o and X_{sat}, while at position 2 both concentrations are essentially zero. The gas stream moves at velocity V_g downstream relative to the fixed bed, while the adsorption zone moves with a velocity V_{ad} relative to the bed. From a mathematical viewpoint, it is convenient to set up the governing equations based on the adsorption zone as a fixed reference, rather than the adsorption bed. Since V_g is much larger in practice than V_{ad} (by several orders of magnitude), we can let V_g represent the gas stream velocity relative to the fixed zone. Then V_{ad} is the velocity of the bed as it passes through the zone from right to left in Figure 6-6.

First we shall make a mass balance on the pollutant (adsorbate) for the overall adsorption zone between positions 1 and 2. The rate of pollutant transferred into the zone by gas and solid phases must equal the rate transferred out. At position 2 there is no pollutant leaving with the gas phase or entering with the solid phase. Thus all mass transfer occurs at position 1. If $\dot{m}_g$ and $\dot{m}_{ad}$ are the mass flow rates of the gas phase entering and solid phase leaving, respectively, then

$$\frac{\dot{m}_g C_1}{\rho_g} = \dot{m}_{ad} X_1 = \rho_{ad} A V_{ad} X_1 \tag{6-4a}$$

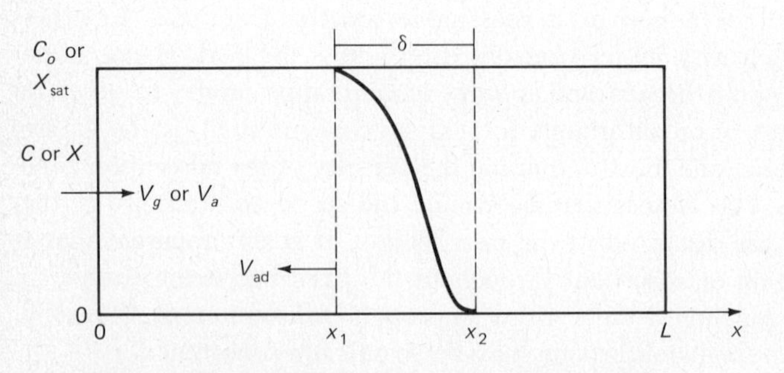

Figure 6-6 General features of an adsorption wave.

where ρ_g is the gas phase density, ρ_{ad} is the apparent density of the granular bed, and A is the cross-sectional area of the bed. Note that

$$\rho_{ad} = \rho_{true}(1 - \phi)$$

where ρ_{true} is the true density of the adsorbent (no void spaces) and ϕ is the fraction of void in the adsorbent. In many adsorption problems involving air pollutants the concentration of pollutant is very dilute. Hence the mass flow rate of carrier gas essentially equals that of the total flow. In addition, the carrier gas is frequently air with a mass flow rate $\dot{m}_a$ and density ρ_a. Making these approximations, we find that Equation (6-4a) can be written symbolically as

$$\frac{\dot{m}_a C_o}{\rho_a} = \rho_{ad} A V_{ad} X_{sat} \tag{6-4b}$$

Furthermore, the inlet concentration C_o and the saturated adsorbent concentration X_{sat} are related by experimental data. For many dilute solutions in the gas phase, the saturation or equilibrium data relating C_e to X_{sat} for adsorption phenomena take on the shape shown in Figure 6-2. Curves of this type are said to be of the Freundlich type, and are represented mathematically by a relation of the type

$$C_e = \alpha X_{sat}^{\beta} \tag{6-5}$$

where α and β are adjustable constants. β is a pure number, while α must have the same units as C. Note that α and β would have different values at different temperatures, even for the same adsorbate and adsorbent. Since we have assumed that equilibrium exists at position 1, although this is not true within the zone, then $C_o = C_e$. If Equation (6-5) is used to eliminate X_{sat} from Equation (6-4b), then we find that

$$\frac{\dot{m}_a C_o}{\rho_a} = \rho_{ad} A V_{ad} \left(\frac{C_o}{\alpha} \right)^{\beta}$$

This equation leads to the following expression for the velocity of the adsorption wave.

$$V_{ad} = \frac{\dot{m}_a}{\rho_a \rho_{ad} A} (\alpha)^{1/\beta} (C_o)^{(\beta - 1)/\beta} \tag{6-6a}$$

or

$$V_{ad} = \frac{V_a}{\rho_{ad}} (\alpha)^{1/\beta} (C_o)^{(\beta - 1)/\beta} \tag{6-6b}$$

We see that the velocity of the adsorption wave is dependent upon the shape of the equilibrium curve (as characterized by α and β), the inlet pollutant gas concentration, the superficial velocity V_a of the air or carrier gas stream, and the apparent density of the adsorption bed.

In order to determine the thickness of an adsorption wave it is necessary to develop a relationship between C and x in the adsorption zone. This is done by examining the transfer of mass from the gas phase for a differential thickness dx of the adsorption zone. As the gas stream passes through dx the pollutant concentration changes by dC. The rate of mass transfer of pollutant gas $d\dot{m}_p$ onto the solid phase is the product of the volume flow rate and the change in concentration, namely,

$$d\dot{m}_p = \frac{\dot{m}_a}{\rho_a} dC$$

Another independent equation for $d\dot{m}_p$ can be developed from the general theory of mass transfer. As might be expected, the mass flow rate is proportional to the difference in pollutant concentration between the actual value in the gas phase and the equilibrium value on the surface of the solid phase. The mass transfer rate also depends upon the volume of the bed, which in this case is $A\,dx$. Lastly, the rate depends upon a film coefficient K for the transfer process. The coefficient K takes into account the film resistance as well as the effective interfacial area of the adsorbent. Values of K vary over a considerable range, but typical values may fall between 5 to 50 s^{-1}. Hence for thickness dx we may also write that

$$d\dot{m}_p = KA(C - C_e)\,dx$$

Equating these two expressions for $d\dot{m}_p$ and using Equation (6-5) to eliminate C_e, we find that

$$-\dot{m}_a\,dC = KA\rho_a(C - \alpha X^{\beta})\,dx$$

Finally, X in the above expression can be eliminated by applying Equation (6-4a) in a general sense to the differential zone of interest. In this situation $\dot{m}_a C/\rho_a = \rho_{ad} A V_{ad} X$. Solving for X, and replacing V_{ad} with Equation (6-6a), we determine that

$$X^{\beta} = \frac{\dot{m}_a C}{\rho_a \rho_{ad} A V_{ad}} = C^{\beta}\alpha^{-1}C_o^{\,1-\beta}$$

The substitution of this result into the expression above leads to the desired relation between x and C, namely

$$-\dot{m}_a\,dC = KA\rho_a\left(C - C^{\beta}C_o^{\,1-\beta}\right)dx$$

or

$$dx = -\frac{\dot{m}_a}{KA\rho_a}\,\frac{dC}{C - C^{\beta}C_o^{\,1-\beta}} \tag{6-7}$$

Note that as the film resistance to mass transfer goes to zero, and K approaches infinity, the adsorption zone thickness becomes zero. Also note that the denominator on the right side of Equation (6-7) becomes zero when $\beta = 1$. In this spatial case it may be shown that the value of C in the main gas phase and C_e inside the film at the solid surface have identical values at a given position. Thus there is no driving force for mass transfer in this situation, and the model predicts an infinite length for the adsorption zone. Since β normally is not zero, the model does predict the zone thickness with reasonable success if the integration limits on Equation (6-7) are modified as noted below.

Integration of Equation (6-7) from x_1 to x_2 (the positions of the trailing and leading edges of the adsorption zone) will yield an expression for the adsorption zone thickness, δ, which equals $x_2 - x_1$. At x_1 the limit on C is C_o and at x_2 the value of C is zero. However, it is convenient from a mathematical viewpoint to alter the right side of Equation (6-7). Rather than use C as the variable, we shall use $\eta = C/C_o$, where η varies from 1 to 0. If the numerator and denominator on the right side of Equation (6-7) is divided by C_o, then δ is found by

$$\frac{\delta K A \rho_a}{\dot{m}_a} = \int_0^1 \frac{d\eta}{\eta - \eta^\beta} = \int_0^1 \frac{d\eta}{\eta(1 - \eta^{\beta - 1})} \tag{6-8}$$

The quantity $\delta K A \rho_a / \dot{m}_a$ is dimensionless in Equation (6-8). The value of the integral on the right side of the equation is undefined for limits of 0 and 1. However, by taking limits close in value to 0 and 1, the integral is defined. For example, we may take η to be within 1 percent of the limiting values, that is, 0.01 and 0.99. This is equivalent to defining an adsorption zone width such that C approaches within 1 percent of its limiting values of 0 and C_o. Based on this arbitrary selection of the integration limits, Equation (6-8) becomes

$$\frac{\delta K A \rho_a}{\dot{m}_a} = 4.595 + \frac{1}{\beta - 1} \ln \frac{1 - (0.01)^{\beta - 1}}{1 - (0.99)^{\beta - 1}} \tag{6-9}$$

The quantity $\dot{m}_a / A\rho_a$ can be replaced by the superficial gas stream velocity, V_a, on the basis of the continuity of flow equation. If one wishes to consider η values within 0.1 percent of the limiting values, then the constant in Equation (6-9) becomes 6.907, and the constants 0.01 and 0.99 are replaced by 0.001 and 0.999, respectively.

The general shape of the concentration versus distance curve is ascertained by integrating Equation (6-7) between the limits of 0.99 and C. The result is

$$x = x_1 + \frac{\dot{m}_a}{K A \rho_a} \left[\ln \frac{0.99 C_o}{C} + \frac{1}{\beta - 1} \ln \frac{1 - (C/C_o)^{\beta - 1}}{1 - (0.99)^{\beta - 1}} \right] \tag{6-10}$$

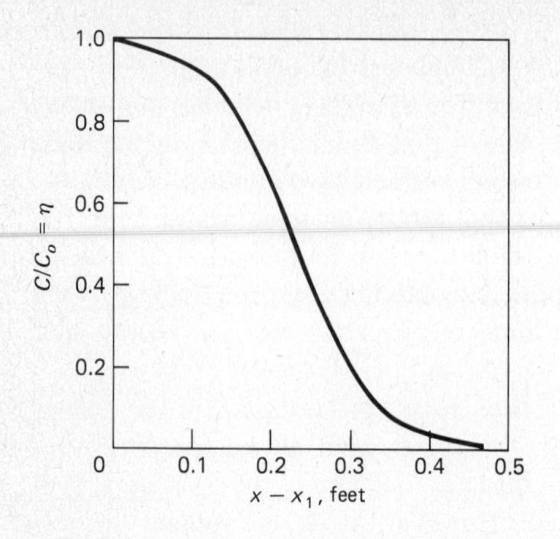

Figure 6-7 General shape of an adsorption wave for $\beta = 2$ and $KA\rho_a/\dot{m}_a = 20$.

This equation is plotted in Figure 6-7 for a β value of 2 and a value of 20 (length^{-1}) for the quantity $KA\rho_a/\dot{m}_a$. In this case δ is 0.46 units in length. The S-shaped nature of the concentration curve is apparent.

Finally, the time of breakthrough can be computed. If we assume that the time required to establish the adsorption zone to its full thickness at the inlet is zero, then

$$t_B = \frac{L - \delta}{V_{ad}} \tag{6-11}$$

This approximation leads to a conservative estimate of the operating time before breakthrough occurs. Also note that the value of δ depends upon an arbitrary choice of the limits of Equation (6-8). For example, Equation (6-9) is based on a C/C_o value of 0.01 at the leading edge of the wave. Thus breakthrough is defined in this case as the situation when C reaches 1 percent of the inlet concentration. Other percentages might be chosen.

Example 6-1

An absorption bed is 0.7 m in length and 12 m^2 in cross-sectional area. The carrier gas, air, has a mass flow rate of 2.2 kg/s and the pollutant concentration is 0.004 kg/m^3 at the inlet. The apparent density of the adsorbent is 350 kg/m^3, the mass transfer coefficient K is 20 s^{-1}, and α and β are 180 kg/m^3 and 2.2, respectively. Determine (a) the speed of the adsorption zone in meters per second, (b) the thickness of the adsorption zone in meters, and (c) the breakthrough time in hours, if the carrier gas temperature and pressure are 30°C and 1.1 bars, respectively.

SOLUTION

(a) The speed of the adsorption zone is found from Equation (6-6a). The density of the carrier gas is required, which in this case is found from the ideal gas equation.

$$\rho_a = \frac{P}{RT} = \frac{1.1 \text{ bars}}{303°\text{K}} \times \frac{29 \text{ kg °K}}{0.08315 \text{ bar·m}^3} = 1.27 \text{ kg/m}^3$$

Now substitution of values into Equation (6-6a) yields

$$V_{\text{ad}} = \frac{2.2}{1.27(350)(12)} \times (180)^{0.455} \times (0.004)^{0.545} = 2.16 \times 10^{-4} \text{ m/s}$$

(b) The thickness of the adsorption zone can be determined from Equation (6-9) where it has been assumed that the limiting values on η are 0.01 and 0.99. Substitution of the proper values yields

$$\delta = \frac{2.2}{20(12)(1.27)}\left[4.595 + \frac{1}{1.2}\ln\frac{1-(0.01)^{1.2}}{1-(0.99)^{1.2}}\right]$$

$$= 7.22 \times 10^{-3}(8.278) = 0.0598 \text{ m}$$

In this particular example the adsorption zone thickness is somewhat less than 10 percent of the length of the bed.

(c) Finally, the time until breakthrough is found from Equation (6-11). On the basis of previous calculations,

$$t = \frac{L-\delta}{V_{\text{ad}}} = \frac{0.7 - 0.06}{2.14 \times 10^{-4}} = 2960 \text{ s} = 0.823 \text{ hr}$$

6-5 REGENERATION OF AN ADSORPTION BED

When breakthrough is reached in an adsorption bed, it is necessary to switch the flow to another unsaturated bed and then proceed with regeneration of the saturated bed. Atmospheric air is sometimes used as the regeneration fluid, and the process is referred to as cold regeneration. However, it is often more effective to regenerate with a relatively hot fluid, since more contaminant will be released per unit of fluid passed through the bed. This tends to concentrate the pollutant gas and makes its disposal by incineration or other techniques easier. Steam or hot air is commonly used for hot regeneration. A disadvantage of hot regeneration is that the bed must then be cooled before adsorption proceeds again. If not cooled, the quantity of pollutant adsorbed per unit mass of adsorbent may be greatly reduced. During regeneration the adsorption zone becomes a desorption zone, and proceeds in the opposite direction. Nevertheless, the basic format of the equation for the speed of the adsorption zone is still valid. In terms of Equation (6-6a) we find that the speed of the desorption zone V_R is given by

$$V_R = \frac{\dot{m}_R}{\rho_R \rho_{ad} A} (\alpha_R)^{1/\beta_R} (C_R)^{(\beta_R - 1)/\beta_R} \tag{6-12}$$

In this relation the subscript R refers to the properties and the flow rate of the regeneration fluid. The values of α and β could be considerably different during desorption as compared to adsorption, due either to the use of a different temperature or a different desorption fluid, or both. Hence the subscript R on α and β is quite important. In addition, the value of C_R at the exit of the bed during desorption is generally different from C_o during adsorption. However, there is one common characteristic of the desorption and adsorption processes. The equilibrium saturation values of X is the same for both processes, since the bed is saturated when desorption begins. Hence C_R and C_o can be related in the following manner. Since

$$C_R = \alpha_R X^{\beta_R} \quad \text{and} \quad C_o = \alpha X^{\beta}$$

then

$$C_R = \alpha_R \left(\frac{C_o}{\alpha} \right)^{\beta_R/\beta} \tag{6-13}$$

Substitution of this relationship into Equation (6-12) leads to

$$V_R = \frac{\dot{m}_R \alpha_R}{\rho_R \rho_{ad} A} \left(\frac{C_o}{\alpha} \right)^{(\beta_R - 1)/\beta} \tag{6-14}$$

If the thickness of the desorption zone is much smaller than the bed length L, then the time for regeneration can be approximated by

$$t = \frac{L}{V_R} = \frac{\rho_R \rho_{ad} A L}{\dot{m}_R \alpha_R} \left(\frac{\alpha}{C_o} \right)^{(\beta_R - 1)/\beta} \tag{6-15}$$

An example calculation of a desorption process, based on the preceding example, is given in Example 6-2.

Example 6-2

The absorption bed analyzed in Example 6-1 is to be regenerated with saturated steam at 150°C. The flow rate of steam is 0.45 kg/s. The values of α and β are 1450 kg/m^3 and 2.4, respectively, and $K = 160\ \text{s}^{-1}$. Estimate the time required for regeneration in hours.

SOLUTION

From the saturation tables for steam the density is found to be 2.55 kg/m^3. The speed of the desorption zone is given by Equation (6-14).

$$V_R = \frac{0.45(1450)}{2.55(350)(12)}\left(\frac{0.004}{180}\right)^{0.583}$$

$$= 0.0609(1.94 \times 10^{-3}) = 1.18 \times 10^{-4} \text{ m/s}$$

The time for regeneration is then approximated by

$$t = \frac{L}{V_R} = \frac{0.7}{1.18 \times 10^{-4}} = 5930 \text{ s} = 1.65 \text{ hr}$$

When the regeneration time is longer than the adsorption time, as is the case above, then three or more beds would be needed in parallel. One would be on the adsorption mode, while two or more would be on regeneration.

As an approximation, the time for saturation to occur in a fixed bed is given empirically by (for adsorption of an organic vapor in air)

$$t_{sat} = \frac{1.3 \times 10^6 W}{QCM} \tag{6-16}$$

where t_{sat} is the time, in hours; W is the mass of adsorbent, in pounds; Q is the volume flow rate of air, in cubic feet per minute; C is the pollutant concentration, in parts per million; and M is the molar mass of the pollutant [11].

This equation might be used to estimate the time of operation of a fixed bed when experimental data based on expected operating conditions are not available. The prediction of the breakthrough curve is a key step in the design of fixed-bed adsorbers. It will be found that the type of equilibrium behavior, as represented by the curves in Figure 6-3, is always a primary criterion for evaluating breakthrough. Methods for estimating the performance of a fixed-bed adsorber appear in the literature [12].

6-6 ABSORPTION

6-6-A Some General Operating Characteristics and Mass Transfer Relations

The source control of gaseous pollutants by liquid scrubbing involves bringing the dirty effluent gas into contact with the scrubbing liquid and subsequently separating the cleaned gas from the contaminated liquid. In the cleaning process the contaminant gas is absorbed into the scrubbing liquid. Absorption is a basic chemical engineering unit operation which in the air pollution control field is referred to as *scrubbing*. Concentration gradients at the liquid gas interfaces serve as driving forces. The process is accelerated by high interfacial surface areas, turbulence, and large mass diffusion coefficients. Absorption is treated extensively in most chemical engineering texts, to which reference can be made for detailed information. Here we discuss the general characteristics of absorbers because of their wide use in controlling such air pollutants as sulfur dioxide, hydrogen sulfide, and light hydrocarbons.

As was mentioned earlier, gas absorption units must provide thorough contact between the gas and the liquid solvent so that interphase diffusion occurs. The rate of mass transfer is proportional to the liquid-gas interface surface area; therefore, absorber units are designed to provide large liquid surface area with a minimum of gas pressure drop. The contact between liquid and gas can be accomplished by dispersing the liquid in the gas or vice versa. Absorbers that disperse the liquid include packed towers, spray towers, and venturi absorbers. We use the packed tower here for illustrative purposes.

The basic arrangement of a counterflow packed tower is shown in Figure 6-8. Dirty gas enters the bottom and the cleaned gas exits at the top of the tower. Clean liquid enters the top of the tower while contaminated liquid is withdrawn from the bottom. The cleaned gas is normally vented through a stack to the atmosphere. The liquid leaving the absorber is either stripped of the contaminant gas and recycled or passed on for further waste treatment or process use. The packing material (inert chemically) is designed to increase the liquid-film surface area; consequently, many geometric shapes are available, each with a unique surface area and associated gas pressure drop. Typical shapes are shown in Figure 6-8(b). Physically, the absorption of a pollutant gas from a moving gas stream into an appropriate liquid stream is quite complex. Basically the transfer process into each fluid stream is accomplished by two mechanisms. The pollutant species is transferred from the bulk of the gas stream toward the gas-liquid interface by turbulent eddy motions. Very close to the interface, however, the fluid motion is essentially laminar and the pollutant must pass the remaining distance by molecular diffusion. On the liquid side of the interface the process is reversed. After absorption at the interface the species diffuses outward toward the bulk liquid stream. After the pollutant travels a short distance into the liquid stream, turbulent or eddy diffusion takes over and transports it into the main body of the liquid.

On the basis of Fick's law for the diffusion of one gas (A) through a second stagnant gas (B) it can be shown [13] that N_A, the molar rate of transfer of A per unit cross-sectional area, is given by the expression

$$N_A = -\frac{D_{AB}(\partial c_A/\partial z)}{1-(c_A/c)} \tag{6-17}$$

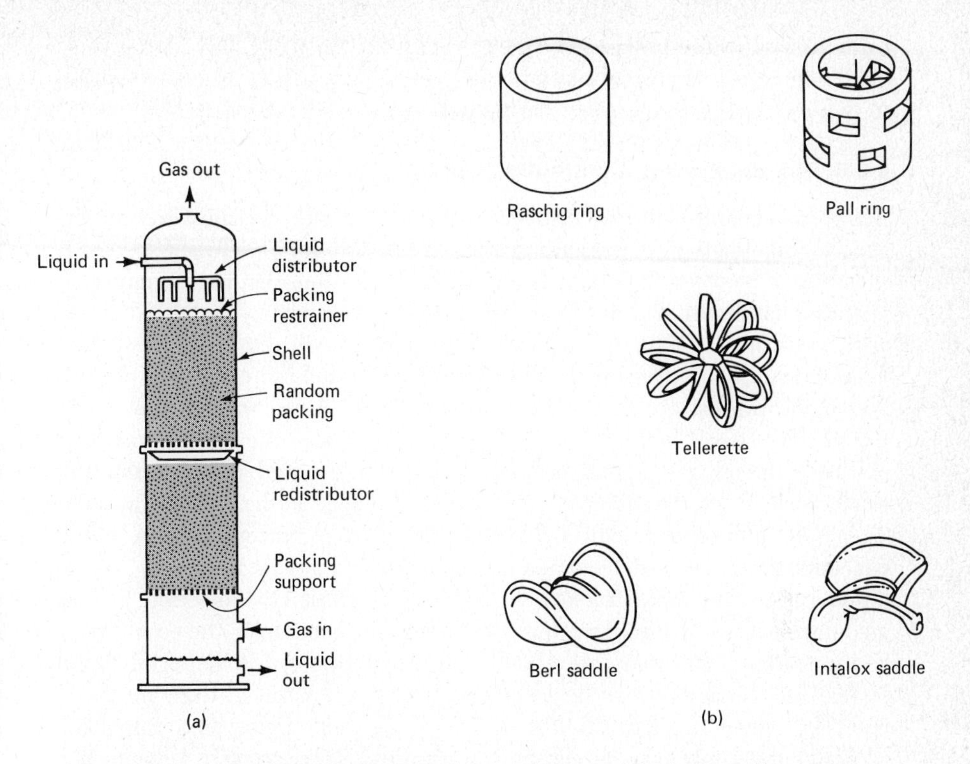

Figure 6-8 (a) Basic arrangement of a counterflow packed absorption tower, and (b) shapes of conventional packing.

where D_{AB} is the molecular diffusion coefficient or diffusivity, in length squared per unit of time; c_A is the molar concentration of species A, in moles per unit volume; c is the molar concentration of the gas mixture, in moles per unit volume; and z is the direction of mass transfer, in units of length.

The values of D_{AB} have been established for a number of binary gas mixtures and are available from various sources [13, 14]. The molecular diffusion coefficient, which is useful in the design of absorption towers, appears in a correlation quoted later in this chapter.

For molecular diffusion of species A through a second liquid the mass transfer rate per unit area is commonly expressed by

$$N_A = \frac{D_L}{z}(c_{A2} - c_{A1}) \tag{6-18}$$

where D_L is the liquid-phase molecular diffusion coefficient for species A through species B, and $(c_{A2} - c_{A1})$ is the concentration difference of species A over the distance z. Unlike D_{AB} in the gas phase, the value of D_L tends to vary considerably with the concentration of the solute A. Nevertheless, typical values of liquid molecular diffusion coefficients for binary mixtures are tabulated in the literature [13, 14]. As in the gas phase discussed above, knowledge of D_L is needed in empirical correlations of data to help evaluate an important absorption-tower design parameter.

6-6-B The Equilibrium Distribution Curve

Before proceeding with the discussion of mass transfer, we must summarize generally the method of presenting equilibrium data for a pollutant species A distributed between liquid and gaseous phases. Consider the situation illustrated in Figure 6-9(a). A constant-pressure system is immersed in an isothermal bath so that the contents of the system will remain at a constant temperature T and pressure P. The liquid solvent and the carrier gas for the absorption process are then introduced into the system. Now a small quantity of solute gas A is injected into the gas phase above the liquid. A portion of the pollutant gas A will be absorbed into the liquid until equilibrium exists. That is, after sufficient time no further change in the concentration of A in the two phases will be noted. By proper instrumentation, these concentrations can be determined and converted into mole fraction x_A in the liquid phase and mole fraction y_A in the gas phase. By injecting further amounts of A in the gas phase into the system and measuring the equilibrium concentrations, the complete equilibrium distribution of A between liquid solvent and carrier gas can be ascertained. A typical result of these measurements is shown in Figure 6-9(b) on a $y_A - x_A$ plot. These data points are representative only for the chosen temperature, pressure, and solvent. The equilibrium distribution curve will shift in position as these variables are altered. It might be noted that in general x_A does not have to equal y_A anywhere along the equilibrium curve (as would be illustrated by a point on the dashed line in the figure).

6-6-C Mass Transfer Coefficients Based on Interfacial Concentrations

When mass transfer occurs in moving liquid and gaseous streams, it becomes difficult to evaluate the separate effects of molecular and turbulent diffusion. An alternate approach is to express the mass transfer rate, N_A, for each phase in terms of a mass transfer coefficient k and a driving force based on the bulk and interfacial concentrations for that phase. For the liquid phase

$$N_A = k_L(c_{Ai} - c_{AL}) = k_x(x_{Ai} - x_{AL}) \tag{6-19}$$

where k_L is the liquid mass transfer coefficient based on concentrations, in length per unit of time; c_{Ai} is the concentration of A in the liquid phase at the interface, in moles per unit of volume; c_{AL} is the concentration of A in the bulk of the liquid phase, in moles per unit volume; k_x is the liquid mass transfer coefficient based on mole fractions, in moles per units of time and

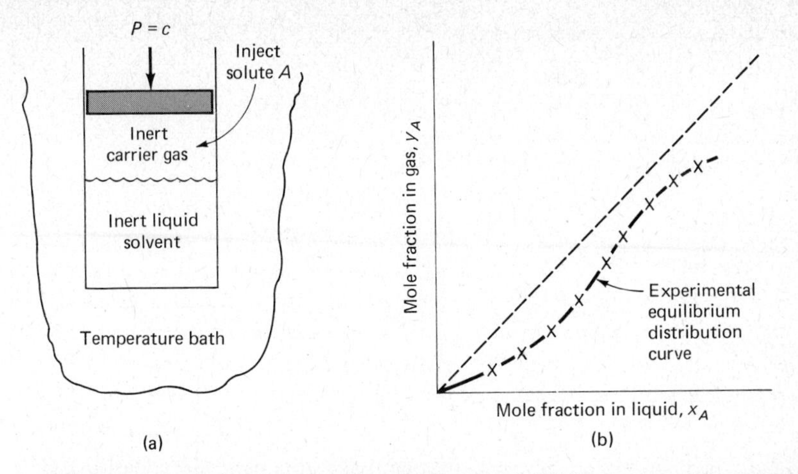

Figure 6-9 (a) A process for determining the equilibrium composition of two phases; (b) a typical equilibrium distribution curve.

length squared; x_{Ai} is the mole fraction of A in the liquid phase at the interface; and x_{AL} is the mole fraction of A in the bulk of the liquid phase.

It may be shown that $k_x = ck_L$. Since we are primarily interested in mass transfer from the gas to the liquid phase, Equation (6-19) has been written for the transfer of gas species A from the interface into the main liquid stream.

In a similar fashion, an expression for mass transfer of solute A from the bulk of the moving gas stream toward the liquid-gas interface is given by

$$N_A = k_G(p_{AG} - p_{Ai}) = k_y(y_{AG} - y_{Ai}) \tag{6-20}$$

where k_G is the gas-phase mass transfer coefficient based on partial pressures, in moles/(length)^2 (time) (pressure unit); p_{AG} is the partial pressure of A in the bulk of the gas phase; p_{Ai} is the partial pressure of A in the gas phase at the interface; k_y is the gas-phase mass transfer coefficient based on mole fractions, in moles/(time) (length)^2; y_{AG} is the mole fraction of A in the bulk of the gas phase; and y_{Ai} is the mole fraction of A in the gas phase at the interface. In this case $k_G = k_y/P$, since $p_i = y_i P$ for an ideal gas mixture.

In gas-liquid absorption processes the pollutant solute A must diffuse out of one phase and into the other through the gas-liquid interface. It is general practice to assume that the two phases are in equilibrium at all points on the surface of contact between the two phases, and that no appreciable diffusional resistance occurs at the actual interface. Consequently, the quantities c_{Ai} and p_{Ai} (or x_{Ai} and y_{Ai}) in Equations (6-19) and (6-20) are directly related by the equilibrium data as represented by the curve in Figure 6-9(b). That is, y_i in Equation (6-20) is taken to be the mole fraction of the solute present in a gas in equilibrium with a liquid with a solute mole fraction x_i, as found in Equation (6-19). The values y_{AG} and x_{AL} in the bulk of the gaseous and liquid streams, respectively, are not equilibrium

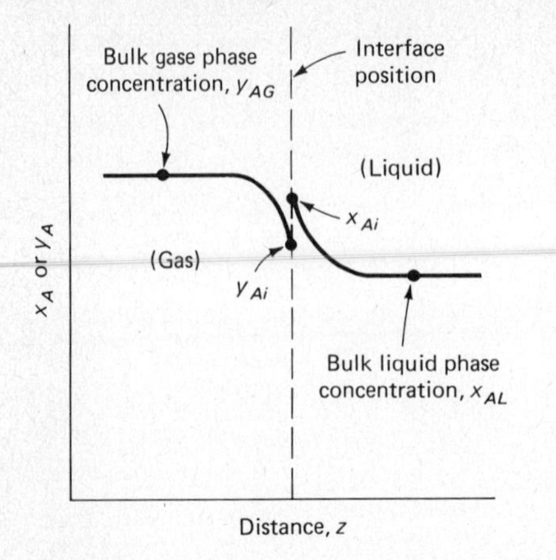

Figure 6-10 Schematic of the concentration variations near the liquid-gas interface.

values. They cannot be, or mass transfer would not occur. The situation described above is shown in Figure 6-10. The gas-phase concentration of solute A rapidly falls off from y_{AG} to y_{Ai} in the vicinity of the interface. At the interface, y_{Ai} and x_{Ai} are equilibrium values. The liquid mole fraction x_A then falls off rapidly until reaching the bulk liquid value of x_{AL}.

In general, this approach to determining N_A is not practical. Values of k_x and k_y are difficult to obtain, owing to the lack of general correlations among experimental data. The other approach would be to attempt to measure the values of y_{Ai} or x_{Ai} experimentally. Unfortunately, any instrument placed in the vicinity of the interface will perturb the equilibrium between the two streams and lead to erroneous values. In order to avoid the difficulties enumerated above, a slightly modified mathematical approach to the "two-resistance" theory is conventionally made. However, the physical and mathematical concepts developed in preceding sections are not wasted. The ideas pertaining to equilibrium data, local coefficients k_x and k_y, and interfacial concentrations will be used in the subsequent discussion.

6-6-D Overall Mass Transfer Coefficients

When mass transfer rates are reasonably low, so that the major resistances to mass transfer still lie in the liquid and gas phases, and not at the interface, it is convenient to express the rate N_A in the following manner:

$$N_A = K_G(p_{AG} - p_A^*) = K_y(y_{AG} - y_A^*) \quad \textbf{(6-21)}$$

where K_G and K_y are defined by these equations as *local overall* mass transfer coefficients. The interpretation of p_A^* and y_A^* is somewhat unusual. By definition, p_A^* is the equilibrium partial pressure of solute A in a gas phase which is in contact with a liquid having the composition c_{AL} of the main body of the absorption liquid. The quantity y_A^* is defined similarly in terms of a liquid with mole fraction x_{AL} of the bulk liquid. The significant difference between y_A^* and the bulk gas-phase mole fraction y_{AG} is illustrated in Figure 6-11 on a y-x diagram. Point P represents the state of the bulk phases of the two fluid streams, y_{AG} and x_{AL}. The point M represents the state (y_{Ai}, x_{Ai}) associated with equilibrium at the interface. By eliminating N_A from Equations (6-19) and (6-20), it is seen that the slope of the line PM is equal to $-k_x/k_y$. From the definition of y_A^* given above, the point (y_A^*, x_{AL}) must lie on the equilibrium distribution curve; it is shown as point C. The distance between points P and C is a measure of the driving force $(y_{AG} - y_A^*)$ in Equation (6-21). The convenience of Equation (6-21) is obvious, since interfacial concentrations are not required. The value of y_A^* is known for any bulk liquid concentration in an absorption tower once the equilibrium data have been determined in the laboratory. The usefulness of Equation (6-21) is usually restricted to the situation where the resistance to mass transfer is primarily in the gas phase, which characterizes the majority of absorption problems in air pollution work. The solubility of the pollutant gas normally determines the liquid that is chosen—that is, a liquid in which the gas is highly soluble. The major physical problem is getting the pollutant to diffuse through the gas phase to the interface; consequently, the gas phase controls the process.

When the liquid phase controls, it is advantageous to use local overall mass transfer coefficients K_L and K_x defined by the following relations:

$$N_A = K_L(c_A^* - c_{AL}) = K_x(x_A^* - x_{AL}) \tag{6-22}$$

The physical interpretation of c_A^* and x_A^* is similar to that used for p_A^* and y_A^*. For example, x_A^* is the equilibrium mole fraction of solute A in the liquid phase which is in contact with a gas which has a solute mole fraction of y_{AG}. This equilibrium condition is given by point D in Figure 6-11. The distance

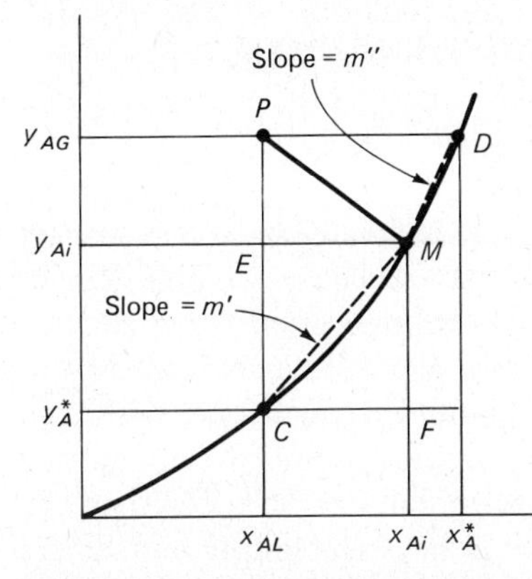

Figure 6-11 Graphical interpretation of mass transfer between gas and liquid phases in terms of local overall coefficients.

between points D and P is a measure of $(x_A^* - x_{AL})$, which is the driving force for mass transfer in the last term in Equation (6-22). Interpretations of the second terms in Equations (6-21) and (6-22) could also be made graphically through the use of a (p, c) diagram instead of a (y, x) plot.

It is important to note that the quantities c_A^*, x_A^*, p_A^*, and y_A^*, are fictitious. That is, they do not represent any actual condition in the absorption process. However, they are related in each case to a real concentration in one of the bulk fluids through the equilibrium data for the two-phase system.

From the geometry shown in Figure 6-11 it is easily seen that

$$y_{AG} - y_A^* = y_{AG} - y_{Ai} + (y_{Ai} - y_A^*)$$

and

$$y_{Ai} - y_A^* = m'(x_{Ai} - x_{AL})$$

where m' is the slope of the chord connecting points C and M. As a result,

$$y_{AG} - y_A^* = y_{AG} - y_{Ai} + m'(x_{Ai} - x_{AL})$$

Substitution of Equation (6-21) and another expression similar to Equation (6-20) yields

$$\frac{1}{K_y} = \frac{1}{k_y} + \frac{m'}{k_x} \tag{6-23}$$

This relation is a good example of why this theory is called a "two-resistance" theory. The term to the left is a measure of the overall resistance to mass transfer, while the two terms on the right represent the gas- and liquid-phase resistances, respectively. That is, the overall resistance is the sum of the resistances in series.

6-7 BASIC DESIGN OF A PACKED ABSORPTION TOWER

6-7-A Mass Balances and the Operating Line

In order to determine some of the operating characteristics of a packed absorption tower, it is necessary to make mass balances on the transfer process. Figure 6-12 is a general schematic of an absorption tower showing the appropriate symbols relative to the gas and liquid streams. In this figure G_m is the total gas flow (carrier gas plus pollutant), in moles per unit time; $G_{C,m}$ is the inert carrier-gas flow rate alone, in moles per unit time; L_m is the total liquid flow rate (solvent plus absorbed pollutant), in moles per unit time; $L_{S,m}$ is just the liquid solvent flow rate alone, in moles per unit time; x is the liquid mole fraction of the pollutant species; y is the gas-phase mole fraction of the pollutant species; X is the liquid-phase mole ratio, in terms of

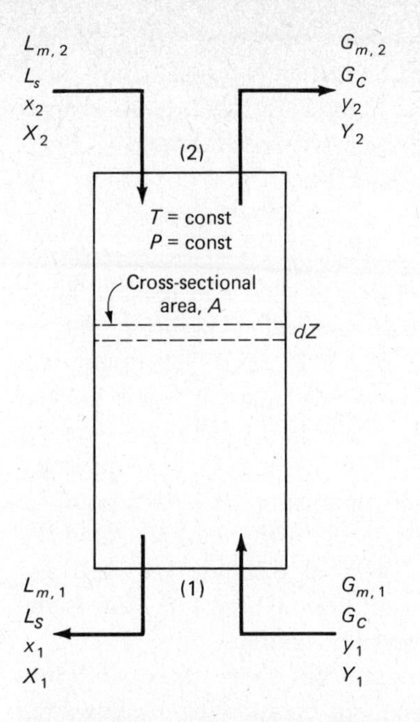

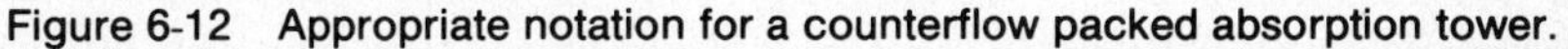

Figure 6-12 Appropriate notation for a counterflow packed absorption tower.

the pollutant species per mole of the liquid solvent; and Y is the gas-phase mole ratio, in terms of the pollutant species per mole of carrier gas.

We have used the subscript m with the G and L symbols to indicate that these rates are on a mole basis. In some equations and correlations developed later, it is required that the gas and liquid rates be on a mass basis, rather than a mole basis. In such cases the subscript m is left off the symbol. It is also important to note the difference between the mole fraction x (or y) and the mole ratio X (or Y). It is easily shown that

$$X = \frac{x}{1-x}, \qquad Y = \frac{y}{1-y}, \qquad x = \frac{X}{1+X}, \qquad y = \frac{Y}{1+Y} \tag{6-24}$$

The conservation-of-mass principle applied to the pollutant species in terms of the total mass flow rates at top and bottom of the column yields, for countercurrent flow,

$$G_{m,1}y_1 + L_{m,2}x_2 = G_{m,2}y_2 + L_{m,1}x_1$$

or

$$G_{m,1}y_1 - G_{m,2}y_2 = L_{m,1}x_1 - L_{m,2}x_2 \tag{6-25}$$

Since the total gas flow rates (and the total liquid flow rates) are different at the top and the bottom of the column, the above equation cannot be simplified further, in general. When plotted on a y-x diagram, this equation would appear as a curve. It is convenient, therefore, to rewrite the conservation equation in terms of the carrier gas and liquid solvent rates, since these quantities are constant throughout the column. This type of balance is accomplished by noting that $G_i y_i$ may be replaced by $G_C Y_i$ and $L_i x_i$ by $L_S X_i$, where i stands for either 1 or 2. Consequently,

$$G_{C,m}(Y_2 - Y_1) = L_{S,m}(X_2 - X_1) \tag{6-26}$$

This is an equation for a straight line on Y-X coordinates, and the line has a slope of $L_{S,m}/G_{C,m}$. Equations (6-25) and (6-26) are called *operating lines* on the y-x and Y-X diagrams, respectively. Figure 6-13 shows representative equilibrium and operating lines on the y-x and Y-X diagrams for a packed absorption tower. Point P on the operating line for each plot represents the bulk-gas and bulk-liquid concentrations of solute species A at some arbitrary position in the column. Every position in the column is related to a point on the operating line between the bottom and the top of the column. As in the discussion pertaining to Figure 6-11, the vertical distance between the operating line and the equilibrium distribution line in Figure 6-13(a) is the overall driving force at any point in the absorption column.

As shown in Figure 6-13, the operating line lies above the equilibrium line for absorption. For a stripping operation in countercurrent flow, the operating line must lie below the equilibrium line in order for the driving force to act from the liquid phase toward the gas phase. Recall that point M in this figure indicates the interfacial composition. For concurrent flow in absorbers and strippers the position of the operating line is quite different from that for countercurrent flow [13].

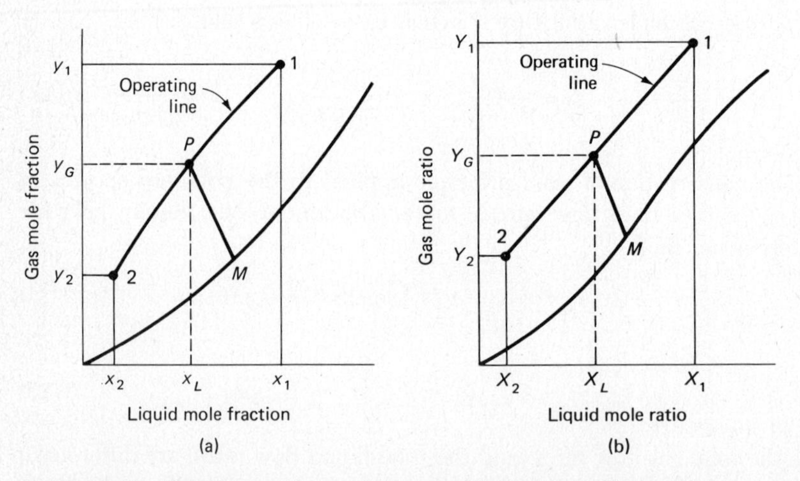

Figure 6-13 Relative positions of the operating and equilibrium lines on x-y and X-Y diagrams for a counterflow packed absorption tower.

6-7-B The Minimum and Design Liquid-Gas Ratio

With reference again to Figure 6-12, the values that are known at the bottom and top of the absorber in air pollution work are $G_{m,1}$, G_C, y_1 (or Y_1), $G_{m,2}$, y_2 (or Y_2), and x_2 (or X_2). That is, everything is known about the gas stream in terms of inlet rates and concentrations, and the outlet gas concentration is specified to meet a certain standard. In addition, the inlet liquid concentration of pollutant A probably is known also. Either fresh, uncontaminated liquid is introduced into the absorber, or the solvent is recycled back to the absorber after reducing the solute concentration to a specified value in a stripper. As a consequence of Equation (6-26), the values of L_S and X_1 (or x_1) are unknown. Thus we have one equation with two unknowns.

This design problem is open-ended; that is, there is no single answer to the "correct" values of L_S and X_1. However, a selection of one of these values obviously fixes the value of the other, through Equation (6-26). In the following discussion, we present one method of determining the design values of these two quantities. Referring to Figure 6-14(b), we note that Y_2, X_2, and Y_1 are known input data. Thus one end of the operating line (point Y_2, X_2 at the top of the absorber) is fixed, and the other end must lie somewhere on the horizontal line Y_1. Three of the indefinitely large number of possible operating lines are shown, with end states (at the bottom of the absorber) of X_{1a}, X_{1b}, and X_{1c}. Note that in going from point $1a$ to $1b$ to $1c$, the value of X_1 increases. Hence the liquid rate L_S must be decreasing in the same direction, in order to satisfy Equation (6-26). It is apparent that the line through X_{1c} represents the minimum possible solvent liquid rate, where the operating line is tangent to the equilibrium line. If the line had any lesser

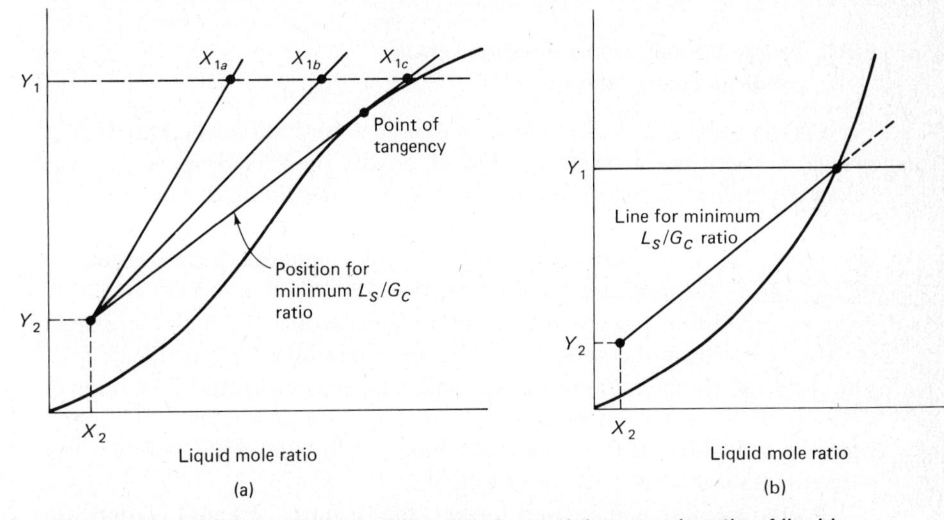

Figure 6-14 Position of the operating line for a minimum mole ratio of liquid solvent to carrier gas, for two different shapes of the equilibrium distribution line.

slope, the operating line would lie below the equilibrium curve for a portion of the tower, and in that region stripping would occur, not absorption. Consequently, the value of L_S associated with the operating line through point (Y_1, X_{1c}) is the minimum solvent rate for the given gas stream conditions.

The minimum rate is highly undesirable. At the point of contact between the operating and equilibrium lines the driving force for mass transfer is zero. Hence it would take an infinitely tall absorber to accomplish the desired separation. Therefore, the design rate must be larger than the minimum. If the rate is just slightly greater than the minimum, the average driving force still will be quite small, and the tower might be inordinately tall. On the other hand, a larger average driving force is attained at the expense of a much larger liquid solvent rate. This means a large recirculation of solvent in most cases, and large pumping losses. However, in this latter case the height of the tower can be relatively small. The ultimate choice lies between these extremes, and must be tempered by economics and other considerations. As a general operating principle, an absorber is typically designed to operate at liquid rates which are 30 to 70 percent greater than the minimum rate.

Frequently the equilibrium curve will be concave upward, and the method of drawing the operating line tangent to the equilibrium line to establish the minimum liquid-gas ratio is not valid. Instead, the minimum ratio now corresponds to the situation where the exit liquid concentration is in equilibrium with the entering gas concentration. Hence the operating line will cut across the equilibrium line at the entering gas concentration. Figure 6-14(b) shows the position of the operating line for a minimum liquid-gas ratio when the equilibrium line is concave upward.

6-7-C Tower Diameter and Pressure Drop per Unit Tower Height

For a given packing and liquid flow rate in an absorption tower, variation in the gas velocity has a marked influence on the pressure drop. As the gas velocity is increased, the liquid tends to be retarded in its downward flow, giving rise to the term "liquid holdup." As the liquid holdup increases, the free cross sectional area for gas flow is reduced. This results in an increase in the pressure drop per unit height, $\Delta P/Z$. At some value of G there is a significant change in the rate of increase of $\Delta P/Z$ with increase in G. This is noted by the bend in the curves at position A on the $(\Delta P/Z)$-G plot shown in Figure 6-15, for various liquid rates. The change in the slope of a line is quite evident at position A, known as the *load point*. Physically, however, it is difficult to see any difference in the appearance of the flow streams within the tower at the load point.

As the gas rate is increased further, the quantity of liquid holdup will increase more rapidly. Ultimately the liquid will tend to fill the entire void

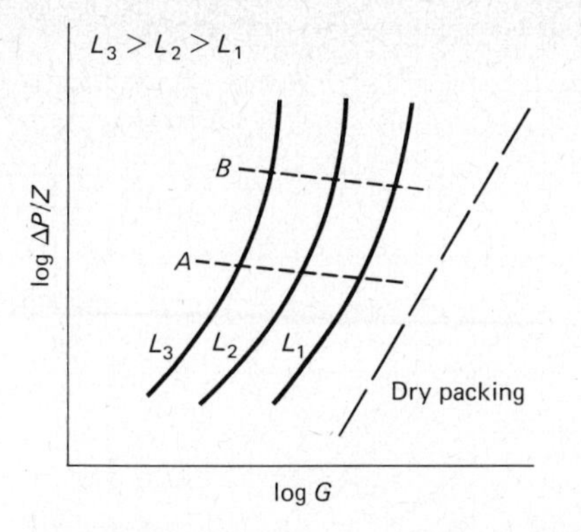

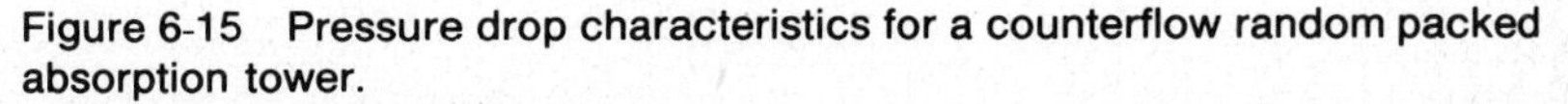
Figure 6-15 Pressure drop characteristics for a counterflow random packed absorption tower.

space, and a layer of liquid may appear on top of the packing. Increased liquid entrainment in the gas stream may also occur. The tower is now said to be flooded, and the second breakpoint B in Figure 6-15 is a good indication of the *flood point*. The danger of operating at this condition is that the gas pressure drop is excessive, and the quantity $\Delta P/Z$ is quite sensitive to flow rates. To avoid this condition it is recommended that an absorption tower operate somewhat below the load point or in the region just above the load point. As a rule of thumb, experience dictates operating at gas velocities which are 40 to 70 percent of those which cause flooding.

The relationship between $\Delta P/Z$ and other important tower variables—liquid and gas rates, liquid and gas stream densities and viscosities, and type of packing—has been extensively studied on an experimental basis. A widely accepted correlation among these parameters is shown in Figure 6-16. As noted below the figure, the correlation is based on a fixed set of units for each of the quantities. One new quantity in the correlation is G', the "*superficial*" *gas mass flow rate*. This is defined as the actual gas flow rate, mass per unit time, divided by the empty cross-sectional area of the tower. A similar definition holds for a *superficial liquid mass flow rate*, L'. The top line in the figure represents the flooding condition. In most conventional packing flooding occurs when the value of $\Delta P/Z$ for the gas phase is in the range of 2 to 3 in. water/ft. The lower limit of loading usually occurs when $\Delta P/Z$ is in the range of 0.5 in. water/ft. All the lines represent different values of the gas-phase pressure drop per unit of height of the tower.

The other new quantity in the correlation is the packing factor, F. This is an experimental quantity which helps to adjust for the differences among packings in their interfacial contact area per unit of tower volume (a) and in their void space (ϵ). Table 6-1 lists values of F for different sizes of various packings. This single correlation allows us to estimate two important tower parameters: the required tower diameter and the expected pressure drop per unit height of the tower. The general procedure is given below.

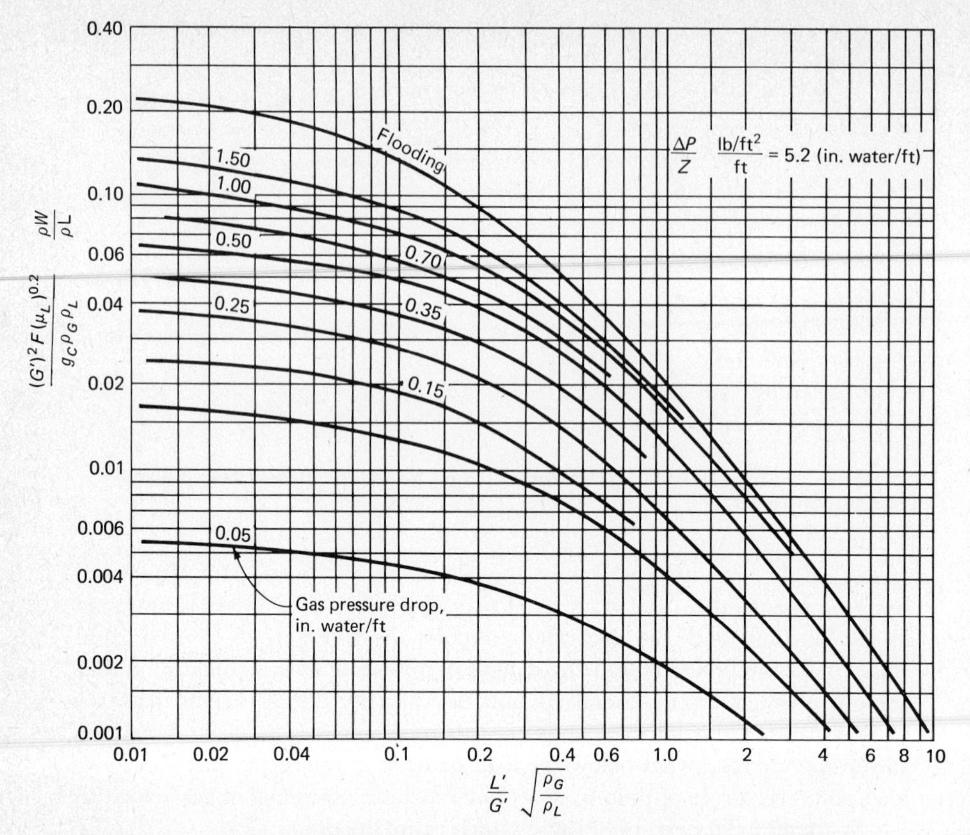

Figure 6-16 Flooding and pressure drop in random packed towers. (Note: If G' and L' are expressed in lb/s·ft^2, then g_c has a value of 32.2 lb$_m$·ft/lb$_f$·s^2. In addition, μ_L must be in centipoises, and ρ_G, ρ_L, and ρ_W in lb$_m$/ft^3.) (Courtesy of the Norton Company.)

Table 6-1 THE PACKING FACTOR, F, FOR RANDOM PACKING

	NOMINAL SIZE (in.)								
PACKING TYPE	$\frac{1}{4}$	$\frac{3}{8}$	$\frac{1}{2}$	$\frac{5}{8}$	$\frac{3}{4}$	1	$1\frac{1}{4}$	$1\frac{1}{2}$	2
Raschig rings									
Ceramic	1600	1000	580	380	255	155	125	95	65
Metal									
$\frac{1}{32}$-in. wall	700	390	300	170	155	115			
$\frac{1}{16}$-in. wall			410	290	220	137	110	82	57
Pall rings									
Plastic				97		52		40	25
Metal				70		48		28	20
Intalox saddles[a]	725	330	200		145	98		52	40
Berl saddles[a]	900		240		170	110		65	45
Tellerettes,[b] plastic, $\frac{3}{4} \times 2$ in. High density: $F = 57$; low density: $F = 65$									

SOURCE: Courtesy of the Norton Company.

[a] Ceramic material.

[b] A. J. Teller and H. E. Ford, *Ind. Engr. Chem.* **50** (1958): 1201.

The abscissa value, $(L/G)\,(\rho_G/\rho_L)^{0.5}$, is calculated first. Where this value intercepts the flooding line on Figure 6-16, move horizontally to the left and read the value of the ordinate, which is the parameter

$$\frac{(G')^2 F(\mu_L)^{0.2}}{g_c \rho_G \rho_L}\left(\frac{\rho_W}{\rho_L}\right)$$

Knowing the value of this overall parameter, we can solve for G', the superficial gas mass flow rate. This, of course, is the value of G' associated with flooding. For proper tower design we use 40 to 70 percent of this flood point value. Now both G and the design value of G' are known. From geometric considerations,

$$\text{tower cross-sectional area,} \quad A = \frac{G}{G'}$$

where G and G' are expressed in a consistent set of units. From the tower area it is a simple matter to evaluate the tower diameter. As a word of caution, it may be necessary to determine the area or diameter at both the top and the bottom of the tower, since they will differ. For a margin of safety the largest value of the area or diameter is chosen for design purposes.

Another empirical correlation found in the literature [15] for the pressure drop in packing when operating below the load point is

$$\frac{\Delta P}{Z} = 10^{-8} m\left[10^{(nL'/\rho_L)}\right]\frac{(G')^2}{\rho_G} \tag{6-27}$$

where ΔP is in pounds per square feet, Z is in feet, and m and n are packing constants (see Table 6-2). Experimental data frequently indicate a wide

Table 6-2 PRESSURE DROP CONSTANTS FOR TOWER PACKING

PACKING TYPE	m	n	L' RANGE (lb/hr·ft²)
Raschig rings			
½ in.	139.00	0.00720	300–8,600
¾ in.	32.90	0.00450	1800–10,800
1 in.	32.10	0.00434	360–27,000
1½ in.	12.08	0.00398	720–18,000
2 in.	11.13	0.00295	720–21,000
Berl saddles			
½ in.	60.40	0.00340	300–14,100
¾ in.	24.10	0.00295	360–14,400
1 in.	16.01	0.00295	720–78,800
1½ in.	8.01	0.00225	720–21,600
Intalox saddles			
1 in.	12.44	0.00277	2520–14,400
1½ in.	5.66	0.00225	2520–14,400

SOURCE: R. E. Treybal. *Mass Transfer Operations*. New York: McGraw-Hill, 1955.

variation in $\Delta P/Z$ for the same packing and the same flow rates. The variation is believed to be due to the variation in packing density in the tower.

Example 6-3

A packed tower is to be designed to remove 95 percent of the ammonia from a gaseous mixture of 8 percent ammonia and 92 percent air, by volume. The flow rate of the gas mixture entering the tower at 68°F and 1 atm is 80 lb·moles/hr. Water containing no ammonia is to be the solvent, and 1-in. ceramic raschig rings will be used as the packing. The tower is to operate at 60 percent of the flood point, and the liquid water rate is to be 30 percent greater than the minimum rate. Determine:

1. The gas-phase flow rates, in pound-moles per hour, for the solute and carrier gas.
2. The mole ratios of the gas and liquid phases at inlet and outlet, and the required water rate in pound-moles per hour.
3. The gas and liquid rates, in pounds per hour, for carrier gas, solute gas, total gas, liquid solvent, solute in liquid, and total liquid.
4. The tower area and diameter.
5. The pressure drop, in pounds per square inch and inches of water, based on two methods.

SOLUTION

1. If we use the symbol A to represent ammonia, then

$$G_{m,C} = 0.92(80) = 73.60 \text{ lb·moles/hr}$$

$$G_{A,1} = 0.08(80) = 6.40 \text{ lb·moles/hr}$$

$$\Delta G_A = 0.95(6.40) = 6.08 \text{ lb·moles/hr}$$

$$G_{A,2} = 0.05(6.4) = 0.32 \text{ lb·mole/hr}$$

2. The gas-phase composition at inlet and outlet is

$$Y_1 = \frac{8}{92} = 0.0870 \text{ lb·mole } A/\text{lb·mole air}$$

$$Y_2 = \frac{0.32}{73.60} = 0.00435 \text{ lb·mole } A/\text{lb·mole air}$$

To determine the remaining composition for the liquid at its exit ($x_2 = X_2 = 0$), it is necessary to evaluate the minimum solvent flow rate. This requires use of equilibrium data for ammonia-air-water mixtures, which are given below for 68°F and 14.7 psia total pressure.

X	0.0206	0.0310	0.0407	0.0502	0.0735	0.0962
Y	0.0158	0.0240	0.0329	0.0418	0.0660	0.0920

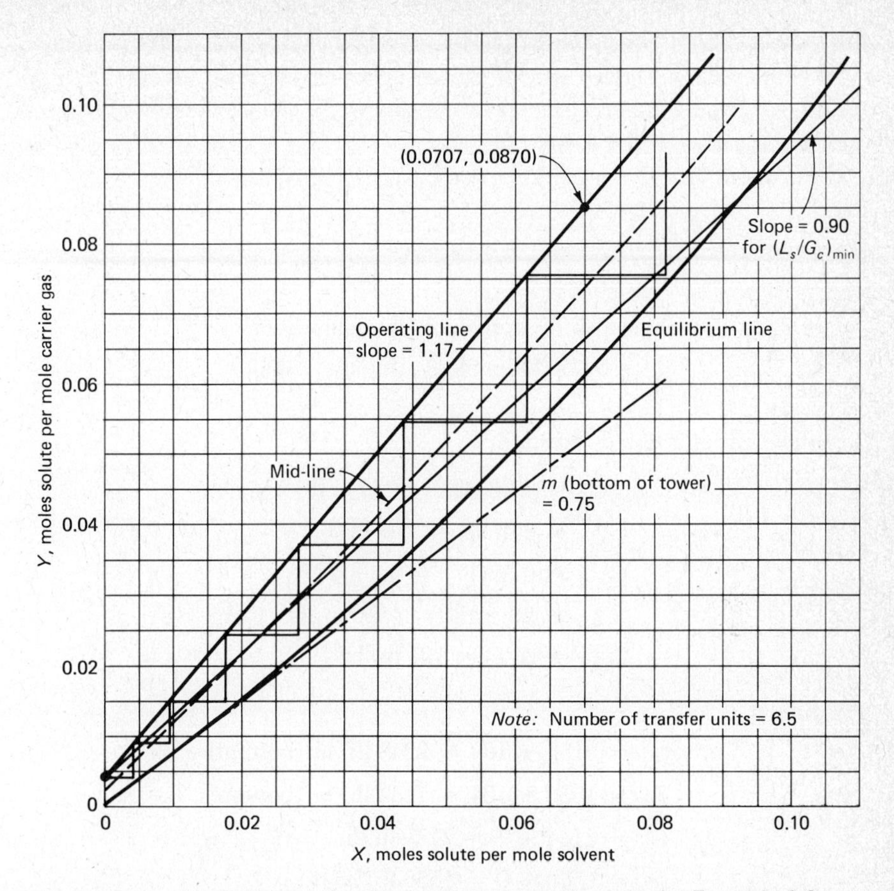

Figure 6-17 Graphical solution for number of transfer units for Example 6-3.

These data are plotted in Figure 6-17. In order to determine the minimum $L_{m,S}/G_{m,C}$ value, a line is drawn from (Y_2, X_2) to the point where the line $Y_1 = 0.0870$ intersects the equilibrium line. On the basis of the diagram,

$$\left(\frac{L_{m,S}}{G_{m,C}}\right)_{\text{minimum}} = \frac{0.0870 - 0.00435}{0.092 - 0} = 0.90$$

Since the liquid rate is to be 30 percent greater than the minimum rate,

$$\left(\frac{L_{m,S}}{G_{m,C}}\right)_{\text{design}} = 1.30(0.90) = 1/1.17 \text{ (mole/mole)}$$

Hence,

$$L_{m,S} = 1.17G_{m,C} = 1.17(73.6) = 86.1 \text{ lb·moles/hr}$$

The composition X_1 can now be found in two ways. First, the operating line with a slope of 1.17 can be drawn from (0.0043,0). The point where it crosses $Y_1 = 0.0870$ is the bottom end of the design operating line, and X_1 can be read directly off the figure. For better accuracy, X_1 can also be found from a material balance on the tower.

$$\frac{L_{m,S}}{G_{m,C}} = \frac{Y_2 - Y_1}{X_2 - X_1} = \frac{0.00435 - 0.0870}{0 - X_1} = 1.17$$

or

$$X_1 = 0.0707 \text{ lb·mole A/lb·mole water}$$
$$x_1 = 0.0660 \text{ lb·mole A/lb·moles solution}$$

3. The gas and liquid rates are

$$G_C = 73.6(29.0) = 2134 \text{ lb/hr}$$
$$G_{A,1} = 6.4(17) = 109 \text{ lb/hr}$$
$$G_{A,2} = 0.32(17) = 5.4 \text{ lb/hr}$$
$$L_S = 86.1(18) = 1550 \text{ lb/hr}$$
$$L_{A,1} = \Delta G_A = 104 \text{ lb/hr}$$

Therefore,

$$G_1 = 2134 + 109 = 2243 \text{ lb/hr (bottom)}$$
$$L_1 = 1550 + 104 = 1654 \text{ lb/hr (bottom)}$$
$$G_2 = 2134 + 5 = 2139 \text{ lb/hr (top)}$$
$$L_2 = 1550 + 0 = 1550 \text{ lb/hr (top)}$$

4. The tower area is based on the flooding correlation of Figure 6-16. This correlation requires knowledge of the gas-phase and liquid-phase densities at the top and bottom of the tower. Since the ammonia content in the liquid phase is quite small, we use the density of pure water, 62.3 lb/ft^3, as the solution density throughout the tower. For the gas phase, on the assumption that the gas mixture acts as an ideal gas and the partial pressure of water can be neglected,

$$\rho = \frac{P}{RT} = \frac{M_m P}{R_u T}$$

where M_m is the average molar mass of the gas and R_u is the universal gas constant. At the top,

$$M_m = \sum y_i M_i = 0.0043(17) + 0.9957(29) = 28.95$$

$$\rho = \frac{28.95(14.7)}{10.73(528)} = 0.0751 \text{ lb/ft}^3$$

At the bottom,

$$M_m = 0.08(17) + 0.92(29) = 28.04$$

$$\rho = \frac{28.04(14.7)}{10.73(528)} = 0.0728 \text{ lb/ft}^3$$

The value of G' at the top and the bottom of the tower can now be ascertained from the flooding correlation. At the top,

$$\frac{L}{G}\left(\frac{\rho_G}{\rho_L}\right)^{0.5} = \frac{1550}{2139}\left(\frac{0.0751}{62.3}\right)^{0.5} = 0.0251$$

From Figure 6-16, the ordinate for flooding condition is found to be 0.20. From Table 6-1 the value of F is 155, and the viscosity of water at 68°F is 1 centipoise. Hence,

$$G'_{\text{flood}} = \left[\frac{0.20(32.2)(0.0751)(62.3)}{155(1)^{0.2}}\right]^{0.5} = 0.439 \text{ lb/s·ft}^2$$

For 60 percent of the flood point,

$$G'_{\text{design}} = 0.439(0.6) = 0.263 \text{ lb/s·ft}^2 = 948 \text{ lb/hr·ft}^2$$

$$A = \frac{G}{G'} = \frac{2139}{948} = 2.26 \text{ ft}^2(\text{top})$$

At the bottom of the tower,

$$\frac{L}{G}\left(\frac{\rho_G}{\rho_L}\right)^{0.5} = \frac{1654}{2243}\left(\frac{0.0728}{62.3}\right)^{0.5} = 0.0252$$

From Figure 6-16 the value of the ordinate is again 0.20. Using the same values of the viscosity and F as before,

$$G'_{\text{flood}} = \left[\frac{0.20(32.2)(0.0728)(62.3)}{155(1)^{0.2}}\right]^{0.5} = 0.434 \text{ lb/s·ft}^2$$

At 60 percent of the flooding condition,

$$G'_{\text{design}} = 0.434(0.6) = 0.260 \text{ lb/s·ft}^2 = 934 \text{ lb/hr·ft}^2$$

The required area at the bottom of the tower then is

$$A = \frac{G}{G'} = \frac{2243}{934} = 2.40 \text{ ft}^2(\text{bottom})$$

As a conservative estimate, the larger area is selected for the tower cross section. As a result, for the tower diameter

$$D = \left(\frac{4A}{\pi}\right)^{0.5} = \left[\frac{4(2.40)}{\pi}\right]^{0.5} = 1.75 \text{ ft} = 21.0 \text{ in.}$$

5. The pressure drop may be determined from the general correlation of Figure 6-16 or from Equation (6-27). On the basis of the correlation the ordinate is (with g_c in $\text{lb}_m\cdot\text{ft}/\text{lb}_f\cdot\text{hr}^2$)

$$\frac{(G')^2 F(\mu_L)^{0.2}}{g_c \rho_G \rho_L} \frac{\rho_W}{\rho_L} = \frac{(934)^2(155)(1)^{0.2}}{4.17 \times 10^8 (0.074)(62.3)} = 0.0702$$

and the horizontal coordinate is (from part 4)

$$\frac{L}{G}\left(\frac{\rho_G}{\rho_L}\right)^{0.5} = 0.0252$$

From Figure 6-16, as an approximate value,

$$\frac{\Delta P}{Z} = 0.65 \text{ in. water/ft} = 0.0234 \text{ psi/ft}$$

An alternative method of estimating the pressure drop is through the use of Equation (6-27). By employing the proper data from Table 6-2 we find that

$$\frac{\Delta P}{Z} = m(10^{-8})(10^{nL'/\rho_L})\frac{(G')^2}{\rho_G}$$

$$= 32.1(10^{-8})(10^{0.00434(659)/62.3})\frac{(934)^2}{0.074}$$

$$= 4.18 \text{ lb per ft}^2/\text{ft} = 0.804 \text{ in. water/ft} = 0.0289 \text{ psi/ft}$$

The answers given by the two methods for the pressure drop per unit height of tower are in reasonable agreement with each other. The overall pressure drop cannot be found until the overall height of the tower is determined. This latter quantity is discussed in the next section.

6-8 DETERMINATION OF AN ABSORPTION TOWER HEIGHT

6-8-A Concept of a Transfer Unit

The required height of a packed absorption tower primarily is determined by the overall resistance to mass transfer between the gas and liquid phases, and the average driving force and interfacial area available for mass transfer. The first two of these parameters may vary considerably along the height of the absorber. Hence some type of averaging technique must be used to predict the overall rate of mass transfer.

Consider a differential height of the absorber, dZ, as shown in Figure 6-12. We define the interfacial area available to mass transfer per unit of volume of the packing by the symbol a. This quantity is a function of the packing in use. In height dZ the total interfacial area open to mass transfer is $a(A\,dZ)$, where A again is the cross-sectional area of the tower. The rate of mass transfer in height dZ of species A then is $N_A A(a\,dZ)$. However, this quantity is also equal to the loss of species A from the gas phase as it passes through section dZ; this loss is given by $d(G_m y)$. Equating these two quantities, we find that

$$N_A A(a\,dZ) = d(G_m y) = A\,d(G'_m y)$$

By recalling that G'_m equals $G_{C,m}/(1-y)$, we can show that the last term may be written as $A\ G'_m\,dy/(1-y)$. With this substitution and the use of Equation (6-21), the above equality becomes

$$K_y a(y_{AG} - y^*_A)\,dZ = \frac{G'_m\,dy}{1-y}$$

or

$$dZ = \frac{G'_m\,dy}{K_y a(1-y)(y-y^*)} = \frac{G'_m\,dy}{K_G aP(1-y)(y-y^*)} \tag{6-28}$$

The subscript A has been omitted from the y quantities in this expression. Integration of Equation (6-28) over the variation of gas-phase composition would lead to the value of Z, the required absorber height, when the gas-phase resistance controls. A similar expression exists for the situation where the liquid-phase resistance controls.

One method of solving Equation (6-28) is based on experimental "pilot plant" data. On the basis of small-scale equipment we can determine the overall value of $K_y a$ (or $K_G a$) for the packing and for the gas and liquid rates under consideration. Then the remaining portion of the right side of Equation (6-28) can be integrated from knowledge of the operating line and equilibrium line characteristics.

The above method can be modified to deal with the concepts of the "height of a transfer unit" and the "number of transfer units," but to do so requires modifying the format of Equation (6-28) somewhat. In terms of the quantities y and y^* previously defined for the pollutant species,

$1 - y$ = mole fraction of the nondiffusing species (carrier gas) in the bulk gas stream at any position Z

$1 - y^*$ = mole fraction of the nondiffusing gas species, based on the concentration of the solute in equilibrium with the liquid at position Z

In addition, the overall driving force $y - y^*$ at any location in the tower may be written in the form

$$y - y^* = (1 - y^*) - (1 - y)$$

The log mean value of $(1 - y^*)$ and $(1 - y)$ is defined as

$$(1 - y)_{LM} \equiv \frac{(1 - y^*) - (1 - y)}{\ln[(1 - y^*)/(1 - y)]} \tag{6-29}$$

On the basis of experimental data, it is found convenient to multiply numerator and denominator of Equation (6-28) by $(1 - y)_{LM}$. As a result,

$$dZ = \frac{G'_m}{K_y a(1 - y)_{LM}} \frac{(1 - y)_{LM} dy}{(1 - y)(y - y^*)} \tag{6-30}$$

Although G'_m, $K_y a$, and $(1 - y)_{LM}$ vary along an absorption column, data indicate that the first term on the right side of Equation (6-30) is reasonably constant. This quantity is defined as the *overall height of a transfer unit* based on gas-phase resistance control, H_{tOG}. That is,

$$H_{tOG} \equiv \frac{G'_m}{K_y a(1 - y)_{LM}} \tag{6-31}$$

As a first approximation, then, to the tower height

$$Z = H_{tOG} \int_{y_2}^{y_1} \frac{(1 - y)_{LM} dy}{(1 - y)(y - y^*)} \tag{6-32}$$

The integral itself is defined as the *number of transfer units*, N_{tOG}. Hence

$$Z = H_{tOG} N_{tOG} \tag{6-33}$$

When H_{tOG} is evaluated in a consistent set of dimensions, it will have the dimension of length, as expected. The quantity N_{tOG} is dimensionless.

6-8-B Interpretation and Evaluation of the Number of Transfer Units

A physical interpretation of the number of transfer units in a packed absorption column can be made in the following way. When the value of y is small, then $(1 - y) \approx (1 - y)_{LM}$. As a result, the definition of the number of transfer units reduces to

$$N_{tOG} \equiv \int_{y_2}^{y_1} \frac{(1 - y)_{LM} dy}{(1 - y)(y - y^*)} \cong \int_{y_2}^{y_1} \frac{dy}{y - y^*}$$

When the denominator is relatively constant along the length of the tower, then integration of the last term in the equation above shows that

$$N_{tOG} \cong \frac{\text{total change in concentration}}{\text{average driving force}}$$

As an approximation, then, the number of overall transfer units is a measure of the number of times the average driving force $(y - y^*)$ divides into the overall concentration change in the gas phase $(y_1 - y_2)$. Likewise, it is a measure of the difficulty of absorption. It must be kept in mind that this interpretation is valid only when the principal resistance to mass transfer occurs in the gas phase. If the major resistance lies in the liquid phase, then an equivalent expression can be developed for the number of transfer units, N_{tOL}.

In addition to direct integration to obtain the value of N_{tOG}, there are a number of analytical solutions to the integration process when the operating and equilibrium lines have special characteristics. Instead of these special treatments of the problem [13, 14], a simple graphical method of determining the number of transfer units is illustrated in Figure 6-18. The basis for the technique is the approximate relation

$$N_{tOG} \approx \int_{y_2}^{y_1} \frac{dy}{y - y^*} = \int_{Y_2}^{Y_1} \frac{dY}{Y - Y^*}$$

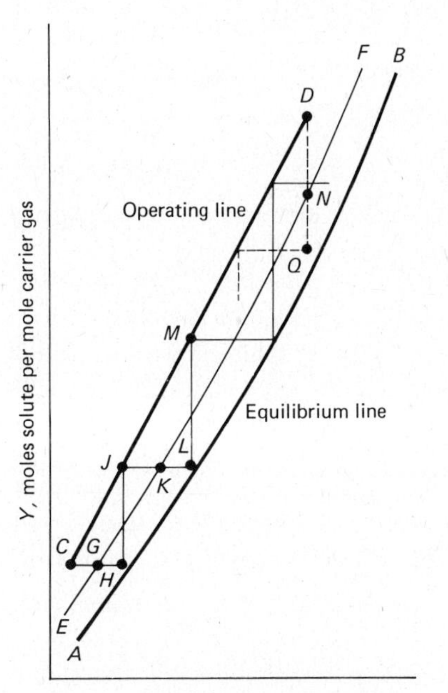

Figure 6-18 Graphical determination of the number of transfer units.

The method allows us to determine the number of times the driving force $Y - Y^*$ will divide into the overall concentration change $Y_1 - Y_2$. Figure 6-18 represents the *Y-X* diagram introduced earlier. The equilibrium curve *AB* typically is not a straight line, but the operating line *CD* is, for reasons discussed previously. Again, the vertical distance between the operating line and the equilibrium line represents the concentration difference or "driving force" which brings about absorption.

When the gas-film resistance controls, a line *EF* is drawn so that all points on the line are located vertically midway between the operating and equilibrium lines. Then, starting at point *C*, a horizontal line is drawn so that $CG = GH$. The position of point *H* may lie to the left or the right of the equilibrium line. From *H* a vertical line is drawn upward to line *CD*, thus locating point *J*. Next draw $JK = KL$. From point *L* another vertical line is drawn upward to line *CD*. Point *M* is established. Steps *CHJ* and *JLM* each represent transfer units. These steps are continued in the same geometrical manner until point *D* is reached or passed to the right by a vertical line. The sum of the whole steps and the last partial step is a measure of the total number of overall transfer units, N_{tOG}.

When the liquid resistance is dominant, we draw the line *EF* so that all points on the line are horizontally midway between the operating and equilibrium lines. Then we start at point *D* on the operating line and draw a vertical line *DQ* such that $DN = NQ$. Then we draw a horizontal line from point *Q* to line *CD*. We continue the procedure to point *C* on the operating line. Again the total number of steps represents the required number of transfer units, N_{tOL}.

Example 6-4

Employing the data and results of Example 6-3, determine the number of overall transfer units by a graphical technique.

SOLUTION

We determine the number of transfer units, N_{tOG}, graphically with the aid of Figure 6-17. From the number of steps measured off, we find that N_{tOG} is roughly $6\frac{1}{2}$.

6-8-C Evaluation of the Height of a Transfer Unit

As noted earlier, the overall height of a transfer unit can be ascertained from pilot plant studies. It is necessary that the small-scale pilot plant duplicate as closely as possible the anticipated operational conditions of the actual equipment. In the absence of direct measurements it is still possible to estimate the value of H_{tOG} by exploiting the similitude among mass transfer operations, in terms of dimensionless correlations. First, recall the relationship

$$\frac{1}{K_y} = \frac{1}{k_y} + \frac{m'}{k_z} \tag{6-23}$$

Multiplying each term by the same quantities, we find that Equation (6-23) may be written as

$$\frac{G'_m}{K_y a(1-y)_{LM}} = \frac{G'_m}{k_y a(1-y)_{LM}} + \frac{mG'_m}{L'_m}\,\frac{L'_m}{k_x a(1-x)_{LM}}\,\frac{(1-x)_{LM}}{(1-y)_{LM}}$$

where m is the average slope of the equilibrium line and replaces m'. By direct analogy to H_{tOG} we define the following quantities:

$$H_{tG} \equiv \frac{G'_m}{k_y a(1-y)_{LM}} \quad \text{and} \quad H_{tL} \equiv \frac{L'_m}{k_x a(1-x)_{LM}} \tag{6-34}$$

where $(1-x)_{LM}$ has the same mathematical interpretation as $(1-y)_{LM}$ introduced by Equation (6-29). Consequently,

$$H_{tOG} = H_{tG} + \left(\frac{mG_m}{L_m}\right)(H_{tL})\frac{(1-x)_{LM}}{(1-y)_{LM}} \tag{6-35}$$

For dilute solutions, Equation (6-35) is approximated to a high degree of accuracy by

$$H_{tOG} = H_{tG} + \left(\frac{mG_m}{L_m}\right)H_{tL} \tag{6-36}$$

This equation permits the evaluation of the overall height of a transfer unit in terms of gas-phase and liquid-phase transfer heights H_{tG} and H_{tL}.

The gas-phase and liquid-phase transfer heights have been correlated over a wide range of experimental data. Generalized equations for H_{tG} and H_{tL} are given below.

$$H_{tG} = \frac{\alpha(G')^{\beta}}{(L')^{\gamma}}\left(\frac{\mu_G}{\rho_G D_G}\right)^{0.5} \tag{6-37}$$

where H_{tG} is the height of a gas transfer unit, in feet; L' is the superficial liquid flow rate, in lb/hr·ft^2; and G' is the superficial gas flow rate, in lb/hr·ft^2. In addition, α, β, and γ are packing constants (see Table 6-3) and the quantity $(\mu_G/\rho_G D_G)$ is the Schmidt number for the gas phase [see Table 6-4(a)]. Also

$$H_{tL} = \phi\left(\frac{L'}{\mu_L}\right)^{\eta}\left(\frac{\mu_L}{\rho_L D_L}\right)^{0.5} \tag{6-38}$$

Table 6-3 CONSTANTS FOR USE IN DETERMINING GAS-PHASE HEIGHT OF A TRANSFER UNIT

PACKING	α	β	γ	GAS-FLOW RATE (lb/hr·ft²)	LIQUID-FLOW RATE (lb/hr·ft²)
Raschig rings					
$\frac{3}{8}$ in.	2.32	0.45	0.47	200–500	500–1500
1 in.	7.00	0.39	0.58	200–800	400–500
	6.41	0.32	0.51	200–600	500–4500
$1\frac{1}{2}$ in.	17.30	0.38	0.66	200–700	500–1500
	2.58	0.38	0.40	200–700	1500–4500
2 in.	3.82	0.41	0.45	200–800	500–4500
Berl saddles					
$\frac{1}{2}$ in.	32.40	0.30	0.74	200–700	500–1500
	0.81	0.30	0.24	200–700	1500–4500
1 in.	1.97	0.36	0.40	200–800	400–4500
$1\frac{1}{2}$ in.	5.05	0.32	0.45	200–1000	400–4500
Partition rings					
3 in.	6.50	0.58	1.06	150–900	3000–10,000
Spiral rings (stacked, staggered)					
3-in. single spiral	2.38	0.35	0.29	130–700	3000–10,000
3-in. triple spiral	15.60	0.38	0.60	200–1000	500–3000

SOURCE: R. E. Treybal. *Mass Transfer Operations*. New York: McGraw-Hill, 1968.

where H_{tL} is the height of the liquid transfer unit, in feet, and μ_L is the liquid viscosity, in lb/ft·hr. In addition, ϕ and η are packing constants (see Table 6-5). Likewise, $(\mu_L/\rho_L D_L)$ is the Schmidt number for the liquid phase [see Table 6-4(b)]. The Schmidt number in either case is a dimensionless value. Example 6-5 will be helpful in illustrating the calculation procedure.

Example 6-5

Employing the data and results of Examples 6-3 and 6-4, determine the overall height of a transfer unit, in feet, and the total height of the tower, in feet. Also compute the total pressure drop on the basis of the $\Delta P/Z$ values found by two methods in part 5 of Example 6-3.

SOLUTION

The height of a transfer unit, H_{tOG}, is found through the use of the generalized correlations given by Equations (6-37) and (6-38). The values of G' and L' used are average values for the overall tower, which in this case are 912 and 668 lb/hr·ft^2, respectively for an area of 2.40 ft^2. On the basis

TABLE 6-4(a) DIFFUSION COEFFICIENTS AND SCHMIDT NUMBERS OF GASES AND VAPORS IN AIR AT 25°C AND 1 ATM

SUBSTANCE	$D(cm^2/s)$	$\mu/\rho D$
Ammonia	0.236	0.66
Carbon dioxide	0.164	0.94
Hydrogen	0.410	0.22
Water	0.256	0.60
Carbon disulfide	0.107	1.45
Ethyl ether	0.093	1.66
Methanol	0.159	0.97
Ethyl alcohol	0.119	1.30
Propyl alcohol	0.100	1.55
Butyl alcohol	0.090	1.72
Benzene	0.088	1.76
Chlorobenzene	0.073	2.12
Ethylbenzene	0.077	2.01

TABLE 6-4(b) DIFFUSION COEFFICIENTS AND SCHMIDT NUMBERS OF LIQUIDS IN WATER AT 20°C

SUBSTANCE	$D[(cm^2/s) \times 10^5]$	$\mu/\rho D$
Carbon dioxide	1.50	570
Ammonia	1.76	570
Chlorine	1.22	824
Bromine	1.20	840
Hydrogen chloride	2.64	381
Hydrogen sulfide	1.41	712
Sulfuric acid	1.73	580
Nitric acid	2.60	390
Methanol	1.28	785
Ethanol	1.00	1005
Propanol	0.87	1150
Butanol	0.77	1310
Glycerol	0.72	1400

SOURCE: J. H. Perry, ed. *Chemical Engineers' Handbook*. New York: McGraw-Hill, 1950.

Table 6-5 CONSTANTS FOR USE IN DETERMINING LIQUID-PHASE HEIGHT OF TRANSFER UNITS

PACKING	ϕ	η	L' RANGE (lb/hr·ft^2)
Raschig rings			
$\frac{3}{8}$ in.	0.00182	0.46	400–15,000
$\frac{1}{2}$ in.	0.00357	0.35	400–15,000
1 in.	0.0100	0.22	400–15,000
$1\frac{1}{2}$ in.	0.0111	0.22	400–15,000
2 in.	0.0125	0.22	400–15,000
Berl saddles			
$\frac{1}{2}$ in.	0.00666	0.28	400–15,000
1 in.	0.00588	0.28	400–15,000
$1\frac{1}{2}$ in.	0.00625	0.28	400–15,000
Partition rings			
3 in.	0.0625	0.09	3000–14,000
Spiral rings (stacked, staggered)			
3-in. single spiral	0.00909	0.28	400–15,000
3-in. triple spiral	0.0116	0.28	3000–14,000

SOURCE: R. E. Treybal. *Mass Transfer Operations*. New York: McGraw-Hill, 1968.

of Equation (6-37), for the gas phase

$$H_{tG} = \frac{\alpha(G')^{\beta}}{(L')^{\gamma}}(\text{Sc})^{0.5} = \frac{7.00(912)^{0.39}}{(668)^{0.58}}(0.66)^{0.5} = 1.88 \text{ ft}$$

where $\alpha = 7.00$, $\beta = 0.39$, and $\gamma = 0.58$ from Table 6-3 and the Schmidt number is given in Table 6-4(a). For the liquid phase

$$H_{tL} = \phi\left(\frac{L'}{\mu_L}\right)^{\eta} \cdot (\text{Sc})^{0.5} = 0.01\left(\frac{668}{2.42}\right)^{0.22}(570)^{0.5} = 0.82 \text{ ft}$$

where the values of ϕ, η, and Sc have come from Tables 6-5 and 6-4(b). Employing Equation (6-36) we find that

$$H_{tOG} = H_{tG} + m\left(\frac{G_m}{L_m}\right)H_{tL} = 1.88 + 0.75\left(\frac{80}{92.3}\right)(0.82) = 2.42 \text{ ft}$$

In the above relation the factor (mG_m/L_m) has been evaluated at the bottom of the tower, where most of the transfer units occur. In Example 6-4 the number of transfer units was found to be $6\frac{1}{2}$. Hence the overall tower height is

$$Z = N_{tOG}H_{tOG} = 6.5(2.42) = 15.6 \text{ ft}$$

The overall pressure drop may be determined from the values of $\Delta P/Z$ which were evaluated by two methods in Example 6-3. Based on these two values,

$$\Delta P = 0.65(15.6) = 10.0 \text{ in. water} = 0.36 \text{ psi}$$

and

$$\Delta P = 0.746(15.6) = 11.5 \text{ in. water} = 0.42 \text{ psi}$$

The answers are in close agreement with each other. The magnitude of the answer is fairly typical of a packed tower operating reasonably far away from the flood point.

Plate or tray towers are also employed as cleaning devices for polluted gas streams. In these towers, stepwise contact between the liquid and gas occurs as shown in Figure 6-19. The design procedure is somewhat similar to that for a packed tower [7, 13, 14].

Estimating the cost of absorption equipment for air pollution control is beyond the scope of the present discussion since so many factors are involved. Several of these factors are:

1. Material to be removed, corrosive or noncorrosive
2. Cost of absorbent

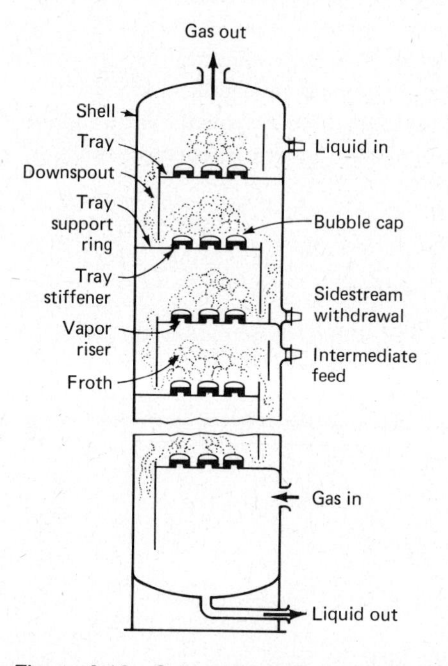

Figure 6-19 Schematic diagram of a tray tower.

3. Type of material for tower (such as an alloy steel)
4. Type of packing material
5. Flow rate of gas being cleaned
6. Value of material removed as solute
7. Recycling of solvent or disposing of waste solvent

Additional discussion of factors in economic design appears in the literature [12].

6-9 FUNDAMENTALS OF CHEMICAL KINETICS

In some cases it may not be feasible to remove the required amount of a specific air pollutant from an exhaust stream by such techniques as adsorption or absorption, in order to meet federal emission standards. Two other general techniques of air pollution control are available. One of these is the chemical alteration of the pollutant species after it is formed, and the process is referred to as incineration. The other technique is to prevent the initial formation of the pollutant by controlling the variables which promote its formation. The proper application of either of these techniques requires an understanding of the fundamentals of chemical kinetics.

The products from combustion processes are responsible for a majority of the contaminants released to the atmosphere. Owing to the wide variation in temperature of combustion gases over a relatively short time period, the rate of reaction becomes as important as the thermodynamics of the reaction. Studies of three of the major air pollutants in our atmosphere—carbon monoxide, sulfur oxides, and nitrogen oxides—have determined that only when we make use of some of the basic concepts from chemical kinetics, the study of reaction rates, does theory begin to match experimental information. Thus in this section we present some of the general theory of chemical kinetics for use in future sections.

6-9-A Rate Equations and Order of Reaction

The rate of chemical reaction is usually stated in terms of the rate of disappearance of reactants or appearance of products. Consider the reaction of interest to be, for example,

$$A + 2B \rightarrow D + 3E$$

Stated in terms of the concentrations of all the reactants, the rate of change of concentration of species A is, in general,

$$\frac{-dC_A}{dt} = kC_A^{\,a}C_B^{\,b}$$

where concentration, C, is expressed in mass per unit volume units. Another common method of symbolically expressing the rate equation is to use brackets around the chemical symbol, rather than the symbol C. That is,

$$\frac{-d[A]}{dt} = k[A]^a[B]^b \tag{6-39}$$

The overall *order* of the reaction described by either rate equation is defined as $(a + b)$. The reaction is of order a with respect to A and of order b with respect to B. The proportionality constant k is a positive number which depends upon temperature (and possibly other variables) but is independent of concentrations. It is frequently called the rate constant or the specific reaction rate. The units of k depend upon the units used for the concentration.

The overall order of a reaction is determined experimentally. Nonintegral values of the order are possible. However, many reactions appear to have integral values for their orders. If the reaction

$$A \rightarrow B$$

is first-order, then

$$\frac{-dC_A}{dt} = kC_A{}^aC_B{}^b \tag{6-40}$$

The solution to this differential equation is

$$C_A = C_A{}^o \exp(-kt) \tag{6-41}$$

where $C_A{}^o$ is the initial concentration of A at time $t = 0$.

An equivalent result may be found by expressing the rate in terms of the formation of the product B. Let x equal the concentration of B at time t. Then the amount of A at time t is simply $(C_A{}^o - x)$. Hence Equation (6-40) becomes

$$-\frac{d(C_A{}^o - x)}{dt} = k(C_A{}^o - x)$$

or

$$\frac{dx}{C_A{}^o - x} = k\,dt$$

Integration yields

$$\ln \frac{C_A{}^o}{C_A{}^o - x} = kt$$

or

$$x = C_A{}^o[1 - \exp(-kt)] \tag{6-42}$$

This equation reduces to Equation (6-41) by replacing x with $C_A{}^o - C_A$.

A second-order reaction may be represented by two different chemical reactions. One of these involves a single reactant, such as

$$2A \rightarrow \text{products}$$

Another possibility involves two different species initially at the same concentration.

$$A + B \rightarrow \text{products}$$

In either case the rate expression is (when $C_A^o = C_B^o$)

$$\frac{-dC_A}{dt} = k_2(C_A)^2$$

Integration of this expression yields

$$\frac{1}{C_A} - \frac{1}{C_A^o} = k_2 t$$

When $C_A^o \neq C_B^o$, a more complex relation develops from the rate equation. In this latter case it may be shown that

$$k_2 t = \frac{1}{C_B^o - C_A^o} \ln \frac{C_B C_A^o}{C_A C_B^o} \tag{6-43}$$

6-9-B Consecutive Set of Reactions

Consider now a consecutive set of first-order reactions, such as

$$A \xrightarrow{k_1} B \xrightarrow{k_2} D$$

where $C_B^o = C_D^o = 0$. The rates are given by

$$\frac{dC_A}{dt} = -k_1 C_A \tag{a}$$

$$\frac{dC_B}{dt} = k_1 C_A - k_2 C_B \tag{b}$$

$$\frac{dC_D}{dt} = k_2 C_B \tag{c}$$

The disappearance of A is expressed by a first-order rate equation. As a result,

$$C_A = C_A^o \exp(-k_1 t) \tag{6-44}$$

Substitution of Equation (6-44) into Equation (b) gives

$$\frac{dC_B}{dt} = k_1 C_A^o \exp(-k_1 t) - k_2 C_B$$

Integration in this case yields

$$C_B = C_A^o \frac{k_1}{k_2 - k_1}[\exp(-k_1 t) - \exp(-k_2 t)] \tag{6-45}$$

The rate of change of species D is

$$\frac{dC_D}{dt} = k_2 C_B = \frac{k_1 k_2 C_A^o}{k_2 - k_1}[\exp(-k_1 t) - \exp(-k_2 t)]$$

Rather than carry out the integration of this equation, the concentration of D at any time can be deduced simply from the relationship

$$C_A + C_B + C_D = C_A^o$$

or

$$C_D = C_A^o - C_A^o \exp(-k_1 t) - C_A^o \frac{k_1}{k_2 - k_1}[\exp(-k_1 t) - \exp(-k_2 t)] \tag{6-46}$$

When k_1 and k_2 are the same order of magnitude, all the terms in Equations (6-45) and (6-46) are important. Significant quantities of species A, B, and D are present at the same time during a portion of the overall process. The general trend in the various concentrations as a function of time is shown by the unbroken lines in Figure 6-20. These lines represent the condition of

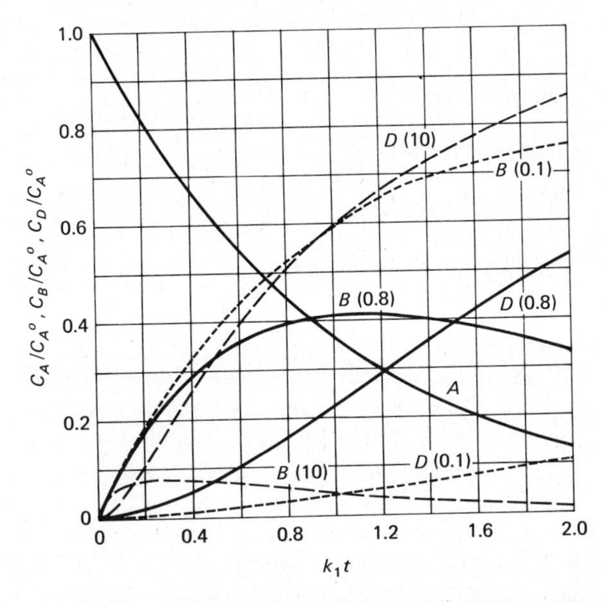

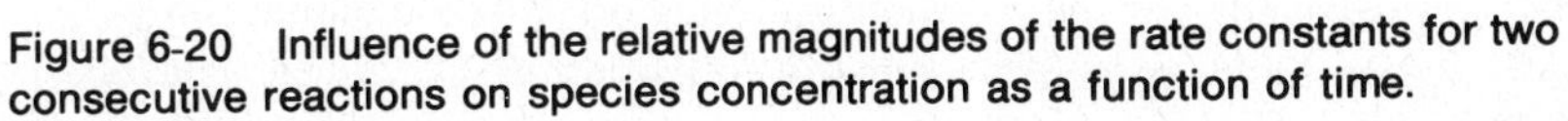
Figure 6-20 Influence of the relative magnitudes of the rate constants for two consecutive reactions on species concentration as a function of time.

$k_2 = 0.8k_1$. Note that the production of D is delayed for a while until a sufficient quantity of B has been built up. This delay period is called the induction period. After sufficient time the rate of D increases while the concentration of B maximizes and begins to fall off. Throughout the time period the concentration of A decays exponentially.

6-9-C Rate-Determining Step

When a number of consecutive and competing reactions occur, the expression for the rate of change of any one of the intermediate species normally is quite complex. In many instances, however, the rate expression can be greatly simplified if some of the rate constants are much larger than others. For example, consider the consecutive set of reactions discussed above, namely,

$$A \overset{k_1}{\rightarrow} B \overset{k_2}{\rightarrow} D$$

If the second reaction is much faster than the first one ($k_2 \gg k_1$), then the general expression

$$\frac{dC_D}{dt} = k_2 C_B = \frac{k_1 k_2 C_A^o}{k_2 - k_1}[\exp(-k_1 t) - \exp(-k_2 t)]$$

reduces to

$$\frac{dC_D}{dt} = k_1 C_A^o \exp(-k_1 t)$$

and the concentration of species D becomes

$$C_D \cong C_A^o [1 - \exp(-k_1 t)]$$

Likewise,

$$C_B \cong C_A^o \left(\frac{k_1}{k_2}\right) \exp(-k_1 t)$$

and this quantity will be relatively small, owing to the ratio k_1/k_2. This latter result is not unexpected. If $k_2 \gg k_1$, the reaction $B \rightarrow D$ will consume species B nearly as fast as it is formed. The concentration-time curves for the condition where $k_2 = 10\,k_1$ are shown as dashed lines in Figure 6-20. Note that the concentration of B is much smaller than that for the condition where $k_2 = 0.8\,k_1$, since we have increased k_2 by a factor of 12.5 over the previous case. The induction period of D is also much shorter in this new case, and its concentration builds up much more rapidly.

Alternatively, if $k_2 \ll k_1$, then the reaction $A \rightarrow B$ dominates. Species B will tend to build up rapidly and the formation of D will be delayed. The equations for C_B and C_D in this case become

$$\frac{C_B}{C_A^o} \cong \exp(-k_2 t) - \exp(-k_1 t)$$

$$\frac{C_D}{C_A^o} \cong 1 - \exp(-k_2 t)$$

Thus the concentration-time profiles for both C_B and C_D are governed by the value of k_2. Figure 6-20 also shows dotted curves for the condition $k_2 = 0.1\,k_1$. This ratio of k_2/k_1 is a factor of 8 less than the original data represented by the unbroken lines. For even smaller values of k_2/k_1, the line for C_D would approach the horizontal axis.

The preceding discussion may be used to illustrate two important concepts with regard to the complex intermediate chemical reactions present in combustion processes. First, the time required for the intermediate products, such as species B above, to be consumed may be considerable. If the reactions are quenched (by suddenly lowering the temperature, for example), the intermediate products may persist for a long time. Some of these intermediate products left to linger in the atmosphere are quite possibly pollutants. Second, although a number of intermediate reaction steps may exist for a given overall reaction, only a very few of these may control the rate of formation or disappearance of any particular species. By comparison of the various rate constants, k, for the intermediate reactions, a number of the reactions may reasonably be ignored. This greatly simplifies the mathematical model for the overall reaction. At the same time, the usefulness of the model in connection with experimental data may be greatly enhanced.

6-9-D The Rate Constant and Activation Energy

As we noted earlier, the rate constant k for any elementary reaction is not a function of concentrations, but of temperature. In some instances this functional relationship with temperature will be weak, while in other instances the value of k will vary several orders of magnitude over a few hundred degrees change in temperature. The rate constant is usually expressed by an Arrhenius-type equation such as the following:

$$k = A \exp\left(\frac{-E}{RT}\right) \tag{a}$$

$$k = AT^n \exp\left(\frac{-E}{RT}\right) \tag{b}$$

where A and n are constants, E is the activation energy for the reaction, and R is the universal gas constant. It is the presence of the exponential factor which has an enormous influence on the variation of k with temperature.

The dependence of k on temperature mathematically is determined by the magnitude of the activation energy, E. The variation of temperature, itself, is small—possibly in the range from 300° to 2500°K for combustion processes. The values of E range typically from a few ten thousands upward toward several hundred thousand, when expressed in joules per gram-mole. It is this variation of E which accounts for the vast differences in how k varies with temperature. To illustrate the point, we shall choose a general example, which involves the formation of species GH by two different elementary reactions, namely,

$$G + H_2 \rightleftharpoons GH + H \qquad \text{(c)}$$

$$H + G_2 \rightleftharpoons GH + G \qquad \text{(d)}$$

The equations for the two rate constants for (c) and (d) follow the format of Equations (a) and (b), and in the forward direction will be assumed to be given by

$$k_c = 6 \times 10^9 T \exp\frac{-24{,}000}{RT} \qquad \text{(e)}$$

$$k_d = 1 \times 10^{14} \exp\frac{-300{,}000}{RT} \qquad \text{(f)}$$

where both values of k are in units of $cm^3/g{\cdot}mole{\cdot}s$, R must be expressed in $J/g{\cdot}mole{\cdot}°K$, and T is in degrees Kelvin. The values of k for reactions (c) and (d) at three typical temperatures for a combustion reaction are shown in the tabulation. In the case of reaction (c) with a relatively low activation energy, the value of k increases simply by a factor of 10. Since the temperature itself increases by a factor of 2, the exponential term actually goes up by a factor of 5. For reaction (d), with a high activation energy, k is altered by roughly 8 orders of magnitude. On the basis of this general example, it is possible to make a generalization. When the activation energy is reasonably small (for example, $< 50{,}000$ J/g·mole), the rate constant k is fairly insensitive to temperature. For activation energies greater than this (and especially beyond 100,000 J/g·mole), the rate constant is extremely sensitive to temperature variations.

TEMPERATURE (°K)	k_c	k_d
1000	3.35×10^{11}	2.14×10^{-2}
1500	1.31×10^{12}	3.58×10^{3}
2000	2.83×10^{12}	1.46×10^{6}

It is important to note that Equations (c) and (d) are written with a double arrow, that is, the reaction is capable of proceeding in either direction. Hence there is a backward as well as a forward rate constant. For reaction (c), this backward rate constant normally would be given the symbol k_{-c}. In determining the net production of species GH by reactions (c) and (d), it would be necessary to examine the effect of both the forward and reverse reactions rates.

The conclusions drawn here have an important bearing on the composition of gases that enter our atmosphere as flue gases from a combustion process. In most combustion processes the gases are maintained at the high temperature level of the flame zone for a few milliseconds or hundredths of a second, and then are cooled by dilution or heat transfer effects. Depending upon the rate of cooling, the temperature of the gas stream may fall quite rapidly. This effect, for example, would quench reaction (d) in terms of its rate, but would have little effect on reaction (c), all other things being equal. Consequently, the rapid quenching of any elementary reaction is dependent, among other things, on the magnitude of its activation energy.

6-10 KINETICS OF CARBON MONOXIDE FORMATION

At the present time, the kinetics of carbon oxidation is not clearly understood. A great deal of additional experimental work needs to be done to achieve a satisfactory elementary reaction mechanism for the formation of CO and CO_2 for hydrocarbon combustion processes. The steps for CO formation must precede those for CO_2, and, unfortunately, the former is probably less well delineated. In most studies of high-temperature oxidation processes of hydrocarbons, such as methane, investigators [16] frequently include the methyl radical (CH_3) reaction with O_2,

$$CH_3 + O_2 \rightleftharpoons HCO + H_2O$$

as a principal intermediate reaction. It is then proposed that carbon monoxide is formed mainly from the reaction

$$HCO + OH \rightleftharpoons CO + H_2O \tag{6-47}$$

although another possibility is

$$HCO + M \rightleftharpoons H + CO + M \tag{6-48}$$

Hydrocarbon flames usually contain appreciable quantities of OH in the flame zone, so the formation of CO is not restricted.

The conversion of CO to CO_2 is now fairly well established [17]. The major reaction, which is intrinsically fast, is considered to be

$$CO + OH \rightleftharpoons CO_2 + H \tag{6-49}$$

Another route is the direct oxidation reaction

$$CO + O_2 \rightarrow CO_2 + O \tag{6-50}$$

However, this step is relatively slow, and probably not too important. Nevertheless, the exclusion of reaction (6-50) has an important ramification. In lean flames an excess of oxygen is always present. Thus it seems puzzling that appreciable quantities of CO are found in the cooled gases. But in the light of elementary kinetics, it appears that direct oxidation is not the preferred path, in spite of excess O_2. Rather, reaction (6-49) dominates. If this is truly the mechanism, then the concentration of OH in various regions of the combustion process controls the extent of CO consumed in those regions.

Although there are a number of elementary reactions which may be considered as contributing to the formation of OH particles, one essential step is

$$H + H_2O \rightleftharpoons OH + H_2 \tag{6-51}$$

This reaction is important because in combination with reaction (6-40) it forms the well-known water-gas reaction, namely,

$$CO + H_2O \rightleftharpoons CO_2 + H_2 \tag{6-52}$$

In the hot zone of premixed flames or turbulent diffusion flames equilibrium is rapidly achieved among the competing reactions. The elementary reactions are equilibrated, and hence an overall reaction like (6-52) is also in equilibrium. Data for the ideal-gas equilibrium constant, K_p, of the water-gas reaction are shown in Table 6-6. On the basis of this thermodynamic consideration alone, CO_2 should be formed from CO when the temperature falls as the gases leave the hot combustion zone. If sufficient O_2 is present, the magnitude of K_p at 298°K indicates nearly complete conversion to CO_2. However, the kinetics of the reaction lead to a much different conclusion.

Consider the elementary reactions (6-49) and (6-51) to be the essential steps of the water-gas reaction (although the total scheme is much more complex [18]). Then a simple kinetic explanation can qualitatively be made for superequilibrium levels of CO in exhaust gases [19]. The forward reaction (6-49) is quite fast, and the rate constant is reasonably independent of temperature. The forward reaction (6-51) is also fast at high temperatures. Around 1500°K the forward-rate constants for the two reactions are comparable. However, reaction (6-51) has a high activation energy (about 85 kJ). Consequently, its rate constant falls rapidly with a temperature decrease. At

Table 6-6 EQUILIBRIUM CONSTANTS FOR THE WATER-GAS REACTION

$CO + H_2O \rightleftharpoons CO_2 + H_2$	T(°K)	T(°F)	K_P
	298	77	1.04×10^5
	500	440	1.38×10^2
$K_p = \dfrac{(p_{CO_2})(p_{H_2}) = [CO_2][H_2]}{(p_{CO})(p_{H_2O}) = [CO][H_2O]}$	1000	1340	1.44
	1500	2240	0.39
	2000	3140	0.22
	2500	4040	0.16

SOURCE: Computed from JANAF (Joint Army Navy Air Force) Thermochemical Tables.

1000°K the constant is around one-thirtieth the value at 1500°K, and at 800°K the constant is about one three-hundredth of the 1500°K value. As a consequence, the reaction

$$H + H_2O \rightarrow OH + H_2$$

is ineffective in producing OH species at a reasonable rate as the gas temperature falls. With a falloff in OH production, the conversion of CO by the reaction

$$CO + OH \rightarrow CO_2 + H$$

would be greatly diminished. It might be suggested that the reaction

$$H + O_2 \rightleftharpoons OH + O$$

is another possible source of OH. This elementary reaction also has a reasonably high activation energy. Hence its rate constant also falls rapidly as the temperature is lowered.

The elementary reaction mechanism discussed above is a provisional one. More work of a theoretical and experimental nature will be required before a comprehensive model of the $CO-CO_2$ system is well established. Nevertheless, this simplified model does illustrate that kinetic studies are useful in gaining insight into and explaining observed phenomena, at least qualitatively. As the model is improved, kinetics can yield more quantitative information which will be useful in predicting the effects of process changes.

6-11 CARBON MONOXIDE EMISSION CONTROL

Carbon monoxide is formed as an intermediate product of the chemical reaction between carbonaceous fuels and oxygen. When an insufficient quantity of oxygen is provided (as in rich mixtures), CO must occur as a final product of the combustion process. In lean mixtures CO occurs for one of two specific reasons: (1) there may be poor mixing of the fuel and air in the reaction zone, so that regions of the zone act fuel-rich, even though the overall mixture is lean; or (2) CO may originate in high-temperature regions of the combustion zone, where chemical equilibrium dictates that dissociation of CO_2 into CO should occur. In Table 6-7 the ideal-gas equilibrium constant K_p for the oxidation of CO to CO_2 is tabulated as a function of temperature. For temperatures above 2000°K the relatively small K_p values indicate that CO_2 dissociation is no longer negligible. Hence the effects of fuel-air ratio, degree of mixing, and temperature may lead to significant CO formation in the hot combustion zone.

Table 6-7 EQUILIBRIUM CONSTANTS FOR THE CO–CO_2 OXIDATION PROCESS

$CO + \frac{1}{2}O_2 \rightleftharpoons CO_2$	T(°K)	T(°F)	K_p
	298	77	1.2×10^{45}
	500	440	1.1×10^{25}
$K_p = \dfrac{p_{CO_2}}{p_{CO}(p_{O_2})^{1/2}}$	1000	1340	1.7×10^{10}
	1500	2240	2.1×10^{5}
	2000	3140	766
	2500	4040	28

SOURCE: Computed from the JANAF Thermochemical Tables.

Example 6-6

Estimate the number of moles of carbon monoxide at equilibrium in an ideal-gas mixture at 2000°K if the initial mixture composition is 1.00 mole of CO, 0.55 mole of O_2, and 2.05 moles of inert N_2 at 1 atm.

SOLUTION

On the basis of the equation and data from Table 6-7, we may write

$$K_p = 766 = \frac{p_{CO_2}}{p_{CO}(p_{O_2})^{1/2}}$$

But the component or partial pressure, p_i, of any species in the mixture is equal to $(n_i/n_T)P$, where P is the total pressure and n_T is the total number of moles present, including inert gases. Consequently, the above equation may be written as

$$K_p = 766 = \frac{n_{CO_2}}{n_{CO}(n_{O_2})^{1/2}}\left(\frac{P}{n_T}\right)^{-1/2}$$

The four unknowns in this expression are related as follows. Since carbon and oxygen atoms must be conserved,

$$n_{CO} + n_{CO_2} = n_{CO,\,init} = 1.00$$

$$n_{CO} + 2n_{CO_2} + 2n_{O_2} = n_{O,\,init} = 2.10$$

In addition,

$$n_T = n_{CO} + n_{CO_2} + n_{O_2} + 2.05$$

If these latter three equations are substituted into the K_p expression in terms of n_{CO}, we find that

$$766 = \frac{(1-a)(3.10+0.5a)^{1/2}}{a(0.05+0.5a)^{1/2}}$$

where the symbol a represents n_{CO}. Solution of this equation leads to the following results:

$$n_{CO} = 0.0097 \text{ mole}$$

$$n_{CO_2} = 0.9903 \text{ mole}$$

$$n_{O_2} = 0.0549 \text{ mole}$$

Hence at the specified temperature and pressure we find that most of the initial amount of CO reacts to form CO_2 when equilibrium prevails. However, at higher temperatures the quantity of carbon monoxide present increases significantly.

As indicated in the preceding section on the kinetics of carbon monoxide formation, the preferred path for oxidation of CO to CO_2 is through the elementary reaction with hydroxyl radical, as given by reaction (6-49). Regardless of the reasons for its formation, removal of the pollutant CO is dependent upon the presence of OH in the reactive mixture. Consequently, thermal quenching of the combustion process normally leads to a superequilibrium level of CO in the exhaust gases. Subsequent removal to levels consistent with ambient air quality standards requires dilution by and further interaction with the ambient atmosphere. Carbon monoxide, unlike some other major gaseous pollutants, does not lend itself to exhaust gas removal techniques. A more productive approach is the control of its formation. This approach is not a simple one, because usually the formation of CO and NO_x must be reduced simultaneously. Unfortunately, the control strategies for each of these gases are basically in conflict. As a consequence, each major type of combustion equipment must be analyzed in terms of these pollutants, and others, on its own merits. For this reason, we discuss the pollution problems from several sources, such as internal combustion engines and gas turbines, separately. The most practical method of reducing the CO emissions from stationary combustion sources is by the proper design, installation, operation and maintenance of combustion equipment.

In the process industries there are a number of noncombustion processes in which carbon monoxide is evolved in large quantities. Steel and petroleum refining are two examples of industries where carbon monoxide from various processes is collected and burned in furnaces or waste heat boilers. As a result, carbon monoxide becomes an economical supplementary source of heat for the plant. In some instances the concentration of CO will be low, but the total quantity of gas evolved may be so large that an appreciable quantity of CO is released. At these lower CO concentrations the gas may not support a flame directly, especially when the initial gas temperature is also low. In such cases catalytic oxidation in an afterburner may be necessary (see Section 6-12). Carbon monoxide may be removed selectively from other gases by scrubbing with specific solutions, such as copper ammonium formate.

6-12 INCINERATION OR AFTERBURNING

Incineration or afterburning is a combustion process used to remove combustible air pollutants (gases, vapors, or odors). It is frequently used in situations where the volume flow rate of waste gas from a process is large but the level of contaminant gas is small. Incineration of objectionable gaseous wastes is a satisfactory pollutant control method from several aspects.

1. Almost all highly odorous pollutants are combustible or are changed chemically to less odorous substances when heated sufficiently in the presence of oxygen. Odorous air pollutants destroyed by incineration include mercaptans, cyanide gases, and hydrogen sulfide.
2. Organic aerosols that cause visible plumes are effectively destroyed by afterburning. Coffee roasters, meat smoke houses, and enamel baking ovens emit such aerosols.
3. Certain organic gases and vapors, if released to the atmosphere, become involved in smog reactions. Flame afterburning effectively destroys these materials.
4. Some industries, such as refineries, produce large quantities of highly combustible waste gases and otherwise dangerous organic materials. The safest method of effluent control is usually by burning in stack flares or specially designed furnaces.

The need for afterburners in many process industries is exemplified by the data of Table 6-8 [20]. We see that solvent evaporation is the major source of hydrocarbon emissions from stationary sources currently, and the total might be expected to increase in the future. Since mobile source emissions are now under strict federal control, solvent evaporation might have been a major contributor to total hydrocarbon emissions if the pollutant sources had remained substantially uncontrolled. Since certain reactive hydrocarbons play an important role in photochemical smog formation, it is necessary that hydrocarbon control techniques be employed at most stationary sources, and that these techniques be optimized in terms of reduction of emissions. Probably the most stringent air pollution code in the United States has been Los Angeles County Rule 66 (now revised as Rule 442). Rule 66 required a facility to eliminate 90 percent of carbonaceous material by incineration according to the formula

$$\frac{HC_{in} - [HC_{out} + (CO_{out} - CO_{in})]}{HC_{in}} \geqslant 0.90$$

Note that the formulation includes carbon monoxide as well as hydrocarbon concentrations. Since a federal ambient air quality standard now exists for carbon monoxide, the inclusion of CO in the code of emission standards of many other localities may be necessary.

The advantages of afterburning are manifold. These include (1) essentially complete destruction of all combustible pollutants when the equipment

Table 6-8 ESTIMATE OF NATIONAL EMISSIONS OF HYDROCARBONS FROM MOBILE AND STATIONARY SOURCES (IN MILLIONS OF POUNDS)

STATIONARY SOURCES	1968	1970	1980
Solvent evaporation	13,400	14,500	22,100
Refuse disposal	3,430	3,510	880
Agricultural burning	3,400	3,400	3,400
Refineries	2,940	3,340	4,300
Gasoline marketing	1,710	1,850	2,470
Carbon black manufacture	1,570	1,450	1,390
Others	2,100	2,200	2,790
Total, stationary sources	28,500	30,250	37,330
Mobile sources	38,300	35,200	11,900

SOURCE: T. Phillips and R. W. Rolke. *Am. Inst. Chem. Engineers Symposium Series No. 137* **70,** 1974.

is properly designed and operated, (2) capability of adapting the equipment to moderate changes in effluent flow rate and concentration, (3) a control effectiveness which is relatively insensitive to the specific gaseous pollutant, (4) absence of performance deterioration (of a thermal-type unit), and (5) possibility of economical waste heat recovery. Among the disadvantages frequently cited are (1) reasonably high capital and operating costs, (2) the necessity of providing collection and ducting equipment in some instances, which adds significantly to the cost, and (3) the possibility of introducing special pollution problems when atoms other than C, H, and O are present in the hydrocarbon. Examples of this latter problem are compounds containing chlorine, nitrogen, and sulfur. The products of incineration in these cases may require further treatment, such as scrubbing, and corrosion problems may arise. In addition, the catalytic type of afterburner may produce products that poison the catalyst bed or that cause the burner to become plugged. In thermal and catalytic afterburners it is necessary to supply additional fuel to achieve a desired temperature level for the oxidation of the waste gas pollutants. Current and possibly persistent fuel dislocations may be an obstacle to obtaining a sufficient supply of gaseous supplementary fuel when required.

Several factors must be known in order to design an incineration process for gaseous contaminants. These include the chemical composition of the contaminants and their concentration level, the inlet waste gas temperature, the volume rate of waste gas to be handled, and the permissible emission levels for the pollutants. With the aid of this information an appropriate incineration technique can be chosen. The three basic types of incineration are classified as (1) direct flame, (2) thermal, and (3) catalytic.

Example 6-7

The waste gas from an industrial process at 450°K contains 1450 ppm of hydrocarbons and 90 ppm of carbon monoxide. After treatment in an afterburner at 900°K the concentrations of hydrocarbons and carbon monoxide are found to be 105 ppm and 280 ppm, respectively. (a) What is the conversion efficiency based on hydrocarbons alone? (b) What is the conversion efficiency based on Los Angeles County Rule 66?

SOLUTION

(a) The conversion efficiency based on hydrocarbons alone is simply

$$\text{efficiency} = \frac{HC_{in} - HC_{out}}{HC_{in}}(100) = \frac{1450 - 105}{1450}(100) = 92.8 \text{ percent}$$

(b) On the basis of Rule 66,

$$\text{efficiency} = \frac{HC_{in} - [HC_{out} + (CO_{out} - CO_{in})]}{HC_{in}}(100)$$

$$= \frac{1450 - [105 + (280 - 90)]}{1450}(100) = 79.7 \text{ percent}$$

On the basis of hydrocarbons alone, the waste gas from the afterburner would meet the 90 percent reduction requirement. However, combustion of hydrocarbons leads to CO formation. This particular afterburner is not designed properly to keep the combined CO and HC levels sufficiently low to meet the standard.

6-12-A Direct Flame Incineration

The design of direct flame afterburners for the incineration of combustible gases and vapors requires the knowledge of the explosive or flammability limits of both the waste materials and the fuel gas in mixtures with air. Such knowledge will show whether a given waste gas will support combustion without additional fuel enrichment. It should be noted that a mixture of combustible material and air within certain concentration limits is explosive. The reaction is most violent when the mixture is slightly fuel-rich. As the mixture is made leaner or richer than this most explosive mixture, the rate of combustion decreases. The concentrations of combustibles on the lean and rich side beyond which a flame will not propagate are known as the lower and upper flammability limits. As a rule of thumb, a stoichiometric (chemically correct) mixture of hydrocarbons and air has a heat of combustion of roughly 100 Btu/scf (3725 kJ/m^3). The lower flammability limit is approximately one-half this value. For safety reasons, to transport gases industrially the concentration of combustibles must be below the lower flammability limit. It is generally recommended that the gas stream have a heating value no more than 10 to 15 Btu/scf (370 to 550 kJ/m^3). [Most insurance regulations require that the waste gas transported to the incinerator be below 25 percent of the lower explosion limit (L.E.L.).] If the gas should be accidentally ignited, then, the increased temperature of the gas stream would not be high enough to sustain combustion.

Direct flame incineration, as the name implies, is a method by which the waste gases are burned directly in a combustor, with or without the aid of an additonal fuel such as natural gas. In some cases the waste gas itself may be a combustible mixture without the addition of air. In other situations, even after addition of air the mixture will be within its flammability limits. If the mixture lies outside its lower flammability limit, the addition of a relatively small amount of fuel frequently will bring it within the limit. In a well-designed burner a waste gas with a heating value of around 85 to 90 Btu/scf (3150 to 3350 kJ/m^3) can be burned successfully without auxiliary fuel. Even this requirement can be circumvented by preheating the gas stream a few hundred degrees, since the lower flammability limit decreases with increasing temperature. In any case, direct flame incineration should be used only in those situations where combustibles in the waste gas contribute a significant portion of the total energy required for combustion. This contribution should be greater than 50 percent of the total heating value required, for economic reasons. This resitriction is even more important in the light of worldwide demands for natural gas and fuel oil.

The *U.S. Bureau of Mines Bulletin 503* (1952) presents data on the combustion limits for many industrially important substances. Examples of these data are presented in Table 6-9. The combustion limits of other gases and vapors appear elsewhere in the literature [21, 22].

One of the problems of direct flame incineration is that flame temperatures in the range of 2500°F are possible. This results in the formation of oxides of nitrogen when sufficient excess air is available and the high gas temperature is maintained for a sufficient time. Thus the combustion process might simply replace one type of pollutant with another. The formation of oxides of nitrogen in combustion processes is discussed in detail in Chapter 8.

A special example of direct flame incineration frequently used in petrochemical plants and refineries is the flare. This is simply an open-ended combustor, usually aimed vertically upward. The flare is used primarily for

Table 6-9 COMBUSTION LIMITS OF SELECTED GASES AND VAPORS IN AIR MIXTURES (IN PERCENT BY VOLUME)

	LOWER	UPPER
Hydrogen	4.	75
Hydrogen sulfide	4.3	45
Carbon monoxide	12.5	74
Natural gas	4.8	13.5
Gasoline	1.4	7.6
Ethyl ether	1.9	48
Ethyl mercaptan	2.8	18

waste gaseous fuels which cannot be disposed of conveniently by other means. It usually requires a pilot to assure continuous burning. Such a combustor must be designed so that the flame will be sustained at various gas flow rates, wind speeds, and fuel compositions. One of the basic observable problems with flares is their occasional smoky or sooty appearance. They burn smokeless only if the hydrogen-carbon ratio, by weight, is above roughly 1 : 3. For example, a methane (CH_4) flare with a hydrogen-carbon ratio of 1 : 3 burns smokeless. On the other hand, acetylene (C_2H_2) with a ratio of 1 : 12 burns sooty. This difficulty can be overcome by steam injection into the flame zone, which effectively increases the turbulence in the burning region and provides better contact of oxygen with the carbon and fuel in the zone. Steam injection also reduces the length of the flame, which helps stabilize the flame in a brisk wind.

6-12-B Thermal Incineration

When the concentration of combustible pollutants is quite low—so low that it would provide only a small fraction of the heating value required to maintain a direct flame—other means of converting the combustibles must be used. One of these methods is thermal incineration, which is often used when the heating value of the waste gas is in the range of 1 to 20 Btu/scf (40 to 750 kJ/m^3). The waste-gas stream typically is preheated in a heat exchanger and then passed through the combustion zone of a burner supplied with supplemental fuel. As a result, the combustibles in the waste-gas stream are brought above their autoignition temperatures and burn with oxygen usually present in the contaminated stream. If sufficient oxygen is not normally available, it is added to the waste-gas flow by means of a blower or fan. A major advantage of the process, shown schematically in Figure 6-21, is that thermal incineration is carried out characteristically with temperatures in the range of 1000° to 1500°F. This temperature range makes design of the combustion chamber less costly, and alleviates the possibility of appreciable formation of the oxides of nitrogen.

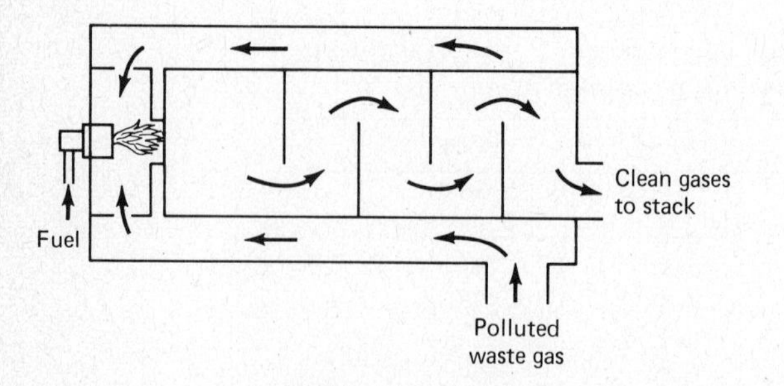

Figure 6-21 Schematic of a thermal incinerator.

An extremely important consideration in the design of thermal incinerators are the "three T's" of combustion, namely, time, temperature, and turbulence. The residence time in the incinerator must be sufficient to permit complete combustion of the combustible material. Normally this is in the range of 0.2 to 0.8 s, with 0.5 s a reasonable guideline. Turbulence refers to the amount of mechanical mixing required to assure complete contact of oxygen with the fuel, and the combustible pollutants with the products of combustion and heat from the flame. Complete mixing is usually much more important in odor control than in general hydrocarbon control, because the escape of the odorous component in concentrations of only a few parts per billion can frequently be detected. Less residence time is required when proper mixing occurs and a short flame is used. An increase in any or all of these three quantities will enhance the possibility of a reasonably complete removal of the combustible waste products. A decrease in any one of the factors will require a concomitant increase in one or both of the remaining factors in order to maintain the same level of pollutant removal. An increase in turbulence by proper burner-duct design is usually the cheapest and easiest alteration of a combustion system. Increased combustion efficiency through increased residence time is appropriate only to a degree, since a longer time means a larger afterburner. Thus first costs of the device could increase appreciably. Higher temperatures can be attained only through the use of more supplemental fuel, which in turn increases the operating cost. In an area of fuel shortages or dislocations, operating temperatures should be minimized as much as possible.

The operating temperatures used in afterburners vary with the nature of the combustible pollutants in the effluent gas. These pollutants can be either general hydrocarbons, carbon monoxide, odor, or a combination of several of these. The approximate average temperature requirements for these pollutants are given in the tabulation [23]:

	AVERAGE TEMPERATURE RANGE	
	(°F)	(°K)
Hydrocarbon oxidation	950–1400	780–1030
Carbon monoxide oxidation	1250–1450	950–1060
Odor control via oxidation	900–1300	750–980

These ranges apply to the short-flame type of burners; for tunnel-type burners, the specific ranges frequently are 100°F higher for the same oxidation process. It should be noted that carbon monoxide oxidation requires temperatures several hundred degrees Fahrenheit higher than hydrocarbon oxidation. If hydrocarbons are only partially oxidized as a result of poor mixing or other design problems, considerable CO may be formed in the afterburner. If the design temperature is not high enough to permit nearly complete oxidation of the carbon monoxide, it may be necessary to allow for a longer residence time or else change the design temperature. Afterburners which operate at the low end of the preferred temperature range for hydrocarbon oxidation frequently exhibit an increasing CO content through the combustor (see Section 6-13-A, and especially Figure 6-28, for a further discussion of this point). Poor hydrocarbon oxidation can also lead to the formation of intermediate oxidation products such as aldehydes, which are objectional pollutants in themselves, and other highly toxic gases.

The temperature maintained in the combustion section of an afterburner has a significant effect on the degree of oxidation of hydrocarbons. For a properly designed afterburner, the efficiency of hydrocarbon oxidation usually falls within the following ranges:

EFFICIENCY	AVERAGE TEMPERATURE RANGE	
(%)	(°F)	(°K)
75–85	1100–1200	870–925
85–90	1150–1250	900–990
90–100	1200–1400	925–1025

These temperature ranges are useful only as guidelines [23], since a number of other factors strongly influence the oxidation efficiency. Among these are the residence time and the degree of mixing, as might be expected, as well as the type of hydrocarbon in the polluted effluent gas stream, the concentration of the hydrocarbons to be controlled, and the type of burner. The accuracy of the measured temperature in the afterburner must also be taken into account. The efficiency of burning CO is also affected by temperature. Experimental measurements indicate that the following ranges of temperature for CO oxidation are reasonably valid:

EFFICIENCY	TEMPERATURE RANGE, SHORT-FLAME-TYPE BURNER		TEMPERATURE RANGE, TUNNEL-TYPE BURNER	
(%)	(°F)	(°K)	(°F)	(°K)
75–90	1250–1350	990–1000	1300–1400	980–1025
90–99	1300–1450	980–1060	1350–1500	1000–1100

It is also found that the change in temperature for a given afterburner design which alters the CO conversion from a moderately low value (for example, 40 to 50 percent) to 95 percent may be on the order of only 50°F.

Table 6-10 lists processes from which the hydrocarbon content of effluent gases can be controlled successfully by thermal incineration [24]. The list is incomplete, and new applications of thermal incineration for controlling odor, hydrocarbons, and carbon monoxide are continually being discovered. Flow rates from 300 to 30,000 scf/min of effluent gas are common in thermal afterburners. Smaller sizes are inherently more costly.

Table 6-10 USES OF THERMAL OXIDATION FOR THE CONTROL OF HYDROCARBONS IN EFFLUENT GAS STREAMS

Adhesive tape curing	Packing house effluents
Brake lining ovens	Paint baking ovens
Coffee roasters	Plastic curing ovens
Core ovens	Printing presses
Cuppola furnace stacks	Solvent degreasing
Fiberglass curing	Textile driers
Lithographing ovens	Varnish burnoff
Meat smokehouses	Varnish kettles
Metal coating ovens	Wire enameling

SOURCE: D. W. Waid. "Afterburners for Control of Gaseous Hydrocarbons and Odor. *Am. Inst. Chem. Engineers Symposium Series No. 137* **70**, 1974.

Costs obviously increase with increase in operating temperature, since additional strength and weight are required. Also, for a given residence time, the size of the combustion chamber is directly proportional to the absolute temperature in the combustion zone, on the basis of ideal-gas behavior.

In general, combustion calculations for afterburners are made to determine the quantity of additional fuel required to attain the desired exhaust gas temperature, the volume of combustion gases or exhaust gases generated and, based upon the latter value, the size of the combustion chamber required to provide the desired contact time in the afterburner. Such calculations require information on the heating value of the fuel supplied to the afterburner. Table 6-11 lists the lower heating values of some common fuels. Illustrative examples of the calculations for direct incineration appear in the literature [6, 7]. Example 6-8 illustrates the required calculations employing the air table found in the appendix.

Example 6-8

A natural-gas-fired circular afterburner is to be designed to incinerate the contaminants discharged from a meat smokehouse and thus eliminate visible emissions an odor. The maximum rate of discharge from the smokehouse is 1200 scfm of gas at 180°F. (Note that the volume rate at 180°F has been corrected to standard conditions of 60°F and 1 atm.) Assume that the contaminated gas has approximately the properties of air, no heating value is assigned to the contaminants due to their low concentration, and a temperature of 1200°F will eliminate the odors. The natural gas supplied to the

Table 6-11 THE LOWER HEATING VALUES OF SOME COMMON FUELS, Δh_c

FUEL (PHASE)	FORMULA	Δh_c (kJ/kg·mole)	Δh_c (Btu/lb·mole)
Methane (g)	CH_4	802,340	345,200
Ethane (g)	C_2H_6	1,427,900	617,000
Propane (g)	C_3H_8	2,044,000	883,000
Carbon monoxide (g)	CO	283,000	121,800

incinerator is 90 percent methane and 10 percent ethane by volume with a lower heating value of 980 Btu/ft^3. The fuel is supplied with 20 percent excess air, with the fuel entering at 60°F and the combustion air entering at 70°F. Also assume that 10 percent of the energy released by combustion of the fuel appears as a heat loss to the surroundings. Determine (a) the mass flow rates and volume flow rates of the combustion air and fuel, (b) the required throat or burner diameter in feet, (c) the large chamber diameter in feet, and (d) the residence time in the large chamber in seconds.

SOLUTION

(a) Refer to Figure 6-22. An energy balance around the entire incinerator is of the form

$$\dot{Q} + \dot{m}_f \Delta h_c = \dot{m}_e(h_{1200} - h_{60})_e - \dot{m}_w(h_{180} - h_{60})_w - \dot{m}_a(h_{70} - h_{60})_a$$

where Δh_c is the lower heating value of the fuel and $\dot{Q}$ is the rate of heat loss. The subscripts e, w, and a represent the exhaust gas, the inlet waste gas, and the inlet air streams, respectively. Since the pollutant concentration in the waste gas is small in this case, we shall assume that the exhaust gas also has the properties of air. From continuity of low considerations,

$$\dot{m}_e = \dot{m}_w + \dot{m}_a + \dot{m}_f$$

In the preceding two equations there are three unknowns, namely, $\dot{m}_a$, $\dot{m}_f$, and $\dot{m}_e$. The third required equation relates the mass of air required per unit mass of fuel. Theoretically, 1 mole of methane requires 9.52 moles of air for complete combustion, and 1 mole of ethane requires 16.66 moles of air. Therefore, for stoichiometric combustion,

$$\frac{\dot{m}_a}{\dot{m}_f} = 0.9(9.52) + 0.1(16.66) = 10.23$$

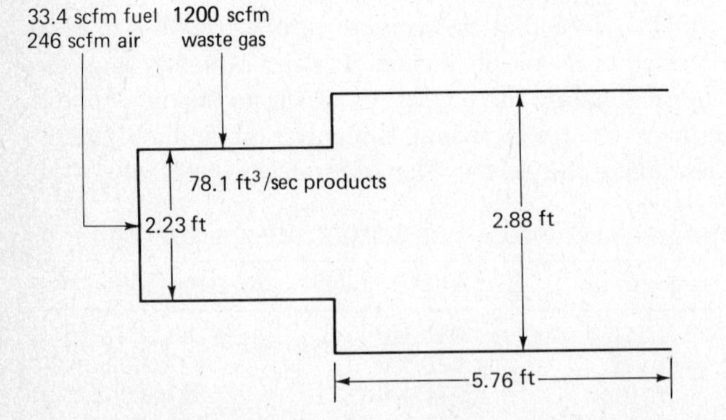

Figure 6-22 Schematic of an afterburner design for Example 6-6.

For 20 percent excess air,

$$\frac{\dot{m}_a}{\dot{m}_f} = 1.2(10.23) = 12.28$$

Before solving the three basic relations simultaneously, we must convert the lower heating value given to a mass basis. The density of the natural gas (which has a molar mass of 17.4) is found from the ideal gas equation.

$$\rho_f = \frac{P}{RT} = \frac{14.7}{520} \times \frac{17.4}{10.73} = 0.0458 \text{ lb/ft}^3$$

Hence the lower heating value becomes

$$\Delta h_c = (980 \text{ Btu/ft}^3) \times (\text{ft}^3/0.0458 \text{ lb}) = 21{,}400 \text{ Btu/lb}$$

In addition, the volume flow rate of waste gas must be converted to a mass basis. The density of air at 60°F and 1 atm is

$$\rho_{\text{air}} = \frac{P}{RT} = \frac{14.7}{520} \times \frac{29}{10.73} = 0.0764 \text{ lb/ft}^3$$

The mass flow rate of contaminated (waste) gas then is

$$\dot{m}_w = 1200 \text{ ft}^3/\text{min} \times 0.0764 \text{ lb/ft}^3 = 91.7 \text{ lb/min}$$

Using the three basic relations and enthalpy data from the air table in the appendix, we find that

$$0.9\left(\frac{\dot{m}_a}{12.28}\right)(21{,}400) = \left(91.7 + \dot{m}_a + \frac{\dot{m}_a}{12.28}\right)(411.8 - 124.3) - 91.7(153.1 - 124.3) - \dot{m}_a(126.7 - 124.3)$$

Thus

$$\dot{m}_a = \frac{23{,}719}{1259.5} = 18.8 \text{ lb/min}$$

and

$$\dot{m}_f = \frac{18.8}{12.28} = 1.53 \text{ lb/min}$$

$$\dot{m}_e = 91.7 + 18.8 + 1.53 = 112.0 \text{ lb/min}$$

In addition,

$$\text{volume rate of fuel} = \frac{1.53}{0.0458} = 33.4 \text{ ft}^3/\text{min}$$

$$\text{volume rate of air} = \frac{18.8}{0.0764} = 246 \text{ ft}^3/\text{min}$$

(b) Gas velocities in afterburner throat regions vary typically from 15 to 30 ft/s to promote mixing of combustion products and contaminated gases. We shall assume velocity of 20 ft/s in this design. The total volume flow rate is required to determine the throat diameter. At 1200°F the density of air is

$$\rho = \frac{P}{RT} = \frac{14.7}{1660} \times \frac{29}{10.73} = 0.0239 \text{ lb/ft}^3$$

Then the volume flow rate of exhaust gas at the exit of the throat is

$$\begin{aligned}\text{volume rate of exhaust gas} &= (112 \text{ lb/min})/(0.0239 \text{ lb/ft}^3)\\ &= 4690 \text{ ft}^3/\text{min} = 78.1 \text{ ft}^3/\text{s}\end{aligned}$$

Evaluation of the throat diameter is now possible.

$$\text{throat diameter} = \left[\frac{4\,(\text{volume rate})}{\pi(\text{velocity})}\right]^{1/2} = \left[\frac{4(78.1)}{\pi(20)}\right]^{1/2} = 2.23 \text{ ft}$$

(c) The combustion chamber diameter is determined by providing adequate residence time with good turbulence levels. If we assume a gas velocity of 12 ft/s in the chamber and a constant temperature of 1200°F, then

$$\text{chamber diameter} = \left[\frac{4(78.1)}{\pi(12)}\right]^{1/2} = 2.88 \text{ ft}$$

(d) If we use a length-to-diameter ratio of 2 for the large combustion chamber, then

$$\text{chamber length} = 2(2.88) = 5.76 \text{ ft}$$

$$\text{residence time} = \frac{5.76 \text{ ft}}{12 \text{ ft/s}} = 0.48 \text{ s}$$

This residence time should be sufficient.

It should be pointed out that the preceding design represents only one of many configurations. For example, a portion of the contaminated gas could be used to provide the theoretical combustion air, as shown in Figure 6-23. A reduction in required fuel gas flow rate might be realized by this arrangement. Detailed design of the gas multiple-port burners and combustion chambers is normally provided by manufacturers of afterburner equipment. References 6 and 7 provide pictures of several different installations.

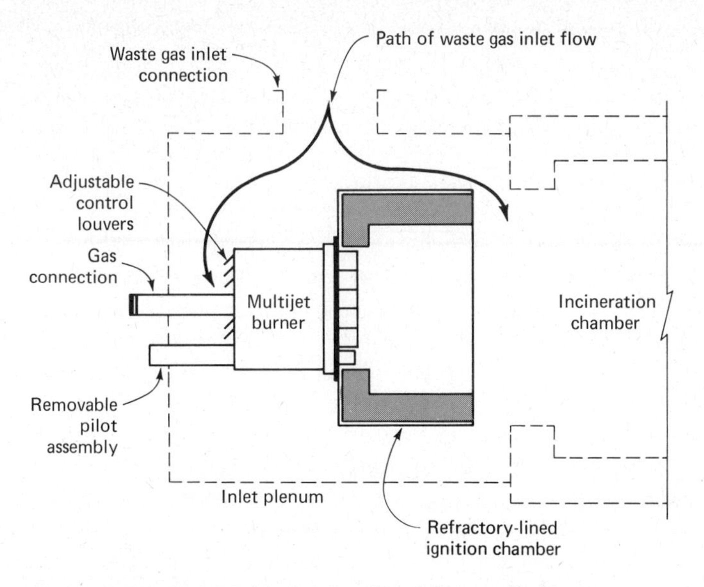

Figure 6-23 Multijet burner for direct flame afterburner.

6-12-C Catalytic Incineration

We have noted that thermal incineration is used in preference to direct flame incineration when the combustible materials in a waste gas are quite low in concentration. A method competitive with thermal afterburning in this situation is catalytic afterburning. A catalyst accelerates the rate of a chemical reaction without undergoing a chemical change itself. As a consequence, the residence times required for catalytic units are much less than those required for thermal units. Instead of the residence or dwell times of 0.3 to 0.9 s typical of thermal afterburners, the time requirements for catalytic action are on the order of a few hundredths of a second. Thermal afterburners may require 20 to 50 times as much residence time as a catalytic unit. In fact, for a catalytic unit we speak of the "space velocity" instead of the residence time. The *space velocity* is defined as the standard cubic feet per hour of gas flow divided by the catalyst volume in cubic feet. The other distinct advantage of catalytic units for an incineration process is that the ignition temperature is lowered. This is really an energy effect. A catalyst lowers the energy requirements of the oxidation process as it takes place on a catalytic site, as opposed to the situation needed for a homogeneous gas-phase reaction. Hence the waste-gas stream does not have to be heated to as high a temperature as in thermal incineration. By passing the waste gases through the catalyst bed the necessary temperature for the exothermic reaction of organic gases and vapors with oxygen may be reduced by as much as 500°F over thermal incineration.

Most waste gases containing combustible pollutants from industrial processes are at a fairly low temperature. Therefore, some type of preheating burner is used to bring the waste gas up to the temperature at which the catalyst will be effective. Temperatures from 600° to 1000°F (590° to 810°K) are typical. This temperature range is below the autoignition temperature for thermal incineration, so the combustibles in the waste gas are merely heated. The combustion reactions occur on the surface of the catalyst. No direct flame structure is seen, but the catalyst surface will glow. The efficiency of catalytic combustion is on the order of 95 to 98 percent, so the effluent gases from the catalytic conversion should be primarily carbon dioxide, water vapor, and nitrogen. A schematic of a catalytic afterburner is shown in Figure 6-24. The blower shown in this configuration will have to be capable of handling the moderate temperatures due to the preheat burner. At these lower temperatures the formation of oxides of nitrogen is not a serious problem, even though the residence time is reasonably long.

Many substances have catalytic properties, but only a few are used for waste-gas treatment. In order to be useful in air pollution control, the substance must be relatively inexpensive, long lasting, able to function at the required temperature, and capable of being formed into a variety of shapes. Successful catalytic beds have been formed into ribbons, rods, beads, pellets, and other shapes. No matter what the specific bed configuration is, it must provide the required catalytic surface area, and maintain uniform density at elevated temperatures to prevent channeling or bypassing. The catalyst matrix should also provide for a moderately low pressure drop and structural integrity and durability. Current practice requires a catalytic surface of

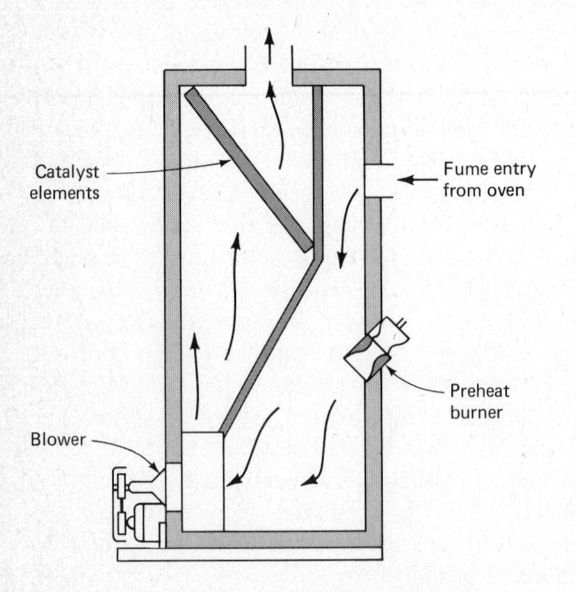

Figure 6-24 A catalytic afterburner design.

approxiamtely 0.2 ft^2/scf/min of waste gases. Some difficult-to-burn materials may require 0.5 ft^2/scf/min. As another basis for design, an 85 to 95 percent conversion of pollutants typically requires from 0.5 to 2.0 ft^3 of catalyst for each 10^6 scf/min of waste gas. Noble metals such as platinum and palladium are used frequently, dispersed on a catalyst support such as alumina. Higher molecular weight hydrocarbons are more easily oxidized than those of lower molecular weight. The catalytic reactivity also varies with molecular structure in the following order: aromatic < branched paraffins < normal paraffins < olefinic < acetylenic.

Catalytic oxidation of gaseous pollutants is attractive because it offers a lower supplemental fuel usage than does thermal afterburning. How much lower depends upon the nature of the gaseous pollutant in the waste-gas stream. The temperature ranges required for three common industrial waste gases are shown in the tabulation [25].

INDUSTRIAL POLLUTANT	CATALYTIC TEMPERATURE RANGE (°F)	(°K)
Solvents	500–850	530–730
Vegetable and animal fats	500–700	530–640
Chemical process exhausts	400–750	480–670

The solvents include toluol, methyl ethyl ketone, xylol, alcohols, and others. Typical examples of contaminants from chemical processes are carbon monoxide, ethylene, ethylene oxide, and propylene.

Catalytic fume incineration is widely used for processes involving paint and enamel bake ovens, varnish kettles, air blowing of asphalt, and phthalic anhydride manufacture. The pressure drop through a catalytic incinerator is usually very low, on the order of several tenths of an inch of water. Thus the operating costs are normally low with the exception of the cost of maintaining the catalyst. The maintenance cost of the catalyst depends upon the nature of the fume. In some cases the waste gas must be cleaned of particulate before it enters the incinerator. Deposition of particulate on the surface of the catalyst bed decreases the available surface area for catalytic action. This lowers the effectiveness of the bed as well as its lifetime. The normal operating life of a catalyst without a particulate deposition problem may be from 3 to 5 yr.

One other problem associated with catalytic incineration is poisoning of the bed by specific contaminants in the waste gas. Materials such as iron, lead, silicon, and phosphorus shorten the life of many catalysts. Sulfur compounds also suppress the effectiveness of some catalysts. Poisoning of the catalyst is a major problem in the search for an effective catalyst for emissions from mobile sources. The application of catalytic units to vehicles is considered in Chapter 10. Finally, it must be recognized that most catalysts have an upper limit in their operating temperature range, which for many is around 1500° to 1600°F (1100°K). As a result, restricting the temperature rise of the waste gas as it passes across the catalyst to 100° to 200°F is usually desirable. If temperature is not restricted, a sudden surge of combustible contaminant above the normal amount might raise the gas temperature sufficiently to cause the catalyst to burn out.

6-12-D Afterburning with Recuperation

Whichever method of waste-gas or fume incineration is used, another important consideration must be noted. Depending on whether catalytic, thermal or direct flame incineration is employed, the temperature of the gases leaving the incinerator may vary from 700° to 2000°F, or more. Thus considerable energy at a relatively high temperature is associated with the waste-gas stream. In addition, owing to the incineration process the gas stream contains essentially only inert gases, since any toxic or corrosive species have been eliminated in the combustion process. Consequently, a clean gas source of high-grade energy is available for use in other processes within the plant.

As noted above, the flue gas temperature from direct flame incineration is typically around 2500°F (1650°K). This high-grade energy is quite useful as a heat source for a waste heat boiler, especially if the volume rate of gas is reasonably high. Other uses must be found for the gases from thermal or catalytic incineration, since the temperature in these cases will be considerably lower (600° to 1500°F or 600° to 1100°K). One of the most obvious and practical uses of the waste-gas stream in this temperature range is to preheat the contaminated gas entering the incinerator. The greater the heating of the contaminated gas toward its autoignition or catalytic ignition temperature, the less the need for supplemental fuel. The heat exchanger required for this operation is called a recuperator or regenerator. A schematic of the overall process is shown in Figure 6-25. One measure of the efficiency of a recuperator is the recuperator effectiveness, defined as

$$\eta_{\text{eff}} = \frac{T_B - T_A}{T_C - T_A} \tag{6-53}$$

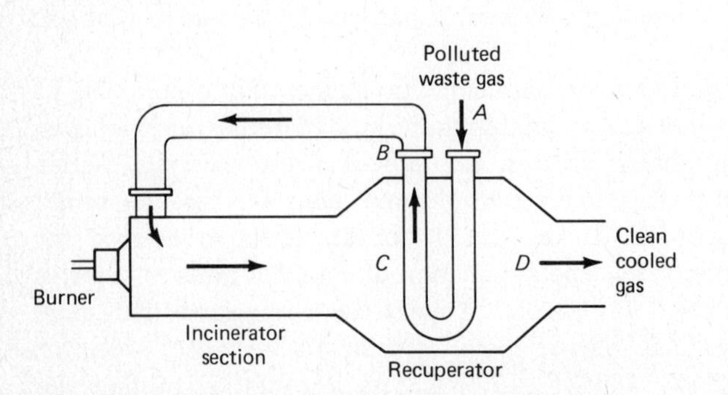

Figure 6-25 Thermal incinerator with recuperator.

where the positions of *A*, *B*, *C*, and *D* are indicated in Figure 6-25. Qualitatively, this is the ratio of the actual temperature rise of the dirty gas to the maximum possible temperature rise based on the desired incineration temperature. A recuperator effectiveness of at least 0.6 to 0.7 (60 to 70 percent) is practical. It may be shown that this is equivalent to a preheat temperature T_B which is at least the average of the incineration temperature T_C and the outlet temperature T_D. As a consequence, a supplemental fuel requirement of only 15 to 35 percent of that required without recuperation may result.

The use of a recuperator to reduce fuel requirements for a thermal or catalytic incinerator is not without economic cost. The reduction of fuel costs must be weighed against an appreciable capital investment. Since the heat transfer coefficients for a process involving transfer of heat from gas phase to gas phase are low, the heat exchanger will need to be relatively large to accommodate the desired amount of heat transfer. The initial cost of the incinerator may, in fact, be almost doubled by the addition of recuperation. Since the clean-gas stream leaving the recuperator may still have a reasonably high temperature, it may be used further as a preheat source for other operations, as shown in Figure 6-26(a). It is not necessary, of course,

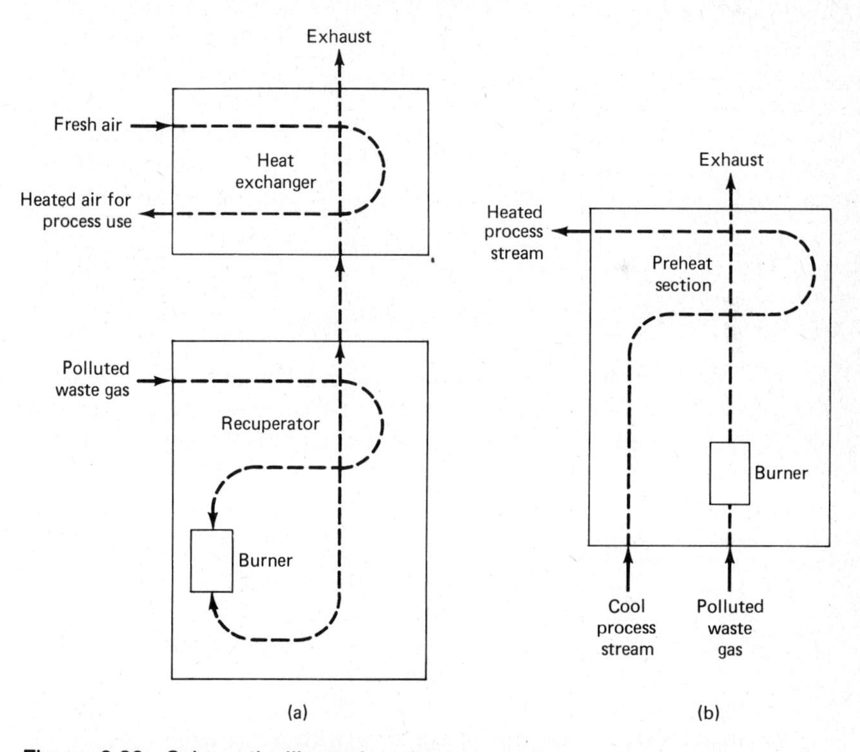

Figure 6-26 Schematic illustrating the use of hot waste gases from an afterburner in auxiliary equipment. (a) Equipment for further use of waste gas leaving the recuperator in preheating another flow stream. (b) Equipment for using hot gases leaving an afterburner to preheat another flow stream.

for the hot gases leaving the incinerator to be used in a recuperator. The available energy might be used directly, as shown in Figure 6-26(b), to preheat other gas or liquid streams in the plant.

6-13 REACTION KINETICS AND CATALYSIS IN AFTERBURNING PROCESSES

6-13-A Some Experimental Reaction Kinetics for Afterburning

As in any oxidation process, the "three T's" of combustion—time, temperature, and turbulence (mixing)—are extremely important in achieving the desired level of hydrocarbon reduction in afterburners. Reaction kinetics determines the temperature level and the time of residence necessary for oxidation of gaseous pollutants before the waste gas is discharged to the environment. In turn, the residence time and temperature directly influence the required physical size of the afterburner and the auxiliary fuel consumption needed, which are the two most important design variables for an afterburner. Hence kinetic data for the oxidation of hydrocarbons and carbon monoxide are requisite information for design purposes.

Experimental data for the oxidation of toluene as a function of temperature and time for a laboratory-scale (5-in. inside diameter) reaction chamber are shown in Figure 6-27 [26]. Note that the concentration scale is logarithmic. The influence of temperature on the oxidation process is apparent. For temperatures above 1450°F (1050°K) essentially complete oxidation of toluene occurs in 0.10 s. (The temperature for these experiments was measured at the outlet of the reaction chamber.) Similar data for methane, hexane, and cyclohexane indicate the same trends for the same range of time and temperature. All the data indicate that only relatively short time periods are required in order to achieve 90 percent or greater hydrocarbon reductions.

On the basis of the experimental data [26] of Figure 6-27 and tests of other hydrocarbons, it has been found that a first-order rate equation is applicable with respect to the hydrocarbon concentration. Tests showed the oxygen concentration to be high enough that the value of $[O_2]$ can be considered constant. The data are fitted by an equation of the type

$$-\frac{d[\mathrm{HC}]}{dt} = k[\mathrm{HC}] = k_o[\mathrm{HC}]\exp\left(\frac{-E}{RT}\right) \tag{6-54}$$

where k_o is the preexponential rate constant and E is the activation energy for the reaction. Table 6-12 lists values of k_o and E derived from the experimental results for four common hydrocarbons and for carbon monoxide oxidation. It was necessary to fit the data with a 1.5 order on the CO concentration. In addition, the experimental data for CO were fitted with separate sets of k_o and E values for temperatures below and above 1400°F

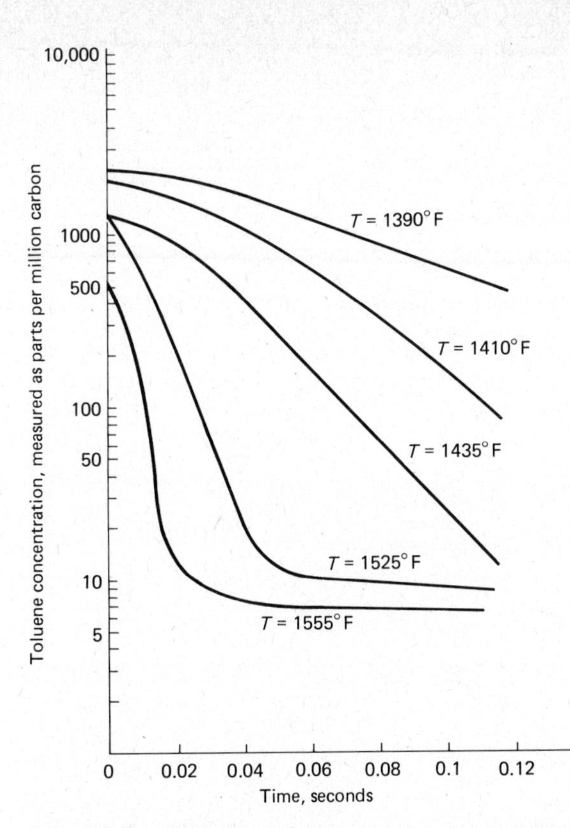

Figure 6-27 Toluene decrease as a function of time and temperature. (SOURCE: K. H. Hemsath and P. E. Susey. *Am. Inst. Chem. Engineers Symposium Series No. 137* **70**, 1974.)

Table 6-12 EXPERIMENTAL RATE DATA FOR REACTION (6-54)

SUBSTANCE	ACTIVATION ENERGY, E (kJ/g·mole)	RATE CONSTANT, k_o
Toluene	245	6.56×10^{13}/s
Hexane	220	4.51×10^{12}/s
Cyclohexane	199	5.13×10^{12}/s
Natural gas	206	1.65×10^{12}/s
Carbon monoxide (> 1400°K)	167	2.5×10^{8}/s/ppm$^{0.5}$
Carbon monoxide (< 1400°K)	418	$\sim 10^{20}$/s/ppm$^{0.5}$

SOURCE: K. H. Hemsath and P. E. Susey. "Fume Incineration Kinetics and Its Application." *Am. Inst. Chem. Engineers Symposium Series No.* 137 **70**, 1974.

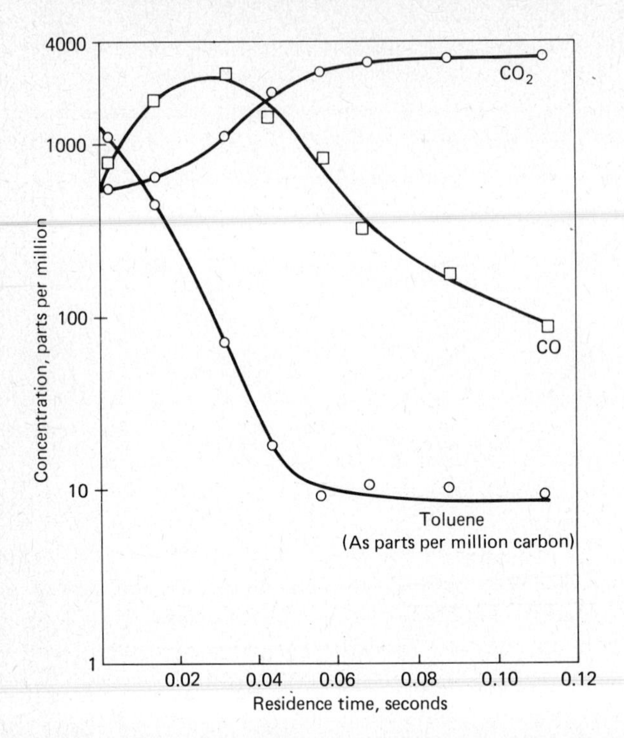

Figure 6-28 Toluene–CO–CO_2 concentrations during incineration at 1525°F. (SOURCE: K. H. Hemsath and P. E. Susey. *Am. Inst. Chem. Engineers Symposium Series No. 137* **70**, 1974.)

(1030°K). An interesting result of these experiments was the difficulty in oxidizing CO compared with the hydrocarbons. Figure 6-28 shows the concentration-time profile for toluene at a temperature of 1525°F (1100°K). Also shown are the profiles of CO and CO_2, which indicate the appearance and subsequent oxidation of CO during the toluene oxidation period. Apparently CO is more difficult to oxidize than hydrocarbons under the same temperature-time conditions, since the other three hydrocarbons showed behavior similar to that indicated in Figure 6-28. Higher temperatures and longer residence time may be required to incinerate CO, compared to hydrocarbons. Hence a situation involving only hydrocarbons may call for an incinerator design different from the design needed when CO is also present.

6-14-B Some Basic Theory on Catalytic Afterburners

Although basic theory on catalytic process pertaining to afterburner design is still under development, several worthwhile points can be considered here. In general, the overall process may be divided into two distinct steps which govern catalytic oxidation: (1) the heat and mass transport processes between the catalytic surface and the gas stream; and (2) the elementary chemical reaction processes at the catalytic sites on the solid surface. At the present stage of development, the theory indicates that for normal design rates for afterburner catalysts, the reaction rate is controlled or impeded by mass diffusion of reactants through the porous substrate of the catalyst. Hence only a fraction of the effective catalyst surface is used to catalyze the oxidation reactions [27]. Generally speaking, the overall rate constant k is related to rate constants for the mass transfer (k_{mt}) and the chemical reaction (k_{chem}) processes by

$$\frac{1}{k} = \frac{1}{k_{\text{mt}}} + \frac{1}{k_{\text{chem}}}$$

As noted earlier in this section, the rate of oxidation of a hydrocarbon is proportional to the concentration of that hydrocarbon in the gas stream, provided at least 2 percent oxygen by volume is present in excess of that required for complete oxidation.

It is found that at reasonably low temperatures, for example, 300° to 500°F (425° to 525°K), the quantity k_{chem} controls. However, as the temperature is increased in order to increase the chemical rate, the quantity k_{mt} controls. This occurs in the temperature range of 600° to 900°F (600° to 750°K). The role of chemical reaction rates versus mass transfer processes and the selectivity of a catalyst is shown by Figure 6-29 [27]. The conversion-temperature curves are shown for various molecular species and solvent hydrocarbons in general for oxidation over a Pt/Al_2O_3 catalyst. This type of catalyst has a high selectivity for hydrogen, with reasonable conversion even below 200°F. High conversions occurs in the 300° to 500°F range. The

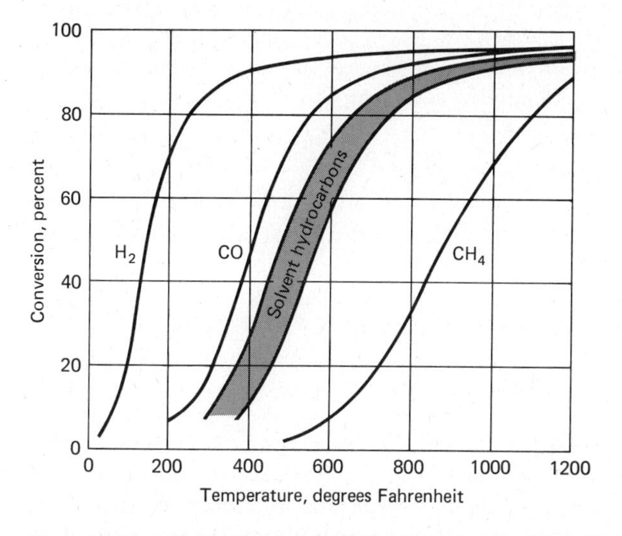

Figure 6-29 Typical temperature-conversion performance curves for various molecular species being oxidized over Pt/Al_2O_3 catalysts. (SOURCE: R. D. Hawthron. *Am. Inst. Chem. Engineers Symposium Series No. 137* **70**, 1974.)

performance curves for most solvent hydrocarbons lie within a narrow band, with significant activity around 400° to 600°F. High conversion requires, however, temperatures in the 750° to 1000°F range. The lower portion of the performance curves, where the conversion percent rises sharply with temperature, is the region where chemical rate processes control. Along the flatter upper portions the mass transfer rate controls. Note that methane is the least reactive to this catalyst system, with high conversions requiring temperatures above 1100°F. Catalysts used in afterburners are usually selected so that they promote oxidation for a wide variety of gaseous, combustible substances. Nevertheless, as indicated by the curves in Figure 6-29, the specific activity of catalysts does vary with the type of molecule to be oxidized. Consequently, the degree of oxidation of various hydrocarbons in a waste stream will differ for a given time-temperature region in the afterburner.

QUESTIONS

1. How is the overall order of a reaction defined?
2. What type of mathematical function relates C_A to $C_A{}^o$ for a first-order reaction?
3. What is meant by a rate-determining step with reference to consecutive reactions?
4. Why is the concept of a rate-determining step important with respect to reactions which include pollutant species in the overall reaction mechanism?
5. Why is the activation energy an important parameter in the study of rates of formation (or depletion) of pollutant species?
6. What intermediate radical species is thought to influence both the formation of CO and the conversion of CO to CO_2 in hydrocarbon oxidation reactions?
7. What are the names of and the differences between the two types of adsorption processes? What is the advantage of one over the other as an air pollutant removal technique?
8. What characteristics of adsorbents are important in air pollution control?
9. What design requirements for adsorption equipment must commonly be met?
10. Why is an adsorption isotherm which is concave upward throughout favorable to adsorption but unfavorable to desorption? How can desorption be made more favorable when it becomes necessary to clean the adsorption medium?
11. What is meant by an adsorption wave? What shape of an adsorption wave is more favorable from a design standpoint?
12. How does one graphically determine the minimum liquid-gas ratio for a packed adsorption tower?
13. What physically is meant by the flood point in a packed absorption tower?
14. Discuss briefly the concept of the height and number of transfer units associated with a packed tower.
15. Distinguish between direct flame and thermal incineration.
16. Compare thermal versus catalytic incineration.

PROBLEMS

6-1. From Figure 6-4(b), determine (a) the values of V_m and c for carbon dioxide at $-78°C$. Then compute the value of V/V_m for the adsorption of CO_2 on silica gel when the CO_2 content of a gas stream at $-78°C$ is (b) 7.5 percent by volume with a total gas pressure of 1.0 bar and (c) 11.4 percent by volume with a total gas pressure of 1.20 bars. The vapor pressure of pure carbon dioxide at this temperature is 1.036 bars.

6-2. From the B.E.T. plot of Figure 6-4(b), estimate the value of V_m for carbon monoxide adsorbed at $-183°C$ on silica gel, in cubic centimeters per gram.

6-3. An adsorption bed 0.80 in length and 1.4 m^2 in cross-sectional area operates at 40°C and with an inlet air stream pressure of 1.2 bars. The mass flow rate of air is 4 kg/min and the concentration of the pollutant in the inlet air stream is 0.0030 kg/m^3. The values of α and β in the Freundlich-type equation for equilibrium concentrations C_e and X_{sat} are 80 kg/m^3 and 1.8, respectively. The bulk density of the adsorbent is 400 kg/m^3, and the mass transfer coefficient K has a value of 25 s^{-1}. On the basis of these data: (a) determine the thickness of the adsorption wave, in meters; (b) determine the velocity of the adsorption wave, in meters per second; and (c) determine the time of breakthrough, in seconds. (d) Prepare a plot of $C_g/C_{g,\text{in}}$ versus $(x - x_1)$. Use $C_g/C_{g,\text{in}}$ values of 0.01, 0.02, 0.05, 0.10, 0.30, 0.50, 0.80, 0.90, 0.95, 0.98, and 0.99.

6-4. The adsorption bed in Problem 6-3 is to be regenerated with saturated steam vapor at 180°C. The flow rate of steam is 10.0 kg/min, and the values of K, α, and β for the process are 120 s^{-1}, 1200 kg/m^3, and 2.2, respectively. Estimate the time for regeneration, in hours.

6-5. An adsorption bed 0.20 m in length and 0.70 m^2 in cross-sectional area operates at 30°C. The inlet air stream pressure is 1.50 bars, and the mass flow rate is 1.50 kg/min. The concentration of the pollutant in the inlet air stream is 2.60×10^{-3} kg/m^3, and the values of α and β in the Fruendlich-type equation for equilibrium concentrations are 4.23 kg/m^3 and 1.5, respectively. The bulk density of the adsorbent is 650 kg/m^3, and the mass transfer coefficient K has a value of 18 s^{-1}. On the basis of these data determine the answers to the same questions asked in Problem 6-3.

6-6. The adsorption bed in Problem 6-5 is to be regenerated with saturated steam vapor at 200°C with a flow rate of 5.0 kg/min. The values of K, α, and β for the process are 110 s^{-1}, 600 kg/m^3, and 2.0, respectively. Estimate the time for regeneration, in hours.

6-7. Reconsider Example 6-1. Determine the time of breakthrough in hours if the integration limits on n are set at 0.001 and 0.999.

6-8. Reconsider Example 6-1. Consider that β is increased from 2.2 to 2.4, and α remains the same. Determine the percent change in time for breakthrough under this new condition.

6-9. With reference to Problem 6-3, determine the percent change in the adsorption velocity if (a) β is changed from 1.8 to 2.2 while α remains constant and (b) α is changed from 80 to 40 kg/m^3 with β remaining the same.

6-10. An adsorption bed is 0.60 m thick and 2.0 m^2 in cross-sectional area. The air mass flow rate is 0.30 kg/s at 1.1 bars and 40°C. The gas phase mass transfer coefficient K is 25 s^{-1} and the values of α and β are 125 kg/m^3 and 2.40, respectively. The actual packed bed density is 400 kg/m^3 and the inlet pollutant concentration is 0.00240 kg/m^3. On the basis of these data, determine the answers to the same questions asked in Problem 6-3.

6-11. An adsorption unit packed with Ambersorb(TM) XE-347 is used to remove vinyl chloride (molar mass of 62.5) from an air stream at 26°C and 1 atm. The mass flow rate of air is 1.9 kg/s and the inlet vinyl chloride concentration is 1000 ppm. The adsorption bed has an area of 5.0 m^2 and is 0.35 m thick. Determine the time of breakthrough in hours based upon the following data: $\alpha = 4.16$, $\beta = 2.95$, $K = 30\ s^{-1}$, and the apparent density of XE-347 is 700 kg/m^3.

6-12. Activated carbon BPL is used to pack the adsorption unit described in Problem 6-11. The operating conditions are the same, but the following data are applicable for activated carbon BPL: $\alpha = 2.62\ kg/m^3$, $\beta = 2.15$, $K = 40\ s^{-1}$, and the bed density is 384 kg/m^3. Determine the time for breakthrough.

6-13. Activated carbon PCB is used to pack the adsorption unit described in Problem 6-11. Employing the same operating conditions as specified in that problem, determine the time of breakthrough by employing the following data for activated carbon PCB: $\alpha = 2.04\ kg/m^3$, $\beta = 2.40$, $K = 40\ s^{-1}$, and the adsorbent bed density is 384 kg/m^3.

6-14. The partial pressure of acetone in a contaminated air stream at 1 atm and 100°C is 100 mm Hg. An adsorption bed of activated carbon is used to remove the acetone. The bed is 3.5 m thick and has a flow area of 7 m^2. The gas flow rate is 1.5 kg/s. Determine the time of breakthrough for this bed based on the following data: $\alpha = 17.3\ kg/m^3$, $\beta = 2.23$, and $K = 40\ s^{-1}$. The apparent density of the adsorbent is 380 kg/m^3.

6-15. The adsorption bed in Problem 6-14 is to be regenerated with saturated steam vapor at 200°C. The flow rate of steam is 2 kg/s, and the values of K, α, and β for the process are 100 s^{-1}, 120 kg/m^3, and 2.4, respectively. Estimate the time for regeneration, in hours.

6-16. A tower packed with $\frac{3}{8}$ in. ceramic Raschig rings is to treat 20,000 ft^3/hr of entering gas. The air entering the tower contains 2.4 percent ammonia by volume, and ammonia-free water is to be used as the absorbent. The tower conditions are 68°F and 1 atm. The ratio of inlet gas flow to inlet liquid flow is (a) 1.10, (b) 1.35, and (c) 1.60 lb of gas per pound of liquid. If the gas flow rate is one-half the flooding rate and 95 percent of the ammonia is to be removed from the inlet gas stream, estimate the required diameter of the tower, in feet, and the pressure drop in inches of water per foot on the basis of Figure 6-16.

6-17. A tower packed with 1-in. ceramic berl saddles is used to treat 30,000 ft^3/hr of entering gas. The air entering the tower contains 1.8 percent ammonia by volume, and ammonia-free water is used as the absorbent. The tower conditions are 68°F and 1 atm. The ratio of inlet gas flow to inlet liquid flow is (a) 1.0, (b) 1.3, and (c) 1.6 lb of gas per pound of liquid. If the gas flow rate is to be 60 percent of the flooding rate, estimate the required diameter of the tower, in feet, and the pressure drop in inches of water per foot of tower height on the basis of Figure 6-16. The ammonia absorbed is to be 96 percent of the initial quantity.

6-18. With reference to the data of Problem 6-16, for a ratio of inlet gas flow to inlet liquid flow of (a) 1.10, (b) 1.35, and (c) 1.60 lb of gas per pound of liquid, determine (1) the actual ratio, L_S/G_C, on a molar basis, (2) the minimum required ratio of solvent to carrier gas on a molar basis, (3) how much greater, in percent, the design liquid flow rate is in comparison with the minimum liquid flow rate, and (4) how many overall gas-phase transfer units are required. The equilibrium data for the solubility of ammonia in water may be expressed as $C = 1.267y$, where C is the pounds of NH_3 per pound of water and y is the mole fraction of ammonia in the gas phase.

6-19. With reference to the data of Problem 6-17, for a ratio of inlet gas flow to inlet liquid flow of (a) 1.00, (b) 1.30, and (c) 1.60 lb of gas per pound of liquid, determine (1) the actual ratio, L_S/G_C, on a molar basis, (2) the minimum required ratio of solvent to carrier gas on a molar basis, (3) how much greater, in percent, the design liquid flow rate is in comparison with the minimum liquid flow rate, and (4) how many overall gas-phase transfer units are required. The equilibrium data for the solubility of ammonia in water may be expressed as $C = 1.267y$, where C is the pounds of ammonia per pound of water and y is the mole fraction of ammonia in the gas phase.

6-20. Reconsider the data of Problems 6-16 and 6-18. For parts (a), (b), or (c), if the packing is $\frac{3}{8}$-in. Raschig rings, determine (1) the value of H_{tG}, in feet, (2) the value of H_{tL}, in feet, (3) the value of H_{tOG}, in feet and (4) the height of the absorption column, in feet.

6-21. Reconsider the data of Problems 6-17 and 6-19. For parts (a), (b), or (c), if the packing is 1-in. berl saddles and the temperature is 77°F, determine (1) the value of H_{tG}, in feet, (2) the value of H_{tL}, in feet, (3) the value of H_{tOG}, in feet, and (4) the height of the absorption column, in feet.

6-22. An absorber is to recover 97 percent of the ammonia in an air-ammonia stream fed to it, using pure water as the inlet absorbing medium. The inlet ammonia content of the air is 24 percent by volume, and the absorber is maintained at 30°C and 1 atm. (a) What is the minimum required ratio of solvent to carrier gas on a molar basis, L_S/G_C? (b) For a water rate which is 50 percent greater than the minimum, how many overall gas-phase transfer units are required? The equilibrium data in terms of C in pounds of ammonia per hundred pounds water and p in millimeters of mercury partial pressure of ammonia at 30°C are:

C_A	1.0	1.5	3.5	5.0	9.0	19.0	30.0
p_A	10	15	35	50	100	250	450

6-23. With reference to Problem 6-22, let the inlet ammonia content be 18 percent and the actual water rate be 40 percent greater than the minimum. Solve for the same two required answers.

6-24. An effluent air stream contains 5.4 percent SO_2 by volume. It is desired to reduce this concentration to 1000 ppm by passing the gas stream through a packed tower before releasing the flow to the atmosphere. The inlet-gas flow rate is 1000 ft^3/min measured at 1 atm and 68°F, and the absorber is counterflow. The equilibrium of SO_2 in water is assumed to be reasonably represented by the relation $y = 30x$. Estimate the minimum amount of water required as a solvent, in cubic feet per minute.

6-25. An effluent air stream from a process contains 4.8 percent SO_2 by volume. It is desired to reduce the concentration to 500 ppm before emitting the gas stream to the atmosphere. This is to be done by passing the gas stream counterflow through a packed tower with water as the absorbing fluid. The inlet-gas flow rate is 100 m^3/min measured at 1 bar and 25°C. The equilibrium relationship of SO_2 in water is assumed to be reasonably represented by the expression $y = 30x$. Estimate the minimum water rate required, in cubic meters per minute.

6-26. A waste gas stream contains 0.045 mole of ammonia per mole of carrier gas. It is desired to design an absorber using water which will reduce the ammonia concentration in the exit gas from the absorber to a maximum value of 600 ppm. The inlet ammonia flow is 2000 ft^3/min when measured at 1 atm and 77°F. The system is to be counterflow, and the operating line slope is to be (a) 33.3 percent and (b) 66.6 percent greater than the slope of the equilibrium line. Assume that the equilibrium line is represented by the relation $y = 0.75x$. Determine (1) the water rate required, in cubic feet per minute, and (2) the number of theoretical transfer units for the packed column.

6-27. A packed absorption tower is to be used to remove methanol (methyl alcohol) from an air stream which contains 10 percent methanol vapor by volume at 104°F and 1 atm. The gas flow rate is 140 lb·moles/hr, and the tower is to remove 96 percent of the methanol. The solvent is water, and it is not to be recirculated. The packing material is 1-in. berl saddles, and the tower is to operate at 50 percent of the flood point under isothermal conditions. The water-methanol equilibrium data at 104°F and 1 atm are as follows:

MOLES METHANOL/MOLE AIR	MOLES METHANOL/MOLE WATER
0.024	0.020
0.046	0.040
0.076	0.070
0.102	0.100
0.128	0.140

The average slope of the equilibrium line, m, may be taken to be 1.125, and neglect the volatility of water. Determine (a) the required water flow rate, in pounds per hour, which is to be 40 percent greater than the minimum for the tower; (b) the required tower diameter, in inches; (c) the tower pressure drop, in inches of water, by two methods; (d) the number of transfer units; and (e) the required tower height, in feet.

6-28. The activation energies for reactions A and B are 40,000 J/g·mole and 200,000 J/g·mole, respectively. Determine the ratio of the rate constant k at 2000°K to k at 1200°K for these two reactions, if $k = A\exp(-E/RT)$.

6-29. The activation energy for the reaction $H + H_2O \rightarrow OH + H_2$ is approximately 85 kJ/g·mole. Show that the rate constant k decreases roughly (a) by a factor of 30 when the temperature drops from 1500° to 1000°K, and (b) by a factor of 390 when the temperature drops from 1500° to 800°K, if $k = A\exp(-E/RT)$.

6-30. Consider that the following two reactions control the disappearance of carbon monoxide in lean flames:

1. $CO + OH \rightleftharpoons CO_2 + H$
2. $H + H_2O \rightleftharpoons OH + H_2$

The forward rate constants for reactions (1) and (2) are:

$$k_{f,1} = 5.6 \times 10^{11} \exp(-545/T)$$
$$k_{f,2} = 8.4 \times 10^{13} \exp(-10{,}100/T)$$

where T is expressed in degrees Kelvin. (a) Determine the ratio of $k_{f,1}$ at 2000°K to that at 1200°K. (b) Determine the ratio of $k_{f,2}$ at 2000°K to that at 1200°K. (c) Determine the ratio $k_{f,2}/k_{f,1}$ at 2000°K and 1200°K.

6-31. For the data of Problem 6-30 determine the same answers except use temperatures of 1800 and 1000°K, rather than 2000 and 1200°K.

6-32. The following data represent some experimental results for the rate constant for the hydrogenation of ethylene over restricted pressure and composition ranges. The units of k are g moles/(s)(atm)(cm^3).

$k \times 10^5$	T(°C)	$k \times 10^5$	T(°C)
2.70	77.	3.03	79.5
0.71	53.3	1.31	64.0
2.40	77.6	0.70	54.5
0.70	52.9	0.323	40.2
0.69	53.7	0.312	39.9

Prepare a plot of $\ln k$ versus $1/T$, where T is in degrees Kelvin. From the plot estimate a value of the activation energy, in kJ/kg·mole.

6-33. The decomposition of NO_2 to NO and O_2 is second order with a rate constant varying with temperature as follows:

T, °K	592	603	627	652	656
k, cm^3/(g·mole)(s)	522	755	1700	4020	5030

(a) Preparc a plot of $\ln k$ versus $1/T$ and on the basis of the plot estimate the value of the activation energy E, in kJ/kg·mole. (b) Using the result of part (a) and the mid-data point above, estimate the value of the pre-exponential factor, in cm^3/(g·mole)(s).

6-34. For the decomposition of N_2O_5, k equals 0.079×10^{-5} s^{-1} at 0°C and 1.29×10^{-4} s^{-1} at 35°C. Estimate (a) the Arrhenius activation energy and (b) the pre-exponential factor for the rate constant.

6-35. For the reaction, $H_2 + I_2 \rightarrow 2HI$, the value of k with respect to HI is 4.45×10^{-5} l/(g·mole)(s) at 283°C and 2.52×10^{-6} l/(g·mole)(s) at 356°C. Determine values of (a) the activation energy in kJ/kg·mole, and (b) the pre-exponential factor in cm^3/(g·mole)(s).

6-36. Rework Example 6-6 with the temperature set at 2500°K rather than 2000°K.

6-37. Rework Example 6-6 with the same data, except with the pressure at 0.10 atm rather than 1.0 atm.

6-38. Rework Example 6-6 with the same data, except with the initial nitrogen content at 5.0 moles rather than 2.05 moles.

6-39. The effluent from a drying oven at 475°K contains 75 ppm carbon monoxide and 1750 ppm hydrocarbons. After the effluent is heated to a temperature of 960°K, the conversion efficiency based solely on hydrocarbons is 98 percent; based on CO and HC it is 90 percent. (a) What is the CO content at the afterburner outlet, in parts per million? (b) What is the CO content in parts per million at the outlet if the CO–HC efficiency is 92 percent at 1000°K?

6-40. The effluent at 470° from a coil coater oven contains 3000 ppm total hydrocarbons, 390 ppm CO, and 320 ppm methane. After being heated in an afterburner to 1230°F, the effluent gas contains 185 ppm total HC, 790 ppm CO, and 140 ppm methane. Determine the conversion efficiency (a) based on total HCs alone, (b) based on total HCs and CO, and (c) based on reactive HCs (excluding methane) and CO.

6-41. A wire enameling oven effluent at 480°K contains 1960 ppm hydrocarbons and 75 ppm CO. After the effluent is heated in an afterburner to 920°K, the concentrations of HC and CO are 75 ppm and 350 ppm, respectively. (a) What is the conversion efficiency based on HCs alone? (b) What is the conversion efficiency based on the Los Angeles County Rule 66?

6-42. A paint-drying oven effluent at 640°K contains 750 ppm hydrocarbons and 25 ppm CO. (a) If the outlet stream from an afterburner at 940°K contains 20 ppm HC and 30 ppm CO, what is the percentage of pollutant conversion in terms of Los Angeles County Rule 66? (b) If the CO content increased to 85 ppm, what would be the percentage of conversion and would the effluent gas satisfy the rule?

6-43. With respect to Example 6-7, (a) if the HC level at the outlet of the afterburner remains the same, to what level, in parts per million, must the CO be reduced at the outlet to satisfy Rule 66, and (b) if the CO level at the outlet can be reduced to 180 ppm by redesign, to what level must the HC also be reduced by the redesign to satisfy Rule 66?

6-44. An afterburner is to be designed to treat 12 m^3/s of air at 1.10 bars and 35°C. The pollutant concentration in the air stream is 0.008 kg/m^3. The fuel specified is methane at an AF ratio of 60, and its enthalpy of combustion is $-50{,}030$ kJ/kg. The reaction chamber that follows the combustion zone is cylindrical. The flow velocity through the reaction chamber is to be 8 m/s, and the residence time for the gas stream is to be 0.6 s. Determine (a) the average temperature of the gas leaving the combustion zone and entering the reaction chamber, in degrees Kelvin, (b) the diameter of the reaction chamber in meters, and (c) the length of the reaction chamber in meters.

6-45. Consider the data of Problem 6-44. A regenerative heat exchanger is added to the flow process, with an effectiveness of 0.60. Assume the reaction chamber temperature must be maintained at 1050°K. On this basis, estimate (a) the AF ratio that now must be used, (b) the temperature of the exhaust gas leaving the regenerator, in degrees Kelvin, and (c) the percent decrease in fuel required compared to the case without a regenerator.

6-46. An afterburner is to be designed to treat 5,000 ft^3/min of polluted air at 15.0 psia and 140°F. The pollutant concentration in the air stream is 0.0005 lb/ft^3. The auxiliary fuel to be used is methane at an AF ratio of 62.0 lb/lb, and the enthalpy of combustion of the fuel is $-20{,}530$ Btu/lb. The cylindrical reaction chamber that follows the combustion zone is to be designed for a flow velocity of 22 ft/s and a residence time of 0.55 s. If the gas streams behave thermodynamically like air, determine (a) the average temperature of the gas leaving the combustion zone and entering the reaction zone, in degrees Fahrenheit, (b) the diameter of the reaction chamber, in feet, and (c) the length of the reaction chamber in feet.

6-47. Reconsider the data of Problem 6-46. A regenerative heat exchanger is added to the flow process, which has an effectiveness of 60 percent. Assume in this new situation that the reaction chamber inlet temperature must be maintained at least at 1350°F. On this basis, estimate (a) the AF ratio that now must be used, (b) the percent decrease in fuel required compared to the case without regeneration, and (c) the temperature of the exhaust gas leaving the regenerator, in degrees Fahrenheit.

References

1. R. L. Duprey. *Compilation of Air Pollution Emission Factors*. PHS Publication 999-AP-42. Durham, N.C.: National Center for Air Pollution Control, 1968.
2. *Reference Book of Nationwide Emissions*. Durham, N.C.: National Air Pollution Control Administration.
3. C. K. Hersh. *Molecular Sieves*. New York: Reinhold Publishing Corp., 1961.
4. "Ms Has Come a Long Way." *Environ. Sci. Tech.* **8** (1974): 106.
5. A. C. Stern. *Air Pollution*. Vol. III, 2nd ed. New York: Academic Press, 1968.
6. *Air Pollution Manual*. Detroit, Mich.: American Industrial Hygiene Association, 1968.
7. *Air Pollution Engineering Manual*, PHS Publication 999-AP-40, 1967.
8. C. L. Mantell. *Adsorption*, 2nd ed. New York: McGraw-Hill, 1951.
9. D. M. Young and A. D. Croswell. *Physical Adsorption of Gases*. London: Butterworth & Co., Ltd., 1962.
10. G. H. Duffey. *Physical Chemistry*. New York: McGraw-Hill, 1962.
11. A. Turk. "Source Control by Gas-Solid Adsorption." in *Air Pollution*, edited by A. Stern. New York: Academic Press, 1968.
12. R. H. Perry and C. H. Chilton, eds. *Chemical Engineers' Handbook*. 5th ed. New York: McGraw-Hill, 1973.
13. R. E. Treybal. *Mass Transfer Operations*. New York: McGraw-Hill, 1968.
14. J. H. Perry. *Chemical Engineers' Handbook*. 4th ed. New York: McGraw-Hill, 1963.
15. M. Leva. *Tower Packings and Packed Tower Design*. Akron, Ohio: The United States Stoneware Co., 1953.
16. D. J. Seery and C. T. Bowman. *Combust. Flame* **14** (1970): 37.
17. F. Dryer et al. *Combust. Flame* **17** (1971): 270.
18. W. Jost et al. *Z. Phys. Chem. N.F.* **45** (1965): 47.
19. H. B. Palmer and J. M. Beers. *Combustion Technology*. New York: Academic Press, 1974.
20. T. Phillips and R. W. Rolke, *Am. Inst. Chem. Engineers Symposium Series No. 137* **70** (1974).
21. *Handbook of Organic Industrial Solvents*. 2nd ed. National Association of Mutual Casualty Companies, 1961.
22. B. Lewis and G. von Elbe. *Combustion, Flames, and Explosion of Gases*. New York: Academic Press, 1961.
23. D. W. Waid. *Thermal Oxidation or Incineration*. APCA Specialty Conference, Odor Control Technology, Pittsburgh, March, 1974.
24. D. W. Waid. "Afterburners for Control of Gaseous Hydrocarbons and Odor." *Am. Inst. Chem. Engineers Symposium Series No. 137* **70**, 1974.
25. R. J. McIlwee. *Catalysis in Air Pollution Control*. APCA Specialty Conference, Odor Control Technology Proceedings, Pittsburgh, March, 1974.
26. K. H. Hemsath and P. E. Susey. "Fume Incineration Kinetics and Its Application." *Am. Inst. Chem. Engineers Symposium Series No. 137* **70** (1974).
27. R. D. Hawthorn. "Afterburner Catalyst—Effects of Heat and Mass Transfer Between Gas and Catalyst Surface." *Am. Inst. Chem. Engineers Symposium Series No. 137* **70** (1974).

Chapter 7
Control of Sulfur Oxides

7-1 INTRODUCTION

One of the most abundant air pollutants emitted in the United States is sulfur dioxide (see Table 1-7). The quantity emitted in 1977 is estimated to have been 30.1 million tons. From Table 7-1, we note that power plant boilers are responsible for over 50 percent of total emissions for 1977. Sulfur is a component of all natural oil and coal with a composition varying from 0.1 to over 5 percent. Thus the total emission of SO_2 varies considerably with the nature or origin of the fossil fuel. The trends of the annual SO_2 average at downtown monitoring stations for some major U.S. cities are shown in Figure 7-1. It is seen that the trend is downward, with significant changes in several instances. This trend is probably due to two factors, namely, the installation of SO_2 control equipment on stacks and the switch to low-sulfur fuels in certain areas.

The SO_2 concentration in large urban areas typically ranges from 0.01 to 0.1 ppm (30 to 300 $\mu g/m^3$) for a 1-hr averaging time. For SO_2 at 25°C and 1 atm,

$$1 \text{ ppm} = 2620\ \mu g/m^3\ SO_2$$

The SO_2 concentration in most urban areas is very localized. Because of their advantageous location with regard to industrial sections and prevailing wind patterns, some sections of a given urban area may easily meet federal SO_2

Table 7-1 ESTIMATED YEARLY EMISSIONS OF SULFUR OXIDES IN MILLION TONS

SOURCE	1971	1974	1977
Transportation	0.7	0.7	0.8
Stationary fuel combustion			
Electric utilities	15.7	17.2	17.6
Industrial	4.0	3.3	3.2
Residential, commercial	1.9	1.6	1.6
Chemical industry	2.0	2.1	1.8
Metals industry	3.8	3.5	2.4
Others	0.2	0.	0.
TOTAL	28.3	28.4	27.4

SOURCE: *National Air Quality, Monitoring, and Emissions Trends Report*, 1977, EPA-450/2-78-052, December 1978.

standards. Other sections of the city often may drastically exceed the federal or local limits because of their unfortunate location.

The federal ambient air quality standards for SO_2 (presented in Chapter 2) are

	STANDARD	
AVERAGING TIME	PRIMARY	SECONDARY
Annual	80 $\mu g/m^3$ (0.03 ppm)	—
24 hr	365 $\mu g/m^3$ (0.14 ppm)	—
3 hr	—	1300 $\mu g/m^3$ (0.5 ppm)

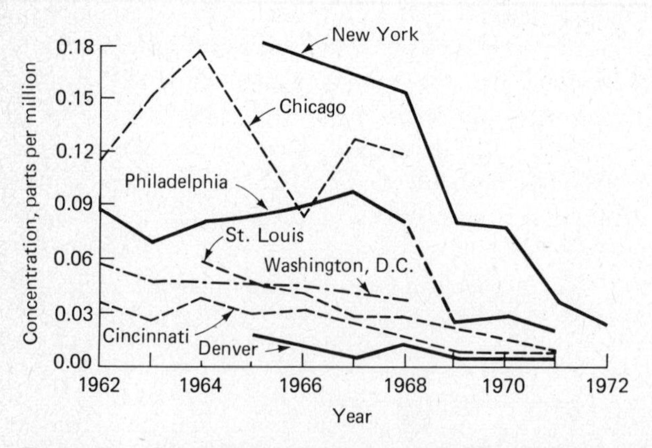

Figure 7-1 Downward trends of annual SO_2 average at downtown stations of Continuous Air Monitoring Program. (SOURCE: R. B. Engdahl. "State of the Air on SO_2 Control," presented at Purdue Industrial Fuel Conference, October, 1974.)

Air quality control regions within the United States can promulgate their own standards, if these standards are acceptable to the EPA. The Metropolitan Indianapolis Intrastate Air Quality Control Region, for example, has the following standards for SO_2 (as of January 1, 1973):

1. 0.015 ppm maximum annual geometric mean.
2. 0.10 ppm maximum 24-hr concentration, not to be exceeded more than 1 day/yr.
3. 0.42 ppm maximum 1-hr concentration, not to be exceeded more than 1 hr/yr.

Note that both the annual mean and the 24-hr value for this air quality region are lower than (that is, more severe than) the federal standards.

In order to meet ambient air quality standards, the U.S. government has established specific emission standards for a number of industries. These are national emission standards and must be met by the industry regardless of its geographic location. For example, consider the national sulfur dioxide emission standards for new steam generating plants. These are presented in Chapter 2. For some plants the emission standard is 1.2 lb SO_2/million Btu of energy released. It is interesting to estimate the percent sulfur permissible in a fuel in order to meet this standard. We assume that coal with a heating value of 10,000 Btu/lb is to be used. Thus a 10^6-Btu heat release requires 100 lb of coal. In addition, 1.2 lb of SO_2 is equivalent to 0.6 lb of sulfur. Hence by this rough estimate a 0.6 percent sulfur coal would just meet the standard. The coal currently being mined in the United States averages around 2.0 or 2.5 percent sulfur. Hence a coal with a sulfur content around 0.5 percent would rank as a very low-sulfur coal. Residual fuel oils have a sulfur content from 1 to 4 percent. In light of the air quality standards and low sulfur requirements discussed above, it is not surprising that control measures must be instituted on fossil-fired combustion processes.

The overall reaction for the formation of sulfur dioxide from sulfur in fossil fuels is simply

$$S + O_2 \rightarrow SO_2 \tag{7-1}$$

The reaction is highly exothermic with a heat release of 296,800 kJ/kg·mole (127,700 Btu/lb·mole) at 25°C. In addition to SO_2, a small quantity of sulfur trioxide, SO_3, is formed in the combustion reaction. In the combustion of fossil fuels the SO_2/SO_3 ratio is typically 40 : 1 to 80 : 1.

Sulfur dioxide can act both as a reducing agent and as an oxidizing agent at atmospheric conditions. The well-established reaction between H_2S and SO_2 is an example of SO_2 acting as an oxidizing agent. This reaction is

$$2H_2S + SO_2 \underset{\text{catalyst}}{\rightarrow} 3S + 2H_2O$$

Of importance to the problems of air pollution is the ability of SO_2 gas to react either photochemically or catalytically with other atmospheric contaminants, forming sulfur trioxide, sulfuric acid, and various salts of sulfuric acid. For example, in the presence of a metallic oxide as a catalyst in the atmosphere,

$$SO_2 + \tfrac{1}{2}O_2 \underset{\text{catalyst}}{\rightarrow} SO_3 \tag{7-2}$$

Many substances, especially particulate matter in the atmosphere and the oxides of nitrogen, act as the catalysts in the above reaction.

Some metal oxides oxidize sulfur dioxide directly to sulfate. As an example,

$$4MgO + 4SO_2 \rightarrow 3MgSO_4 + MgS \tag{7-3}$$

Two other reactions taking place in the atmosphere when high humidity prevails are

$$2SO_2 + 2H_2O + O_2 \underset{\text{catalyst}}{\rightarrow} 2H_2SO_4 \tag{7-4}$$

$$SO_3 + H_2O \rightarrow H_2SO_4 \tag{7-5}$$

The first oxidation reaction above occurs rapidly in the presence of metal salts, such as the sulfates and chlorides of iron and manganese. These metal salts act as catalysts for the reaction. The droplets of sulfuric acid formed, in addition to being highly corrosive, tend to reduce visibility. These reactions are also partially responsible for the acid rain phenomenon discussed in Section 1-7-H.

7-2 THERMODYNAMICS AND KINETICS OF SULFUR OXIDE FORMATION

The thermodynamics and kinetics of SO_x formation in homogeneous flame processes were extensively reviewed [1] in 1972. On a fundamental level, the formation processes for SO_2 and SO_3 cannot be explained quantitatively by the simple overall reactions

$$S + O_2 \rightarrow SO_2 \tag{7-1}$$

$$SO_2 + \tfrac{1}{2}O_2 \rightleftharpoons SO_3 \tag{7-2}$$

It appears, for example, that the monoxide, SO, is important kinetically in the oxidation scheme. There are indications that the species S_2O may also be important under certain circumstances. Since both SO and S_2O are unstable and reactive at atmospheric temperatures, they are not found among the normal products of combustion processes.

The conventional sulfur oxides found as stable products resulting from high-temperature combustion of sulfur-bearing fuels are SO_2 and SO_3. The thermodynamic equilibrium relationship between them is expressed by reaction (7-2). Table 7-2 summarizes values of the ideal-gas equilibrium constant

Table 7-2 EQUILIBRIUM CONSTANTS FOR THE SO_2-SO_3 OXIDATION REACTION

$SO_2+\frac{1}{2}O_2 \rightleftharpoons SO_3$	T(°K)	T(°F)	K_p
	298	77	2.6×10^{12}
	500	440	2.6×10^{5}
$K_p = \dfrac{p_{SO_3}}{p_{SO_2}(p_{O_2})^{1/2}}$	1000	1340	1.8
	1500	2240	3.8×10^{-2}
	2000	3140	5.6×10^{-3}

SOURCE: Computed from JANAF Thermochemical Tables.

K_p for this reaction as a function of temperature. From the K_p values found in the table it is apparent that equilibrium favors SO_3 at low temperatures and SO_2 at high temperatures in lean mixtures (equivalence ratio less than unity). On this basis one would expect to find small amounts of SO_3 in the actual flame zone, and large amounts in the cooled flue gases as the equilibrium shifts with temperature. Hedley [2] has shown that the opposite effect is observed in power plant furnaces. In the flame zone the SO_3 level exceeds the predicted equilibrium value, while in the cooled flue gas near the exit the SO_3 concentration is far below the equilibrium value for that temperature.

This anomalous behavior of SO_3 is due to the role of kinetics in determining the rate of formation of various intermediate species in the elementary chemical steps. It is proposed that the monoxide, SO, is an important intermediate product in the reaction scheme. To shorten the discussion we assume that some mechanism produces SO from the sulfur-bearing molecules early in the reaction zone. The next step is to examine the elementary reactions by which SO is oxidized first to SO_2, and then ultimately to SO_3. The major SO_2 formation reactions are thought to be

$$SO + O_2 \rightarrow SO_2 + O \tag{7-6}$$

and

$$SO + OH \rightarrow SO_2 + H \tag{7-7}$$

Note that these reactions are also producing O and H atoms, which are highly reactive and may enter the overall reaction scheme later. Also, it is important to keep in mind that the sulfur oxidation reactions are occurring simultaneously with the oxidation of the hydrocarbon fuel. This latter process is producing large quantities of intermediate species such as O, H, and OH, which are available to the sulfur oxide reactions.

Our main concern is the formation and removal of SO_3. The major formation reaction is the three-body process

$$SO_2 + O + M \rightarrow SO_3 + M \tag{7-8}$$

where M is a third body which is an energy absorber. Three-body processes are intrinsically fairly slow; however, this reaction will proceed rapidly in the hot reaction zone because the concentration of O atoms is at a maximum there. The major steps for the removal of SO_3 are thought to be the following:

$$SO_3 + O \rightarrow SO_2 + O_2 \tag{7-9}$$

$$SO_3 + H \rightarrow SO_2 + OH \tag{7-10}$$

$$SO_3 + M \rightarrow SO_2 + O + M \tag{7-11}$$

The last step, which is simply the reverse of reaction (7-8), is a thermal decomposition process.

In the hot reaction zone, where high levels of O atoms occur, it is reasonable to assume that reactions (7-8) and (7-9) dominate the overall reaction mechanism. Applying basic kinetic concepts, we may write for the formation and removal rates of SO_3,

$$\frac{d[SO_3]}{dt} = k_8[SO_2][O][M] - k_9[SO_3][O]$$

The maximum SO_3 concentration is found by setting $d[SO_3]/dt = 0$. As a result,

$$[SO_3]_{max} = \frac{k_8[SO_2][M]}{k_9}$$

A numerical value for $[SO_3]_{max}$ has been estimated [3] on the basis of estimates of k_8, k_9, and $[M]$ which appear in the literature. From these values it is found that the maximum SO_3 level is between 1 and 5 percent of the SO_2 concentration. This is higher than that predicted from equilibrium considerations, but substantially agrees with measurements [2].

In hydrocarbon-rich flames (equivalence ratio greater than unity) the removal reaction is (7-10), since O atom concentration is much lower under this condition. If the rate equation for SO_3 concentration is now written in terms of reactions (7-8) and (7-10), and then set equal to zero, we find that under fuel-rich conditions,

$$[SO_3]_{max} = \frac{k_8[SO_2][M][O]}{k_{10}[H]}$$

The ratio $[O]/[H]$ controls the value of the maximum SO_3 concentration in this circumstance. Since this ratio is small, the conversion of SO_2 to SO_3 is suppressed significantly under rich conditions.

As the products of combustion leave the hot reaction zone the temperature drops and atom concentrations decay rapidly by recombination. Without appreciable atom concentrations, the formation and removal steps for SO_3, such as reactions (7-8), (7-9), and (7-10), become negligible. Some SO_3 may thermally decompose by reaction (7-11) as the temperature decreases. However, this reaction has a large activation energy, so its rate becomes unimportant once the temperature has fallen significantly. As a consequence, kinetics predicts superequilibrium levels of SO_3 in the flame zone of lean hydrocarbon mixtures, followed by a modest drop in concentration as the gases cool. In the cool gases the level is far below that calculated on the basis of equilibrium. The predicted final level of SO_3 is a few percent of the SO_2 concentration. This agrees with industrial experience, where SO_3/SO_2 values typically lie between 1 : 40 and 1 : 80. In fuel-rich combustion SO_3 levels are negligible.

Thus the main oxide of sulfur formed in combustion processes is SO_2, in spite of thermodynamic forces which should push the reaction strongly toward SO_3. The eventual conversion of SO_2 to SO_3 in our atmosphere normally would be a slow process, owing to small rates of conversion at atmospheric temperatures. However, once SO_2 is in the atmosphere, the rate is affected by heterogeneous catalysis on the surface of suspended particles. This leads to an increased rate of SO_2 conversion to SO_3. Unfortunately, it also leads to an increase in sulfate aerosol formation, and these aerosols are a human health hazard. A similar situation exists when oxidation catalysts are used to control CO and hydrocarbon emissions from automotive engines. These catalysts speed up the oxidation of SO_2 to SO_3 in the exhaust, and sulfate aerosol formation is greatly enhanced.

7-3 GENERAL CONTROL METHODS

The data of Table 7-1 indicate that nearly three-fourths of sulfur dioxide emissions originate from the combustion of fossil fuels. Consequently, the control methods discussed here are those specifically applicable to fossil-fuel combustion, such as found in steam power plants. Possible control methods for other industries such as sulfuric acid plants, smelters, and paper plants may be found in other references [4].

Four possible methods or alternatives may be used to reduce sulfur dioxide emissions from fossil-fuel combustion. These potential methods are:

A. Change to low-sulfur fuel
   1. Natural gas
   2. Liquefied natural gas
   3. Low-sulfur oil
   4. Low-sulfur coal
B. Use desulfurized coal and oil
C. Build tall stacks to increase atmospheric dispersion
D. Use flue-gas desulfurization systems

The first three methods will be discussed in this section. Flue-gas desulfurization is the topic of Section 7-4.

7-3-A Low-Sulfur Fuels

The present known reserve of natural gas in the continental United States is roughly a 15-year supply at the current rate of usage. The financial incentive provided by short supply will probably stimulate gas-well drillers to increase their efforts and thus extend that supply somewhat. In addition, the availability of natural gas from the Alaskan north shore by pipeline to the midcontinental United States would help alleviate the impending shortage. Nevertheless, a switch to natural gas to lessen SO_2 emissions is not a viable alternative.

Liquefied natural gas (LNG) from foreign suppliers is a potential solution. However, for utility use the cost will be much higher than that of other alternatives. Also, it seems impractical to place additional reliance for clean fuel on foreign markets.

Natural or desulfurized oils appear to offer real long-term potential to small industries in general and to utilities located on coastal waters for dealing with SO_2 control problems. Roughly 40 percent of the crude oil produced in the United States has a sulfur content between 0 and 0.25 percent. However, it is the residual fuel oil left over from refining processes that is used as fuel for power utilities. The average sulfur content of residual fuel oil obtained from domestic crude oil is nearly 2 percent [5]. For East Coast utilities residual oil with a 1 percent sulfur content is commonly achieved by blending Libyan crude oil with other high-sulfur residual oils. Although low-sulfur crude oil is available from foreign sources, a coherent energy policy for the future must be based on domestic sources.

Reserves of low-sulfur coal (1 percent or less) in the United States are estimated to exceed one trillion tons. Unfortunately, getting this coal out of the ground is neither quick nor cheap. Approximately 62 percent of this coal is west of the Mississippi, whereas 90 percent of the coal-consuming utility power plants are located east of the Mississippi. Table 7-3 lists the sulfur content of major bituminous coal reserves in the continental United States [6].

7-3-B Desulfurization Processes for Coal and Oil

Sulfur occurs in coal in two principal forms, organic and inorganic. The inorganic compound, known as iron pyrite (FeS_2), is present as discrete particles and is thus amenable to physical removal by gravity washing. The percentages of pyritic and organic sulfur vary with mining location. Typically, the pyritic form is 40 percent or less, and water washing may reduce the total sulfur content by one-third. Sulfur in the organic form is chemically bound in the coal. Consequently, more complex and costly chemical processes are required to remove organic sulfur. For example, such a process might involve coal gasification, or the conversion of coal to a synthetic oil or solid material.

Table 7-3 SULFUR CONTENT OF THE MAJOR BITUMINOUS COAL RESERVES IN THE UNITED STATES (IN BILLIONS OF TONS)

LOCATION	SULFUR CONTENT (%) <1	1–3	>3
Colorado	62	—	—
Illinois	—	21	119
Indiana	—	16	19
Eastern Kentucky	22	7	—
Western Kentucky	—	5	32
Missouri	—	—	79
New Mexico	11	—	—
Ohio	—	13	38
Pennsylvania	1	50	7
Utah	22	—	4
West Virginia	47	45	9
Wyoming	13	—	—
TOTAL	194	191	322

Several other coal gasification techniques are under development. The differences between the processes involve the method of feed introduction, the method of supplying heat for the gasification step, the method of achieving reactant contact by fixed or fluidized beds, and the use of different gasifying mediums. Nearly all possible combinations of ways to gasify coal have been studied, and the specific processes mentioned above are only a few of the most promising.

It is also possible to produce synthetic liquid fuels (and other types of petroleum products such as alcohols, ketones, and waxes) from coal [7]. In general, the carbon-to-hydrogen ratio in coal is much higher than in oil. Hence additional hydrogen must be added to the process if a liquid product is to result. Consequently, coal liquefaction usually involves the production of hydrogen from coal by a gasification process as an intermediate step, if hydrogen is not available from another source.

The desulfurization of natural oil can be accomplished by commercially available processes. Although desulfurization is expensive, there is no technological deterrent. Nevertheless, it is important for economic and political reasons that the United States find other fuel sources to supplement its dwindling domestic supply of oil and gas. As we note here preliminary studies indicate that synthetic gaseous and liquid fuels can be made from coal and can be essentially free of sulfur. It has been estimated [8] that the cost of desulfurization of coal, although expensive, may be the same order of magnitude as that for desulfurization of stack gases. Consequently, both technologies will undoubtedly develop concurrently, and each may have specific advantages in a given situation.

The reserves of coal in the United States are incredibly large. A recent survey [9] indicates that the recoverable stripping reserve with an overburned of less than 150 ft is 128 billion tons. This is roughly a 200-yr supply at 1970 production rates. Over 10 times this amount lies at a depth of less than 3000 ft. If solution to sociological problems in the mining industry can be found, coal will increase in importance as an energy source. This lends additional impetus to the development of technology to desulfurize coal that contains more than 0.5 percent sulfur.

7-3-C Tall Stack Dispersion

A control method involving dispersion from tall stacks may seem archaic, in light of a federal emission standard for steam power plants of 1.2 lb SO_2/MB input, or smaller. Nevertheless, there is still considerable controversy over the use of this technique as opposed to flue-gas desulfurization. The control method is based on natural dispersion at high elevation so that ground-level concentrations are acceptable at all times (see the dispersion calculations in Chapter 4). High stacks have been widely used in England with acceptable results. The TVA Cumberland power station in Tennessee has twin 1000-ft concrete chimneys, while the International Nickel Company has built one that is 1250 ft high. If tall stacks can provide adequate ground-level protection, then it is a moot point whether an emission standard or effective atmosphere dispersion is the better control method in some geographical locations.

7-3-D Flue-Gas Desulfurization Systems

Progress in developing satisfactory desulfurization processes for flue gases has been extremely slow because of the complexity and magnitude of the problem. For example, a 1000-MW electric power plant using coal containing between 2.5 and 3 percent sulfur will discharge 1.7 to 2 million ft^3/min of flue gas with an SO_2 content between 0.2 and 0.3 percent by volume. Many sizes of power plants exist, and the technical and economic feasibility of most desulfurization processes are closely related to plant size and location. It seems highly unlikely that a single desulfurization method will be developed that is capable of controlling effluents from all types of sources. The control techniques to be employed depend upon such factors as boiler size, configuration, load pattern, geographical location, and the like. The most promising SO_2 removal methods [5] will be discussed in the following section.

7-4 FLUE-GAS DESULFURIZATION PROCESSES

Sulfur dioxide removal processes may be grouped according to two classifications: (1) throwaway or regenerative, and (2) wet or dry. Throwaway processes are those in which a solid waste product is formed which must be discarded. As a result, fresh chemicals also must be continually added. In regenerative processes, as the name implies, the chemistry is such that the removal agents can be continually regenerated in a closed-loop system. Wet or dry processes are differentiated simply by whether or not the active removal agent is contained in a liquid solution. The removal system typically involves the use of absorption, adsorption, or catalytic processes.

In 1980 the Environmental Protection Agency (EPA) published in the *Federal Register* a new standard for SO_2 emissions from fossil-fueled electric utility steam generating units for which construction began after September 18, 1978. The standard, which is based both on the heating value of the fuel as well as the sulfur content (see Table 2-6), has the following maximum limitation on SO_2 emissions.

$$\text{Solid fossil fuel: } 1.2 \text{ lb } SO_2/10^6 \text{ Btu}$$

For comparative purposes, flue gases generated from fuel with a 3 percent sulfur content would need to have about 75 to 80 percent SO_2 removed to meet these standards. The major processes described or referred to below are being developed primarily for use in the power-utility field. However, there is also a real demand for solutions to the problem of industrial-scale boiler emissions of SO_2. Utilities typically emit SO_2 at rates 10 to 30 times greater than those for industrial boilers. Because of these smaller sulfur dioxide emission rates, the recovery processes that apply to utilities are much less economical and practical on the industrial scale. This is especially true if the recovery process is fairly complex, and if a usable by-product is required to make the process economically attractive. Consequently, two different technologies may develop for SO_2 removal systems for flue gases, depending upon the size of the unit requiring control. By mid-1977 there were scrubbers of some type installed on 50,000 MW of the then current 225,000 MW of coal-fired capacity in the United States.

A brief comparison of some of the major processes under development for SO_2 removal from stack gases appears in Table 7-4. The processes in this table are divided into three classes: throwaway and regenerative—both of which are wet scrubbing processes—and dry processes. Since SO_2 is an acidic gas, almost all of the scrubbing processes use an aqueous solution or slurry of alkaline material. The throwaway designs usually dispose of the removed sulfur in some form of calcium-type waste sludge. As a result, the alkali makeup required is considerable. The product of the regenerative processes is usually sulfur or sulfuric acid, and the alkali solution is recycled. Hence little makeup is generally required in regenerative systems. Throwaway processes frequently can be used to remove particulate (fly ash) as well, if the system is enlarged to accommodate the particulate. In some regenerative processes a high-efficiency particulate collector, such as an electric precipitator, must precede the SO_2 removal equipment, because particulates are not acceptable in its operation.

Table 7-4 DESCRIPTIONS OF SOME SULFUR DIOXIDE REMOVAL PROCESSES

PROCESS GENERICS	PROCESS OPERATIONS	ACTIVE MATERIAL	KEY SULFER PRODUCT
A. Throwaway scrubbing processes			
1. Lime or limestone	Slurry scrubbing	CaO, $CaCO_3$	$CaSO_3/CaSO_4$
2. Sodium	Na_2SO_3 solution	Na_2CO_3	Na_2SO_4
3. Double alkali	Na_2SO_3 solution, regenerated by CaO or $CaCO_3$	$CaCO_3/Na_2SO_3$ or CaO/NaOH	$CaSO_3/CaSO_4$
4. Magnesium-promoted lime/limestone	$MgSO_3$ solution, regenerated by CaO or $CaCO_3$	$MgO/MgSO_4$	$CaSO_3/CaSO_4$
B. Regenerative scrubbing processes			
1. Magnesium oxide	$Mg(OH)_2$ slurry	MgO	15% SO_2
2. Sodium	Na_2SO_3 solution	Na_2SO_3	90% SO_2
3. Citrate	Sodium citrate solution	H_2S	Sulfur
4. Ammonia	Ammonia solution, conversion to SO_2	NH_4OH	Sulfur (99.9%)
C. Dry processes			
1. Carbon adsorption	Adsorption at 400°K, reaction with H_2S to S, reaction with H_2 to H_2S	Activated carbon/H_2	Sulfur
2. Spray dryer	Absorption by sodium carbonate or slaked lime solutions	$Na_2CO_3/Ca(OH)_2$	Na_2SO_3/Na_2SO_4 or $CaSO_3/CaSO_4$

Missing from Table 7-4 is the dry limestone injection method, which was one of the earliest tested. Finely ground limestone or calcinated limestone is injected with the coal into a furnace. The SO_2 is absorbed as calcium sulfites and sulfates in the high-temperature (1400° to 2000°F) section of the furnace. Cyclones or precipitators may be used to remove this solid particulate as well as the coal fly ash. Unfortunately, results indicate that less than 20 percent of the limestone is utilized, and only 20 to 40 percent of the SO_2 is removed. Experiments indicate that over 200 percent of the stoichiometric quantity of limestone is required to remove 50 percent of the SO_2. Therefore, the particulate removal equipment must be at least twice the normal capacity of that for a conventional furnace. In addition, disposal of the increased quantity of solid waste presents a major problem. Some increase in boiler tube erosion was also encountered with this method. One of the major advantages of the limestone injection method is that it can be applied to existing boilers without requiring major changes. Although the dry limestone direct-injection process in conventional boilers is not practical, there are several alternative dry limestone processes under active investigation. These are discussed in Section 7-4-A.

7-4-A Dry Limestone Scrubbing in Fluidized Beds

Fluidized-bed coal combustion may become an attractive, but elaborate, alternative to the conventional means of burning coal in a boiler. By 1980 the process was under extensive study in the pilot-plant state. One way of making the process environmentally clean is to add dry limestone to the pulverized coal. The sulfur oxides formed in the hot (600–1000°C, 1100–1800°F) bed react with the limestone to form calcium sulfate. This waste product, along with the ash left from the combustion process, can be removed by an electrostatic precipitator, for example. Such a process removes more than 90 percent of the sulfur oxides. In addition, due to the relative low temperatures in such a fluidized bed, the nitrogen oxide emissions are typically less than half that associated with conventional coal combustion. However, the process requires a mass ratio of limestone to coal of about 1 : 4, even though the sulfur content may only be 3 percent. Hence the amount of solid waste product is large; a 1000-MW fluidized-bed power plant might produce around 800,000 tons of used stone per year which must be disposed. One disadvantage of fluidized-bed combustion is that current boilers cannot be altered to use the process.

A regeneration process developed by the Argonne National Laboratory in 1978 may reduce the waste problem by 75 percent. Experiments indicate that stone can be regenerated and reused through about 10 cycles without a serious deterioration of its SO_2 removal effectiveness. The regeneration process passes carbon monoxide gas (obtained by partial combustion of coal) over the calcium sulfate waste. The products are carbon dioxide, sulfur dioxide, and calcium oxide. CO_2 is released to the atmosphere, the SO_2 is concentrated so that it can be converted to elemental sulfur as a by-product, and the calcium oxide (lime) is reused in the fluidized bed.

Another way to reduce the use of limestone, and hence the amount of waste product, without the recycle process, is to modify the limestone in such a way as to increase its effectiveness in removing sulfur dioxide. Research has shown [10] that the excess limestone usually required (three or four times the stoichiometric quantity) could be significantly reduced by the addition of calcium chloride, which is an inexpensive mineral. At 700°C, where the maximum transformation occurs, the conversion of limestone to sulfate exceeds 90 percent when $CaCl_2$ is used, compared to 20–25 percent for untreated limestone. At this same temperature more than 90 percent of the SO_2 in a gas containing 0.35 volume percent SO_2 was removed in bench tests. In these tests the modified limestone was maintained in a fluidized bed which was independent of the combustion process. Hence the method is applicable also to the flue gases from conventional coal combustion. The $CaCl_2$ content used was roughly 2 mole percent, and SO_2 removal was much better when the SO_2 content in the gas stream was less than 0.5 percent (5,000 ppm). On the basis of recent investigations, dry limestone scrubbing may again appear attractive, in this case in conjunction with fluidized beds.

7-4-B Lime and Limestone Scrubbing

The lime wet scrubbing process was originally developed in England by Imperial Chemical Industries (ICI) in the 1930s. Sulfur dioxide removal efficiencies of over 90 percent were attained in pilot plant operation. In modern processes the flue gas is scrubbed with a 5 to 15 percent slurry of calcium sulfite/sulfate salts which also contains amounts of lime (CaO) and limestone ($CaCO_3$). The SO_2 reacts with the slurry to form additional sulfite and sulfate salts. The solids (along with coal fly ash) are continuously separated from the slurry and discharged into a settling pond. The remaining liquor is recycled to the scrubbing tower after fresh lime or limestone has been added. A schematic of the process is shown in Figure 7-2.

Although the overall mechanism for the process is not completely understood, the following reactions are thought to occur [9]:

$$CaO + H_2O \rightarrow Ca(OH)_2$$

$$Ca(OH)_2 + CO_2 \rightarrow CaCO_3 + H_2O$$

$$CaCO_3 + CO_2 + H_2O \rightarrow Ca(HCO_3)_2$$

$$Ca(HCO_3)_2 + SO_2 + H_2O \rightarrow CaSO_3 \cdot 2H_2O\downarrow + 2CO_2$$

$$CaSO_3 \cdot 2H_2O + \tfrac{1}{2}O_2 \rightarrow CaSO_4 \cdot 2H_2O\downarrow$$

The many attendant process problems include chemical scaling, corrosion, erosion, and solid-waste disposal. The latter may seriously restrict the number of utilities that can employ the lime-limestone scrubbing technique. A

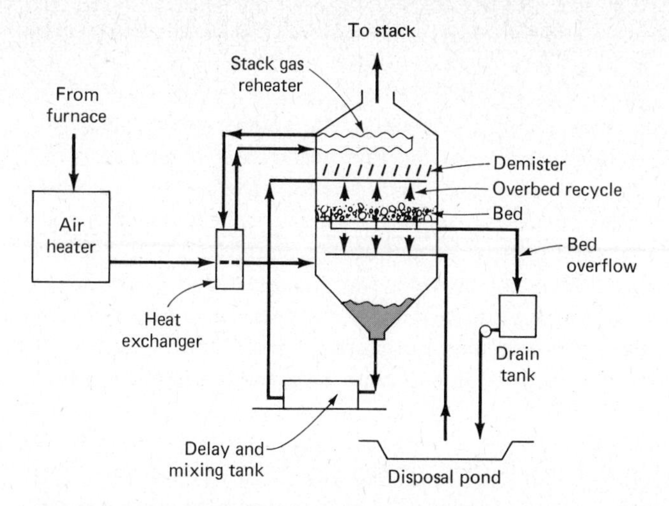

Figure 7-2 Schematic of SO_2 lime scrubbing removal system.

sizable disposal area is required, preferably adjacent to the plant. The Kansas City Power and Light Company currently uses a 160-acre pond for waste disposal. Another drawback of the process is the necessity for reheating the clean flue gas. As a result of the scrubbing process, the flue gases are cooled to around 120°F. In order to have the necessary buoyancy to exit from a stack at the required velocity, the gases must be reheated to 250° to 300°F. Reheating is accomplished, as shown in Figure 7-2, by installing a gas cooler before the scrubber and a gas stack heater after the scrubber. Thus two units must be added to the gas flow system, which increases the equipment investment and also increases the system pressure drop, so that additional fan capacity must be provided.

It has been reported [11] that by mid-July, 1978, there were over two dozen utility flue gas desulfurization units classified as operational in the United States based on lime or limestone slurry scrubbing. The measured or guaranteed removal efficiencies on these units ranged from 50 to 90 percent, for coal sulfur contents of 0.3 to 5 percent. Due to scaling and other inherent problems associated with this process, improvement on the basic limestone scrubbing process is desirable. One possible modification is discussed in the next section.

7-4-C Wet Limestone Scrubbing Modified with Magnesium Sulfate

One method of overcoming the scaling and low SO_2 removal efficiency of limestone scrubbing is by modifications in equipment or operation. A more basic approach is to modify the solution chemistry. It is possible to absorb SO_2 as a soluble sulfate in the scrubbing liquor, rather than an insoluble calcium sulfite or sulfate. The calcium salt then can be precipitated in a tank external to the scrubber. The use of magnesium sulfate in the scrubber liquid leads to increased SO_2 absorbing capacity (in excess of 90 percent) and virtual elimination of scaling in the scrubber [12]. The power consumption may also be decreased by as much as 50 percent. The scrubber can also be used to remove a high percentage of the fine particulate which has passed through the electrostatic precipitator, which precedes the scrubber. Another major advantage is that an unlimited turndown ratio (the ratio of gas and liquid flow rates) is possible since the scrubber is simply a horizontal open chamber with several stages of liquid spray nozzles at the top of the chamber. A simplified flow diagram of the process is shown in Figure 7-3.

There are four major reactions of interest. First, sulfur dioxide and water form H_2SO_3. This quantity reacts with the ion pair $MgSO_3$ accordingly.

$$H_2SO_3 + MgSO_3 \rightarrow Mg^{+2} + 2HSO_3^{\,-}$$

In the presence of calcium carbonate the $MgSO_3$ is regenerated:

$$Mg^{+2} + 2HSO_3^{\,-} + CaCO_3 \rightarrow MgSO_3 + Ca^{+2} + SO_3^{\,-2} + CO_2 + H_2O$$

This latter reaction takes place in an external reactor-holding tank which is part of the scrubbing loop. Finally, the precipitation reaction in the magnesium-promoted limestone slurry is

$$Ca^{+2} + SO_3^{\,-2} + \tfrac{1}{2}H_2O \rightarrow CaSO_3 \cdot \tfrac{1}{2}H_2O(s)$$

Some of the sulfite in liquor is oxidized to sulfate. Hence gypsum ($CaSO_4 \cdot 2H_2O$) is also precipitated. It is possible to oxidize the calcium sulfite

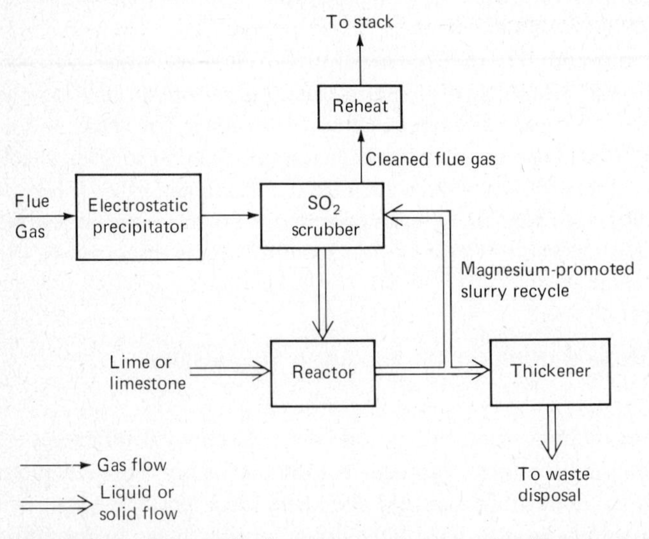

Figure 7-3 Schematic of a magnesium-promoted limestone slurry desulfurization process.

to gypsum if such a product is desired. The function of the magnesium sulfate is to assure that an appreciable concentration of the ion pair $MgSO_3$ is present in the slurry. Generally, 0.3 to 1.0 molar $MgSO_4$ solutions are used in the scrubber.

A 908-MW unit of the Pennsylvania Power Company at the Bruce Mansfield generating station is to have such a scrubbing system installed on it. Construction of the SO_2 removal equipment, using the Kellogg-Weir scrubbing system, began in 1979. Approximately 95 percent of the fly ash will be removed in electrostatic precipitators which precede the scrubbers. The scrubbers will operate with five stages of spray nozzles and a gas flow rate of about 22 ft/s (7 m/s). Reheating of the flue gas leaving the scrubber by an oil burner will raise the flue gas temperature by roughly 35°F (20°C).

The positive effect of the addition of magnesium to limestone/lime slurries on the SO_2 removal efficiency has been confirmed by other data. Tests were held between 1973 and 1977 on a slipstream of the TVA Shawnee Power Station near Pacucah, Kentucky. The test results are based on the use of a Turbulent Contact Absorber, and the absorbing liquor contained 8 to 15 weight percent solids. Without magnesium present, the percent removal varied from 60 to 75 percent, depending upon the pH. For a Mg^{+2} concentration around 8000 ppm, and the same pH range, the removal efficiency ranged from 82 to 94 percent. This amounts to about a 25 to 35 percent increase in SO_2 removal efficiency when magnesium is added.

7-4-D Magnesium Oxide Scrubbing

In the magnesium oxide scrubbing process, MgO in the slurry functions in the same manner that limestone or lime does in the lime scrubbing process. The critical difference between the processes is that magnesium oxide scrubbing is regenerative, whereas lime scrubbing is generally considered a throwaway process. All MgO processes involve scrubbing with a $Mg(OH)_2$ slurry. Absorption of SO_2 by the slurry leads to the formation of magnesium sulfite (or sulfate). Separation and calcination of this solid regenerates MgO and produces a stream of 10 to 15 percent sulfur dioxide. (Coke or some other reducing agent added in the calcining step reduces any sulfate present.) The regenerated magnesium oxide is returned to the scrubber. A process schematic is shown in Figure 7-4.

The regenerative process requires that fly ash be removed prior to passage of the flue gas through the scrubbing unit. The SO_2 stream from the calciner is most frequently used to produce sulfuric acid. Sulfur production is possible, but would be expensive since the SO_2 stream is fairly dilute. A distinct advantage of the system is that several power stations could ship the loaded absorbent to a central regeneration plant for MgO recovery. If a sufficient local market existed for all the acid produced, the economics of the system would be greatly improved.

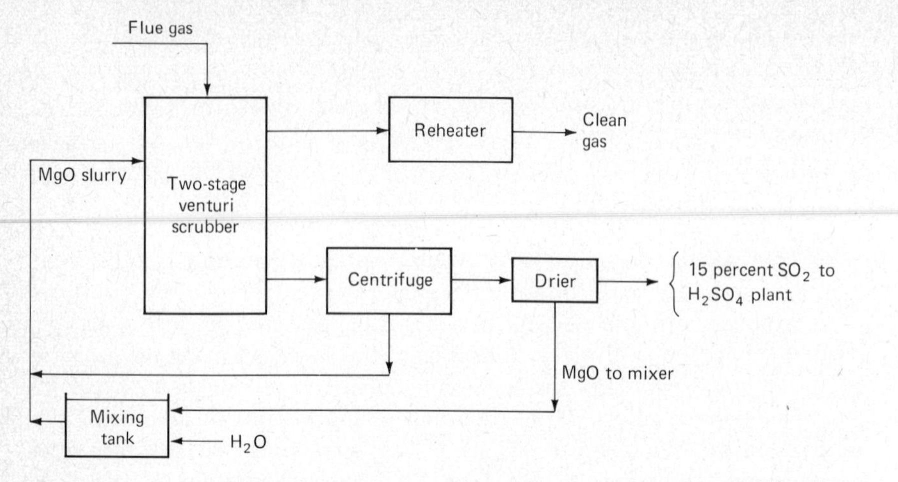

Figure 7-4 Simplified schematic of the magnesium oxide flue gas desulfurization process.

Demonstration projects, partially funded by EPA, have been carried out at the Mystic station of Boston Edison and the Dickerson station of Potomac Electric and Power. The former is a 150-MW unit operating on 2.5 percent sulfur oil; the latter is a 100-MW unit operating on 3 percent sulfur coal. The Mystic station completed a 2-yr study in May, 1974. During the early months of 1974 the system availability was 85 percent. Boiler tube failures accounted for 75 percent of the shutdowns, with the scrubber accounting for the rest. The system achieved 91 percent SO_2 removal efficiency, and plugging, scaling, and erosion were negligible. Both the Mystic and the Dickerson systems are Chemico-Basic Processes using venturi scrubbers. Study has proved the capability of calcining sulfites and sulfates back to active magnesium oxide. The reactivated absorbent showed 90 percent removal efficiency, compared with 95 percent for virgin absorbent. In 1974, the added cost was $1.40 per barrel of oil burned. In comparison, it would cost an extra $3.00 per barrel to use 0.5 percent sulfur oil rather than the 2.5 percent sulfur oil in actual use. Experiments with coal-fired units have shown them to be as good as, if not better than, the oil-fired unit discussed above. Another magnesium oxide scrubbing system, by United Engineers, has been set for the Eddystone station of the Philadelphia Electric Company. This 120-MW unit operates on 2.5 percent sulfur coal.

Because the MgO scrubbing system is regenerative, it does have one disadvantage: heat for calcining is required in addition to the heat already required to raise the temperature of the flue gases for proper buoyancy in the stack. Under optimum conditions, the sales of a product such as sulfuric acid would offset the additional cost of calcining. A major advantage of the system is that, because it is a closed loop, it produces no wastes and hence no waste disposal problem [13].

7-4-E Single Alkali Scrubbing

The main competition in the field of alkali absorption is between sodium and ammonia [14]. Clear solutions of either are excellent absorbents for SO_2. Thus the problems associated with slurry scrubbing (using alkaline earths) are avoided. In addition, the regeneration step can be carried out at a relatively low temperature in a liquid system. One advantage that sodium scrubbing (usually with NaOH or Na_2SO_3 solutions) has over ammonia is that the cation is nonvolatile. Fume development is a problem in all ammonia scrubbers. Both processes, however, produce an unwanted but unavoidable side product in the scrubber: sodium sulfate in the one case and ammonium sulfate in the other. For regeneration (or for sale as a fertilizer), ammonium sulfate is a much more desirable side product and thus gives ammonia scrubbing the advantage [15].

One of the best known sodium-based methods is the Wellman–Power Gas system. The EPA co-sponsored a project with Northern Indiana Public Service for a 115-MW unit at Gary, Indiana, with a startup date of mid-1977. A sodium sulfite (Na_2SO_3) solution scrubs the SO_2 from the flue gas and sodium bisulfite ($NaHSO_3$) is formed. The basic reaction is

$$SO_2 + Na_2SO_3 + H_2O \rightarrow 2NaHSO_3$$

The sodium sulfite is regenerated in an evaporator-crystallizer through the application of heat. A concentrated 90 percent SO_2 stream is produced at the same time. The overall regeneration reaction is

$$2NaHSO_3 \underset{\text{heat}}{\rightarrow} Na_2SO_3(s) + H_2O + SO_2(\text{conc.})$$

In this particular case, the SO_2 stream goes to an Allied Chemical plant where elemental sulfur is formed by the Claus process. Sulfuric acid production follows. For lower sulfur coal, a 200-ppm SO_2 concentration in the exit-stack stream is expected. One objection to this process, referred to previously, is the necessity of disposing of sodium sulfate crystals. The installation cost of the system was estimated to be \$120/kW, with a \$0.60/MBtu operating cost. A process schematic is shown in Figure 7-5.

Actual performance data at the NIPSCO plant in Gary indicated an average SO_2 removal efficiency of 91 percent in late 1977. The system also did a good job of final dust removal, with particulate emissions below 0.08 $lb/10^6$ Btu. Operating cost goals were met, and the sulfur produced reached 99.9 purity. The consumption of sodium carbonate, used for sodium makeup, was also below maximum specifications during the test period. The world's largest application of the Davy/Wellman Lord scrubbing and recovery system for SO_2 control is built at the San Juan station of Public Service of New Mexico, located in the four corners area of New Mexico. The two-year construction program was completed in mid-1978. The project cost \$120

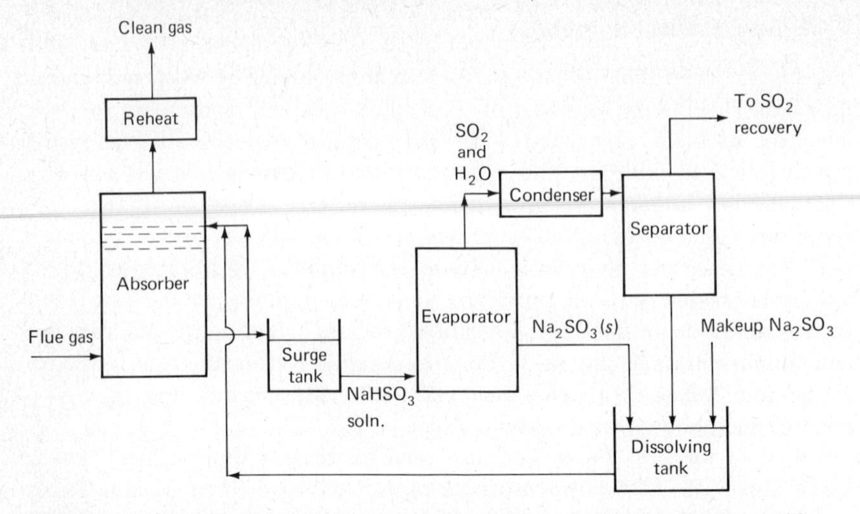

Figure 7-5 Simplified schematic for single alkali scrubbing of flue gas with regeneration.

million for the two 330-MW units. The scrubbers are preceded by electrostatic precipitators designed for 99.9–99.7 percent efficiency. Sodium carbonate is used for makeup, and an Allied Chemical recovery process produces sulfur for sale. About 2300 lb/hr of sulfur should be available from each 330-MW unit.

Sodium scrubbing at the Reid Gardner station of the Nevada Power Company began in early 1974 on two 125-MW units. A scrubber on a third 125-MW unit started up in mid-1976. Sulfur dioxide control is achieved by a sodium carbonate throwaway system designed by Combustion Equipment Associates. The interesting feature of this plant is that a low sulfur (0.5 to 1 percent) coal is burned. Combustion gases first pass through a CEA venturi scrubber, where most of the particulate is removed. The gases then pass through an absorption tower where they come in contact with the sodium carbonate solution. Over 90 percent of the SO_2 and most of the remaining particulate is removed here. Reduction of the SO_2 content from 350 ppm to as low as 12 ppm has occurred. The state limitation on the plant is 50 ppm. The sodium sulfite-sulfate residue and fly ash are dumped in 45-acre sealed ponds. Cost was 9.5 million dollars for retrofitting the original 250-MW plant.

Ammonia scrubbing is under active investigation, and a successful demonstration on a 30-MW slipstream of a 250-MW plant near Paris, France, was carried out in mid-1976 [16]. In the French IFP process the stack gas pollutants are first captured in an aqueous ammonia liquor in a scrubbing section. In a second reaction or conversion section the liquor is changed into ammonia, sulfur dioxide, and water. A portion of the SO_2 is then converted into H_2S, and the H_2S and SO_2 in a 2 : 1 ratio is converted into elemental sulfur by the Claus process. The ammonia is recovered and recycled back to the scrubber section. In 1977 the estimated installation cost was $60/kW, and operating costs were cited as 3.3 mils/kW-hr (neglecting income from sulfur sales).

7-4-F Double Alkali Scrubbing

A sodium single-alkali throwaway system was discussed in the preceding section. This process is quite expensive if the sulfur content of the coal is high, since costly chemicals are not being regenerated. However, this feature can be overcome in a double or dual-alkali system. The primary scrubbing loop usually involves a sodium oxide or sodium hydroxide solution which combines with sulfur dioxide to form primarily sodium sulfite (Na_2SO_3). The spent scrubbing liquor is removed to a secondary loop where lime (a second and inexpensive alkali) is added in a reactor. The precipitated calcium sulfite/sulfate is separated in a thickening tank, and final drying is accomplished in a rotary drum filter. The final dried product is suitable for a landfill. The expensive sodium alkali solution is continuously regenerated in the reactor and recycled to the primary loop absorber. Combustion Equipment Associates and the Envirotech Corporation are among the suppliers of a double alkali system. A general schematic of a double-alkali scrubbing system is shown in Figure 7-6.

CEA installed dual alkali systems became operational at the Colstrip station of the Montana Power Company in late 1975 and early 1976. Both units are 360 MW and measured removal efficiencies for SO_2 are in the 70 to 75 percent range [11]. Annual average availability was around 87 percent for the two units. Other units by this company and other firms are in operation or in the planning-construction stage at this time.

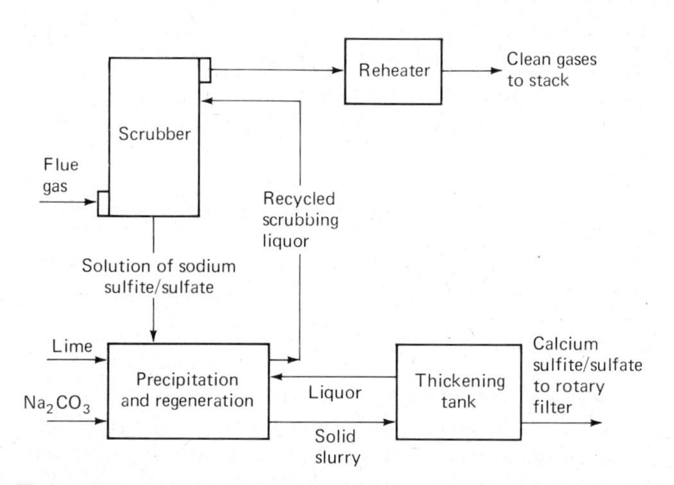

Figure 7-6 Schematic of double alkali scrubbing process.

7-4-G Citric Acid Scrubbing

The absorption of SO_2 in an aqueous solution which contains sodium citrate (with some additional citric acid and sodium thiosulfate) is another possible method for flue gas desulfurization. This approach, known as the Citrate Process, is applicable to sulfur dioxide emissions from power, refining, and metallurgical industries. It was developed by the United States Bureau of Mines in the late 1960s for specific application to the nonferrous smelting industry. After nearly a decade of small-scale testing, a three-year pilot test was completed in 1976 on a base metal smelter. On this basis the overall process design was improved and a commercial size demonstration of citrate technology was begun [17]. The testing of a 156,000-scfm flue gas stream from a coal-fired utility power plant owned by the St. Joe Minerals Corporation near Pittsburgh, Pennsylvania, was to commence during 1978.

From an economics viewpoint, it is claimed [18] that the Citrate Process is applicable to flue gases containing from 500 ppm to 5 percent by volume SO_2. The removal efficiency achieved is at least 80 to 90 percent, and pilot plant tests have shown values exceeding 99 percent. One of the major advantages of this process is that elemental sulfur is the primary product, which is easily stored and directly marketable. The purity of the sulfur product reaches 99.5 percent, and hence is of sufficient quality to use as feedstock for a number of industrial processes. No significant sludge disposal is required and the organic acid reagent is nontoxic and biodegradeable. Thus the environmental impact of the overall process is minimal. The process has a good capacity for short-term overloads of SO_2, and the absorption process is free of scaling and plugging problems. The liquid-to-gas ratio (approximately 4 to 7 gpm per 1000 acfm) is relatively low and results in reduced pumping and fan head losses. Overall, both electrical power and process steam demand is low. As a result only about 3 percent of a utility output is required to operate the Citrate Process.

It is imperative that particulates be removed before the absorption stage in the Citrate Process. In the absorber the removal of SO_2 is enhanced by the buffering property of citric acid, which maintains the pH around 4.5. (A buffered solution is one to which large amounts of acid or base can be added with only a small resultant change in hydrogen ion concentration.) For combustion gases a 0.25 *M* citric acid solution is required for pH control. Solution temperature is the other major operating variable having the greatest effect on SO_2 removal efficiency. The SO_2 absorption is controlled by a reaction which forms bisulfite, namely,

$$SO_2(g) + H_2O(l) \rightleftharpoons HSO_3^- + H^+$$

There are no solids formed in the absorber.

A major portion of the overall process is devoted to regeneration of the rich citrate solution leaving the absorber. This regeneration is achieved by the liquid phase reaction of the bisulfite with hydrogen sulfide (H_2S). The overall regeneration reaction is given by

$$H^+ + HSO_3^- + H_2S \rightarrow 3S(s) + 3H_2O$$

The elemental sulfur formed in this reaction is concentrated and removed from the solution by air flotation. The resulting slurry is further treated to produce a liquid sulfur product. The lean citric acid solution from the flotation process is returned to the SO_2 absorber. Finally, H_2S must be available for the conversion of bisulfite to elemental sulfur. If not obtainable from some other source in the plant, it can be produced by the reaction of some of the recovered sulfur product with a reducing gas, such as hydrogen, methane, methanol, or carbon monoxide. Roughly two-thirds of the sulfur produced is converted to H_2S for use in the sulfur precipitation step.

The economics for the citrate process has been estimated [18] on the basis of 1977 dollars. For a 1000-MW coal-fired power plant, the total capital cost would run around 73 million for recovered sulfur in the amount of 63,000 tons. The total direct operational cost (without capital charges applied and sulfur product credited) would run around 16 million. The unit operational cost is 2.4 mils/kW-hr.

7-4-H Dry Scrubbing

Prior to 1980 the removal of SO_2 by absorption was usually carried out by a general technique known as wet scrubbing. Wet scrubbing simply implies that the SO_2 leaves the absorption equipment in a solution or slurry, and further treatment is necessary in order for the sulfur to end up in a suitable land-fill operation or in some form of marketable product. Hence considerable equipment is needed to complete the process, once the SO_2 has been removed in the scrubber operation. An alternative to this is a process known as dry scrubbing, or spray-dryer absorption. As in wet scrubbing, the SO_2 is removed by coming into intimate contact with a suitable absorbing solution. Typically this might be a solution of sodium carbonate or slaked lime. In dry scrubbing, however, the solution is pumped to atomizers, which create a spray of very fine droplets. The droplets mix with the incoming flue gas in a very large chamber, and subsequent absorption leads to the formation of sulfites and sulfates within the droplets. Nearly simultaneously the sensible heat of the flue gas, which enters the chamber at around 300°F (150°C), evaporates the water in the droplet. Thus a dry powder is formed before the gas leaves the spray dryer; hence the name—dry scrubbing. The temperature of the gas leaving the spray dryer is now around 50°F (30°C) above its dewpoint.

Another major consideration in dry scrubbing is that the fly ash in the flue gas is not removed prior to the spray dryer. Hence the flue gas leaving the spray dryer contains a particulate mixture of both reacted products and fly ash. This particulate mixture commonly is very similar in particle size and physical characteristics to the fly ash itself. Thus conventional particulate removal systems can be used to complete the overall process. Although electrostatic precipitators might be used, the design typically involves a baghouse employing teflon-coated fiberglass. Expected particulate emissions are expected to be less than 0.01 gr/ft^3 (0.02 g/m^3). The use of a baghouse instead of an ESP is advantageous in terms of SO_2 removal. The flue gas from the dryer passes through unreacted material collected on the bags, and additional SO_2 removal may occur. A major advantage of the spray-dryer system (in comparison to many wet scrubbing processes) is the use of much less equipment and the concomitant lowering of maintenance and operating costs. In addition, one is able to collect both fly ash and waste material simultaneously, which is not possible in some wet scrubbing processes. In practice both the first-stage spray dryer and the second-stage fabric filter are designed so that it is not necessary to shut down the system for normal maintenance.

The first commercial dry scrubbing system installed in the United States is scheduled for operation in mid-1981 at the Coyote station near Beulah, North Dakota. The 410-MW unit burns lignite with an average sulfur content of 0.78 percent and roughly 7 percent ash [19]. Sodium carbonate is used as the sulfur removal reactant. The operating pressure drop across the FGD equipment and associated ductwork is anticipated to be around 13 in. H_2O (32 mbar), and the liquid-to-gas ratio to be around 0.3 gal/1000 scf of flue gas. The turnkey installation cost for the emissions control system is $78/kW in 1977 dollars. A second dry scrubbing system is to be installed on two 440-MW units at the Antelope Valley station of Basin Electric Power Cooperative near Bismarck, North Dakota [20]. The units are scheduled to be operational in April, 1982, and November, 1983. The station uses pulverized lignite firing, and a slaked lime slurry will be used to remove SO_2 in the spray dryer. A baghouse will be used to capture the fly ash and the dry powder containing calcium sulfites and sulfates. The inlet SO_2 concentration of roughly 800 ppm in the flue gas will be lowered to 300 ppm, which is sufficient to meet the North Dakota standard of 0.78 lb SO_2/million Btu.

The dry-scrubbing units in North Dakota are throwaway systems. The viability of a new system to regenerate the sodium scrubbing agent and produce a sulfur product will be tested at 100-MW demonstration program scheduled for the Huntley station of Niagara Mohawk Corporation. Using Rockwell International's Aqueous Carbonate Process, the plant construction is expected to be completed by mid-1981. The solids from the spray-drying process are collected by an ESP, combined with some coal, and fed to a reactor at 1800–1900°F (1250–1310°K) and then to a series of columns. The result is the regeneration of the sodium carbonate and the production of hydrogen sulfide, which is converted to sulfur. A major advantage of the reducing process is the use of coal, rather than an expensive gas such as natural gas. The projected cost is $51 million.

7-4-I Other Methods and Costs

Other methods of desulfurizing flue gas, involving both wet and dry systems, are possible. Included are adsorption on charcoal, scrubbing with sulfuric acid followed by crystallization with limestone, and organic scrubbing. Foster-Wheeler is involved with German firms in a demonstration effort which uses activated carbon to remove SO_2. The \$4.1 million program, demonstrated at a utility coal-fired boiler, involves conversion of SO_2 to elemental sulfur. Joy Manufacturing has announced (1979) a recovery process, called the Cooper process, which removes 99 percent or better of SO_2 and produces potassium sulfate, a valuable fertilizer product. A process called SULF-X, patented by Pittsburgh Environment and Energy Systems, Inc., uses iron sulfide (FeS) instead of limestone to remove sulfur dioxide from stack gases. Elemental sulfur is the only by-product of the process. As of 1979 the process was to be adapted to a 3-MW coal-fired plant in Pittsburgh. The reader is referred to the current literature for additional information. References 11, 14, 21, and 22 provide summaries of many of the desulfurization processes in use or under investigation.

Because of the rapid change in prices in the mid-1970s, it is difficult to predict the installation and operating costs for sulfur dioxide removal units in power plants. In early 1979 the installation cost on new units probably ranged from \$100 to \$150/kW. Costs for retrofitting on existing units probably were around \$125 to \$175/kW. For a 500-MW station, the cost would be in the \$40 to \$60 million range. The cost of the power plant alone would run about \$500 to \$650/kW. Thus a sulfur dioxide removal system accounts for a sizable fraction of the total cost of a new power plant requiring emission controls. The operating cost in this same time period might run upward of 4.5 mils/kW-hr, 50 cents/MBtu, or \$12/ton of coal.

In summary, it can be stated that no single method of removing SO_2 from large quantities of flue gas has a clear-cut advantage over the others. All of the proposed systems require extensive modification of the basic power plant design. Some systems cannot be retrofitted to existing units. In some cases, relatively large land areas are required to accommodate the additional equipment for the SO_2 removal process. Additional land area is required if a throwaway system is employed. It is highly likely that a number of processes may prove fruitful, tailored to a particular situation involving land availability, the sulfur content of the fuel, presence of a local market for recovery products such as sulfur or sulfuric acid, as well as other factors.

QUESTIONS

1. What are the objections to the lime scrubbing throwaway process for removing SO_2 from stack gases?
2. What are the advantages of the lime scrubbing throwaway process?

3. What processes would you recommend for removing SO_2 from the flue gases of an industrial oil-fired furnace with a steam rate of 100,000 lb/hr?
4. In what ways can a coal-fired industrial heating plant meet the 1970 federal emission standards? Why are these methods feasible?
5. What are the objections to the use of tall stacks to meet the 1970 federal emission standards?
6. Discuss the impact that the tonnage of sulfur or sulfuric acid recovered from SO_2 in flue-gas treatment would have upon the chemical industry if all coal- and oil-fired power plants were required to use such treatment processes.
7. If all coal-burning power plants were required to use sulfur recovery processes, how would you suggest disposing of the surplus sulfur or sulfuric acid?

PROBLEMS

7-1. A new power plant is being designed to burn coal having a 3 percent sulfur content and a heating value of 11,000 Btu/lb. What is the minimum efficiency that a sulfur dioxide removal system can have for the plant to comply with an emission standard of 1.2 lb/10^6 Btu?

7-2. A new power plant is designed to burn coal with a 4.1 percent sulfur content and a heating value of 10,800 Btu/lb. What is the minimum efficiency that a sulfur dioxide removal system can have in order for the plant to comply with an emission standard of 1.2 lb/10^6 Btu?

7-3. A No. 5 fuel oil having a maximum sulfur content of 2.0 percent and a heating value of 148,000 Btu/gal is being considered as a fuel for an industrial boiler. If the specific gravity of the oil is 0.953, what is the minimum efficiency required of a sulfur dioxide scrubber system in order for the plant to meet an emission standard of 0.8 lb/10^6 Btu for a new plant?

7-4. Consider an initial gaseous mixture which contains 2500 ppm of SO_2 and 60 ppm of SO_3, plus 3.0 percent by volume of O_2. If the temperature is 1000°K, estimate the equilibrium concentrations of SO_2 and SO_3 based on thermodynamic considerations alone. The pressure is 1 atm.

7-5. Limestone ($CaCO_3$) is used in a throwaway limestone scrubbing system for a powerplant burning coal containing 3.60 percent sulfur and 7.70 percent ash. The SO_2 removal efficiency must be 85 percent to meet federal emission standards for SO_2. Determine (a) the stoichiometric pounds of limestone required per pound of sulfur in the coal, (b) the pounds of limestone required per ton of coal if 30 percent excess limestone is used, (c) the pounds of sludge produced per ton of coal if the sludge is 60 percent water and 40 percent $CaSO_4 \cdot 2H_2O$, plus ash.

7-6. A 500-MW electric power plant burns coal having 4.0 percent sulfur and a heating value of 11,500 Btu/lb. The plant thermal efficiency is 33 percent and the ash content of the coal is 6.5 percent. Determine (a) the required SO_2 removal efficiency if the federal SO_2 emission standard is satisfied, (b) the tons of limestone ($CaCO_3$) required per day if the limestone scrubbing slurry contains 30 percent excess limestone, and (c) the tons per day of sludge produced (including ash) by the throwaway process if the sludge is 60 percent water and 40 percent $CaSO_4 \cdot 2H_2O$ and ash, by weight.

References

1. C. F. Cullis and M. F. R. Mulcahy. "The Kinetics of Combustion of Gaseous Sulfur Compounds." *Combust. Flame* **18** (1972): 225.
2. A. B. Hedley. "Factors Affecting the Formation of Sulfur Trioxide in Flame Gases." *J. Inst. Fuel* **40** (1967): 142.
3. H. Palmer and J. M. Beers. *Combustion Technology*. New York: Academic Press, 1974.
4. *Air Quality Criteria for Sulfur Oxides*, AP-50. Washington, D.C.: Public Health Service, 1969.
5. *Control Techniques for Sulfur Oxide Air Pollutants*, AP-52. Washington, D.C.: Public Health Service, 1969.
6. A. V. Slack. *Sulfur Dioxide Removal for Waste Gases*. Noyes Data Corp., 1971.
7. C. W. Ertel and J. T. Metcalf. "New Fuels, Old Coal." *Mech Engr*. **94** (March 1972).
8. H. C. Messman. "Desulfurize Coal?" *Chem. Tech.* **1** (February 1971).
9. *Geological Survey Bulletin 1322*. U.S. Geological Survey, 1970.
10. G. Van Houte. "Desulfurization of Flue Gases in Fluidized Beds of Modified Limestone." *J. Air Pollu. Control Assoc*. **28**, no. 10 (1978): 1030–1033.
11. W. H. Megonnell, "Efficiency and Reliability of Sulfur Dioxide Scrubbers." *J. Air Pollu. Control Assoc*. **28**, no. 7 (1978): 725–731.
12. W. J. Raymond and A. G. Sliger. "The Kellogg Weir Scrubbing System." *Chem. Engr. Progress* **74**, no. 2 (1978): 75–80.
13. *Sulfur Oxide Removal from Power Plant Stack Gas: Magnesia Scrubbing*. PB-222 509, TVA. National Technical Information Service, 1973.
14. A. V. Slack. "Removing SO_2 from Stack Gases." *Environ. Sci. Tech*. **7** (1973): 110.
15. *Sulfur Oxide Removal from Power Plant Stack Gas: Ammonia Scrubbing*. PB-196 804, TVA. National Technical Information Service, 1970.
16. "Regenerative SO_2 Removal Process." *Environ. Sci. Tech*. **11**, no. 1 (1977): 22–23.
17. W. I. Nissen et al. "Citrate Process for Flue Gas Desulfurization, A Status Report." Proceedings of the Flue Gas Desulfurization Symposium, EPA-600/2-76, May, 1977.
18. R. S. Madenburg et al. "Industrial Application of Citrate FGD Technology." 71st Annual Meeting, Air Pollu. Control Assoc., Houston, Texas, June, 1978.
19. O. B. Johnson et al. "Coyote Station—First Commercial Dry FGD System." 41st Annual American Power Conference, Chicago, Ill., April, 1979.
20. R. A. Davis et al. "Dry SO_2 Scrubbing at Antelope Valley Station." 41st Annual American Power Conference, Chicago, Ill., April, 1979.
21. G. T. Rochelle. "A Critical Evaluation of Processes for the Removal of SO_2 from Power Plant Stack Gas." Annual Meeting, Air Pollution Control Association, June, 1973.
22. "SO_2 Removal Technology Enters the Growth Stage." *Environ. Sci. Tech*. **6** (August 1972).

Chapter 8

Control of Oxides of Nitrogen from Stationary Sources

8-1 INTRODUCTION

The stable gaseous oxides of nitrogen include N_2O (nitrous oxide), NO (nitric oxide), N_2O_3 (nitrogen trioxide), NO_2 (nitrogen dioxide), and N_2O_5 (nitrogen pentoxide). An unstable form, NO_3, also exists. Of these, the only ones present in the atmosphere in any significant amount are N_2O, NO, and NO_2. Thus these three are potential contributors to air pollution. Nitrous oxide (N_2O) is an inert gas with anesthetic characteristics. Its ambient concentration is 0.50 ppm, which is considerably below the threshold concentration for a biological effect. In addition, it has a balanced environmental cycle which is independent of the other oxides of nitrogen.

Nitric oxide (NO) is a colorless gas and its ambient concentration is usually far less than 0.5 ppm. At these concentrations its biological toxicity in terms of human health is insignificant. However, nitric oxide is a precursor to the formation of nitrogen dioxide and is an active compound in photochemical smog formation as well. Hence it initiates reactions that produce air pollutants. Consequently, the control of NO is an important factor in reducing air pollution. Nitrogen dioxide (NO_2) is a reddish brown gas and is quite visible in sufficient amounts. A concentration of 1 ppm of NO_2 probably would be detected by the eye. The NO_2 ambient air quality standard in California initially was set at 0.25 ppm (on an hourly average), primarily on the basis of visibility effects. The toxicological and epidemiological effects of NO_2 on human beings are not completely known. As far as we know, NO_2 is not a primary pollutant in the sense that it directly affects human health, unless the concentration is extremely high. In the past, the threshold limit in ambient air has been considered to be around 5 ppm for daily exposure. A threshold level for physiological derangement of approximately 1.5 ppm NO_2 has been suggested [1]. The environmental hazard of nitrogen dioxide is primarily associated with the pulmonary (respiratory) effects of the pollutant. Healthy individuals exposed to concentrations of NO_2 of from 0.7 to 5.0 ppm for 10 to 15 min have developed abnormalities in pulmonary airway resistance. Exposure to 15 ppm of NO_2 causes eye and nose irritation, and pulmonary discomfort is noted at 25 ppm for exposure of less than 1 hr. However, these results are probably without medical significance with regard to present atmospheric concentrations [1], since they occurred only at nitrogen dioxide levels of from 5 to 20 times that of ambient air. The significance of current atmospheric levels of NO_2 with regard to human health is inadequately known.

The primary federal air quality standard for nitrogen dioxide on an annual basis is 100 $\mu g/m^3$, or 0.05 ppm. In process gases and in the ambient atmosphere both NO and NO_2 are usually present in significant amounts but varying proportions. In general discussions they are usually lumped together under the generic formula NO_x. When actual data are presented, however, the mass or volume values frequently are quoted on an "equivalent NO_2" basis.

8-2 SOURCES AND CONCENTRATIONS OF NO_x

Well over 90 percent of all the man-made nitrogen oxides that enter our atmosphere are produced by the combustion of various fuels. On a nationwide basis, roughly one-half of the NO_x is from stationary sources, while the remainder is from mobile sources such as spark-ignition and compression-ignition engines in automobiles and trucks. For 1977, Table 8-1 lists estimated nationwide NO_x emissions from stationary and mobile sources. We see that fuel combustion in stationary sources, including electric utilities, accounts for about 56 percent of the total 25.4 million tons/yr, with about 40 percent attributed to mobile sources [2].

On a global basis, the man-made emission rate of NO_x is not of grave concern at the present time. Nitrogen oxides are an essential part of the

Table 8-1 ESTIMATED NO_x EMISSIONS IN THE UNITED STATES IN 1977

SOURCE	THOUSANDS OF TONS	PERCENT
Transportation	10,100	39.8
Fuel combustion, stationary sources	14,300	56.3
Industrial processes	780	3.1
Solid waste disposal	110	0.4
Miscellaneous	110	0.4
	25,400	100.0

nitrogen cycle in nature. Nitrogen dioxide is hydrolyzed to nitric acid in the atmosphere, which in turn is precipitated out as nitrates. These return to the earth's surface as fertilizer for organic growth. Among the possible atmospheric reactions for the formation of nitric acid are:

$$O_3 + NO_2 \rightarrow NO_3 + O_2$$
$$NO_3 + NO_2 \rightleftharpoons N_2O_5$$
$$N_2O_5 + H_2O \rightarrow 2HNO_3$$
$$OH + NO_2 \rightarrow HNO_3$$

Nitric acid is a major contributor to acid rain, as discussed in Chapter 1. It is estimated that the NO_x formed in this natural cycle amounts to 500×10^6 tons/yr. This is 20 times the total national emission rate in 1977. As a result of this cycle the average background concentration of NO_x, in nonpolluted environments, is around 1 ppb. In contrast, urban metropolitan areas have concentrations that frequently average 40 to 80 ppb or higher. Maximum concentration of oxides of nitrogen in major urban areas can reach the 0.3- to 1.4-ppm levels, for averaging times of 1 day or less. The average level of NO_x in the atmosphere of smog-ridden areas of California is around 0.25 ppm, with maximum concentrations reaching 3.5 ppm. Although data from other countries are not extensive, hourly values of 0.11 to 0.25 ppm have been reported for London and several Japanese cities [3, 4, 5].

From the preceding data we can see that air pollution associated with NO_x is a local problem. It depends on the emission rates of sources in the local area. On the basis of current medical knowledge and ambient NO_x concentrations, NO_x is not considered a health hazard, per se. The real danger posed by NO_x at the concentrations found in metropolitan areas lies in its role in photochemical reactions leading to smog formation. These atmospheric reactions lead to the formation of chemical compounds that do have a direct adverse effect on human beings and plants. In some situations, NO_x may be present in a high enough concentration, yet not react to form smog because other necessary conditions for the reaction are absent. However, nearly every major city in any technologically advanced country at times now experiences the effects induced by the presence of NO_x.

At an emission source the concentration of oxides of nitrogen is much higher than ambient values. For example, the NO_x concentration in the flue gas from a gas-fired domestic water heater is 10 ppm or less, while the level in the flue gas from the steam boiler of a power plant may reach 500 to 1000 ppm. The actual quantities of NO_x produced by any given industry can be quite large. For example, a 750-MW gas- or coal-fired steam power plant produces around 7500 to 9500 lb of NO_x (NO_2 equivalent) per hour. The total flue-gas flow rate for the plant would be about 80×10^6 scf/hr. From such combustion processes the NO_x is the exhaust stack gas would be 90 percent or more NO, and the rest NO_2. The type of fuel used can alter the emission rate significantly. Generally, the NO_x emissions, based on the same energy released, increase in the order gas-oil-coal. This trend is illustrated in Table 8-2, which compares the fuels for different types of applications [6]. In about 1970, nationwide data for stationary sources showed that coal and oil combustion each accounted for 35 to 45 percent of NO_x emissions, while gas contributed about 15 percent. These values continue to change as specific fuel allocations are altered in the light of new energy and air pollution problems. Because of the differences in NO_x emissions from the combustion of coal, oil, and gas, the standards of performance for new stationary sources promulgated by EPA are different for different fuels. As an example, the standards of performance for new steam electric power plants, which went into effect in February, 1980, are (based on 30-day rolling average):

FUEL TYPE	$lb/10^6$ Btu	$kg/10^6$ kJ
Gaseous	0.20	0.09
Liquid	0.30	0.13
Subbituminous coal, shale oil, and fuels derived from coal	0.50	0.21
Anthracite or bituminous coal	0.60	0.26

These values are 2-hr averages, expressed as NO_2 equivalent. In contrast to Table 8-2, which indicates typical NO_x emissions on an energy-release basis, Table 8-3 lists estimates of average emission factors for NO_x based on the amount of fuel consumed [7].

The quantity of NO_x dispersed into the atmosphere can be reduced two ways. The primary method is control over the reaction which produces the pollutant. As a second possibility, the pollutant might be removed after it is formed. These two approaches will be discussed in Section 8-4. The control methods for NO_x emissions require an understanding of the main factors responsible for NO_x formation. This in turn requires a general understanding of the basic chemisty, thermodynamics, and kinetics of the formation reactions. The vast majority of NO_x emissions come from the burning of fuels in stationary and mobile devices. Hence, we limit our attention in this chapter to combustion reactions. The major noncombustion source of NO_x is in the manufacture and use of nitric acid.

There are two sources of nitrogen which contribute to the formation of oxides of nitrogen in combustion reactions. The inevitable source in fuel-air reactions is the air itself, which contains molecular nitrogen and oxygen in a molar ratio of roughly 3.75 : 1. In addition, evidence indicates that fuels

Table 8-2 COMPARISON OF THE EFFECT OF FUEL TYPE ON NO_x EMISSIONS, EXPRESSED AS lb $NO_x/10^9$ Btu RELEASED

	HOUSEHOLD AND COMMERCIAL	INDUSTRY	ELECTRIC UTILITIES
Natural gas	110	205	375
Fuel oil	80–480	480	690
Coal	340	840	840

Table 8-3 EMISSION FACTORS FOR NITROGEN OXIDES

SOURCE	AVERAGE EMISSION FACTOR
Coal	
Household and commercial	8 lb/ton of coal burned
Industry and utilities	20 lb/ton of coal burned
Fuel oil	
Household and commercial	12 to 72 lb/1000 gal of oil burned
Industry	72 lb/1000 gal of oil burned
Utility	104 lb/1000 gal of oil burned
Natural gas	
Household and commercial	116 lb/million ft^3 of gas burned
Industry	214 lb/million ft^3 of gas burned
Utility	390 lb/million ft^3 of gas burned
Gas turbines	200 lb/million ft^3 of gas burned
Waste disposal	
Conical incinerator	0.65 lb/ton of waste burned
Municipal incinerator	2 lb/ton of waste burned
Mobile source combustion	
Gasoline-powered vehicle	113 lb/1000 gal of gasoline burned
Diesel-powered vehicle	222 lb/1000 gal of oil burned
Aircraft: conventional	23 lb/flight per engine
fan-type jet	9.2 lb/flight per engine
Nitric acid manufacture	57 lb/ton of acid product

SOURCE: NAPCA. *Control Techniques for Nitrogen Oxides from Stationary Sources*, AP-67. Washington, D.C.: HEW, 1970 [7].

which contain nitrogen in their structure may be responsible for an appreciable fraction of the nitrogen that eventually appears as NO_x. Fuel oil and coal are of primary concern as contributors, since natural gas is essentially free of nitrogen-type compounds. From a theoretical viewpoint, the bond energy of N≡N in molecular nitrogen is much greater than that of a C—N bond in an organic compound. Thus, oxygen should preferentially attack the weaker C—N bond. Laboratory studies [8] have shown that the addition of organic nitrogen compounds to CO-oxidizer flames leads to NO_x formation in the absence of molecular nitrogen in the oxidizer. These studies covered a temperature range of 860° to 1150°K (1100° to 1600°F).

8-3 THERMODYNAMICS OF NO AND NO_2 FORMATION

In order to interpret data correctly, make suggestions for modification of current equipment, and design new equipment, one must have a basic understanding of the thermodynamics and kinetics of nitrogen-oxygen reactions, especially at high temperatures. The two overall reactions of concern are those which produce nitric oxide, NO, and nitrogen dioxide, NO_2. The equilibrium reactions are

$$N_2 + O_2 \rightleftharpoons 2NO \tag{8-1}$$

$$NO + \tfrac{1}{2}O_2 \rightleftharpoons NO_2 \tag{8-2}$$

The expression for the equilibrium constant K_p for reaction (8-1), and typical values for K_p based on the JANAF Thermochemical Tables, are shown in Table 8-4. The value of K_p is extremely small ($< 10^{-4}$) for temperatures below 1000°K (1350°F). Up to this temperature, then, the partial pressure of NO and hence the equilibrium quantity of NO would be very small. Above 1000°K appreciable amounts of NO could be formed. Table 8-5 lists the theoretical amount of NO at equilibrium for two special situations. The third-column data are for nitrogen and oxygen initially present in a 4 : 1 ratio with no other gases present. The fourth column is for a 40 : 1 ratio, but in a product gas mixture roughly representative of hydrocarbon combustion with 10 percent excess air. The values are only an approximation to actual combustion conditions, since the presence of CO_2 and H_2O in the exhaust gases has been neglected. However, the values do indicate two points. First, equilibrium NO formation increases rapidly with increasing temperature. [Note that the flame temperature of hydrocarbon reactions is in the vicinity of 3000° to 3500°F (1900° to 2200°K), for air concentrations slightly above stoichiometric.] Second, the equilibrium values of NO are of the same order of magnitude as the measured values quoted earlier for a steam power plant (that is, 500 to 1000 ppm).

Table 8-4 EQUILIBRIUM CONSTANTS FOR THE FORMATION OF NITRIC OXIDE FROM MOLECULAR OXYGEN AND NITROGEN

	T		
$N_2 + O_2 \rightleftharpoons 2NO$	(°K)	(°F)	K_p
$K_p = \frac{(p_{NO})^2}{(p_{N_2})(p_{O_2})}$	300	80	10^{-30}
	1000	1340	7.5×10^{-9}
	1200	1700	2.8×10^{-7}
	1500	2240	1.1×10^{-5}
	2000	3140	4.0×10^{-4}
	2500	4040	3.5×10^{-3}

SOURCE: JANAF Thermochemical Tables, Dow Chemical Company [9].

Table 8-5 TYPICAL NO COMPOSITION AT VARIOUS TEMPERATURES FOR THE EQUILIBRIUM REACTION $N_2 + O_2 \rightleftharpoons 2NO$, AS A FUNCTION OF INITIAL N_2/O COMPOSITION

T			
(°K)	(°F)	$4N_2/O_2$ (ppm)	$40N_2/O_2$ (ppm)
1200	1700	210	80
1500	2240	1,300	500
1800	2780	4,400	1650
2000	3140	8,000	2950
2200	3500	13,100	4800
2400	3860	19,800	7000

Table 8-6 EQUILIBRIUM CONSTANTS FOR THE NITRIC OXIDE OXIDATION REACTION TO NITROGEN DIOXIDE

$NO + \frac{1}{2}O_2 \rightleftharpoons NO_2$	T (°K)	T (°F)	K_p
$K_p = \dfrac{p_{NO_2}}{(p_{NO})(p_{O_2})^{1/2}}$	300	80	10^6
	500	440	1.2×10^2
	1000	1340	1.1×10^{-1}
	1500	2240	1.1×10^{-2}
	2000	3140	3.5×10^{-3}

SOURCE: JANAF Thermochemical Tables, Dow Chemical Company [9].

The values of K_p for reactions (8-2) are listed in Table 8-6 [9]. The equilibrium constant, K_p, for the formation of NO_2 decreases with increasing temperature. Hence the formation of NO_2 is favored at low temperatures, but NO_2 dissociates back into NO at higher temperatures. As the temperature rises above 1000°K, the formation of NO_2 becomes increasingly less likely under equilibrium conditions.

In actual combustion processes, reactions (8-1) and (8-2) occur simultaneously. The predicted equilibrium concentrations of NO and NO_2 in a typical flue gas (3.3 percent O_2, 76 percent N_2) are shown in Table 8-7. On the basis of these thermodynamic data alone, the following situation exists for the overall reactions involving NO and NO_2. At room temperature little NO or NO_2 is formed. Whatever NO is formed is converted to NO_2. In the range of 800°K (980°F) the quantities of NO and NO_2 present are still negligible, although the amount of NO now exceeds that of NO_2. However, at conventional combustion temperatures (> 1500°K or 2250°F) appreciable NO formation is possible, with negligible amounts of NO_2 present.

Nevertheless, all combustion gases eventually are cooled. In an electric power plant, for example, cooling is due to heat transfer from the gas to the steam in the boiler, superheater, economizer, and air preheater. The decreasing temperature theoretically would shift the equilibrium composition of NO

Table 8-7 PREDICTED EQUILIBRIUM COMPOSITIONS OF NO AND NO_2 AT VARIOUS TEMPERATURES FOR THE SIMULTANEOUS REACTIONS $N_2 + O_2 \rightleftharpoons 2NO$ AND $NO + \frac{1}{2}O_2 \rightleftharpoons NO_2$ FOR AN INITIAL COMPOSITION OF 3.3 PERCENT O_2 AND 76 PERCENT N_2

T (°K)	T (°F)	NO (ppm)	NO_2 (ppm)
300	80	1.1×10^{-10}	3.3×10^{-5}
800	980	0.77	0.11
1400	2060	250	0.87
1873	2912	2000	1.8

and NO_2. If excess oxygen is present as the gas cools, the conversion of NO into NO_2 is preferred [reaction (8-2)]. Thermodynamics predicts, then, that the somewhat cooled flue gas leaving a combustion process will consist primarily of NO_x in the form of NO_2.

In actuality, this is not what happens. Although the ambient or global form of the pollutant NO_x is NO_2, the flue gases leaving most combustion processes contain predominantly NO. Roughly 90 to 95 percent of the NO_x emitted in combustion processes appears in the form of NO. As we have seen, NO in flue gases is thermodynamically unstable as it enters the environment and its temperature drops. However, both the decomposition of NO into N_2 and O_2 and the reaction of NO with O_2 to form NO_2 are kinetically limited. As the temperature drops below 2300° to 2400°F (1550°K) the rates of these reactions are very small. Hence the final concentrations of the oxides of nitrogen in the flue gas are essentially frozen at the values found during their formation at higher temperatures. As we noted above, at lower temperatures (below 1500°K) the amount of NO_2 formed would be fairly small. Thus the NO formed in combustion gases at a high temperature is carried out into the ambient atmosphere. The bulk of the oxidation reaction for NO into NO_2 occurs in the atmosphere over a period of time dictated by the kinetics of the reaction. Decomposition of NO into N_2 and O_2 is limited by the high activation energy (~375 kJ/g·mole) of the reaction, which limits the rate. Thus the NO formed at high temperatures preferentially forms NO_2 at lower temperatures, rather than decomposing.

The kinetic reaction mechanism for the formation of NO_2 from N_2 is not as simple and straightforward as that given by the overall reactions (8-1) and (8-2). First, there is evidence [10] that substantial NO_2 may be formed in the preheat zone preceding the major flame zone. Some of this decomposes into NO in the flame zone, and additional NO is formed by some other mechanism in or above the flame zone. The study from which this information was obtained involved a methane-air flame, with two-thirds of the stoichiometric air premixed with the fuel.

Second, the overall reactions are achieved by a complicated set of kinetic reactions which may involve a number of intermediate species. The rate at which NO is formed, for example, is generally considered [11] to be controlled by the reaction

$$N_2 + O \rightarrow NO + N$$

Supplementing this reaction, we might also find the following reactions important in the overall reaction scheme:

$$O_2 + N \rightarrow NO + O$$
$$NO + O \rightarrow NO_2$$
$$O_2 + M \rightleftharpoons 2O + M$$
$$N_2 + M \rightleftharpoons 2N + M$$

where M is a nonreacting body. Thus it must be recognized that although we are concerned specifically about the formation of NO and NO_2 in combustion reactions, the chemistry involved is complex and still open to question.

The preceding discussion has emphasized the importance of two parameters in NO_x formation—temperature and time. Basically, at high temperatures ($> 3000°F$ or $1900°K$) both thermodynamics and kinetics favor NO formation. Hence, peak temperatures should be avoided. Two ways of accomplishing this are:

1. Avoid high heat-release rates.
2. Achieve high heat-removal rates.

One possible result of these methods would be to limit the residence time at peak temperatures to a minimum. If the temperature in the combustion zone can be lowered, then reaction rates become the limiting factor. In this case, the temperature-time profile of the reacting mixture controls the ultimate composition of the flue gas as it leaves the equipment. At these lower temperatures, since equilibrium is not attained, the kinetics of formation is more important than the kinetics of decomposition. Figure 8-1 shows two possible temperature-time profiles for a gas element passing through a combustion zone. The upper profile is undesirable since the element is maintained at a high temperature for a considerable time. With a high heat removal rate, for example, the lower, more desirable, profile might be achieved. In this latter case the rate of formation of NO would be suppressed, and a lower amount of nitrogen dioxide would ultimately be formed.

A third parameter, which is intimately connected to the temperature, is the oxygen content of the reacting mixture. The air-fuel ratio determines to a large extent the peak temperature. In addition, excess oxygen must be present to react with nitrogen. Although the stoichiometric air-fuel ratio leads essentially to the maximum possible combustion temperature, the maximum equilibrium NO_x concentration occurs at a percentage of stoichiometric air somewhat higher than 100 percent. This trend is illustrated by

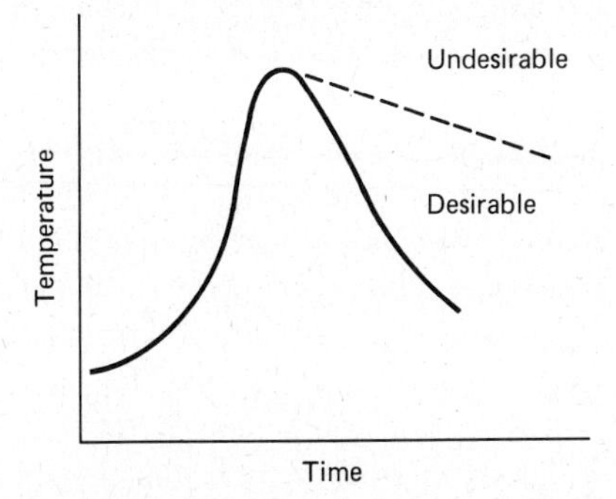

Figure 8-1 Possible temperature-time profiles for a gas element passing through a combustion zone.

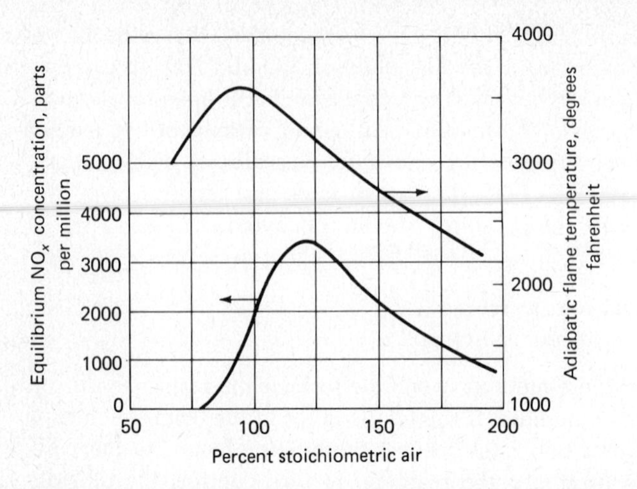

Figure 8-2 Calculated NO_x equilibrium values in methane-air flames.

Figure 8-2 for methane combustion. At equilibrium, the NO_x concentration is a maximum around 115 percent stoichiometric air, or 15 percent excess air. Although the equilibrium constant for NO formation is somewhat lower because of the lower temperature, the presence of an increased oxygen concentration leads to more NO at equilibrium. If the implication of this figure is qualitatively correct for combustion reactions in general, then one approach to NO_x control is combustion at near or substoichiometric air. This leads to problems with other air pollutants in the combustion process. However, operation within the rich fuel-air ratio zone is the basis for one approach to NO_x control.

8-4 KINETICS OF NITRIC OXIDE FORMATION IN COMBUSTION PROCESSES

As noted in Section 8-3, the amounts of NO and NO_2 formed in various combustion systems are not adequately explained by a thermodynamic analysis alone. It is necessary to take into account the rate of reaction as well. Application of chemical kinetics to combustion processes that contain nitrogen and oxygen from atmospheric air has helped elucidate the structure of the chemical reaction system which leads to the formation of the nitrogen oxides. Although the general reaction mechanism is quite complex and some of the details are still subjects of controversy, considerable advancement in the understanding of the complex chemical process has been made since the mid-1960s.

Interest in predicting nitric oxide emissions from mobile and stationary sources has lead to the formulation of various analytical models for NO formation in combustion processes. An inherent feature of such models is a reaction (kinetic) mechanism for NO formation from atmospheric nitrogen. The basic model presently in use had its origin in the work of Zeldovich and coworkers [12, 13] around 1946. Once oxygen atoms are formed by the process

$$O_2 \rightleftharpoons 2O \tag{8-3}$$

then the primary reactions of interest according to Zeldovich's free radical chain mechanism are

$$O + N_2 \rightleftharpoons NO + N \tag{8-4}$$

$$N + O_2 \rightleftharpoons NO + O \tag{8-5}$$

It is interesting to note that although atomic oxygen is formed from the dissociation of diatomic oxygen, as indicated by Equation (8-3), atomic nitrogen is not considered to be formed from molecular nitrogen by dissociation. Instead, atomic nitrogen is formed by means of the reaction expressed in Equation (8-4). The reason for the significant amount of atomic oxygen formed, in comparison to atomic nitrogen, by dissociation becomes apparent when the K_p values for the two dissociation reactions are compared as a function of temperature, as shown below.

TEMPERATURE, °K	2000	2200	2400	2600	2800
$\log_{10} K_p$ for $O_2 \rightleftharpoons 2O$	−6.356	−5.142	−4.130	−3.272	−2.536
$\log_{10} K_p$ for $N_2 \rightleftharpoons 2N$	−18.092	−15.810	−13.908	−12.298	−10.914

Note that the K_p value for the latter reaction is from 10^{-8} to 10^{-12} smaller than that for O_2 dissociation. Hence its effect can be neglected. However, note that the K_p values for O_2 dissociation are small; consequently the actual amount of atomic oxygen formed is extremely small, even at elevated temperatures found in flame zones.

It is assumed that the basic combustion reactions are equilibrated prior to the onset of NO formation in the postcombustion zone of flames. Another elementary reaction frequently added to the list is

$$N + OH \rightleftharpoons NO + H \tag{8-6}$$

However, reactions (8-4) and (8-5) are by far the most important in terms of NO formation in lean and moderately rich flames ($\phi \leqslant 1.2$). Reaction (8-6) is of some importance in fuel-rich flames for $\phi \geqslant 1.2$, where ϕ is the equivalence ratio for the fuel-air mixture. Reaction (8-4) is generally rate-controlling, since breaking the N_2 bond is the most difficult step in the Zeldovich mechanism.

8-4-A Qualitative Study of NO Kinetics

A qualitative evaluation of the kinetics which control NO formation in hydrocarbon-air combustion processes may be made by examining the Zeldovich mechanism, as represented by reactions (8-4) and (8-5). Of these two, reaction (8-4) is thought to be the controlling formation reaction. The removal of NO is due primarily to the reverse of reaction (8-5). The rate constants for these two elementary reactions for the formation and consumption of NO are

$$k_4 = 1.4 \times 10^{14} \exp\left(\frac{-315.5}{RT}\right), \ \mathrm{cm^3/g{\cdot}mole{\cdot}s}$$

$$k_5 = 6.4 \times 10^{9} T \exp\left(\frac{-26.15}{RT}\right), \ \mathrm{cm^3/g{\cdot}mole{\cdot}s}$$

$$k_{-5} = 1.6 \times 10^{9} T \exp\left(\frac{-161.5}{RT}\right), \ \mathrm{cm^3/g{\cdot}mole{\cdot}s}$$

where R is in kJ/g·mole·°K and T is in degrees Kelvin. Note that both reactions have large activation energies, so that the rates of these two reactions will be relatively small at low temperatures. In addition, the O atom concentration which also controls the reaction rates will be negligible at low temperatures. Thus kinetics predicts a very low rate of formation of NO at low temperatures and thermodynamics predicts a low equilibrium concentration.

At high temperatures NO will form rapidly if sufficient O atoms are present. In fact, in the hottest zone of the flame a superequilibrium level of O atoms exists. As a consequence, the concentration of NO increases quickly as hydrocarbon combustion provides the right conditions for the oxidation of atmospheric nitrogen. However, as the gases cool rapidly on leaving the hot flame zone the NO concentration is "frozen" because its removal is dependent upon the reverse of reaction (8-5). The rate of this reaction falls significantly with a decrease in temperature because of its large activation energy and its dependency on O atoms, which now recombine in the cooling gases to form O_2. Although the equilibrium constant, based on thermodynamic considerations, would indicate a shift from NO to N_2 and O_2 as the temperature falls, this shift is kinetically limited. Thus significant NO leaves with the flue gases and enters the atmosphere.

8-4-B Quantitative Study of NO Kinetics

On the basis of the elementary reactions, an equation for the rate of formation of nitric oxide can be developed. It is extremely important, however, that the mechanisms and assumptions used in the development be kept clearly in mind. Rates based on different mechanisms for gas-phase combustion can be significantly different. These differences in NO formation rates can be related directly to differences in O atom concentration and in temperature profiles predicted by the various combustion mechanisms [14].

One basic approach to the prediction of NO formation rates is first to restrict the development to reactions (8-4) and (8-5), namely,

$$O + N_2 \rightleftharpoons NO + N \qquad (8\text{-}4)$$

$$N + O_2 \rightleftharpoons NO + O \qquad (8\text{-}5)$$

In terms of the general theory presented in Section 6-9, the net rate of formation of NO through reaction (8-4) would be

$$\frac{d[\mathrm{NO}]}{dt} = k_4[\mathrm{O}][\mathrm{N_2}] - k_{-4}[\mathrm{N}][\mathrm{NO}]$$

where k_4 and k_{-4} are the forward and reverse rate constants, respectively, for Equation (8-4). The order of the reaction with respect to each species has been taken as unity. In terms of reactions (8-4) and (8-5) the general rate equation for NO is

$$\frac{d[\mathrm{NO}]}{dt} = k_4[\mathrm{O}][\mathrm{N_2}] - k_{-4}[\mathrm{N}][\mathrm{NO}] + k_5[\mathrm{N}][\mathrm{O_2}] - k_{-5}[\mathrm{O}][\mathrm{NO}] \tag{8-7}$$

The next step is to express some of the quantities on the right in terms of other variables.

First, it is usually assumed that a steady concentration of N atoms exists. This is a fairly standard procedure whenever a species is present in very small amounts compared with the major species. This latter requirement is fulfilled by N atoms in ordinary combustion processes, especially when $\phi \leqslant 1.2$. In general, from reactions (8-4) and (8-5),

$$\frac{d[\mathrm{N}]}{dt} = k_4[\mathrm{O}][\mathrm{N_2}] - k_{-4}[\mathrm{NO}][\mathrm{N}] + k_{-5}[\mathrm{NO}][\mathrm{O}] - k_5[\mathrm{N}][\mathrm{O_2}]$$

In steady state, $d[\mathrm{N}]/dt = 0$. Hence

$$[\mathrm{N}]_{\mathrm{SS}} = \frac{k_4[\mathrm{O}][\mathrm{N_2}] + k_{-5}[\mathrm{NO}][\mathrm{O}]}{k_{-4}[\mathrm{NO}] + k_5[\mathrm{O_2}]} \tag{8-8}$$

The use of Equation (8-8) in Equation (8-7) yields, after manipulation and cancellation of terms,

$$\frac{d[\mathrm{NO}]}{dt} = 2[\mathrm{O}]\frac{k_4[\mathrm{N_2}] - \left(k_{-4}k_{-5}[\mathrm{NO}]^2/k_5[\mathrm{O_2}]\right)}{1 + \left(k_{-4}[\mathrm{NO}]/k_5[\mathrm{O_2}]\right)} \tag{8-9}$$

One final simplification remains. If the equilibrium constants for reactions (8-4) and (8-5) are K_4 and K_5, respectively, then

$$K_4K_5 = \left(\frac{k_4}{k_{-4}}\right)\left(\frac{k_5}{k_{-5}}\right) = \frac{[\mathrm{NO}]^2}{[\mathrm{N_2}][\mathrm{O_2}]} \equiv K_{p,\mathrm{NO}} \tag{8-10}$$

where $K_{p,\mathrm{NO}}$ is the equilibrium constant for the reaction

$$\mathrm{N_2} + \mathrm{O_2} \rightleftharpoons 2\mathrm{NO}$$

The value of $K_{p,\mathrm{NO}}$, of course, is accurately known. Substitution of Equation

(8-10) into Equation (8-9) leads to

$$\frac{d[\mathrm{NO}]}{dt} = \frac{2k_4[\mathrm{O}][\mathrm{N_2}]\{1 - ([\mathrm{NO}]^2/K_{p,\mathrm{NO}}[\mathrm{N_2}][\mathrm{O_2}])\}}{1 + (k_{-4}[\mathrm{NO}]/k_5[\mathrm{O_2}])} \tag{8-11}$$

This is one possible format of the rate equation for the formation of nitric oxide from atmospheric nitrogen in combustion processes.

The determination of the amount of nitric oxide formed at any time t is found by integrating Equation (8-11). The value of $[N_2]$ is usually set equal to its equilibrium value at the given temperature. In addition, it is necessary to develop an expression for [O]. The assumption is usually made that [O] may be set equal to its equilibrium value $[O]_e$ in the hot products for the reaction

$$\tfrac{1}{2}\mathrm{O_2} \rightleftharpoons \mathrm{O}$$

This assumption appears to be more valid the hotter the flame and the leaner the mixture. The value of $[O]_e$ is obtained from the equilibrium constant $K_{p,\mathrm{O}}$ for the above reaction. Hence

$$K_{p,\mathrm{O}} = \frac{p_{\mathrm{O}}}{(p_{\mathrm{O_2}})^{1/2}} = \frac{[\mathrm{O}]_e RT}{[\mathrm{O_2}]_e^{1/2}(RT)^{1/2}} = \frac{[\mathrm{O}]_e(RT)^{1/2}}{[\mathrm{O_2}]_e^{1/2}}$$

or

$$[\mathrm{O}]_e = \frac{[\mathrm{O_2}]_e^{1/2} K_{p,\mathrm{O}}}{(RT)^{1/2}} \tag{8-12}$$

Each quantity on the right side of Equation (8-12) is normally known.

Through the use of Equations (8-10) and (8-12), Equation (8-11) may be written as

$$\frac{dY}{dt} = \frac{M(1 - Y^2)}{2(1 + CY)} \tag{8-13}$$

$$M = \frac{4k_4 K_{p,\mathrm{O}}[\mathrm{N_2}]^{1/2}}{(RT)^{1/2}(K_{p,\mathrm{NO}})^{1/2}}$$

$$C = \frac{k_{-4}(K_{p,\mathrm{NO}})^{1/2}[\mathrm{N_2}]^{1/2}}{k_2[\mathrm{O_2}]^{1/2}}$$

$$Y = \frac{[\mathrm{NO}]}{[\mathrm{NO}]_e}$$

If we regard the postcombustion zone where NO is formed as a constant-temperature region, then Equation (8-13) may be integrated directly [15]. The result is

$$(1 - Y)^{C+1}(1 + Y)^{C-1} = \exp(-Mt) \tag{8-14}$$

This equation relates the fraction of NO formed (Y) to the time t in terms of parameters C and M.

It may be demonstrated mathematically that the relation between Y and t does not depend strongly on values of C, which are representative of postflame combustion conditions. The important parameter that heavily influences Equation (8-14) is M. The expression that defines M may be written as

$$M = \frac{4k_4 K_{p,\mathrm{O}} P^{1/2}}{RTK_{p,\mathrm{NO}}}$$

On the basis of experimental k_4 data and K data from the JANAF tables, the value of M has been estimated [15] as

$$M = 5.7 \times 10^{15} T^{-1} P^{1/2} \exp\left(\frac{-58{,}400}{T}\right)$$

with M in units of second^{-1}, P in atmospheres, and T in degrees Kelvin. It is seen that the effect of pressure on M is slight. However, there is a very strong temperature dependence. This is extremely significant in terms not only of the total NO produced in time t, but also of the rate of formation of nitric oxide. For a pressure of 1 atm, the table below lists values of M as a function of T.

T		
(°K)	(°F)	M(s^{-1})
1800	2780	0.0244
2000	3140	0.56
2200	3500	7.27
2400	3860	60.9
2500	4040	154

On the basis of this model, values of M in the range of 0.1 to 100 probably are of major importance, in terms of practical combustion temperatures at atmospheric pressure. Although its influence is small, C values typically lie between 0 and 1.

A general plot of Equation (8-14) is shown as Figure 8-3 for the range of M and C values discussed above. It should first be noted that C has little effect on $[\mathrm{NO}]/[\mathrm{NO}]_e$ for a given M, as indicated earlier. The significance of M (or T), however, is striking. For temperatures in the range of 2200° to 2400°K (3500° to 3860°F), appreciable NO would be formed in less than 0.1 s. Only when the temperature is considerably less than 2000°K (3140°F) will the amount of nitric oxide formed be small after a 1-s interval. This trend is especially important for stationary combustion sources, such as those in steam power plants. In these cases the gas velocity is fairly small. As a result a fluid element remains in the hot postcombustion zone for an appreciable time (in power plant boilers this may be on the order of several seconds).

In Section 8-3 we noted that the temperature-time profile of the reacting mixture controls the NO composition in a flue gas. In the absence of equilibrium, the kinetics of formation is the important factor. This is confirmed by Figure 8-3, where the slope of any line measures the formation *rate* at that condition. At any given time, t, the rate is significantly different for the various temperatures (or M values). As an example, the tabulation below lists the rate of formation, dY/dt, at $t = 10^{-2}$ s for a C value of unity, on the basis of Equation (8-13). These results indicate an order of magnitude change in dY/dt for approximately a 200°K change. (These same rates are valid at $t = 10^{-1}$ s for M values of 0.1, 1, and 10.) To prevent the ultimate formation of a large NO concentration, the rate of formation must be kept as low as possible. This requires that the temperature be reasonably low in the region of the flame. If the rate of formation of nitric oxide is relatively low initially, then there may be sufficient time to quench the reaction further and prevent a buildup in NO concentration. This quenching might be accomplished, for example, by dilution with cooler gases or by heat transfer to solid surfaces.

	T		
M	(°K)	(°F)	dY/dt (s^{-1})
1	2040	3210	0.498
10	2230	3550	4.76
100	2450	3950	30.25

On the basis of equilibrium data from Table 8-5 and kinetic data from Figure 8-3, we can estimate the time required to reach various NO concentration levels at a given temperature. The results of this estimate are summarized in Figure 8-4. It should be emphasized that this is a log-log plot. Hence the range of values on the coordinates is many-fold. The rapid buildup of nitric oxide toward the equilibrium level is apparent at high temperatures (> 2000°K). (These curves are based on an initial mixture of only nitrogen and oxygen in a 40 : 1 ratio. This mixture merely approximates that of real combustion gases, since CO_2 and H_2O concentrations have been ignored. In addition, the postflame N_2/O_2 ratio depends upon the initial air-fuel ratio supplied to the combustion process. Other data of this type may be found elsewhere [16].)

Consider the situation where it is desired to keep the nitric oxide level less than 100 ppm. The table below summarizes data taken from Figure 8-4,

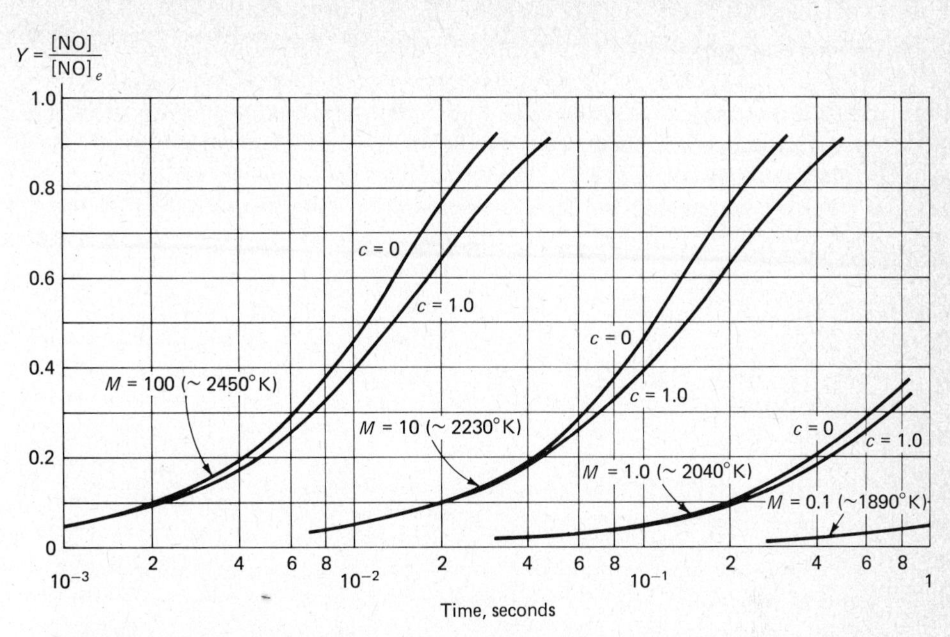

Figure 8-3 Concentration-time profiles at various temperatures for a 40 : 1 ratio of N_2/O_2 reacting to form NO.

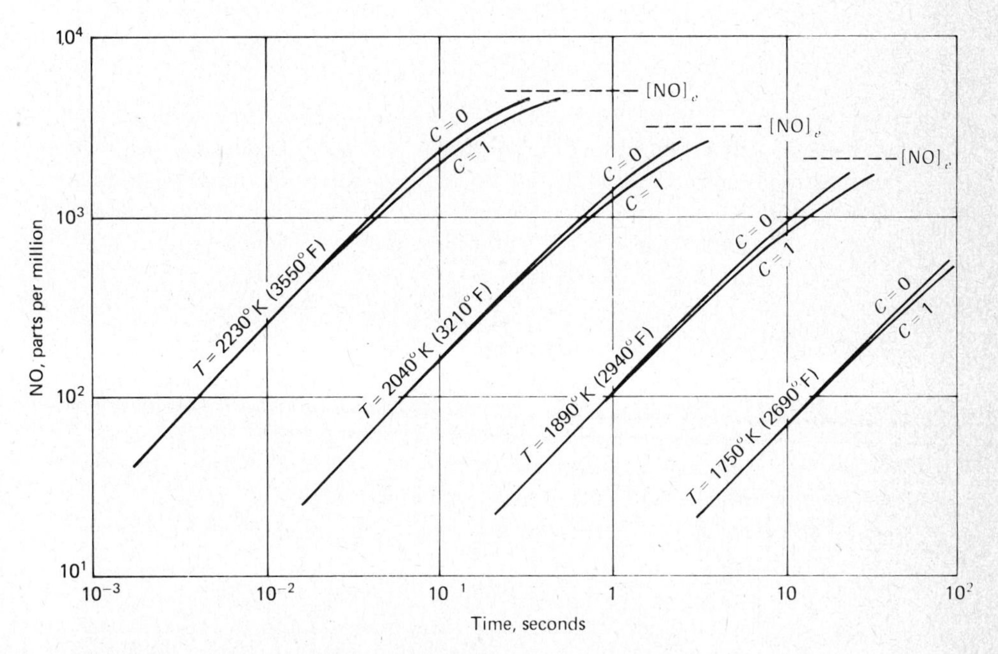

Figure 8-4 Concentration of NO formed from a 40 : 1 ratio of N_2/O_2 as a function of time at various temperatures.

for a C value of unity. The drastic effect of temperature on the time to reach the given NO level is obvious. A decrease from 2230° to 1890°K allows a 250-fold increase in time before the 100-ppm limit is exceeded. Thus the control of the temperature-time profile in the postcombustion process is extremely important if nitric oxide formation is to be minimized. Also, any "hot spots" within the combustion gases will have an adverse effect on the final, average NO concentration of the gas.

T		
(°K)	(°F)	t (ms)
2230	3550	4
2040	3210	70
1890	2940	1000

It must be kept in mind that the preceding data are semiquantitative, based as they are on the following major assumptions:

1. Reactions (8-4) and (8-5) are the elementary steps which control the formation of nitric oxide.
2. The temperature remains constant in the postflame region.
3. The initial composition is a 40 : 1 ratio of solely molecular nitrogen and oxygen, respectively.

With regard to the first assumption, the inclusion of reaction (8-6) in the mechanism is important in fuel-rich mixtures. However, it can be shown that this inclusion only modifies (8-11) by adding the term k_6[OH] to the denominator, where k_6 is the forward-rate constant for reaction (8-6). Occasionally, studies show nitric oxide formation rates which exceed those predicted by the Zeldovich mechanism. However, these observed rates are consistent with reactions (8-4) and (8-5) if account is taken of nonequilibrium oxygen atom concentration.

We assume fixed temperature in order to generate the generalized curves shown in Figures 8-3 and 8-4. In practice, of course, the temperature may be dropping as a result of heat transfer or dilution effects. Consequently, the true final NO composition depends heavily upon the temperature-time profile in the postcombustion gases. The effect of initial composition in the postflame region has already been discussed. Nevertheless, the preceding discussion has presented some of the basic or underlying causes for NO formation in combustion processes involving atmospheric air. Combustion control methods are based on an understanding of these basic factors.

8-5 NO_x FORMATION FROM FUEL NITROGEN

The preceding section outlined the kinetics of formation of nitric oxide from atmospheric molecular nitrogen by means of the Zeldovich mechanism. Recent studies [17, 18] have indicated that chemically bound nitrogen in hydrocarbon fuels, termed fuel nitrogen, is another major source of NO formation. The heterocyclic ring compounds of pyridine, piperidine, and quinoline are the most common ones found in oil. Coal contains both chain and ring nitrogen-bearing compounds [19]. Natural gas is essentially free of such components. It is significant that in most of these nitrogen-bearing organic compounds, the bonds between the nitrogen atom and the rest of the molecule are considerably weaker than the N—N bond in molecular nitrogen (~940 kJ/g·mole). Hence it is not surprising that fuel nitrogen can contribute large quantities of NO in combustion processes.

The reactions involved in the kinetics of formation of NO from fuel nitrogen are not clearly known. Experimental data do indicate, however, that the oxidation of fuel nitrogen is rapid. The time-scale is on the order of that of the main combustion reactions. Experiments have also shown that NO concentrations in excess of equilibrium occur in the vicinity of the flame front [20, 21]. Several theories have been proposed. These include (1) the use of cyanide (CN) as an intermediate [17, 22], (2) the release of atomic nitrogen as bonds are broken [17], and (3) a partial-equilibrium mechanism [23].

Some experimental results [17, 24] have indicated from 20 to 80 percent fuel nitrogen conversion to NO_x. More recent measurements [22] with kerosene containing 0.5 percent pyridine indicate fuel nitrogen conversion approaching 100 percent. For an equivalence ratio of around 0.7, the NO concentration increased from roughly 50 to 400 ppm with the addition of pyridine under good fuel-air mixing conditions. For an equivalence ratio around 1.0, the increase was from 150 to 800 ppm in the exhaust. Consequently, there is ample evidence that fuel nitrogen does play a role in NO_x formation when coal or oil is burned. The kinetic mechanism for this process is being actively investigated.

8-6 COMBUSTION CONTROL METHODS FOR NO_x FROM STATIONARY SOURCES

On the basis of thermodynamic and kinetic considerations discussed previously, some insight is gained into possible methods of reducing NO_x concentrations in flue gases by combustion control. In general, the main parameters that affect NO_x formation are temperature, residence time, concentrations of various species, and extent of mixing. From an experimental viewpoint, the factors that control NO_x formation include (1) air-fuel ratio, (2) combustion air temperature, (3) extent of combustion-zone cooling, and (4) furnace-burner configuration. Consideration of these basic design factors leads to the combustion techniques known as *flue-gas recirculation* and *two-stage combustion*. Type of fuel is also a major influence on performance.

Before examining data, several comparative values should be reviewed. First, on the basis of kinetic information presented earlier, the temperature above which the rate of NO formation becomes significant is about 1800°K (2800°F). Second, NO_2 concentrations frequently are reported in parts per million. To convert from ppm to lb $NO_2/10^6$ Btu, we apply the relation

$$NO_2\left(\frac{\text{lb}}{10^6\ \text{Btu}}\right) = \frac{46.0(FG)(PPM)}{(MW_{fg})(HHV)} \tag{8-15}$$

where FG is the pounds of dry flue gas per pound of fuel, PPM is the NO_2 concentration in parts per million, MW_{fg} is the molar mass of the flue gas (on a dry basis), and HHV is the higher heating value in British thermal units per pound.

Lastly, the emission standards for new steam-generating units of more than 250×10^6 Btu/hr heat input are

1. Gas-fired, 0.20 lb/10^6 Btu
2. Oil-fired, 0.30 lb/10^6 Btu
3. Coal-fired, 0.60 lb/10^6 Btu

These emissions are expressed as NO_2. On a parts-per-million basis, these values are approximately 175 ppm for gas-fired, 230 ppm for oil-fired, and 490 ppm for coal-fired units.

8-6-A Effect of Excess Air

Fundamentally, the presence of excess air affects both the temperature and the oxygen concentration of gases in the postcombustion zone. The results for pulverized coal firing in an experimental unit are shown in Figure 8-5

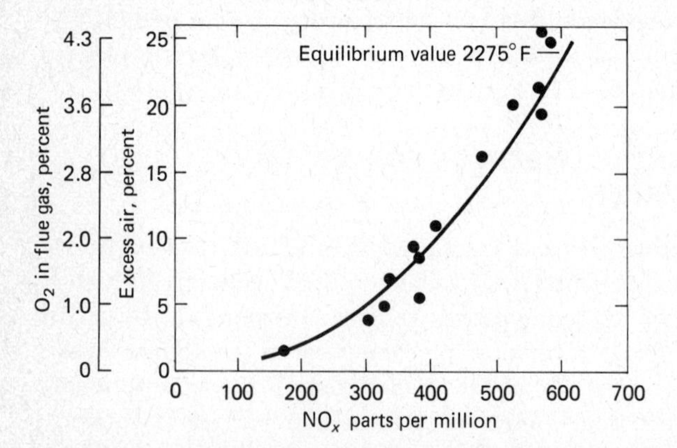

Figure 8-5 Nitric oxide formation as a function of excess air for pulverized coal firing. (SOURCE: D. Bienstock. 1972 Industrial Coal Conference, Purdue University, October, 1972.)

[25]. In this specific case NO_x decreased from 550 to 575 ppm at 25 percent excess air to as low as 175 ppm at 1.4 percent excess air. (Although not shown, this 70 percent overall decrease in NO_x was accompanied by a decrease in carbon combustion efficiency. It fell from 99.5 percent efficiency at 25 percent excess air to around 92 percent at 2 percent excess air.) The general trend would be the same for oil or natural gas as a fuel. Note that the most pronounced lowering of the NO_x concentration is at low (< 10 percent) excess air. This occurs in spite of the fact that the flame temperature increases as the inlet air concentration approaches the stoichiometric value. The rapid dropoff in oxygen concentration in this case dominates over the temperature effect.

8-6-B Effect of Combustion Air Temperature

In many industrial operations waste heat frequently is available to help preheat air entering a combustion process. The use of an air preheater in large-scale power plants is a typical example. Although this process leads to an appreciable energy savings, the added energy increases the flame temperature. Thus NO_x emissions increase. Data from full-sized boiler tests [40] indicate a threefold increase in NO_x emissions when combustion air is preheated from 80° to 600°F. A significant portion of the increase occurred in the upper range of temperatures (from 450° to 600°F), as might be expected from kinetic considerations. The effect of change of combustion turbulence due to change in air density with temperature was not segregated from the data.

8-6-C Effect of Combustion-Zone Cooling

Since temperature has such a major influence on nitric oxide formation, an effective method of control is cooling of the primary flame zone by heat transfer to surrounding surfaces. It would be anticipated that the higher the rate of heat release per effective surface area of the combustion chamber, the higher the temperature of the flame zone, and thus the higher the NO_x emissions. This is confirmed by data [26] on front-fired boilers using natural gas, oil, and pulverized coal. Figures 8-6(a), (b), and (c) shows the trends. The data for the oil and coal units have been corrected for the expected contribution of fuel nitrogen. Hence all three correlations are based on NO_x from atmospheric nitrogen. In addition, an effort was made to take into account the type and extent of wall deposits for coal firing, since these influence the heat transfer to the walls. In terms of the desirability of high heat-removal rates, coal and oil have an advantage over gas. The flames of these first two fuels are usually luminous. Hence the radiative heat transfer is enhanced and lower temperatures are maintained. As a result, the amount of NO_x formed in gas firing can be greater than in oil firing at the same heat release. Nevertheless, gas-fired units usually have lower NO_x concentrations.

It is suspected that the contribution of fuel nitrogen in oil and coal offsets other factors.

8-6-D Effect of Furnace-Burner Configuration

Burner configuration plays an important role in NO_x control. The highly turbulent cyclone-type coal burner, for example, leads to high NO_x concentrations. It is an example of a high heat-release rate device, which should be avoided. Tangential firing (units where heat sinks are in close proximity to the burner flame) has led to reported NO_x reductions of 50 to 60 percent over conventional firing techniques.

The effect of burner location in power plant boilers is illustrated in Figure 8-6(c) for pulverized coal operation. The front-fired type of boiler has all burners on a single wall, while the Turbo-fired® boiler is a modification of an opposed-firing furnace. Similar trends exist for natural-gas and oil operation. Burner location is important due to the type of flame produced and the degree of turbulence involved.

8-6-E Flue-Gas Recirculation

A major method employed in NO_x control from stationary sources (and mobile sources) is flue-gas recirculation. A portion of the cooled flue gas is injected back into the combustion zone. This additional gas acts as a thermal sink and reduces the overall combustion temperature. In addition, the oxygen concentration is lowered. Both of these effects favor a reduction in NO_x emissions. One disadvantage is the increased cost of duct work, since large volumes of gas are involved. However, the technique does give the operator an additional element of control, since the percent recirculation can be varied over some restricted range. In amounts of up to 25 percent, the recirculated gas has a negligible effect on flame development. Recirculation, however, appears to be relatively ineffective in reducing NO_x formed from fuel nitrogen [18].

Results of laboratory studies by Exxon of the effect of recirculation in oil-fired domestic furnaces are shown in Figure 8-7. A 50 percent recirculation, for example, at 15 percent excess air leads to a potential reduction of NO_x of roughly 60 percent. Similar percentage reductions occur at higher percentages of excess air. Figure 8-8 is based on natural-gas-fired units [27]; and the same trend has been noted elsewhere [26]. Table 8-8 records the effect of flue gas recirculation on NO_x formation for pulverized coal combustion [25]. The effect of flue-gas recirculation in natural-gas and oil-fired units is much more dramatic than in coal units, owing to the nature of the combustion processes.

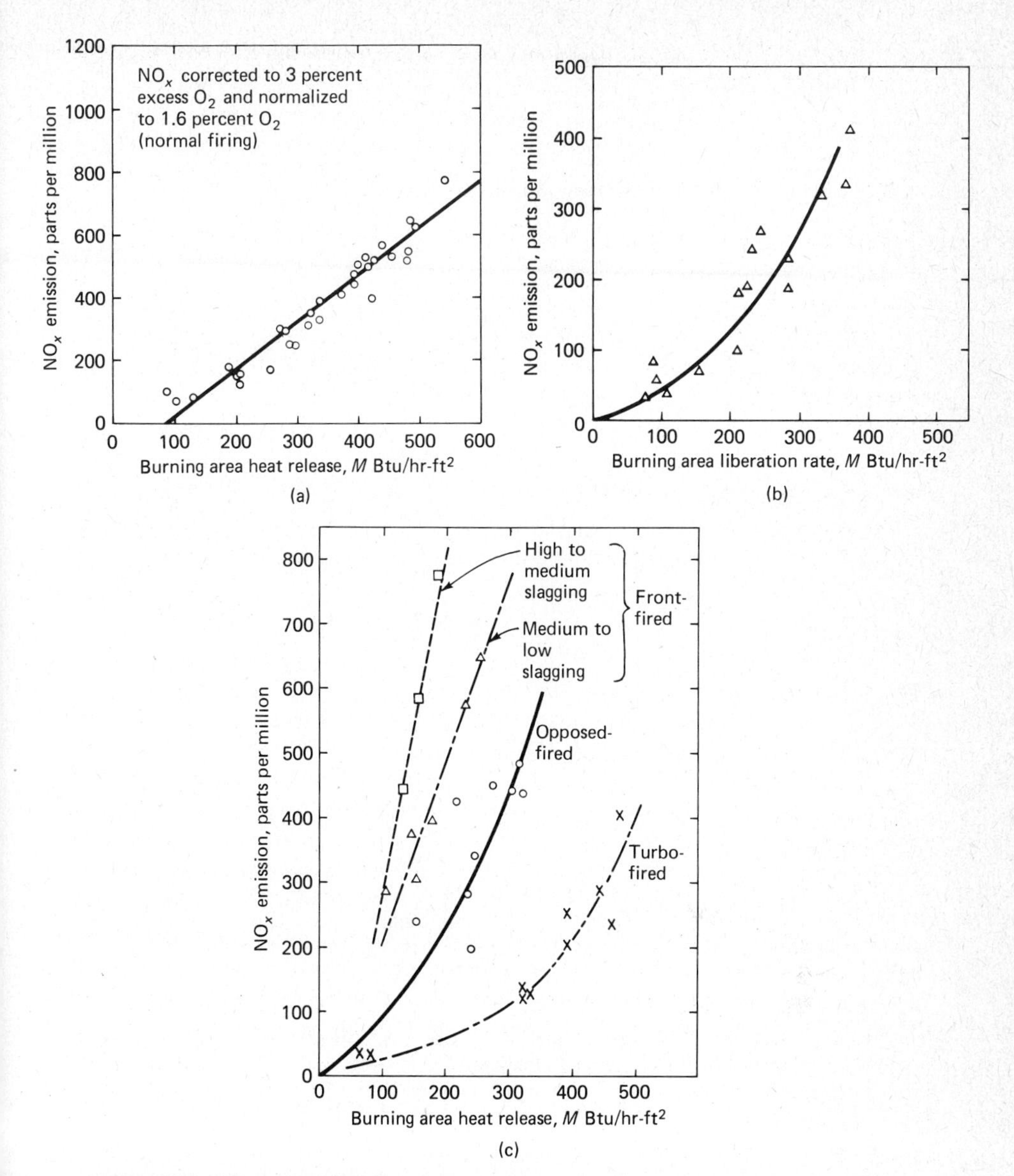

Figure 8-6 NO$_x$ emissions from atmospheric nitrogen as a function of the heat release rate per unit surface area. (a) Natural gas front-fired. (b) Oil front-fired. (c) Pulverized coal firing. (SOURCE: A. H. Rawdon and R. S. Sadowski. "An Experimental Correlation of Oxides of Nitrogen Emissions from Power Boilers Based on Field Data." Paper No. 72-WA/Pwr-5, ASME Annual Winter Meeting, New York, November, 1972.)

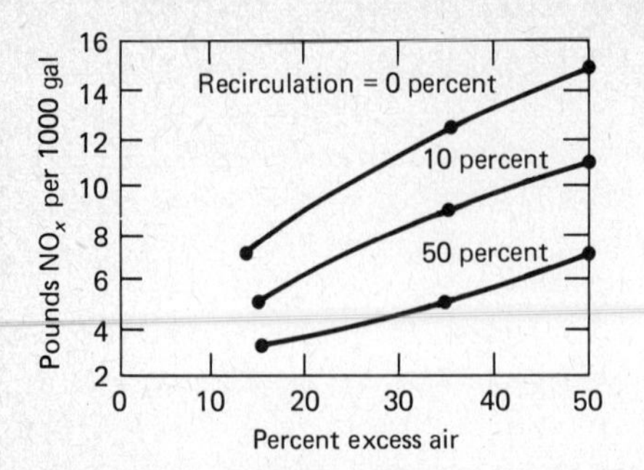

Figure 8-7 Combined effect of recirculation and excess air on NO_x emissions.

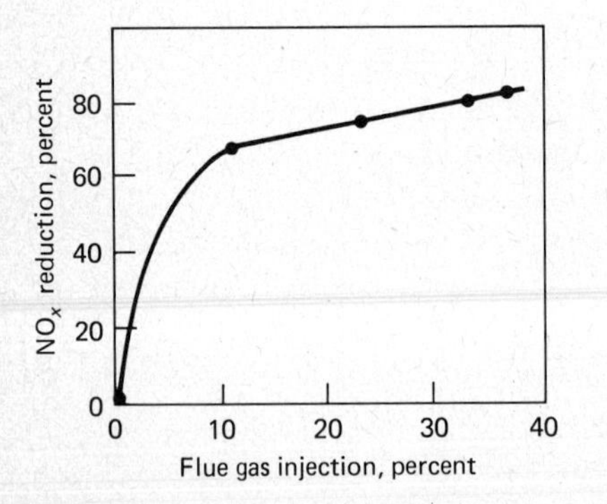

Figure 8-8 Effect of flue-gas recirculation on NO_x emissions, natural-gas firing, 7.5 percent excess air, no preheat. (SOURCE: R. E. Sommerlad, R. P. Welden, and R. H. Pai. "Nitrogen Oxides Emissions–An Analytical Evaluation of Test Data." 33d Annual Meeting, The American Power Conference, Chicago, April, 1971.)

Table 8-8 THE EFFECT OF FLUE-GAS RECIRCULATION ON NO_x FORMATION FOR COAL COMBUSTION

EXCESS AIR (%)	RECIRCULATION VOLUME (%)	QUANTITY OF NO_x (lb $NO_x/10^6$ Btu)
20	0	0.80
20	10	0.75
20	15	0.72
20	20	0.65
15	0	0.70
15	15	0.67
15	20	0.61

SOURCE: D. Bienstock. "Control of NO_x Emissions in Coal Firing." 1972 Industrial Coal Conference, Purdue University, October, 1972.

8-6-F Two-Stage or Off-Stoichiometric Combustion

Earlier we noted the beneficial effect of low excess air in NO_x formation. To take advantage of this effect, a firing method termed two-stage combustion can be employed. In this case the fuel (oil or gas) and air are burned at near-stoichiometric conditions. A fuel-rich, reducing atmosphere for a portion of the flame region might be used. Typically, 85 to 95 percent of the total air requirements (which normally range from 110 to 130 percent of stoichiometric values) is introduced in the burner with the fuel. Because of incomplete combustion, and possibly the concomitant increase in radiant heat transfer, the gas temperature is held down in this first stage of combustion. This lowers NO formation. Second, oxygen is lacking for NO formation in this stage. Complete burnout, which consumes the CO and hydrocarbons left in the first stage, is accomplished by injecting secondary air downstream. By this time the gas temperature is low enough so that NO formation is kinetically limited. This method leads to a much more desirable temperature-time profile. Off-stoichiometric firing, which uses alternating fuel-rich and fuel-lean burners, is a variant optimization of staged combustion.

Figure 8-9 represents data compiled by the Bureau of Mines [25] on two-stage coal combustion. Reduction of combustion air to 95 percent of stoichiometric in the first stage proved very beneficial in reducing NO_x emissions. A similar reduction for natural-gas and oil fuels is shown in Figures 8-10(a) and 8-10(b), respectively [28]. This figure also illustrates the combined effect of two-stage combustion with the flue-gas recirculation of 15 percent. At full load the NO_x concentration was reduced from 275 to 90 ppm by the addition of recirculation. The original value was 1425 ppm without either technique in use. On the basis of these data alone, flue-gas recirculation and two-stage combustion are less effective for oil than for natural gas. Nevertheless, the NO_x concentration was cut in half for the operation with oil as a fuel.

Two-stage combustion is less readily adaptable to coal-fired units. In the absence of excess air, hazardous fuel-air distributions may occur. This leads to unburned fuel and possibly increased carbon monoxide emissions. One method of attack, though, to achieve much lower average combustor bed temperatures for coal-fired units is through the use of a fluidized bed. Because of the very high heat transfer rates with a fluidized bed, the overall temperature is lowered (in the range of 1600° to 1700°F). Under this temperature condition, the NO_x concentration from molecular nitrogen should be negligible. However, current evidence indicates that as much NO_x is formed in fluidized beds as in conventional equipment. More research will be required before the process is understood well enough to realize its potential in reducing NO_x formation. Fluidized beds offer other potential advantages, which will be discussed under NO_x removal methods.

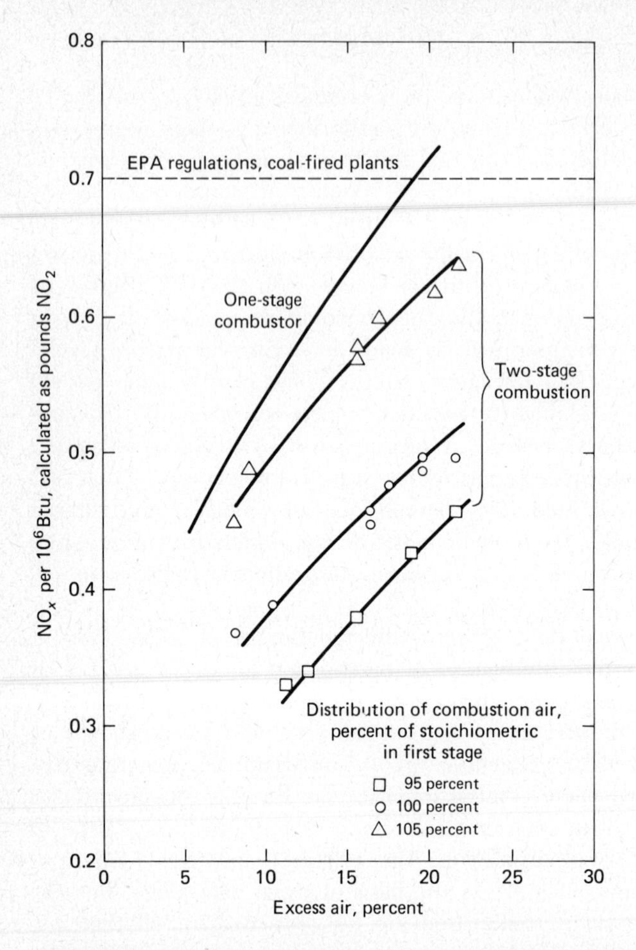

Figure 8-9 Nitrogen oxide formation in two-stage coal combustion. (SOURCE: D. Bienstock. "Control of NOx Emissions in Coal Firing." 1972 Industrial Coal Conference, Purdue University, October, 1972.)

In summary, the techniques of combustion modification appear to be an attractive approach to the control of oxides of nitrogen. Basically, all these control methods are predicated on two theoretical principles:

1. Achieve a minimum residence time of the combustion gases at peak flame temperatures, through (a) directly lowering the peak flame temperature; (b) increasing the rate of heat removal; and (c) avoiding high heat-release rates.
2. Limit the availability of reactants through use of low excess air.

Any particular control method may take advantage of several of the above means, and synergistic effects may occur. Attractive possibilities include two-stage combustion, low excess air firing, flue-gas recirculation, redesign to

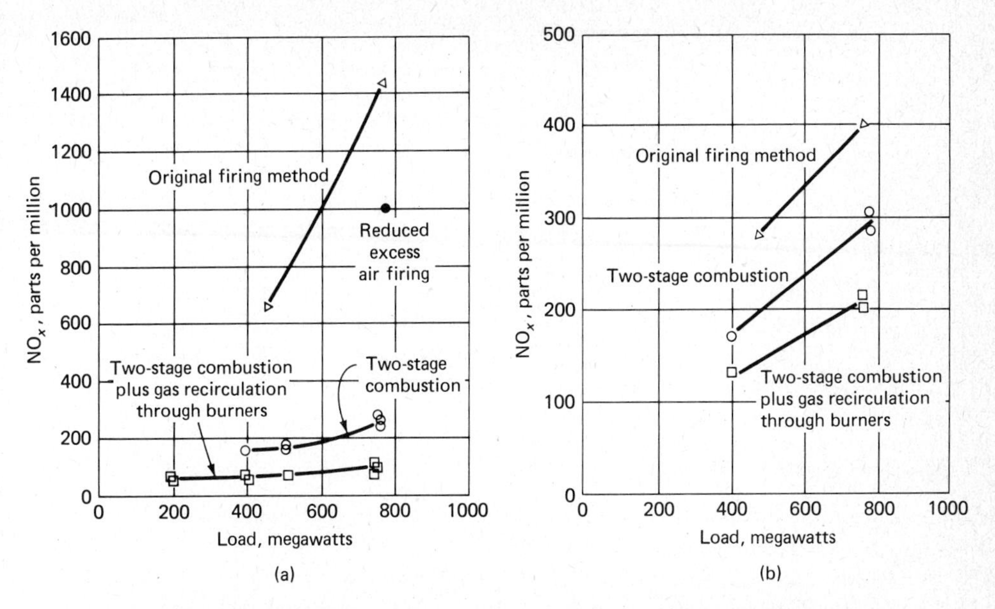

Figure 8-10 Effects of nitric oxide control methods. (a) Natural-gas fuel; (b) oil fuel. (SOURCE: W. H. Barr, and D. E. James. "Nitric Oxide Control—A Program of Significant Accomplishments." Paper No. 72-WA/Pwr-13, ASME Winter Meeting, New York, 1972.)

burner and firing techniques, and fluidized bed combustion. The type of fuel also greatly influences the rate of NO_x emissions.

8-7 FLUE-GAS CONTROL METHODS FOR NO_x

In some combustion processes it may not be feasible, for various reasons, to use direct combustion controls to attain desirable levels of NO_x emissions. In these cases it will be necessary to remove the oxides of nitrogen from the cooled flue gases before they are released to the atmosphere. A number of methods for NO_x removal are currently being investigated.

Control of NO_x emissions by flue-gas treatment is a formidable task. One reason is that the volume rate of flue gas requiring treatment is enormous at any given installation. For example, the gas from a 1000-MW power plant is released at a rate of roughly 10^8 scf/hr. On a comparable basis, the quantity of NO_x released is about 10,000 lb/hr. So while the concentration of NO_x is relatively low (200 to 1500 ppm), the tonnage rate can be sizable. (The concentration of NO_x may be a factor of 3 lower than typical SO_2 values in an untreated flue gas, on a volume basis.) A second factor is tied directly to the previous point. The large tonnages of NO_x plus treating solutions constitute a major waste disposal problem when adsorption or absorption techniques are used. In most of these cases flue-gas treatment should be considered only if useful products can be recovered from the waste solution and the adsorbing or absorbing medium is recycled.

Among the possible removal techniques for oxides of nitrogen are catalytic decomposition, catalytic reduction, absorption, and adsorption.

8-7-A Catalytic Decomposition

The direct decomposition of NO into N_2 and O_2 would be highly desirable. Numerous data are available for a variety of catalysts. Even though the kinetics of the reaction mechanism is in dispute, data to date indicate the decomposition reaction to be extremely slow. No catalyst has been found which provides sufficient activity at reasonable temperatures. Removal of NO from flue gases by such a technique is not feasible at this time. References 29, 30, and 31 discuss work of this type.

8-7-B Catalytic Reduction

A reaction with another compound reduces NO to molecular nitrogen. Two types of reduction must be considered—selective and nonselective. In selective reduction, the added reactant preferentially reduces NO_x. In non-selective reduction, the excess O_2 must be consumed first. Selective reduction is to be preferred, since it minimizes the amount of reactant required.

Selective reduction can be carried out with H_2, CO, NH_3, or H_2S as the reactant gas, and with a suitable catalyst [32]. When CO is used, the proper catalyst permits overall reactions of the type

$$2CO + 2NO \rightarrow CO_2 + N_2$$

and

$$4CO + 2NO_2 \rightarrow 4CO_2 + N_2$$

rather than the combustion reaction

$$2CO + O_2 \rightarrow 2CO_2$$

The use of CO has the disadvantage that any amount which is unreacted adds to the general CO pollution of the atmosphere. Using hydrogen is preferable to using CO. However, some catalysts for H_2 are not effective in the presence of CO (present from the main combustion reaction). The catalysts found to be most effective are the platinum-type metals.

The reduction of NO by ammonia has long been known. It has been practiced to a limited extent in nitric acid plants. Temperature control for reactions such as

$$6NO + 4NH_3 \rightarrow 5N_2 + 6H_2O$$

$$6NO_2 + 8NH_3 \rightarrow 7N_2 + 12H_2O$$

must be maintained in the range of 400° to 600°F. At higher temperatures the ammonia is oxidized to NO_x, which compounds the problem. Below

400°F ammonium nitrate is formed. The Gibbs function change for these reactions in the given temperature range is highly negative. Thus the reactions are very favorable in terms of equilibrium yields. For example, an experiment [32] involving NH_3, NO, and O_2 in a 1 : 1 : 10 ratio was carried out at 150°C. When the mixture was passed over a platinum catalyst, the NO concentration was reduced by a factor of 100. Unfortunately, the platinum group catalysts are easily poisoned by sulfur, which must be held to less than 1 ppm. Claims have been made for other catalysts which would make this a promising technique. Ammonia injection also holds forth the possibility of simultaneous removal of NO_x and SO_2.

The Claus process is a well-known process for converting SO_2 into elemental sulfur by the reaction

$$SO_2 + 2H_2S \rightarrow 3S + 2H_2O$$

It is proposed that a similar reaction would remove NO.

$$NO + H_2S \rightarrow S + \tfrac{1}{2}N_2 + H_2O$$

The combined removal of SO_2 and NO by H_2S addition is under study. As in reduction using CO, the utilization of H_2S must be complete, since it is a pollutant by itself.

In nonselective reduction the NO_x is converted to N_2 simply by supplying sufficient reactant to reduce the excess oxygen as well. Typical reactants are hydrogen or methane, CH_4. When methane is used, the overall NO_x reduction reactions are

$$CH_4 + 4NO_2 \rightarrow 4NO + CO_2 + 2H_2O$$

and

$$CH_4 + 4NO \rightarrow 2N_2 + CO_2 + 2H_2O$$

These reactions compete with the methane combustion reaction

$$CH_4 + 2O_2 \rightarrow CO_2 + 2H_2O$$

These reactions are highly exothermic. Hence the gas temperature could become quite high as a result of the reduction process. The actual value depends upon the quantity of pollutant reduced. When noble metal catalysts, such as platinum or palladium, are used, the flue-gas temperature must be above some minimum value to ensure ignition. For CH_4 this temperature is around 850°F (730°K). The rate of conversion is high, even with relatively high space velocities. Nonnoble metal catalysts also can be used (such as supported copper oxide), but data on their performance are limited. One source [33] claims nearly complete reduction of NO by hydrogen at temperatures of 550° to 750°F and space velocities in excess of 50,000 vol/hr/vol. (Space velocity is defined as volume rate of gas flow per volume of catalyst.)

The catalyst was 75 percent nickel oxide on alumina. Catalysts for NO reduction are currently being used for automotive emission control. If nonselective catalytic reduction is to be used for stationary combustion sources, then stoichiometric (or even substoichiometric) air must be used in the primary combustion zone. Otherwise the cost of the reducing reactant becomes prohibitive. For example, a coal- or oil-fired unit with moderate excess air typically has around 3 percent O_2 in the flue gas. It would require about 15 to 20 percent more (equivalent) reducing agent just to remove the oxygen, plus that required to convert the NO_x. Low primary air, on the other hand, leads to CO formation in the primary combustion zone. This CO can react with NO_x, further reducing the amount of reductant added downstream. There is also some promise that nonselective reduction with nonnoble metal catalysts can reduce both NO_x and SO_2 simultaneously.

8-7-C Absorption

The oxides of nitrogen can be absorbed by water, hydroxide and carbonate solutions, sulfuric acid, organic solutions, and molten alkali carbonates and hydroxides. When aqueous alkaline solutions, such as NaOH and MgOH, are used, complete removal requires that one-half of the NO must first be oxidized to NO_2 (or NO_2 gas added to the gas stream). Best absorption occurs when the NO/NO_2 molar ratio is 1 : 1, which indicates that absorption of the combined oxide, N_2O_3, is the most favorable. The absorption of NO_x by alkaline solutions has been confirmed during desulfurization of power plant emissions by such solutions. In the desulfurization process, apparently about 10 percent of the NO was oxidized to NO_2 before the flue gas reached the scrubber. The scrubber then removed about 20 percent of the total NO_x, in equal parts of NO and NO_2.

The absorption of nitric oxides by strong sulfuric acid has long been known. The compounds formed in this case are violet acid, $H_2SO_4 \cdot NO$, and nitrosyl sulfuric acid, $NOHSO_4$. The latter is very stable in concentrated acid. The reaction is

$$NO + NO_2 + 2H_2SO_4 \rightleftharpoons 2NOHSO_4 + H_2O$$

Any moisture in the flue gas is picked up by the acid, and this moisture drives the above reaction to the left. This problem area can be reduced by operating at an elevated temperature ($> 250°F$) so that the vapor pressure of water in the solution is equal to the partial pressure of water in the flue gas. The use of sulfuric acid to remove SO_2 and NO_x is currently under active investigation.

8-7-D Adsorption

The use of char (activated carbon) to adsorb oxides of nitrogen has been studied extensively [34]. Char has a high adsorption rate and capacity, compared to other materials. However, regeneration may be a problem. A potential fire and explosion hazard may be another difficulty with this material since O_2 is usually present in most stack gases. Manganese oxide and alkalized ferric oxides show technical potential. However, sorbent attrition is a major technological stumbling block. The technique does not show much promise at the present time.

In summary, methods such as catalytic reduction, catalytic decomposition, absorption, or adsorption may provide some limited means for the reduction of NO_x emissions from flue gases associated with stationary combustion sources. However, the limited reactivity of nitrogen oxides and the vast quantities of gas to be handled place a severe restriction on the overall technique of flue-gas cleaning. For stationary sources the most promising methods are those involving the control of the combustion process itself.

QUESTIONS

1. List the oxides of nitrogen which are present in sufficient quantities to contribute to air pollution.
2. Why is the reduction of nitric oxide an important factor in reducing air pollution?
3. List the major sources of the oxides of nitrogen.
4. At what temperature does the formation of NO in a combustion reaction become significant?
5. How does the temperature of the reaction influence the quantity of NO_2 formed based upon equilibrium conditions?
6. Why are the quantities of NO and NO_2 in the gases emitted from combustion processes so different from those predicted by equilibrium chemistry?
7. What are the three basic parameters that must be controlled in order to control the quantity of NO emitted from a combustion process?
8. What are the factors that control the formation of NO_x in combustion processes?
9. Using data presented in Section 8-6, predict the reduction in NO_x expected when 20 percent flue-gas recirculation is used with (a) 20 percent excess air and (b) 15 percent excess air.
10. What is meant by two-stage combustion?
11. Why is the removal of NO_x from flue gas a formidable task?
12. List the possible removal techniques for controlling the emission of oxides of nitrogen.
13. Select one of the catalytic reduction processes for removing NO_x from combustion gases and tell why you think it is the best method.
14. Under what conditions should the adsorption method be employed to control NO_x emissions?
15. On the basis of equilibrium constant data, what conditions increase the quantity of NO formed by combustion?
16. Based on the data presented in Figure 8-4, estimate the time required to keep the nitric oxide level less than 200 ppm. Compare this result with data for 100 ppm presented in Section 8-4.

17. What are the combustion control methods that can be employed to reduce the emission of the oxides of nitrogen?
18. Why is the catalytic reduction of NO not considered a good method of NO control for steam power plants?
19. Is it possible to combine NO and SO_2 removal into one system applicable to coal-burning power plants?

PROBLEMS

8-1. Estimate by means of the emission factor table in this chapter the NO_x emissions, in tons, from the burning of waste for a city of 100,000 population. A municipal incinerator without controls will be used. Assume that each person generates 5 lb of waste per day, and that industrial sources within the city generate another 2.5 lb of waste per capita per day.

8-2. A 1000-MW power plant may operate with (a) coal with a heating value of 11,000 Btu/lb, (b) fuel oil with a heating value of 18,000 Btu/lb and a specific gravity of 0.86, and (c) natural gas with a heating value of 1000 Btu/scf. If the plant thermal efficiency is 38 percent, determine the NO_x emissions for these three fuel sources, in tons per day, on the basis of emission factors.

8-3. A semianthracite coal has the following ultimate analysis of basic elements in percent by weight: sulfur, 0.6; hydrogen, 3.7; carbon, 79.5; nitrogen, 0.9; oxygen, 4.7; and an ash content of 10.6 percent. The coal is burned with 20 percent excess air beyond that required to change S, H, and C to SO_2, H_2O, and CO_2. If (a) 50 percent and (b) 20 percent of the bound nitrogen is converted to NO, and none of the atmospheric nitrogen is oxidized to NO, what would be the parts per million of NO in the flue gas?

8-4. Consider an initial gaseous mixture of 8.0 percent CO_2, 12 percent H_2O, 75.0 percent N_2, and 5.0 percent O_2 by volume. If the mixture is maintained at (a) 1200°K, (b) 1500°K, and (c) 2000°K, estimate the equilibrium concentration of NO in parts per million, if the only reaction to be considered is N_2 and O_2 reacting to form NO.

8-5. Consider an initial gaseous mixture composed of 10.0 percent CO_2, 10.0 percent H_2O, 70.0 percent N_2, and 10.0 percent O_2, by volume. If the mixture is heated to (a) 1500°K and (b) 2000°K at 1 atm, predict the equilibrium concentration of NO in parts per million, if the only reaction to be considered is $N_2 + O_2 \rightleftharpoons 2NO$.

8-6. On the basis of equations in Section 8-4-A, determine the values of k_4 and k_{-5} at (a) 600 and 1200°K and (b) 500 and 1000°K.

8-7. Estimate the ratio $[NO]/[NO]_e$ based on kinetic considerations in Section 8-4-B for (a) a value of $M = 70$ and $C = 0.5$ and for $t = 0.01$, 0.04, and 0.10 s. (b) What temperature corresponds to $M = 70$, in degrees Kelvin?

8-8. Consider Problem 8-7, except that $M = 50$.

8-9. Consider Problem 8-7, except that $M = 30$.

8-10. Show that when reaction (8-6) is included as a reaction in the Zeldovich mechanism, then

$$\frac{d[NO]}{dt} = \frac{2[O][N_2]\left\{1 - \left([NO]^2/K_{p,NO}[N_2][O_2]\right)\right\}}{1 + k_{-4}[NO]/k_5[O_2] + k_6[OH]}$$

where $N + OH \rightarrow NO + H$ would be reaction (6) according to the simplified designation system used in Section 8-4-B.

8-11. A residual fuel oil with an average hydrocarbon formula of $C_{10}H_{20}$ also contains (a) 0.3 percent and (b) 0.5 percent by weight of nitrogen chemically bound in the fuel. If all of this nitrogen is converted to NO, and the $C_{10}H_{20}$ component is burned with 60 percent excess air, calculate the NO concentration in the flue gas, in parts per million, assuming no conversion of nitrogen in the atmospheric air to NO and complete combustion of the hydrocarbon.

8-12. The NO emission standard in the exhaust gas from an oil-fired unit is roughly 230 ppm. Consider an oil with a formula of $C_{10}H_{20}N_x$. The oil is burned with 50 percent excess air beyond that required to change C and H to CO_2 and H_2O. Assume that 50 percent of the nitrogen bound in the fuel is converted to NO, and no nitrogen (N_2) in the atmospheric air used for combustion is converted. What is the maximum percent by weight of nitrogen permitted in the fuel without exceeding the emission standard?

8-13. Through the use of Equation 8-15 determine the NO_x emission in pounds per million Btu for the combustion of methane gas (CH_4) with 10 percent excess air if the NO_x concentration is reported as (a) 100 ppm and (b) 500 ppm.

8-14. The NO_x emission from a power boiler is 0.65 lb/10^6 Btu input. Assume that the fuel is pure carbon, and that 20 percent excess air is used in the combustion process. Determine the NO_x emitted in parts per million.

8-15. Determine the concentration in ppb of atomic oxygen that exists at equilibrium for an initial mixture of 97 percent N_2 and 3 percent O_2 by volume and for temperatures of (a) 2000, (b) 2200, and (c) 2400°K.

References

1. E. Goldstein. "Reevaluation of the Air Quality Standard for Nitrogen Dioxide." *Calif. Air Environment* 4, no. 2 (Winter 1974).
2. "New Focus." *J. Air Pollu. Control Assoc.* **24** (January 1974): 65.
3. C. F. Barrett and L. E. Reėd. *Intntl. J. Air Water Pollu.* **9** (1965): 347.
4. S. Hayushi. *Kuki Seijo* (Tokyo) **4** (1964): 8.
5. O. Tadu. *J. Sci. Labour* (Tokyo) **42** (1966): 263.
6. R. L. Duprey. *Compilation of Air Pollution Emission Factors*, AP-42. U.S. Public Health Service, 1968.
7. NAPCA. *Control Techniques for Nitrogen Oxides from Stationary Sources*, AP-67. Washington, D.C.: HEW, 1970.
8. J. T. Shaw and A. C. Thomas. "Oxides of Nitrogen in Relation to the Combustion of Coal." Paper presented at Conference on Coal Science, Prague, June 10–14, 1968.
9. Joint Army Navy Air Force (JANAF) Thermochemical Tables. Dow Chemical Company, PB-168370. U.S. Government Clearinghouse, 1965.
10. R. B. Rosenberg and D. H. Larson. Basic Research Symposium, Institute of Gas Technology, Chicago, 1967.
11. L. S. Caretto, R. F. Sawyer, and E. S. Starkman. "The Formation of Nitric Oxide in Combustion Processes," University of California (Berkeley) Report No. TS-68-1, 1968.
12. Y. B. Zeldovich. *Acta Physicochim. USSR* **21** (1946): 577.

13. Y. B. Zeldovich, P. Y. Sadovnikov, and D. A. Frank-Kamenetskii. *Oxidation of Nitrogen in Combustion* (translated by M. Shelef). Academy of Sciences of USSR, 1947.
14. C. T. Bowman and A. S. Kesten. "Kinetic Modeling of Nitric Oxide Formation in Combustion Processes." Fall Meeting, Western Section of the Combustion Institute, October, 1971.
15. A. A. Westenberg. "Kinetics of NO and CO in Lean, Premixed Hydrocarbon-Air Flames." *Combust. Sci. and Tech.* **4** (1971): 59.
16. *Air Quality Criteria for Nitrogen Oxides*, AP-84. Washington, D.C.: EPA, 1971.
17. G. B. Martin and E. E. Berkau. "An Investigation of the Conversion of Various Fuel Nitrogen Compounds to Nitrogen Oxides in Oil Combustion." 70th National American Institute of Chemical Engineers Meeting, Atlantic City, N.J., August, 1971.
18. D. W. Turner, R. L. Andrews, and C. W. Siegmund. "Influence of Combustion Modification and Fuel Nitrogen Content on Nitrogen Oxide Emissions from Fuel Oil Combustion." Winter American Institute of Chemical Engineers Meeting, San Francisco, December, 1971.
19. H. M. Spiers. *Technical Data on Fuels*. 6th ed. World Power Conference, 1962.
20. C. P. Fenimore and G. W. Jones, *J. Phys. Chem.* **65** (1961): 298.
21. D. I. Maclean and H. G. Wagner. 11th International Symposium on Combustion, 1967.
22. J. P. Appleton and J. B. Heywood. 14th International Symposium on Combustion, 1973.
23. C. T. Bowman. 14th International Symposium on Combustion, 1973.
24. B. Bartok et al. "Basic Kinetic Studies of Modeling of Nitrogen Oxide Formation in Combustion Processes." 70th National Meeting, American Institute of Chemical Engineers, Atlantic City, N.J., 1971.
25. D. Bienstock. "Control of NO_x Emissions in Coal Firing." 1972 Industrial Coal Conference, Purdue University, October, 1972.
26. A. H. Rawdon and R. S. Sadowski. "An Experimental Correlation of Oxides of Nitrogen Emissions from Power Boilers Based on Field Data." Paper No. 72-WA/Pwr-5, ASME Annual Winter Meeting, New York, November, 1972.
27. R. E. Sommerlad, R. P. Welden, and R. H. Pai. "Nitrogen Oxides Emission—An Analytical Evaluation of Test Data." 33rd Annual Meeting, The American Power Conference, Chicago, April, 1971.
28. W. H. Barr and D. E. James. "Nitric Oxide Control—A Program of Significant Accomplishments." Paper No. 72-WA/Pwr-13, ASME Annual Winter Meeting, New York, November, 1972.
29. S. Sourirajin and J. L. Blumenthal. *Actes Congr. Intern. Catalyse* **2** (1960): 2521.
30. J. M. Fraser and F. Daniels. *J. Phys. Chem.* **20** (1952): 22.
31. C. H. Riesz et al. "Catalytic Decomposition of Nitric Oxide." Air Pollution Foundation Report No. 20, 1957.
32. H. C. Andersen, W. J. Green, and D. R. Steele. "Catalytic Treatment of HNO_3 Plant Tailgas." *Industrial and Engineering Chemistry* **53**, no. 3 (1961): 199–204.
33. Dupont, British Patent No. 799,159, 1958.
34. G. Badjai, H. K. Orbach, and F. C. Riesenfeld. *Ind. Engr. Chem.* **50** (1958): 1165.

Chapter 9
Atmospheric Photochemical Reactions

9-1 INTRODUCTION

Among the most discomforting features of life in metropolitan areas of industrialized nations is the frequent presence of smog. Smog is the name usually applied to that form of air pollution which arises from the interaction of sunlight with various constituents of the atmosphere. Smog is characterized chemically by a relatively high level of oxidants, which irritate the eyes and throat, damage plants, and cause rubberlike products to crack. Odor and decreased visibility are also characteristic of smog conditions.

9-2 THERMODYNAMICS OF PHOTOCHEMICAL REACTIONS

A number of reactions that occur in our atmosphere would not be possible unless a relatively large amount of energy were available from some source. For example, the dissociation of molecular oxygen requires about 500 kJ/g·mole. Energy in such a concentrated dose simply is not available from the low-temperature gases of the atmosphere. Nevertheless, solar radiation does provide this quantity of energy. The photons that are emitted by the sun have an energy, $h\nu$, where ν is the frequency associated with a particular photon, and h is Planck's constant with a value of 6.62×10^{-34} J·s. A continuous spectrum of frequencies is involved, so that a wide range of energies is available. Since the wavelength of light is inversely proportional to its frequency ($\lambda = c/\nu$), a relatively large energy photon is one with a small wavelength. The infrared region of radiation is roughly 1 to 100 μm (1 $\mu\text{m} = 10^{-6}$ m $= 10^4$ Å). The value of $h\nu$ (hc/λ) is found to be roughly 125 kJ/g·mole for λ of 1 μm. (The g·mole used here represents the number of photons equal to Avogadro's number, that is, 6.0225×10^{23} photons.) At 100 μm, the energy of a light quantum obviously is 1.25 kJ/g·mole. Photons of this energy range may heat a gas or excite its rotational or vibrational energy modes. They will not, however, excite the electronic motions. To break a C—C bond or a C—H bond requires about 350 and 420 kJ/g·mole, respectively. These bonds, as well as the O—O bond discussed earlier, require photons of wavelength much shorter than those of the infrared region. Taking the visible spectrum to have limits of 4000 to 8000 Å (0.4 to 0.8 μm), we find that photons at the outer limits have energies of about 290 to 145 kJ/g·mole. Thus we are getting into the range where molecules are activated electronically by solar radiation. In the ultraviolet region of 0.4 to 0.2 μm, the respective energies of photons are 290 to 580 kJ/g·mole. Hence most photochemical reactions require light energy in the near ultraviolet or lower visible parts of the spectrum.

Although the absorption of a photon of energy may bring about a number of possible changes to a molecule, in atmospheric photochemical studies we are primarily interested in the phenomenon of photochemical dissociation. Fundamentally, photochemical dissociation can be considered a two-step process. The absorption of a photon of energy by species A leads to an excited state, A^*.

$$A + h\nu \rightarrow A^*$$

This is followed by the dissociation of A^* into two products, such as

$$A^* \rightarrow B + C$$

An excited state is normally very unstable, so that the second reaction occurs quickly after the formation of A^*. Either or both B and C may be highly reactive. Hence they bring about a chain of reactions which culminate in the undesirable end product of photochemical smog.

9-3 MONATOMIC OXYGEN AND OZONE FORMATION

In the upper atmosphere (above 50 mi) high-energy photons ($\lambda \sim 0.2$ μm) attack molecular oxygen, as given by the reaction

$$O_2 + h\nu \rightarrow 2O \qquad \text{(a)}$$

As a result, oxygen exists almost solely as monatomic O in this region. At lower levels the monatomic oxygen undergoes a number of reactions. Two of these are recombination to form O_2 and, more importantly, combination with O_2 to form ozone, O_3, according to the reaction

$$O + O_2 + M \rightarrow O_3 + M \qquad \text{(b)}$$

where M is any energy-accepting third body. Ozone itself undergoes photochemical change (with $\lambda \sim 0.2 - 0.29\ \mu m$) so that

$$O_3 + h\nu \rightarrow O_2 + O \tag{c}$$

The end result is that an ozone layer is created above the earth, with its greatest concentration (0.03 ppm) in the region between 10 and 20 mi above the earth.

Now, by turbulence and diffusion in the atmosphere, some of the substances released by man-made sources are exposed to upper atmospheric conditions. Any oxides of nitrogen carried into the upper atmosphere, or formed there, will probably return to earth as nitric acid as a result of the oxidation effect in the ozone layer.

This ozone layer has several interesting effects. Owing to the absorption characteristics of ozone, it acts as a filter to ultraviolet radiation trying to reach the earth's surface. Consequently, some photochemical activity is reduced at lower altitudes. From the standpoint of man, that reduction is essential, independent of smog problems. The human body as it is now constituted could not survive the ultraviolet radiation from the sun if this radiation were not attenuated by the ozone layer.

9-4 ROLE OF OXIDES OF NITROGEN IN PHOTOOXIDATION

In areas where emissions from heavy industry combine with emissions from mobile sources, the atmosphere receives large quantities of SO_2 and NO, which are reducing agents, and hydrocarbons, which generally have no oxidizing power. In the absence of sunlight the SO_2 and NO would slowly be converted into sulfates and nitrates; the conditions now prevalent in large urban areas would not exist. However, because, in addition to SO_2 and NO, hydrocarbons and sunlight are present in urban atmospheres, we have the problem of air pollution.

As nitric oxide (NO) is released by stationary and mobile sources, it tends to oxidize to NO_2 via the reaction

$$2NO + O_2 \rightarrow 2NO_2 \tag{d}$$

If the concentration were 1000 ppm, the near-stoichiometric conversion of NO to NO_2 would occur in a few seconds. During the mixing of exhaust and flue gases in the atmosphere, however, the concentration of NO is lowered significantly. If 1 ppm of NO exists in the atmosphere, the half-life (50 percent conversion) of nitric oxide will be more like 100 hr. The half-life will be greater for smaller concentrations. However, if ozone (O_3) is present, the conversion is extremely rapid even at low concentrations. For respective NO and O_3 concentrations of 0.1 ppm, about 20 s is required for total oxidation.

The synthesis of ozone near the earth's surface does not proceed by means of reactions (a) and (b). The photodissociation of O_2 by reaction (a) requires solar energy in the region of 0.2 μm. However, solar radiation less than 0.29 μm does not reach the tropopause, having been removed largely by reactions with ozone in the stratosphere. Hence ozone in the lower atmosphere must be produced by some other mechanism. Formation usually is attributed to the nitrogen dioxide photolytic cycle. Nitrogen dioxide is highly active photochemically. For radiation below 0.38 μm the gas dissociates according to the reaction

$$NO_2 + h\nu \rightarrow NO + O \tag{e}$$

This is one of the most important photochemical reactions in the lower atmosphere, since it produces the highly active monatomic oxygen O. As we noted earlier O combines with O_2 (in the presence of a third body) to form ozone via reaction (b). Ozone then oxidizes nitric oxide to nitrogen dioxide. In summary, the nitrogen dioxide photolytic cycle may be represented by

$$NO_2 + h\nu \rightarrow NO + O \tag{e}$$

$$O + O_2 + M \rightarrow O_3 + M \tag{b}$$

$$O_3 + NO \rightarrow NO_2 + O_2 \tag{f}$$

The initial NO_2 for reaction (e) is formed by reaction (d), that is, the direct but slow oxidation of NO to NO_2. We noted earlier that approximately 5 percent of the NO leaving a stationary combustion source has already been converted to NO_2 before it leaves the stack. All three reactions noted above are very fast, and the combination would tend to maintain at steady state a constant and low level of ozone. A simplified model of the nitrogen dioxide photolytic cycle is shown in Figure 9-1. It represents reactions (e), (b), and (f). The position of these reactions in the cycle is noted.

Numerous other reactions involving species containing nitrogen and oxygen arc possible. Suggested reactions include the following:

$$O + NO_2 \rightarrow NO + O_2 \tag{g}$$

$$O + NO_2 + M \rightarrow NO_3 + M \tag{h}$$

$$O + NO + M \rightarrow NO_2 + M \tag{i}$$

$$NO_3 + NO \rightarrow 2NO_2 \tag{j}$$

$$NO_2 + O_3 \rightarrow NO_3 + O_2 \tag{k}$$

$$NO_3 + NO_2 + M \rightarrow N_2O_5 + M \tag{l}$$

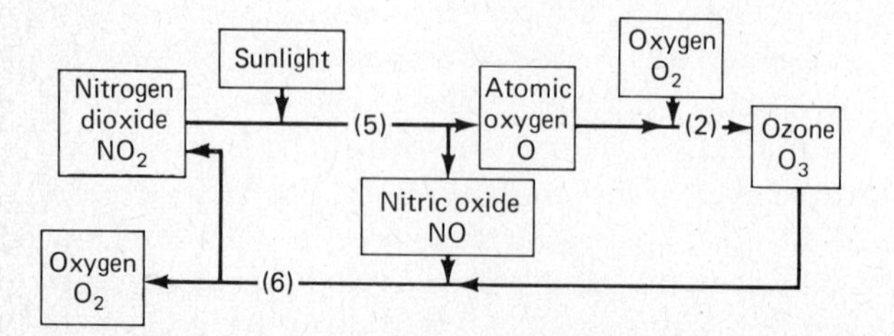

Figure 9-1 Atmospheric nitrogen dioxide photolytic cycle.

Some of these species eventually would be removed from the series of reactions. For example, in the presence of water-vapor droplets in the atmosphere we find that

$$4NO_2 + 2H_2O + O_2 \rightarrow 4HNO_3 \quad \text{(m)}$$

Nitrogen dioxide also hydrolyzes in the gas phase as

$$3NO_2 + H_2O \rightleftharpoons 2HNO_3 + NO \quad \text{(n)}$$

Reaction (n) is an equilibrium reaction. The nitric acid from either of these reactions may react further to form nitrate salts.

The general mechanism described above would be essentially correct if the species indicated in the cycle of events did not interact with other species in the atmosphere. However, atmospheric measurements indicate that the reaction mechanism must be much more complex than that provided by the nitrogen dioxide photolytic cycle. First, actual ozone concentrations in urban atmospheres may reach 0.2 to 0.5 ppm for 1-hr peak averages. This is an order of magnitude greater than the concentration predicted by the steady-state model of Figure 9-1. Second, the daily variation of NO, NO_2, and O_3 is broad and complex, and cannot be explained by the simple model. The diurnal variation of these three species in Los Angeles is shown in Figure 9-2. Except early in the morning (7 to 10 A.M.), the concentrations of O_3 and NO are not high (< 0.01 ppm) at the same time. Even in this early time period the O_3 is increasing while NO is decreasing. Generally the concentrations of O_3 and NO_2 are high at the same time. This latter statement is in agreement qualitatively with Equation (9-1), which represents

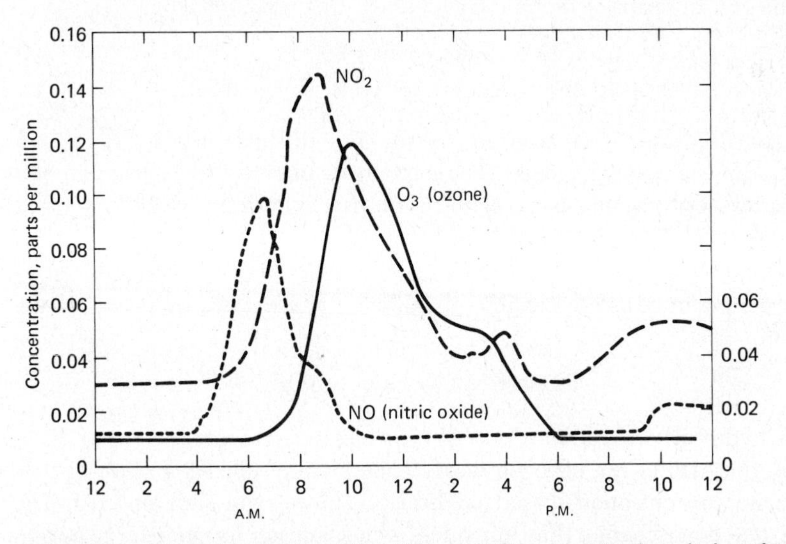

Figure 9-2 Diurnal variation of NO, NO_2, and O_3 concentrations in Los Angeles, July 19, 1965. (SOURCE: NAPCA. Air Quality Criteria for Photochemical Oxidants, AP-63. Washington, D.C.: HEW, 1970.)

the steady-state kinetics of reactions (b), (e), and (f), namely,

$$[O_3] = \frac{kI[NO_2]}{[NO]} \tag{9-1}$$

In this expression I represents the intensity of light. For the atmosphere, measurements indicate that kI is less than approximately 25 $\mu g/m^3$ (0.015 ppm). Since O_3 concentrations greater than 0.1 ppm are observed, Equation (9-1) requires that $[NO_2]/[NO]$ must be 10 or greater. Generally, atmospheric measurements such as those of Figure 9-2 support this trend. Another important result stems from Equation (9-1). The ozone concentration in the atmosphere increases in proportion to the amount of NO oxidized to NO_2. Hence the large quantity of ozone observed in urban atmospheres, compared with the amount predicted by the nitrogen dioxide photolytic cycle, can be understood only if an alternative mechanism for the oxidation of NO to NO_2 can be established.

9-5 HYDROCARBONS IN ATMOSPHERIC PHOTOCHEMISTRY

The increased complexity of atmospheric reactions is brought about by the presence of hydrocarbons. The photochemistry of the lower atmosphere has not been completely delineated. However, we might mention at this point two basic mechanisms which lead to the increased NO_2 and O_3 concentrations noted in Figure 9-2. First, a small part of the atomic oxygen formed by reaction (e) is capable of reacting with various organic compounds to form organic and inorganic free radicals. The same is true of ozone formed by reaction (b). In terms of olefinic compounds these reactions might be

$$O + \text{olefin} \rightarrow R\cdot + RO\cdot \tag{o}$$

$$O_3 + RCH{=}CHR \rightarrow RCHO + RO\cdot + HCO\cdot \tag{p}$$

where R·, RO·, and HCO· are free radicals. The aldehyde, RCHO, formed in reaction (p) is a pollutant, itself. The next important step is the reaction of a free radical with molecular oxygen to form peroxy radicals, (ROO·), such as

$$R\cdot + O_2 \rightarrow ROO\cdot \tag{q}$$

These peroxy radicals are capable of oxidizing NO to NO_2 by means of the reaction

$$ROO\cdot + NO \rightarrow NO_2 + RO\cdot \tag{r}$$

Hence hydrocarbon reactions of this type enhance the production of NO_2, beyond that due to the nitrogen dioxide photolytic cycle. As a result of this, the ozone concentration [based on Equation (9-1)] will shift upward. The above reaction scheme thus provides a mechanism for increasing ozone concentration. Reaction (r) overshadows reaction (f) in the oxidation of NO to NO_2, and hence reaction (f) does not play a dominant role in reducing the ozone concentration. Thus reaction (b) produces ozone, but little is used up in the morning hours of an urban atmosphere. An additional source of O_3 may be by the reaction of peroxy radicals with O_2, such that

$$ROO\cdot + O_2 \rightarrow RO\cdot + O_3$$

This type of reaction could also contribute to the rapid rise in ozone concentration in midmorning. However, it has been suggested [1] that the reaction above may be of minor importance, and that the influence of radical species on rapidly converting NO to NO_2 is sufficient to account for the O_3 buildup.

Finally, we can redraw Figure 9-1 to include the interaction of hydrocarbons with the nitrogen oxide photolytic cycle. A simplified representation is shown in Figure 9-3. Oxygen atoms attack various hydrocarbons (especially olefins and substituted hydrocarbons). Ozone can also oxidize hydrocarbons, but the rates of reaction are considerably slower than that of hydrocarbon oxidation by atomic oxygen. The oxidized compounds and free radicals then react with NO to form more NO_2. Since a significant portion of the NO now reacts with hydrocarbon species (upper path in Figure 9-3), less is available for reaction with O_3 (lower path). This upsets the consumption of O_3 by NO, so that the O_3 level increases. Under the same circumstances, the NO_2 level also increases, rather than remaining relatively constant, since an additional source of NO_2 formation is present. Hence the concentration of NO_2, which normally should decrease because of its initial photodissociation by sunlight, actually increases significantly. NO, in turn, is depleted. Figure 9-2, which was referred to earlier, is in agreement with the general timing of events predicted by this chemical model. The events shown in Figure 9-3, of course, do not exhibit the complexity of the actual system. However, it is a useful scheme to explain qualitatively the rapid buildup of NO_2 and the increase in ozone level.

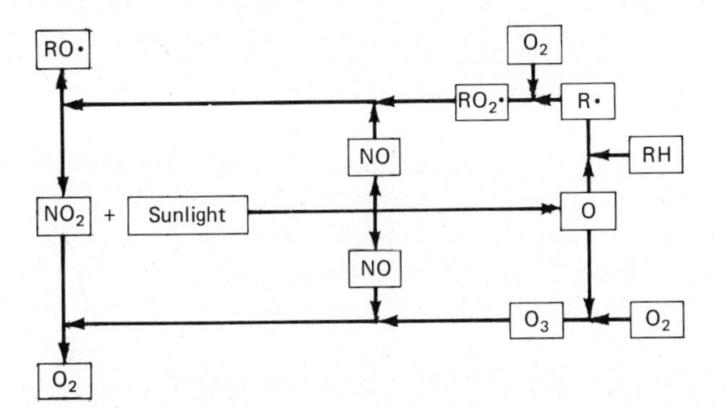

Figure 9-3 Interaction of hydrocarbons with atmospheric nitrogen oxide photolytic cycle. (SOURCE: NAPCA. Air Quality Criteria for Photochemical Oxidants, AP-63. Washington, D.C.: HEW, 1970.)

As a second mechanism for free radical formation, many organic compounds in the air are themselves photochemically active. These include aldehydes, ketones, peroxides, and acyl nitrates. The absorption of solar radiation by these compounds will contribute to the production of free radicals, which in turn will be responsible for many new compounds. As an example, aldehydes undergo photodissociations such as

$$RCHO + h\nu \rightarrow R\cdot + HCO\cdot$$

By whatever initiating mechanism, the presence of free radicals leads to the complex chemistry of smog formation. They are present in extremely dilute amounts, say, less than 10 ppb. However, their dilution leads to long half-lives. That is, they may persist for minutes or hours. Hence they have sufficient time to enter into a number of reactions.

One of the major problems in the study of photochemical smog has been the identification of those chemical species of major significance. The reactions listed below are additional ones which are frequently cited. However, the list is not intended to be complete, nor is it to be inferred that all those listed will eventually be found to play a dominant role.

$$RCO_2\cdot + NO \rightarrow RCO\cdot + NO_2$$
$$RCO\cdot + O_2 \rightarrow RCO_3\cdot$$
$$RCO_2\cdot + O_2 \rightarrow RO_2\cdot + CO_2$$
$$RCO_3\cdot + NO \rightarrow NO_2 + RCO_2\cdot$$
$$RCO_3\cdot + NO_2 \rightarrow RCO_3NO_2 \text{ (peroxyacyl nitrate)}$$
$$RO\cdot + NO \rightarrow RONO$$
$$RO\cdot + RH \rightarrow ROH + R\cdot$$
$$RH + O \rightarrow R\cdot + OH$$
$$RO\cdot + RCO\cdot \rightarrow \text{ketone} + \text{alcohol}$$

Several of the above reactions are chain-terminating steps. These reactions are but a few of the many that might be included in a generalized kinetic mechanism [2].

The major point to be noted in the above reactions is that frequently the same radical appears as a product in one reaction and as a reactant in another. Hence a relatively small quantity of various free radicals may be responsible for the formation of a sizable quantity of pollutants. The chain of reactions is initiated by the photooxidation of hydrocarbons by atomic oxygen. This latter species is formed, as emphasized earlier, by the photodecomposition of nitrogen dioxide. The major products from these photochemical reactions are aldehydes, ketones, CO, CO_2, organic nitrates, and oxidants. Oxidants include ozone, NO_2, peroxyacyl-type nitrate compounds, possible other types of organic peroxy and hydroperoxide compounds, and hydrogen peroxide (H_2O_2).

One other mechanism is frequently cited [3, 4] as a possible source of large radical concentrations in urban atmospheres. In this case atomic oxygen reacts with water to form hydroxyl radicals, OH. A short chain reaction is then initiated by the hydroxyl radical as it reacts with both ozone and carbon monoxide such that

$$OH + O_3 \rightarrow HO_2 + O_2$$
$$OH + CO \rightarrow CO_2 + H$$

The hydrogen atom immediately acts to reform the hydroperoxyl radical, HO_2, by means of

$$H + O_2 + M \rightarrow HO_2 + M$$

This chain reaction is completed by the oxidation of nitric oxide by the hydroperoxyl radical to nitrogen dioxide via the reaction

$$HO_2 + NO \rightarrow NO_2 + OH$$

Chain termination reactions include

$$OH + OH \rightarrow H_2O + O$$
$$HO_2 + OH \rightarrow H_2O + O_2$$

This series of reactions includes those of chain initiation, propagation, and termination. Note that the mechanism not only provides an oxidation mechanism for NO to NO_2 but also removes carbon monoxide from the atmosphere. Formaldehyde might also be produced by a similar chain reaction which begins with the reaction of hydroxyl radical with methane. The interesting feature of this mechanism is that carbon monoxide plays a dominant role in the chain-initiating step. Until the early 1970s carbon monoxide was considered relatively unreactive with respect to photochemical smog formation. Data [5] from an irradiated smog chamber indicate, however, that carbon monoxide accelerates the oxidation of nitric oxide and the formation of ozone. The initial composition within the chamber during a specific test included 3 ppm of isobutene, 1.5 ppm of nitric oxide, less than 0.04 ppm of NO_2, and a relative humidity of 70 percent. Comparative data were taken for initial CO concentrations of 0 and 100 ppm. The presence of CO markedly accelerated the disappearance of the olefin, the appearance of ozone, and the conversion of NO to NO_2. An ozone concentration of 0.5 ppm was reached in roughly 2 hr with CO initially present, while nearly 3 hr were required when CO was absent initially. A similar time reduction occurred for the peak concentration of NO_2 within the chamber. If these smog chamber results are a valid approximation to atmospheric reactions, then we must surmise that the presence of carbon monoxide in the atmosphere in the early morning may not only accelerate the oxidation of NO to NO_2 but also hasten the appearance of oxidants in the atmosphere.

The effect of CO concentration on photooxidation processes as described above was obtained for a CO level of 100 ppm. However, such levels are not usually experienced in urban atmospheres. For example, it is reported [6] that the geometric mean carbon monoxide level for Los Angeles was around 10 ppm for hourly values, and the largest daily maximum hourly average reported for CO was 37 ppm. Interpolation of results from the 100-ppm level to intermediate levels of less than 50 ppm would be difficult. In addition, the measurements were carried out in the presence of a single hydrocarbon. Subsequent measurements [7] at CO levels of 25 and 50 ppm have yielded somewhat different results. This later investigation used a mixture of hydrocarbons thought to be representative of the Los Angeles region (total HC concentration of 1.96 ppm), 0.97 ppm of nitric oxide, and 0.05 ppm of nitrogen dioxide. The ozone yield was measured after 500 min of irradiation. For CO levels less than 35 ppm it was found that ozone yield would be relatively unaffected and that the formation rate of NO_2 would be slightly increased, in comparison with atmospheres having zero carbon monoxide concentration. The overall conclusion was that CO would have negligible effect on photochemical smog formation at concentrations of CO, NO_x, and hydrocarbons commonly found in urban environments.

9-6 OXIDANTS IN PHOTOCHEMICAL SMOG

For the purpose of air pollution studies, the term oxidant generally refers to those substances which can oxidize potassium iodide or other chemical reagents. It is often difficult to obtain a correlation among the methods of measuring oxidant concentration because of the variation in the relative response of different oxidants to the technique in use. Oxidants in the atmosphere include ozone, nitrogen dioxide, hydrogen peroxide (H_2O_2), peroxyacyl-type nitrate compounds, and possibly other types of organic peroxy and hydroperoxide compounds.

One of the organic oxidants which is a major pollutant is peroxyacetyl nitrate, PAN. Although its structure was fairly well established by the mid-1960s, research in the late 1960s led to a more definitive structure. Its basic formulation is now considered to be

$$CH_3-\overset{\overset{\displaystyle O}{\|}}{C}-O-O-NO_2$$

This compound is a strong eye irritant and causes plant damage. Concentrations of PAN at Riverside, California, for example, have reached 60 ppb. Another strong eye irritant identified in 1968 [8] is peroxybenzoyl nitrate, PBN, having the structure

$$C_6H_5-\overset{\overset{\displaystyle O}{\|}}{C}-O-O-NO_2$$

It has been reported that PBN is about 100 times more eye irritating than

PAN. Research seems to indicate that nearly every hydrocarbon (with the possible exception of methane) is capable of being photooxidized in the presence of NO_x to some form of oxidant. The only requirements are that the hydrocarbon-NO_x ratio be high enough and the irradiation time long enough. This indicates that a sizable reduction in hydrocarbons released might lead to a significant reduction in oxidant concentration in urban areas.

The national ambient air quality standard for photochemical oxidants expressed as ozone (corrected for SO_2 and NO_2) is 240 $\mu g/m^3$ (0.12 ppm) averaged over a 1-hr period. This limit is not to be exceeded for a 1-hr period more than once a year.

The effect of photochemical oxidants on human health is difficult to appraise, since other factors affecting health frequently occur simultaneously. For example, daily mortality of the aged in Los Angeles correlates well with days of high temperature and high oxidant concentrations. For lower values of these two parameters, no consistent relationship prevails [9]. Temperature alone is a severe environmental stress. Whether a synergistic effect occurs with high oxidant levels has not yet been ascertained. Even with chronic diseases such as emphysema, bronchitis, and asthma, the correlation between symptoms and abnormally high oxidant levels is not significant. The most widespread and common symptom clearly associated with increasing oxidant levels is acute eye irritation. Ozone is not a primary eye irritant. However, eye irritation is associated consistently with oxidant levels exceeding 200 $\mu g/m^3$ (0.1 ppm). With a federal oxidant standard of 240 $\mu g/m^3$, there is no margin of safety between the standard and the onset of eye irritation.

Although the ambient air quality standard for photochemical oxidants is 240 $\mu g/m^3$ for a 1-hr average, this value is frequently exceeded in areas such as the Los Angeles basin. Consequently, states must have a contingency plan by which various degrees of warning are provided to the populace when certain levels of oxidant are reached. As of March 13, 1974, EPA set new levels for such warnings. They are as follows:

Alert	200 $\mu g/m^3$ (0.1 ppm) for 1-hr average
Warning	800 $\mu g/m^3$ (0.4 ppm) for 1-hr average
Emergency	1000 $\mu g/m^3$ (0.5 ppm) for 1-hr average
Significant harm	1200 $\mu g/m^3$ (0.6 ppm) for 1-hr average

By "significant harm" is meant that level of photochemical oxidant concentration which never should be reached. Any contingency plan should include abatement actions such as shutting down offices, businesses, and industries, as well as prohibiting the use of vehicles when the significant harm level is approached. Health studies have shown that 1000 $\mu g/m^3$ of oxidants can cause breathing difficulty and chest pains even in healthy individuals. More susceptible persons are likely to experience even more adverse effects. From January, 1970, to June, 1973, the new significant harm level was exceeded 18 times in the Los Angeles area. During the July 25, 1973, air pollution emergency episode in Los Angeles, hourly averages of 1120 and 1260 $\mu g/m^3$ were recorded.

9-7 HYDROCARBON REACTIVITY

The presence of hydrocarbons is essential to the formation of photochemical smog. However, not every hydrocarbon will react with the other constituents of the atmosphere to produce to the same degree the chemical species that are responsible for effects such as plant damage, visibility reduction, and eye irritation. The differences in the end result are explained in terms of the *reactivity* of hydrocarbons. Several different measures of reactivity appear in the literature. Among the experimental data used for evaluation are hydrocarbon consumption rate, nitric oxide oxidation, oxidant formation, aerosol formation, eye irritation, and plant damage [10].

The results based on hydrocarbon reaction rates and nitric oxide oxidation are generally in good agreement. Branched or straight-chain olefins with interval double bonds form the most reactive class. The trialkylbenzenes,

Table 9-1 REACTIVITIES OF HYDROCARBONS BASED ON ABILITY TO PARTICIPATE IN PHOTOOXIDATION OF NITRIC OXIDE TO NITROGEN DIOXIDE

	RANKING	
HYDROCARBON	ALTSHULLER AND COHEN	GLASSON AND TUESDAY
2, 3-Dimethylbutene-2		10
2-Methyl-2-butene		3
trans-2-Butene[a]	2	2
Isobutene	1	
Propylene	1	0.5
1, 3, 5-Trimethylbenzene	1.2	1.2
m-Xylene	1	0.9
1, 2, 3, 5-Tetramethylbenzene	0.9	0.7
o- and *p*-Xylene	0.4	0.4
o- and *p*-Diethylbenzene	0.4	0.4
Ethylene	0.4	0.3
Toluene	0.2	0.2
Benzene	0.15	0.04
3-Methylheptane	0.15	
n-Heptane		0.2
2, 2, 4-Trimethylpentane	0.15	0.15
Butanes		0.1
Acetylene	0.1	
Ethane		0.03
Methane		0.01

SOURCE: A. P. Altshuller and J. J. Bufalini. *Environ. Sci. Tech.* 5 (1971): 39–64. Reprinted with permission of the American Chemical Society.

[a] Two ranking scales adjusted to give same value on both scales for *trans*-2-butene, for comparative purposes.

tetraalkylbenzenes, and olefins with terminal double bonds (except ethylene) rank next, followed by the dialkylbenzenes, aldehydes, and ethylene. Toluene is less reactive than ethylene, and the paraffinic hydrocarbons, acetylene, and benzene are less reactive than toluene. Table 9-1 summarizes this trend for two sets of data. The alkylbenzenes, which rank high on the hydrocarbon reactivity list, also have a high eye irritation potential. On the basis of reactivity scales such as these, we find that the olefins from automotive sources account for 75 to 85 percent of the overall reactivity of the hydrocarbons released from such a source.

9-8 DAILY HISTORY OF POLLUTANTS IN PHOTOCHEMICAL SMOG

A typical graph of the variation of pollutant concentrations versus time for a day of high eye irritation in downtown Los Angeles is shown in Figure 9-4. In the early morning hours the hydrocarbon content of the air begins to increase and nearly doubles by 8 A.M. The raw materials for photochemical reactions in this case are due to increased traffic. Around dawn the concentrations of CO and NO likewise increase, for the same reason. We note that NO_2 is generated at a substantial rate by 7 A.M. (several hours after dawn) as a result of the reactions involving NO and the hydrocarbons. Shortly afterward, the NO concentration drops to a very small value, as the NO_2 reaches its peak. Note also that the disappearance of NO coincides with the rise in ozone concentration. The ozone is formed during photooxidation of hydrocarbons in the presence of nitrogen dioxide.

As early afternoon approaches, this maximum concentration of ozone is reached, after which it declines gradually. During a severe smog in the Los Angeles basin, for example, a 0.5-ppm concentration of ozone may be reached. This represents 500 tons of ozone present in the air at a given time.

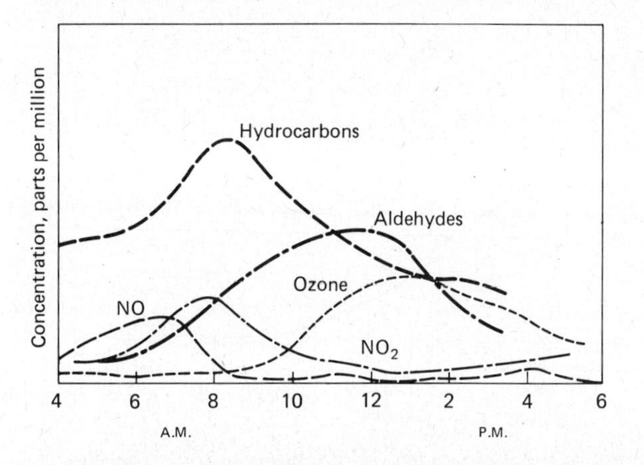

Figure 9-4 Average concentration during smog in downtown Los Angeles: hydrocarbons, aldehydes, and ozone for 1953—1954; NO and NO_2 for 1958. (SOURCE: Based on data of the Los Angeles Air Pollution Control District.)

Good ozone formers include the diolefins, olefins, aldehydes, and alcohols. Of major significance is that these groups of compounds form ozone at different rates. Consequently, as the more reactive species are consumed, less reactive species take over the role of forming ozone. The net result, as shown in Figure 9-4, is the growth of ozone over a considerable length of time.

Nitrogen dioxide diminishes as the ozone reaches its peak. A slight positive perturbation in nitric oxide occurs around 4 to 5 P.M., as late afternoon traffic injects more of the substance into the atmosphere. The NO scavenges most of the ozone still present. Aldehydes are one of a group of compounds which are irritants. The data of Figure 9-4 indicate a general increase in aldehydes throughout the morning hours, and a slow reduction in the afternoon. Throughout the preceding discussion the influence of temperature has been ignored. Some investigations indicate that a change of 40°F will increase the overall rates of reaction by a factor of 2 to 4. This may help explain why smog formation is more prevalent on warmer days, other factors being equal.

The trends established in Figure 9-4 for downtown Los Angeles are replicated by smog-chamber studies. Figure 9-5 shows the variation of concentrations with time within an irradiated smog chamber which initially contained propylene and nitric oxide [11]. The rapid decline in NO, followed

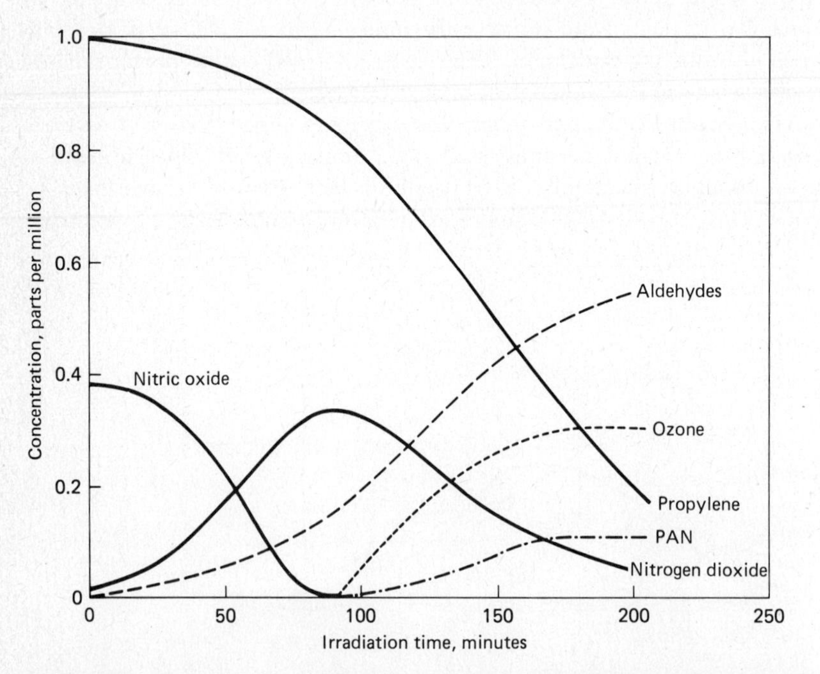

Figure 9-5 Typical concentration changes in a photochemical smog reaction. (SOURCE: W. G. Agnew. "Automotive Air Pollution Research." Proc. Royal Soc. **307A** (1968): 153. Data courtesy of the General Motors Research Laboratories.)

by the rise in NO_2 and O_3 concentrations, is again observed. The subsequent formation of PAN is especially to be noted.

In summary, photochemical smog begins with the photodissociation of nitrogen dioxide and the concomitant production of atomic oxygen. In the absence of hydrocarbons, nitrogen dioxide is regenerated about as fast as it is photolyzed. At the same time, small concentrations of O, O_3, NO, and N_2O_3 are reached. Once organic vapors are injected into the system, nitrogen dioxide actually begins to accumulate, probably as a consequence of the following situation. Hydrocarbons react with the oxygen atoms to form free radicals. Chain reactions then commence, which involve molecular oxygen, so that nitric oxide is consumed. As a result, nitrogen dioxide is produced faster than it photodissociates. At the same time the complex chain reactions lead to other products which act as irritants to animals and plants.

Although the complete chemistry of photochemical smog formation is not known, it is apparent that the oxides of nitrogen catalyze, in a sense, reactions between hydrocarbons and oxidants. The alleviation of the effects of smog may require a substantial reduction in NO_x produced in urban areas. At the same time it is necessary to control the release of hydrocarbons from numerous mobile and stationary sources.

9-9 OXIDATION OF SULFUR DIOXIDE IN POLLUTED ATMOSPHERES

An important physical effect of air pollution is reduced visibility. This reduction in visibility is generally caused by the scatter and absorption of radiation by aerosol particles in the atmosphere. While the quantitative contribution of sulfur oxides is not known, sulfuric acid mist and other sulfate particulate matter are recognized as important sources of scattering. Such particulates arise from the complex oxidation processes described above for the interaction of atmospheric pollutants. Photodissociation of sulfur dioxide released from combustion processes into the atmosphere is not possible, since it requires wavelengths shorter than those reaching the lower atmosphere. Hence any photochemistry of SO_2 can involve only molecular reactions of the electronically excited states of SO_2.

In terms of the photochemistry of the SO_2-NO_x system, the addition of NO_x to low concentrations of SO_2 seems to produce varying results. Some authors report an enhancement of the photooxidation of SO_2, while others report a hindrance, as measured by the formation of aerosol. Recall that the photodissociation of NO_2 (Section 9-4) leads to atomic oxygen and ozone. Thus SO_2 could be a secondary competitor with NO and NO_2 for the oxygen atom, by the reaction

$$SO_2 + O + M \rightarrow SO_3 + M$$

The effectiveness of the reaction increases as the SO_2/NO_x concentration increases. There is experimental evidence for this effect [12]. For NO_x and

SO_2 concentrations of 20 pphm (typical of photochemical smog), kinetic estimates indicate that the SO_2-O reaction rate would be about a factor of 10 less than the NO_x-O rate. Not only the SO_2-NO_x system but also the SO_2-hydrocarbon system in the absence of NO_x has been studied. Available data indicate that the reactions of SO_2 in the presence of hydrocarbons at common atmospheric levels are not significant in the formation of aerosols.

In any polluted atmosphere SO_2, NO_x, and hydrocarbons are usually present simultaneously. Experimental work in this area indicates that the irradiation of straight-chain and cyclic olefins in the presence of NO_x and SO_2 leads to significant aerosol formation [13]. One set of data is reproduced in Table 9-2 [14], where the photometer reading indicates the relative presence of aerosols. The effect of SO_2 on aerosol formation is quite apparent. Paraffinic hydrocarbons appear to yield little or no aerosol in the presence of NO_x and SO_2, while data on aromatics are meager and inconclusive [13]. While the data in the table are for a fixed NO_x-SO_2 ratio, experiments indicate that the rate of disappearance of SO_2 is dependent on the ratio of the reactants. In one set of experiments the initial concentrations of SO_2 and NO are held constant at 0.1 and 1 ppm. When the olefin concentration is decreased from 3 to 0.5 ppm, the rate of disappearance of SO_2 increases. This indicates that low concentrations of olefins can be quite effective in photooxidizing SO_2 in the presence of NO. For the situation of fixed olefin and NO concentrations and variable SO_2 quantities, the data are contradictory to date. Some studies indicate that light scattering is directly proportional to SO_2 concentration, while others observe an increase in rate of consumption of SO_2 as the SO_2 concentration is decreased from 1 to 0.1 ppm. It should also be pointed out that the disappearance of SO_2 and the formation of aerosol do not normally coincide [15]. Data indicate that SO_2 disappearance occurs first, followed later by the buildup and maximization of aerosol as measured by light scattering. The decay of SO_2 and the buildup

Table 9-2 EFFECT OF SO_2 ON AEROSOL FORMATION FOR IRRADIATED NITROGEN DIOXIDE-HYDROCARBON MIXTURES

HYDROCARBON	PHOTOMETER READING[a]	
	OLEFIN-NO_2	OLEFIN-NO_2–SO_2
Ethylene	0.1	2.2
1-Butene	0.05	2.85
2-Butene	0.05	3.0
1-Pentene	0.1	3.05
1-Hexene	0.05	3.35
2-Hexene	0.25	3.05
3-Heptene	0.45	3.35
2, 4, 4-Trimethyl-1-pentene	3.3	3.3

SOURCE: M. J. Prager et al. *Ind. Engr. Chem.* **52** (1960): 521. Reprinted with permission of the American Chemical Society.

[a] Reactant concentrations: olefin, 10 ppm; NO_2, 5 ppm; SO_2, 2 ppm.

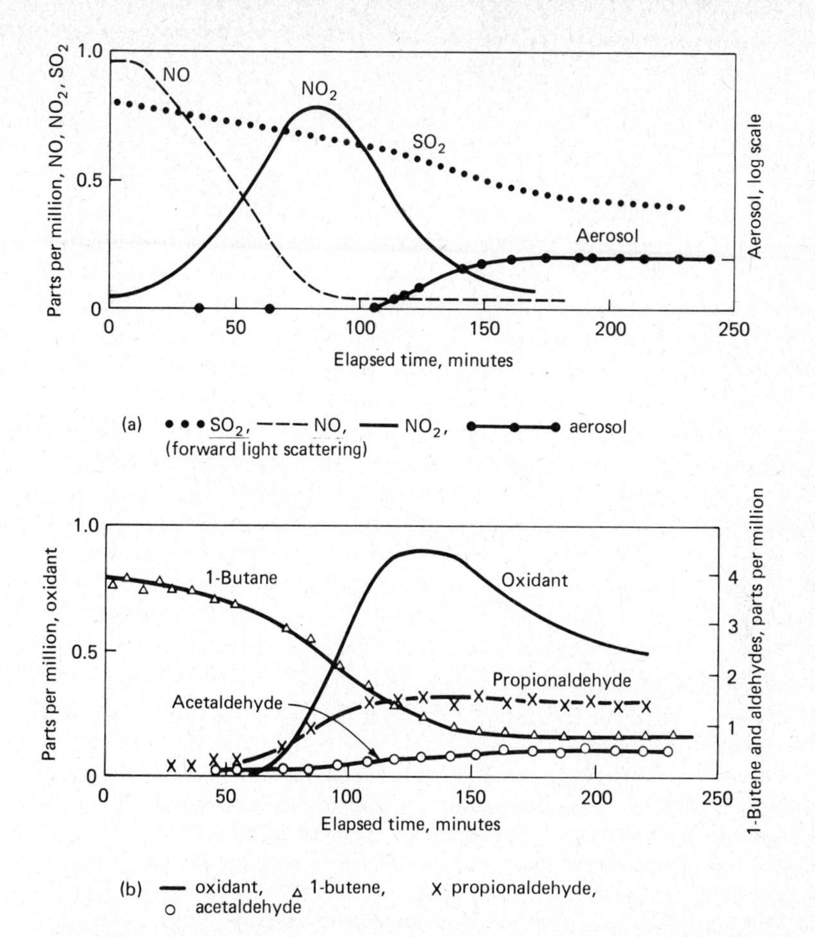

Figure 9-6 Smog profile for a system with an initial composition of 4 ppm 1-butene, 1 ppm NO, and 0.75 ppm SO_2, under dry conditions. (SOURCE: J. Harkins and S. W. Nicksic. *J. Air Pollu. Control Assoc.* 15 (1965): 218.)

of aerosol in a smog chamber under dynamic conditions is shown in Figure 9-6 for the 1-butene, NO, and SO_2 system [16]. The aerosol is not observed until after the NO_2 peak has been reached and the NO concentration has fallen to a low value. At this point the production of O atoms is also at a maximum. Buildup of particulate continues until the ozone maximum is reached, after which it levels off. Hence aerosol accumulates rapidly once ozone is produced, on the basis of a number of tests.

In summary, experiments generally indicate that the rate of disappearance of SO_2 and the formation of aerosol increase when SO_2 is photo-oxidized in the presence of NO_x and olefinic hydrocarbons. Apparently some intermediate species in the hydrocarbon-NO_x-ozone system is effective in enhancing the oxidation of SO_2. At the same time, SO_2 does appear to have the general effect of slowing down photochemical smog reactions, which as yet is unexplained. Consequently, urban regions will be affected differently in terms of photochemical smog and aerosol formation. The magnitude of the problem will vary with the relative concentrations of NO_x, hydrocarbons, and SO_2 emitted into the atmosphere from stationary and mobile sources.

Finally, some comment must be made on the effect of H_2O concentration on the photooxidation of SO_2, and on aerosol identification. Several sets of experiments indicate that aerosol formation (as measured by light scattering) decreases with increase in relative humidity at constant SO_2 concentration. This effect is most noticeable at low relative humidity. Little work has been done on determining the chemical identity of aerosols. The aerosol is H_2SO_4 from the simple SO_2-O_2 system. For the SO_2-NO_x system the aerosol more than likely is sulfate also. Aerosols produced by irradiation of the lower olefins in the presence of NO_x and SO_x are not of organic composition [16]. Higher olefinic hydrocarbons appear to produce carbonaceous aerosol also, such as organic acids. The chemical identification of these aerosols is extremely important, since they are of such a size as to enter the respiratory tract easily.

QUESTIONS

1. Why is photochemical smog generated over a city?
2. What four factors or chemical species must be present for the formation of photochemical smog?
3. What is the time lag, in minutes, between the emission of NO_x and hydrocarbons and the buildup of eye irritants in the atmosphere?
4. What are the major eye irritants in photochemical smog?
5. List the hydrocarbons that have the highest reactivity relative to smog formation.
6. Why is methane not considered a polluting hydrocarbon?
7. What is the major result of the photochemical oxidation of SO_2 in the presence of NO_x?

PROBLEMS

9-1. The photodissociation of SO_2 into SO and O atoms requires 565 kJ/g·mole. What wavelength of light is required, in microns? Why does this process not occur in the lower atmosphere?

9-2. The rate constant at 25°C for the reaction $NO_2 + O_3 \rightarrow NO_3 + O_2$ is 3.3×10^{-17} cm^3/molecule·s. The concentrations of NO_2 and O_3 at a given time are 0.05 and 0.02 ppm, respectively. Determine the rate of formation of the intermediate species, NO_3, in parts per million per second, at 1 bar pressure.

9-3. The rate of formation of nitrogen dioxide, NO_2, in the atmosphere may be estimated from the rate equation, $d[NO_2]/dt = k[NO]^2[O_2]$, where k has a value of 10^4 l^2/g·mole^2·s. If at a given location the concentration of NO is 0.05 ppm, and oxygen comprises 21.0 percent of the atmosphere by volume, calculate the rate of formation of NO_2 in (a) molecules/l·s, and (b) g·moles/l·s.

9-4. The rate constant for the reaction $O_3 + NO \rightarrow NO_2 + O_2$ is 5×10^{-14} cm^3/molecule·s. If the concentrations of O_3 and NO are each 0.1 ppm at a given time, determine the rate of formation of NO_2 in parts per million per second.

References

1. R. J. O'Brien. "Photostationary State in Photochemical Smog Studies." *Environ. Sci. Tech.* **8** (1974): 579.
2. T. A. Hecht, J. H. Seinfeld, and M. C. Dodge. "Further Development of Generalized Kinetic Mechanism for Photochemical Smog." *Environ. Sci. Tech.* **8** (1974): 327.
3. H. Levy. "Normal Atmosphere: Large Radical and Formaldehyde Concentrations Predicted." *Science* **173** (1971): 141.
4. B. Weinstock. *Science* **166** (1969): 224.
5. K. Westberg, N. Cohen, and K. W. Wilson. "Carbon Monoxide: Its Role in Photochemical Smog Formation." *Science*, **171** (1971): 1013.
6. R. I. Larsen. *J. Air Pollu. Control Assoc.* **19** (1969): 24.
7. W. A. Glasson. *Environ. Sci. Tech.* **9**, no. 4 (1975): 343.
8. J. M. Heuss and W. A. Glasson, *Environ. Sci. Tech.* **2** (1968): 1109.
9. J. McCarroll. "Photochemical Oxidants and Human Health: An Epidemiologic Appraisal." *Calif. Air Environ.* **4** (Winter 1974).
10. A. P. Altshuller and J. J. Bufalini. "A Review of Photochemical Aspects of Air Pollution." *Environ. Sci. Tech.* **5** (1971): 39–64.
11. W. G. Agnew. "Automotive Air Pollution Research." *Proc. Royal Soc.* **307A** (1968): 153.
12. S. Jaffee and F. S. Klein. *Trans. Faraday Soc.* **62** (1966): 2150.
13. M. Bufalini. "Oxidation of Sulfur Dioxide in Polluted Atmospheres—A Review." *Environ. Sci. Tech.* **5** (1971): 685.
14. M. J. Prager, E. R. Stephens, and W. E. Scott, *Ind. Engr. Chem.* **52** (1960): 521.
15. L. Reckner and W. E. Scott. American Chemical Society National Meeting, New York, September, 1963.
16. J. Harkins and S. W. Nicksic. *J. Air Pollu. Control Assoc.* **15** (1965): 218.

Chapter 10
Mobile Sources

10-1 INTRODUCTION

Serious air pollution was detected as early as 1943 in the Los Angeles area. By 1948 the problem was so severe that the state legislature passed a law for the formation of air pollution control districts with the power to curb emission sources. Early attempts to reduce eye irritation by controlling emissions from stationary sources such as open burning, steel mills, and refineries failed. It soon became apparent that the air pollution in Los Angeles was of a composition different from that of such cities as London and Pittsburgh, and therefore could not be controlled by reducing emissions from stationary sources alone. Consequently, a research program was undertaken by the state of California to establish the cause of the air pollution problem. Professor A. J. Haagen-Smit [1] first demonstrated that hydrocarbon compounds react with the oxides of nitrogen when irradiated by sunlight to form eye-irritating oxidants such as PAN and ozone. Those reactions are discussed in detail in Chapter 9. Since 1953 much research and study have been devoted to the subject of air pollutants generated by the combustion of hydrocarbons. The literature on the subject is extensive, yet much more knowledge is necessary to deal with the problem effectively.

10-2 EMISSION STANDARDS FOR AUTOMOBILES

On a national basis in 1977, highway vehicles were responsible for 75 percent of the CO, 35 percent of the hydrocarbons, and 29 percent of the

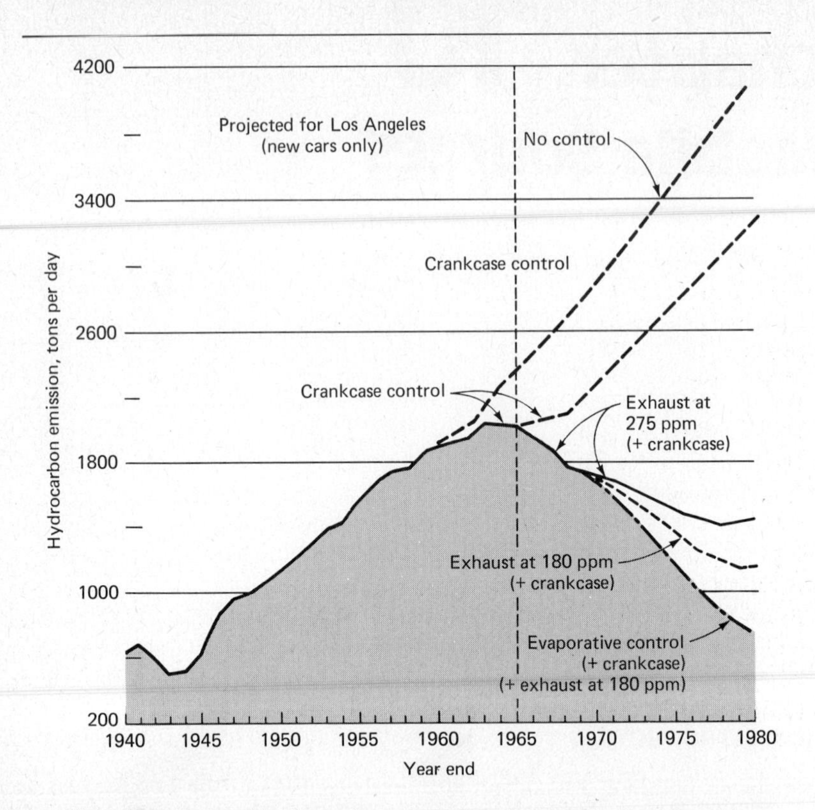

Figure 10-1 Forecasts of hydrocarbon emissions by cars in the city of Los Angeles. (SOURCE: *SAE Journal* 76, no. 12 (1968): 45. Reprinted by permission of the Society of Automotive Engineers, Inc.)

oxides of nitrogen [46]. In 1965 there were estimated to be 80 million cars in operation in the United States. That number may increase to 145 million by 1985. The effect of these cars upon the quantities of air pollutants is illustrated in Figure 10-1, which presents the estimated hydrocarbon emission as a function of the calendar year. Similar diagrams are available for CO and NO_x. From Figure 10-1 it can be seen that in the absence of control, the hydrocarbon emissions would have increased by a factor of over 2 by 1980.

Early analysis of the automobile with no emission controls shows the sources of emissions to be as follows:

SOURCE	POLLUTANT (%) CO	HC	NO_x	PARTICULATE
Exhaust	100	62	100	90
Crankcase emission		20		10
Fuel tank evaporation		9		
Carburetor evaporation		9		

Table 10-1 FEDERAL EXHAUST EMISSION AND FUEL ECONOMY STANDARDS FOR PASSENGER CARS AND LIGHT-DUTY TRUCKS

	PASSENGER CARS				LIGHT DUTY TRUCKS			
YEAR	HC (g/mi)	CO (g/mi)	NO_x (g/mi)	F/E (mpg)	HC (g/mi)	CO (g/mi)	NO_x (g/mi)	F/E (mpg)
1968	3.2	33						
1971[a]	4.6	47	4.0					
1974	3.4	39	3.0					
1977	1.5	15	2.0					
1978	1.5	15	2.0	18	2.0	20	3.1	
1979	1.5	15	2.0	19	1.7	18	2.3	17.2[b]
1980	0.41	7.0	2.0	20	1.7	18	2.3	16.0
1981	0.41	3.4[c]	1.0	22	1.7	18	2.3	18.0
1982[d]	0.41	3.4	1.0	24	0.50	4.5	1.2	20.0

[a] Test method changed in 1971.
[b] Fuel economy data for 2-wheel drive trucks.
[c] Manufacturer may apply for waiver, not to exceed 7.0 g/mi.
[d] Data for trucks are estimated values.

As an additional complication, the mode of vehicle operation has a marked effect upon the emissions. In 1960, the Motor Vehicle Pollution Control Board of the State of California was created to establish specifications on vehicle exhaust and evaporative emissions. The first automotive emission requirement was for the reduction of crankcase blow-by. The California Motor Vehicle Pollution Control Board adopted a resolution requiring that a positive crankcase ventilation system be installed on all new cars sold in California beginning with 1963 models. The result of this requirement is shown in Figure 10-1 by the crankcase control line.

From Figure 10-1 it can be seen that any attempt to reduce air pollution in 1980 to the 1947 level, a level deemed satisfactory, will necessitate more drastic controls to reduce the quantity of hydrocarbons. Consequently, various emission standards for new cars have been established in subsequent years by the control board. The federal government, by an amendment to the Clean Air Act in 1965, specifically authorized the writing of national standards for emissions from all motor vehicles sold in the United States. A summary of legislation and projections for passenger-car and light-duty truck exhaust emissions is presented in Table 10-1. To gain some understanding of the many factors involved in achieving the projected emission standards, it is advantageous to examine in some detail the phenomena responsible for the formation of the various pollutants.

10-3 GASOLINE

Modern gasolines are blends of varying quantities of paraffins, olefins, naphthenes, and aromatics compounded to give the desired characteristics (of starting, accelerating, and so on) when burned in automobile engines.

The composition varies from company to company and from one geographical region to another. For example, a gasoline designed for use in the winter of Minnesota will have a larger percentage of highly volatile constituents than will a gasoline for use in Florida in the summer. Examples of the hydrocarbon families are shown in Figure 10-2. These include saturated straight and branched chains, and aromatic types. The hydrocarbons have different boiling points, ignition temperatures, and combustion characteristics.

Extensive tests have shown that different gasolines when burned in a given engine will yield different unburned hydrocarbon exhaust products. Furthermore, the exhaust products vary from one engine to another when operating with the same gasoline. Over 200 different hydrocarbons have been identified in the exhaust products of one engine burning a given gasoline. The combustion phenomena occurring in an internal combustion engine are so complex that it has been impossible to develop a chemical reaction model which successfully predicts the composition of the exhaust products.

As an additional complicating factor, it has been found that the different unburned hydrocarbons have varying potentials for forming eye-irritating smog. This potential, as discussed in Chapter 9, is termed *reactivity*. Thus

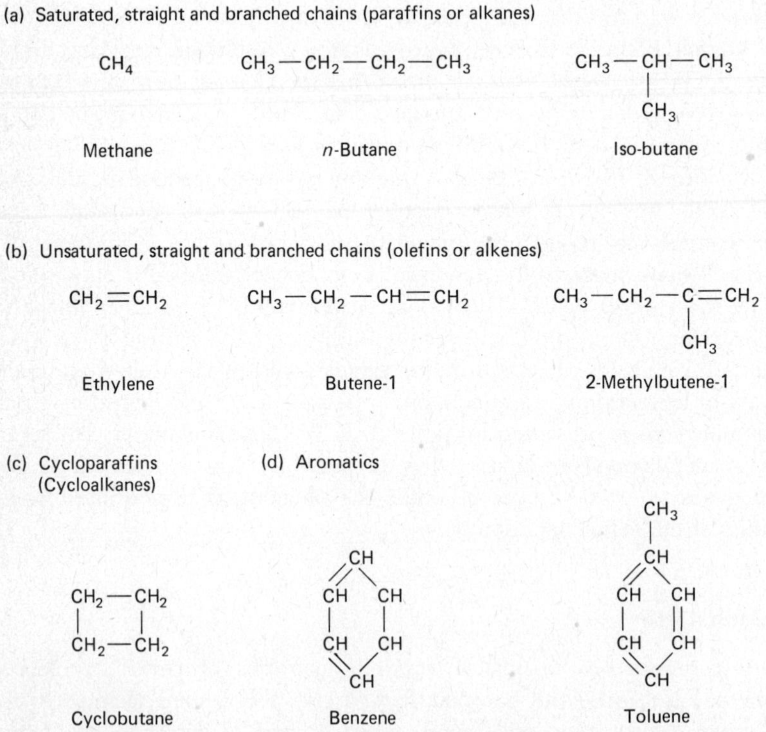

(a) Saturated, straight and branched chains (paraffins or alkanes)

CH_4 — Methane

$CH_3{-}CH_2{-}CH_2{-}CH_3$ — *n*-Butane

$CH_3{-}CH(CH_3){-}CH_3$ — Iso-butane

(b) Unsaturated, straight and branched chains (olefins or alkenes)

$CH_2{=}CH_2$ — Ethylene

$CH_3{-}CH_2{-}CH{=}CH_2$ — Butene-1

$CH_3{-}CH_2{-}C(CH_3){=}CH_2$ — 2-Methylbutene-1

(c) Cycloparaffins (Cycloalkanes)

$CH_2{-}CH_2$, $CH_2{-}CH_2$ (four-membered ring) — Cyclobutane

(d) Aromatics

(CH)$_6$ ring — Benzene

CH_3 substituted ring (C, CH)$_6$ — Toluene

Figure 10-2 Examples of members of general hydrocarbon families.

one hydrocarbon has a higher reactivity than another, as illustrated in Table 9-1. The reactivity of methane has been shown to be near zero. In light of the preceding discussion, it is not surprising that one of the methods proposed to reduce hydrocarbon emissions was to change fuels. Various compositions of gasolines were found to yield approximately the same total quantities of unburned hydrocarbons.

10-4 ORIGIN OF EXHAUST EMISSIONS FROM GASOLINE ENGINES

In the conventional spark ignition gasoline engine, a mixture of air and gasoline provided by the carburetor is inducted into the cylinder through the intake manifold and intake valve during the intake stroke, and then is compressed and ignited by a spark from the spark plug during the compression stroke. The mixture burns, and the products of combustion expand as the piston travels downward during the expansion stroke. The combustion products are exhausted from the cylinder through the exhaust valves and manifold to the exhaust system during the exhaust stroke. An estimate of the theoretical quantity of air required to burn the fuel completely may be obtained by writing the complete chemical reaction equation employing a theoretical fuel to represent the actual gasoline blend of hydrocarbons. As an example, on a molar basis,

$$C_7H_{13} + 10.25O_2 + 38.54N_2 \rightarrow 7CO_2 + 6.5H_2O + 38.54N_2 \quad (10\text{-}1)$$

and on a mass basis such as pounds or kilograms,

$$97(\text{lb})C_7H_{13} + 328(\text{lb})O_2 + 1080(\text{lb})N_2 \rightarrow 308(\text{lb})CO_2 + 1170(\text{lb})H_2O + 1080(\text{lb})N_2 \quad (10\text{-}2)$$

The complete or theoretical combustion, as shown by the above reactions, is defined as the complete conversion of carbon to CO_2 and hydrogen to H_2O. This is also frequently termed the *stoichiometric* reaction. It is useful to define a mixture ratio, or air-fuel ratio (A/F). This is the ratio of the mass of air required per unit mass of fuel for combustion. That is,

$$\text{mixture ratio} = A/F \equiv \frac{\text{mass of air}}{\text{mass of fuel}} \quad (10\text{-}3)$$

For comparative purposes we find that the stoichiometric air-fuel ratio is quite useful. In terms of Equation (10-2) this is

$$(A/F)_{\text{stoich}} = \frac{328 + 1080}{97} = 14.5$$

This value is typical of the stoichiometric air-fuel ratios for many individual hydrocarbons or hydrocarbon mixtures. The fuel-air ratio (F/A) is the reciprocal value.

The ratio of the mass of air actually supplied to the mass of fuel may be larger or smaller than the stoichiometric ratio. It is common practice to employ the equivalence ratio, ϕ, defined by Equation (10-4) to express the actual air-fuel ratio.

$$\phi \equiv \frac{(A/F)_{\text{stoich}}}{(A/F)_{\text{actual}}} = \frac{(F/A)_{\text{actual}}}{(F/A)_{\text{stoich}}} \tag{10-4}$$

When ϕ is less than 1, more air is being supplied than is required for complete combustion or an excess of air exists and the mixture is referred to as a lean mixture. Conversely, when the value of ϕ is greater than 1, the mass of air supplied is less than that required for complete combustion of the fuel and the mixture is said to be rich (an excess of fuel).

The values presented in reaction (10-1) indicate that 52.04 moles of products are formed by the combustion of 1 mole of fuel. If 0.10 percent of the fuel is unburned and is exhausted with the combustion products as unburned hydrocarbon, the exhaust will contain approximately 20 ppm of unburned hydrocarbons. Should the quantity of unburned fuel be 1 percent, the exhaust would contain roughly 200 ppm. The original emission standard for 1970 was 180 ppm of unburned hydrocarbons. Thus to satisfy the federal emission standards established from 1970 onward, practically complete combustion of the fuel must be attained. The detailed analysis of the combustion of hydrocarbons present in gasoline shows that many (25 to 100) separate competing chemical reactions may occur simultaneously and that the products of combustion vary depending upon the pressure and temperature existing at the time considered. A further complicating factor is the rate at which the chemical reactions occur, since the combustion process is a dynamic phenomenon. There may not be sufficient time for the complete combustion of some of the hydrocarbon species.

As was mentioned in the introductory paragraphs of this section, the combustion process is initiated by a spark. The combustion flame front then travels outward in all directions through the unburned mixture toward the walls of the combustion chamber. The surfaces of the combustion chamber are either air- or water-cooled. Consequently, the combustible mixture is cooled by contact with these cooler surfaces. This cooling action may lower the temperature of the air-fuel mixture in this region to such an extent that the flame goes out or is quenched before all of the fuel present is burned. This phenomenon is aggravated in combustion chambers which have a large surface-to-volume ratio. In any case, a film of unburned hydrocarbons will exist along the wall of the cooled cylinder, causing a certain amount of unavoidable flame quenching. These unburned hydrocarbons are removed from the film along the cylinder walls and are exhausted with the combustion gases as the piston and its sealing surfaces (piston rings) move along the cylinder during the exhaust stroke.

Carbon monoxide, like unburned hydrocarbons, results from the incomplete combustion of the fuel. Consequently, those conditions that promote or enhance complete combustion tend to reduce the quantity of carbon monoxide in the exhaust gas of the engine. A major factor is the air-fuel ratio. Test results presented in Figure 10-3 show that as the air-fuel ratio increases from 10 to 16, the CO in the exhaust gas decreases from 12 percent to nearly zero. Thus one method of reducing the CO emission is to operate the engine with lean-mixture ratios. The value of the air-fuel ratio also exerts a major influence upon the quantity of unburned hydrocarbons emitted by a given engine. The results of experiments, as presented in Figure 10-3, show that the value of unburned hydrocarbons in the exhaust gas first decreases as the air-fuel ratio is increased from 11 to approximately 16, and then increases as the air-fuel ratio increases further to 22.5. The increase in unburned hydrocarbons is attributed to what is known as misfire. The mixture is so lean that combustion does not always proceed from the ignition spark.

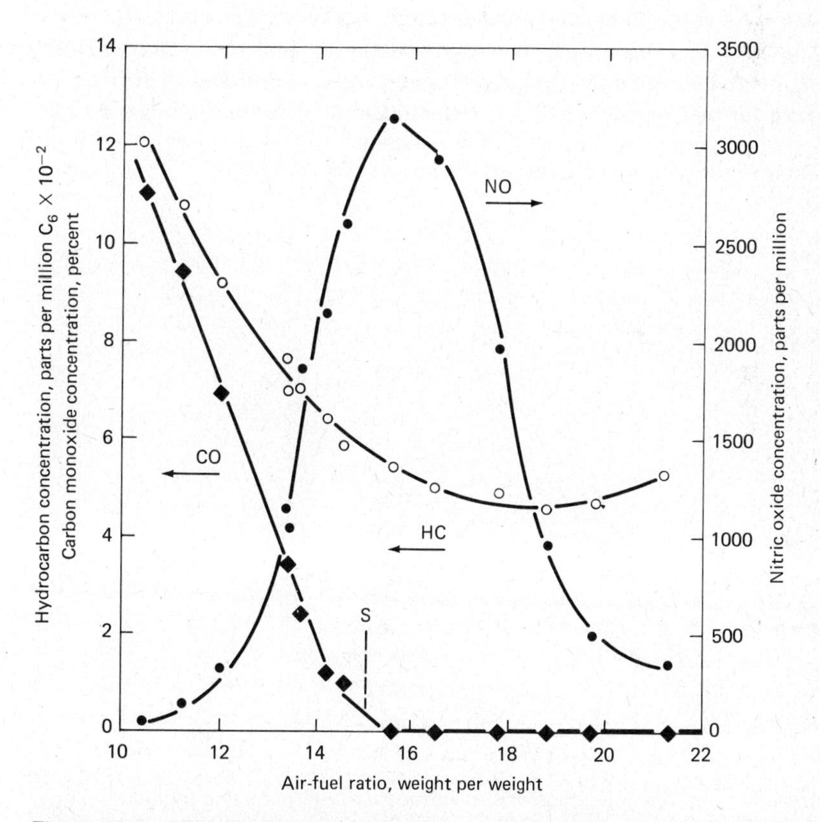

Figure 10-3 The effects of air-fuel ratio on hydrocarbon, carbon monoxide, and nitric oxide exhaust emissions. (SOURCE: W. G. Agnew. Research Publication GMR-743, General Motors Corp., 1968.)

The oxides of nitrogen are formed basically by reactions between atmospheric oxygen and nitrogen inducted into the engine. The major component of the oxides of nitrogen formed in the combustion zone is nitric oxide, NO. The quantities of the oxides of nitrogen formed are complicated functions of temperature, pressure, time of reaction, and the quantities of the reactants present. A basic discussion of these points is presented in Chapter 8. The salient features of that discussion apply to the study of internal and external combustion engines. As indicated in the discussion in Chapter 8 and corroborated by many experimental results, the following fact is well established: nitric oxide is formed within the combustion chamber during the time period that the maximum temperature exists, and it persists in above-equilibrium quantities during expansion and exhaust. The influence of the air-fuel ratio upon the quantity of NO in the exhaust gas of a typical gasoline engine is presented in Figure 10-4. Note that the maximum amount of NO is obtained at an air-fuel ratio near stoichiometric.

Comparing the data presented in Figures 10-3 and 10-4, we can see that as the air-fuel ratio is increased from around 11 : 1, the unburned hydrocarbons and carbon monoxide concentrations decrease. The concentration of NO increases to a maximum at a value of the air-fuel ratio which gives a minimum of unburned hydrocarbons and carbon monoxide. A further increase in air-fuel ratio results in a reduction of NO but an increase in HC.

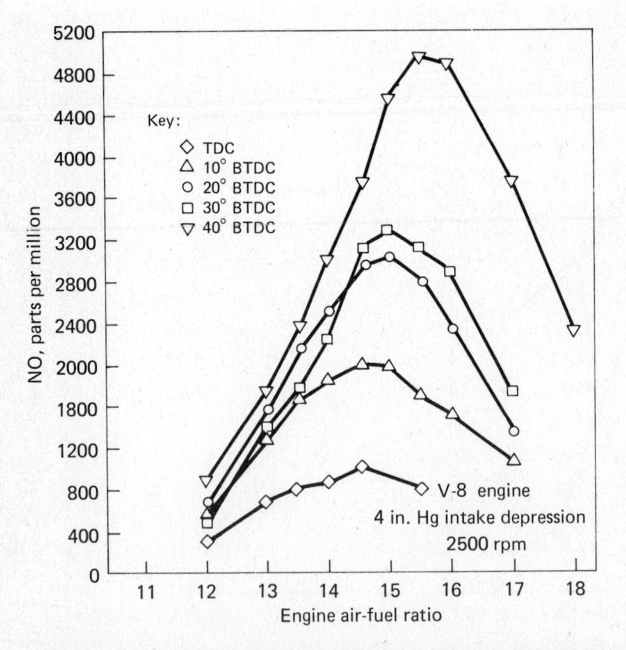

Figure 10-4 Nitric oxide emissions at various air-fuel ratios and ignition timings. (SOURCE: R. M. Campau and J. C. Neerman. From *Vehicle Emissions*, Part II. Selected Soc. Automotive Engineers Papers, 1963–1966. Reprinted by permission of the Society of Automotive Engineers, Inc.)

Thus it may be concluded that minimum quantities of HC, NO, and CO cannot be attained concurrently by changes in the air-fuel ratio alone.

Many factors other than those previously mentioned influence the emissions from a given spark ignition engine. A detailed discussion of these factors appears in other sources [2, 3] as well as in the publications of the Society of Automotive Engineers. However, it is informative to examine briefly a limited number of the factors which exert strong influence upon the emissions from gasoline engines, namely: (1) carburetion, (2) spark timing, and (3) surface-to-volume ratio.

CARBURETION

The function of the carburetor is to provide the engine with a homogeneous mixture of fuel vapor and air at the mixture ratios required for satisfactory engine operation under the wide range of conditions. These include cold starting, warmup, acceleration, part load, full load, idle, and the like. The liquid fuel is metered through a number of small holes called jets. The fuel flow rate is a function of the pressure drop across the jet as well as the shape and size of the individual jet. Because of a number of factors, such as manufacturing tolerances, carburetor adjustment, basic carburetor design, and physical properties of specific gasolines, it is not always possible for the carburetor to provide the engine with the correct vaporized fuel-air mixture. The result is an increase in exhaust gas emissions.

SPARK TIMING

The conventional gasoline engine with no air pollution control equipment normally operates with the ignition spark occurring some place between 15 and 20 degrees before the piston reaches top dead center. This leads to what is commonly referred to as good "drive-ability." The influence of spark timing (degrees of advance) upon hydrocarbons and nitric oxide in the exhaust gases is shown in Figures 10-4 and 10-5. Note that both hydrocarbons and nitrogen oxide can be reduced by retarding the spark (reducing the amount of spark advance). Unfortunately, as is indicated in Figure 10-5, retarding the spark below a certain point results in loss of power and engine "drive-ability."

SURFACE-TO-VOLUME RATIO

As has been mentioned before, the ratio of the surface area to the volume of the combustion chamber has an influence upon the quenching of the combustion reactions. The relation between the surface-to-volume ratio and the unburned hydrocarbons in the exhaust for typical automobile engines is presented in Figure 10-6. Note that the amount of unburned hydrocarbons is doubled as the surface-to-volume ratio increases from roughly 4 to 8.

In summary, it may be concluded that some reduction in exhaust emissions can be obtained by (1) improved carburetors, (2) spark retardation, and (3) low surface-to-volume ratio of the combustion chamber.

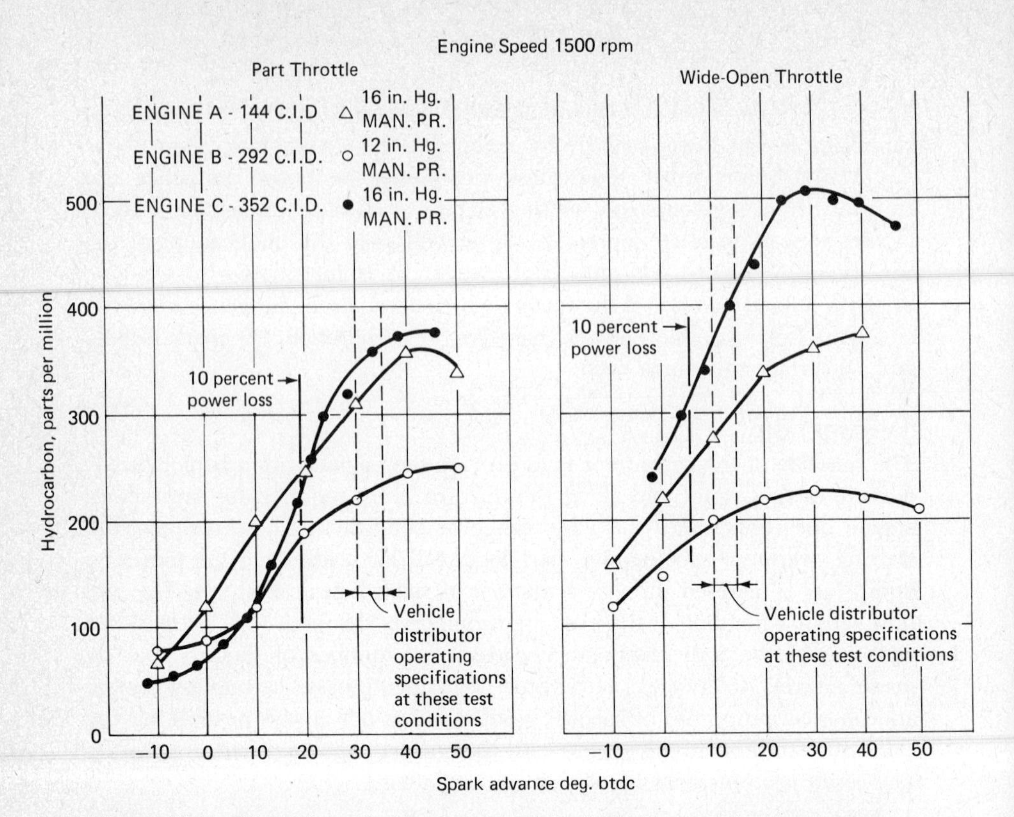

Figure 10-5 Effect of spark advance on exhaust gas hydrocarbon concentration. (SOURCE: D. F. Hagen and G. W. Holiday. Soc. Automotive Engineers Paper 486-C, 1962. Reprinted with permission of the Society of Automotive Engineers, Inc.)

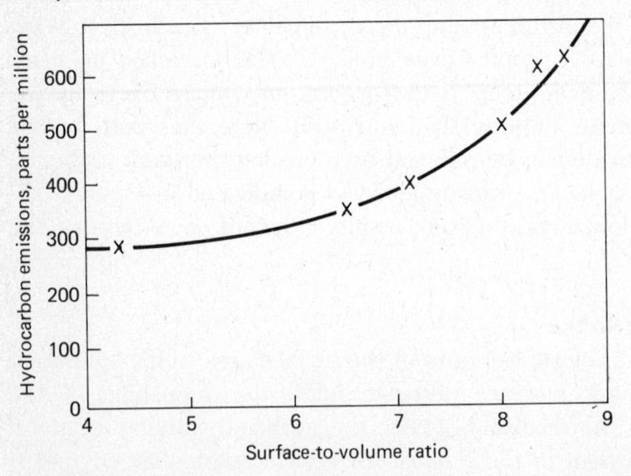

Figure 10-6 Results of the California chassis dynamometer schedule measurements of hydrocarbons emissions for engines having various values of combustion chamber surface-to-volume ratios. (SOURCE: C. E. Scheffler. From *Vehicle Emissions*, Part II. Selected Soc. Automotive Engineers Papers, 1963–1966. Reprinted by permission of the Society of Automotive Engineers, Inc.)

10-5 CRANKCASE AND EVAPORATIVE EMISSIONS

Since the flow rate of pollutants (unburned hydrocarbons and carbon monoxide) from the crankcase and fuel systems is at least an order of magnitude lower than the rate of exhaust flow, incineration can be employed as a common means of treating these combustible pollutants when the engine is running. Crankcase blow-by occurs only during engine operation; thus, direct incineration of all blow-by is the control method used. The general features of the system used are shown in Figure 10-7. The combustion gases that leak past the piston rings into the crankcase are metered through a positive crankcase ventilation valve (PCV) into the intake manifold of the engine and burned in the combustion chamber. The fresh air for crankcase ventilation is drawn from the air cleanser so that if the blow-by exceeds the capacity of the PCV valve, the backflow out the crankcase air intake will be entrained in the air flow to the carburetor.

A second source of hydrocarbon emissions is the evaporation loss from carburetor and fuel-tank vents. Because of the high volatility of gasoline and the inability of carburetors accurately to meter fuel streams containing large proportions of vapors in the liquids, carburetor venting is employed. In the modern carburetor, these vapors are vented internally so that they are consumed when the engine is running. However, during hot soak periods when the engine is shut off after a period of operation, the vapors from the carburetor can enter the atmosphere. The vapors from both the fuel tank and carburetor depend upon the volatility of the fuel used and the ambient temperature. The more volatile the fuel, and the higher the outside temperature, the larger the losses.

A storage system can be used to collect and contain these vapors until they can be purged and incinerated during subsequent engine operation; this readies the storage system for reuse. A system using the regenerative storage principle is shown in Figure 10-8. In this system, the carburetor is internally

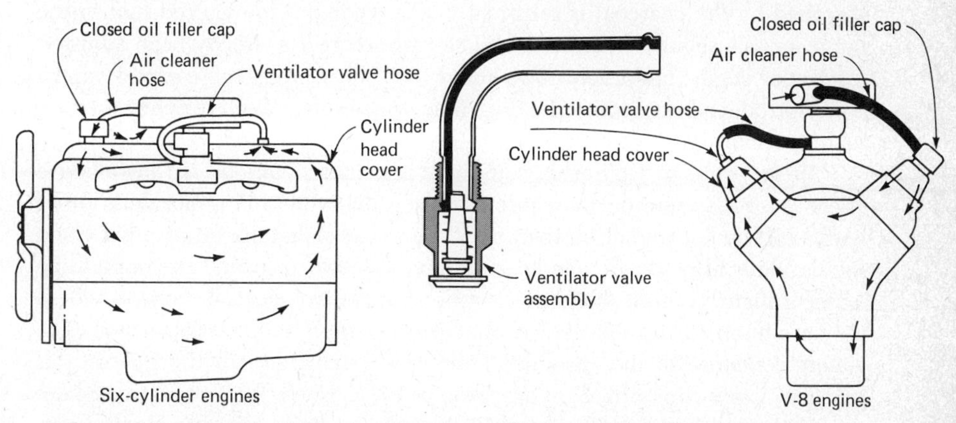

Figure 10-7 Closed positive crankcase ventilation system. (Courtesy of the Chrysler Corporation.)

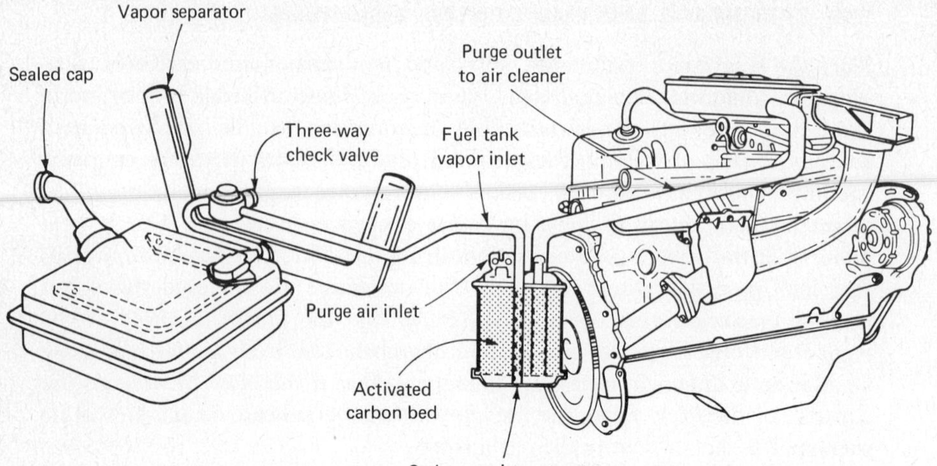

Figure 10-8 Fuel evaporative control systems. (SOURCE: W. M. Brehob. "Control of Emission Sources." Eighth Conference on Air Pollution Control, Purdue University, 1969.)

vented so that its evaporative emissions are stored in the engine intake manifold and air cleaner. The vapor from the fuel tank is collected, separated from any entrained liquid, and stored in a canister of activated carbon. When the engine is operated after a period of nonoperation, purge air is drawn through the carbon bed, purging the activated carbon of the absorbed hydrocarbon vapors. The purge air from the carbon canister is ducted into the air cleaner; thus, the hydrocarbon vapors are incinerated in the combustion chamber.

On cursory examination, it might seem that this system would result in a gain in fuel economy since the vapors that normally escape into the atmosphere are burned in the engine. However, this is not the case as the fuel absorbed by the charcoal is returned to the engine as unmetered fuel in the purge air as a slightly rich mixture. The carburetor has already been adjusted for proper engine operation using liquid fuel. Thus, when these excess vapors are reintroduced on engine starting, they will result only in a richer mixture than is necessary.

In the state of California some gasoline filling stations are installing a vapor control system on their pumps. The automobile filling nozzle is fitted with a soft gasket that effectively seals the automobile tank fill opening when the nozzle is in place. As one hose delivers gasoline, a second hose operating under a slight vacuum ducts the gasoline vapors to a central location where they are bubbled through the gasoline in the supply tank of the station. The vapors condense in the gasoline. Thus this system reduces the quantity of gasoline vapor (unburned hydrocarbons) discharged into the atmosphere during the automobile filling process.

10-6 EMISSION REDUCTION BY FUEL CHANGES

The spark ignition engine can be made to operate on fuels other than gasoline. In general, when gaseous fuels are employed in engines designed to operate on gasoline, major reductions in emissions are obtained with moderate (6 to 20 percent) losses in power. Lee and Wimmer [4] tested propane, methane, steam reformed hexane, and gasoline in a single cylinder engine under conditions that were optimum for each fuel. An estimated duty cycle simulating traffic service was used. They found reductions in CO of 98%, in HC of 87 to 99%, and in NO_x of 53 to 87% when the emissions of the gaseous fuels were compared to the emissions obtained with gasoline.

The emission and performance characteristics of liquefied petroleum, LP-gas have been investigated by Baxter, Leck, and Mizelle [5], the California Air Resources Board, and others employing a large variety of engines and vehicles. In all cases major reductions in CO, NO_x and HC emissions were obtained. Natural gas, either liquefied or gaseous, also is a potential low-emissions fuel for spark ignition engines. Several fleets of vehicles have been employed in natural gas fuel evaluation programs. McJones and Carbeil [6] obtained reductions of approximately 90% in CO, 91% in HC, and 85% in NO_x when natural gas was used in place of gasoline in 1969, in 4-, 6-, and 8-cylinder cars.

One of the major disadvantages of using the gaseous fuels, LP-gas, or liquefied natural gas is the complicated pressurized fuel supply system and the limited range of operation between fuel refillings. At present, because of the limited fuel space in most passenger cars, the range of a car using LP-gas is from 100 to 200 mi. The range for natural gas is less. The fuel filling procedure is more complicated with LP-gas than with gasoline.

Additional fuels receiving consideration are methanol, ammonia, and hydrogen. Experimental results [7] indicate that the thermal efficiency is much higher and the NO_x emissions are lower when a 350 in.^{3} V-8 engine was operated on hydrogen rather than gasoline. The prices of LP-gas, liquefied natural gas, methanol, hydrogen, and others are uncertain at this time. However, the projected costs appear to be competitive with gasoline, should large quantities be required by the motoring public.

10-7 EMISSION REDUCTION BY ENGINE DESIGN CHANGES

The conventional spark ignition engine prior to 1968 was developed primarily to provide suitable driving characteristics, such as smoothness, rapid acceleration, and high performance (large output per unit size); thus, it is not surprising that high emissions resulted. Reduced emissions can be obtained with the conventional engine by operating with lean air-fuel mixtures and retarded spark settings; unfortunately, drive-ability tends to deteriorate.

The surface-to-volume ratio can be decreased by increasing the displacement per cylinder, by increasing the stroke compared to the bore, by decreasing the compressions ratio, and by changing the combustion chamber shape. All of these modifications have severe practical limitations, since when carried to extremes, they cause a reduction in specific power and an increase in fuel consumption, engine roughness, and valve-gear complexity.

Retarding the spark tends to reduce the emission of unburned hydrocarbons by completing combustion later in the expansion stroke and reducing the time between combustion and exhaust, causing a higher temperature to exist at exhaust. This promotes additional chemical oxidation of the exhaust products. For all cases, except extreme overadvance, retarding the ignition timing causes a marked reduction in hydrocarbon emission. Again the relationship between emission reduction and performance must be pointed out. The degree of spark retardation must be carefully controlled to minimize emissions while minimizing the adverse effects on performance, economy, and drive-ability.

One method of providing the desired air-fuel mixture ratio to each cylinder is by the use of a fuel injection system. In one form of this system, known as the stratified-charge engine [8], the piston has a recess in the center of the piston face so that a small cavity is formed between the flat cylinder head and the piston, as shown in Figure 10-9. During the intake stroke, air is inducted into the cylinder with a swirling motion imparted to it by the intake manifold, and compressed, retaining its swirling motion, into the aforementioned small cavity. Near the end of the compression stroke, liquid fuel is injected tangentially into the cavity counterflow to the swirling air. The mixture is ignited by a conventional spark plug. Expansion and

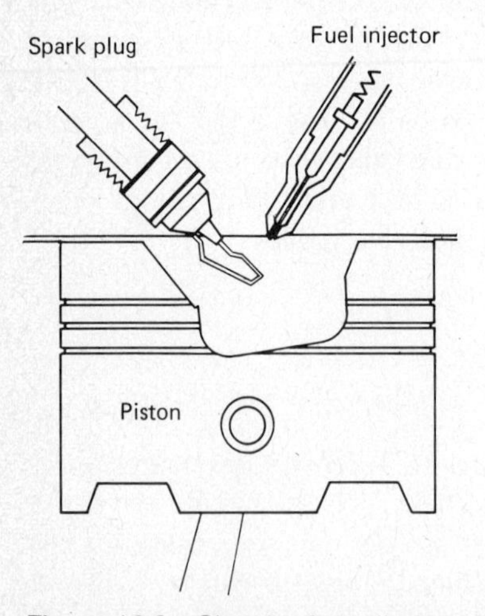

Figure 10-9 Sketch of a cylinder of the Ford stratified-charge engine. (SOURCE: A. Simko. "Emission Reduction Potential of Stratified Charge Combustion." Technical Meeting, Combustion Institute, Ann Arbor, Mich., March, 1971.)

exhaust occur as in a conventional spark ignition engine. It has been claimed that nonleaded and low-octane fuels can be used.

A second method of reducing exhaust emissions is termed, by General Motors Corporation, the combustion control system. The intake air is preheated to obtain a more constant density and the carburetor is changed to give both a higher idle speed and leaner mixtures. The spark timing is retarded. In some models, the spark is retarded during light accelerations by means of a pressure switch located in the hydraulic circuit of the transmission [9]. It is claimed that hydrocarbon emissions are reduced without penalizing full-throttle accelerations or high-gear cruising economy. A reduction in the emission of unburned hydrocarbons may also be obtained by injecting air into the exhaust port of each engine cylinder where it mixes with the hot exhaust products and promotes oxidation of the hydrocarbons and carbon monoxide [9]. Carburetion is tailored for optimum emission reduction, fuel economy, and performance. Increased engine cooling capacity, increased thermostat temperature setting, and thermal modulation of the spark retard are provided. The injected air is provided by a separate belt-driven air pump.

Since, as was pointed out in Chapter 8, the quantity of the oxides of nitrogen are a function of the flame temperature, one method of reducing NO emission is by reducing flame temperature. Such a reduction can be accomplished by either water injection or exhaust-gas recirculation. Nicholls et al. [10] obtained a 60 percent reduction in nitric oxide emission with a water-fuel ratio of 1, as shown in Figure 10-10. For water injection rates greater than a water-fuel ratio of 1.25, a deterioration of engine performance was observed. A major reduction of NO_x emissions can be obtained [11] by

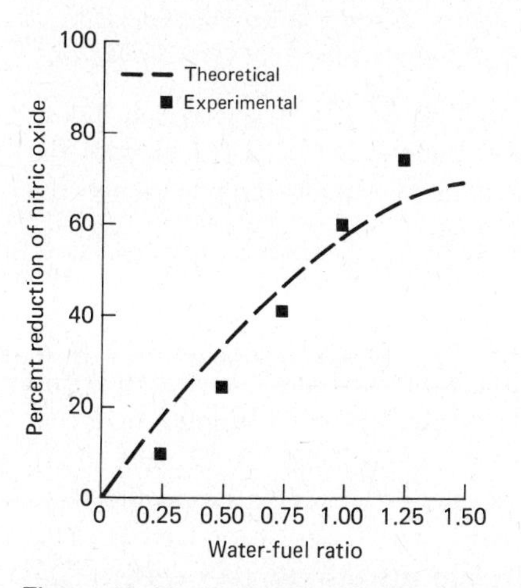

Figure 10-10 Reduction of nitric oxide emissions with water injection, for an equivalence ratio of unity. (SOURCE: J. E. Nicholls, I. S. El Messiri, and H. K. Newhall. Soc. Automotive Engineers Paper 690018, 1968. Reprinted with permission of the Society of Automotive Engineers, Inc.)

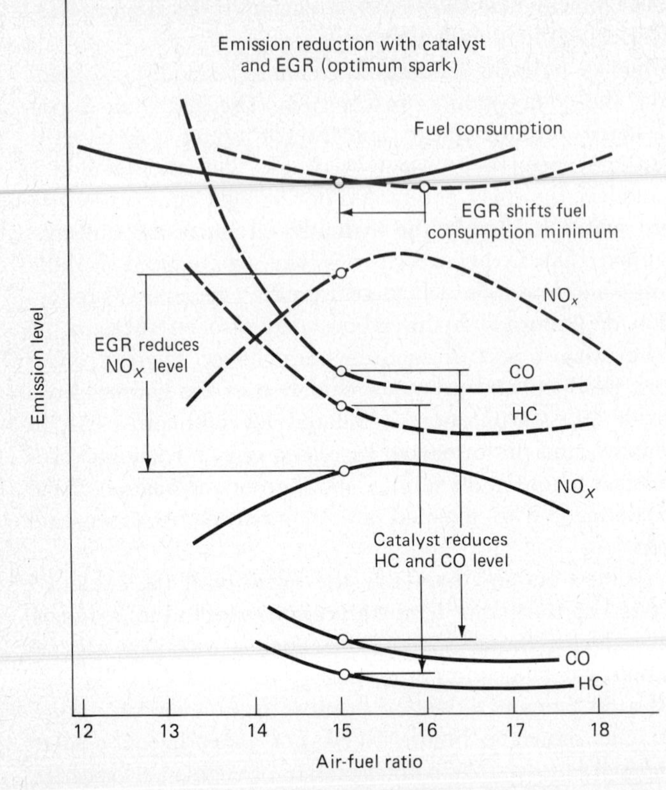

Figure 10-11 Effect of EGR and catalytic converter on NO*x*, HC, and CO emissions in a spark ignition engine. (SOURCE: "Fuel economy trends and catalytic devices." R. C. Stempel and S. W. Martens. *Automotive Fuel Economy, Progress in Technology Series No. 15*, Soc. of Automotive Engineers, 1976.)

employing the recirculation of exhaust products (EGR). As seen from Figure 10-11, the use of EGR reduces NO_x but, unfortunately, also increases the emissions of HC and CO. This latter problem may be overcome through the use of a catalytic converter.

10-8 EXTERNAL REACTORS

When the compromises in engine design (required to meet exhaust emission standards by internal engine modifications only) become too great, external reactor systems can be used. In one system the conventional exhaust manifold is replaced by a much larger exhaust manifold reactor fabricated from high-temperature materials, as shown in Figure 10-12. Secondary air supplied by a separate air pump is injected into the exhaust port close to the exhaust valve of each cylinder. The hot combustible exhaust products are oxidized by the additional air, resulting in a decrease in the hydrocarbon and CO concentrations in the exhaust. It has been found necessary to operate the

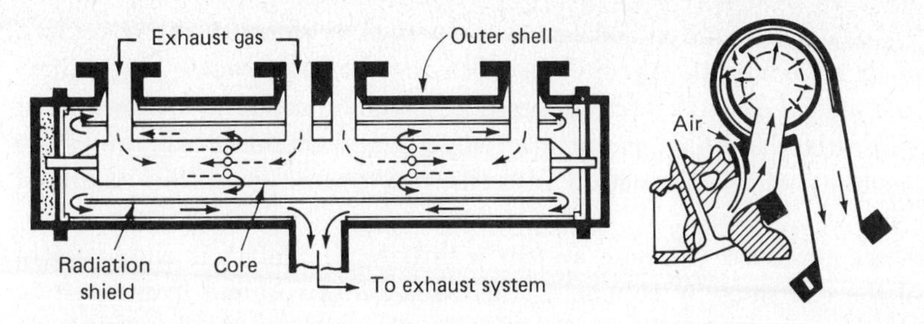

Figure 10-12 Exhaust manifold reactor. (SOURCE: D. J. Patterson et al. Soc. Automotive Engineers Paper 730201, 1973. Reprinted with permission of the Society of Automotive Engineers, Inc.)

engine fuel-rich to provide sufficient combustible species in the engine exhaust to assure thermal ignition and sustained combustion in the reactor. The performance of individual thermal reactors is a function of reactor design, gas composition, and temperature. Patterson et al. [12] report that exhaust temperature must be at least 1300°F (1000°K) to obtain ignition in the reactor and that higher temperatures are desirable. Kadlec et al. [13] show that steady-state reactor temperatures of 1500°F (1100°K) and above are required to attain approximately 85 percent and above conversion of unburned species in the engine exhaust. Glass et al. [14] obtained exhaust emission levels of 0.3 percent CO, 10 ppm HC, and 150 ppm NO when testing a 281 in.^{3} engine operating fuel-rich with 11 percent exhaust-gas recirculation. Fuel consumption data indicating a fuel penalty of 16 percent were obtained from extended runs on a mileage accumulation dynamometer.

Three of the major problems encountered with thermal reactors are: (1) special materials are required to withstand the high temperatures and cyclic modes of operation, (2) the reactor requires a larger space if sufficient time is provided for the incineration of the unburned species, and (3) when the engine misfires, an increased quantity of unburned hydrocarbons enter the reactor, resulting in possible overheating of the reactor. It should be pointed out that a limited amount of incineration of HC and CO can be obtained by injecting air into the exhaust port near the exhaust valve, thus permitting oxidation of the unburned species to occur in the exhaust manifold.

External catalytic reactors may be employed to lower hydrocarbon, carbon monoxide, or oxides of nitrogen emissions. Unfortunately, to obtain a decrease in HC and CO, the gas stream passing through the catalyst must be oxygen-rich, whereas to obtain a decrease in NO_x, the stream must be reducing, that is, deficient in oxygen. Originally two catalytic beds were used. In one system employing dual catalytic chambers, the engine is supplied with an air-fuel mixture sufficient fuel-rich (air-fuel ratio of 14.5–14.7 to 1) so as to provide a reducing atmosphere directly from the engine to the first catalytic bed. The engine exhaust contains approximately 70 g/mi CO and 8 g/mi NO_x. The oxides of nitrogen are reduced according to the chemical reaction equations discussed in Section 8-7-B. The large quantity of CO is required because of the very slow chemical reaction rates. Secondary air is mixed with the gases leaving the first reactor before they enter the second catalytic bed. In the second reactor, catalytic combustion of the hydrocarbon and CO occurs at a temperature sufficiently low to reduce the possibility of the formation of additional quantities of the oxides of nitrogen.

In an improved catalytic system a three-way catalyst is employed to control the emissions by enhancing the oxidation of CO and hydrocarbons, and reduction of NO_x, simultaneously. To obtain the desired exhaust gas composition leaving the catalyst bed, the gases entering the three-way catalyst from the engine must have a specific composition. The very narrow mixture ratio "operating window" of a platinum-rhodium three-way catalyst is shown in Figure 10-13. A slight change in either A/F or catalytic temperature will result in unacceptable exhaust emissions, therefore a sophisticated A/F control system is required for the engine. It is estimated that a computerized system sensing ten engine operating parameters will be required on the 1985 gasoline powered vehicles in order to satisfy both the emission standards and fuel economy requirements. The questions of sources of supply of the required quantities of catalytic materials, of catalytic life, and of required maintenance have not been completely answered at this time.

The effectiveness of a catalytic converter required to attain the projected 1982 emission standard is illustrated by Example 10-1.

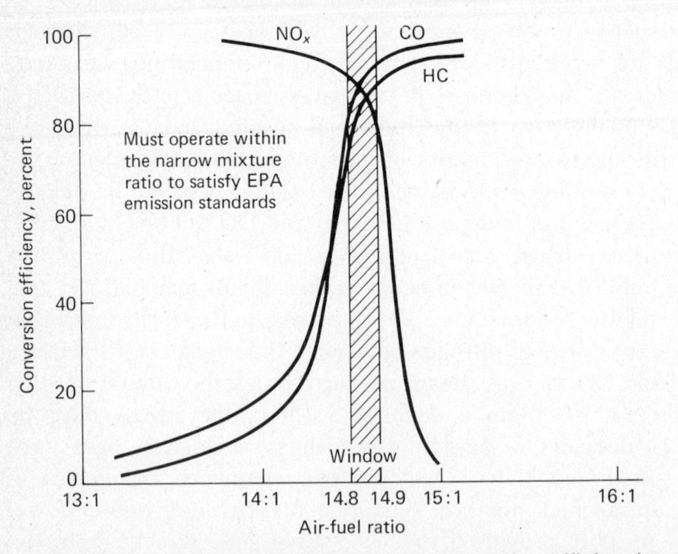

Figure 10-13 Effect of air-fuel ratio on conversion efficiencies of a three-way catalyst. (SOURCE: "Closed loop carburetor emission control systems." G. W. Niepoth, J. J. Gumbleton, and D. R. Haefner. 71st Annual Meeting, Air Pollution Control Association, June, 1978. Courtesy of General Motors Corp.

Example 10-1

The flow rate of NO_x measured at the exhaust manifold of a 6-cylinder engine is 6.9 g/mi. Calculate the effectiveness of a catalytic converter required to satisfy the projected 1982 federal emission standards.

SOLUTION

From Table 10-1 the federal emission standard is 1.0 g/mi. Consequently,

$$\text{effectiveness} = \frac{6.9 - 1.0}{6.9} \times 100 = 85.5\%$$

Successful operation of current catalytic reactors requires engines that are properly tuned and the use of nonleaded gasolines to prevent poisoning of the catalysts. If the engine exhaust is too hot or contains excessive unburned hydrocarbons as the result of an improperly adjusted carburetor or faulty spark plugs, the catalyst material can be severely damaged, even to the point of failure.

10-9 STRATIFIED-CHARGE ENGINES

The stratified-charge gasoline engine is designed to provide precision control of fuel combustion, a feature not possible in the conventional engine. In the engine being developed by the Ford Motor Company [15], air alone enters the cylinder during the intake stroke. The intake passages are so designed that they swirl the air as they carry it to the center of the cup in the dished or recessed head of the piston. Near the end of the compression stroke, fuel is injected into the air close to the center of the piston with a swirling motion counter to that of the air. Ignition is provided by a spark plug with electrodes extending into the center of the cup in the piston head, as shown in Figure 10-9. Thus, the flame travels outward from the center, burning the combustion mixture, which theoretically is surrounded by air only. By concentrating the fuel-air at the center of the combustion chamber, the engine can operate at a much leaner overall mixture ratio. Also, the fuel injection system provides more precise control of the air-fuel ratio than is possible with the conventional carburetor. One version of the stratified-charge engine as developed by the Ford Motor Company [16] using some exhaust gas recirculation gave the following exhaust emissions in grams per mile: HC, 0.23; CO, 0.2; and NO_x, 0.76. The overall air-fuel ratio used was approximately 20/1, and the fuel economy was 21 mpg on a highway cycle.

In a different version of the stratified-charge engine being developed by Honda [17], a lean fuel-air mixture provided by a carburetor is compressed in the cylinder of the engine. Simultaneously, a fuel-rich mixture provided by another carburetor is compressed in a small combustion chamber connected to the main cylinder by a small passage. See Figure 10-14. A spark from a conventional spark plug initiates combustion in the fuel-rich mixture. The

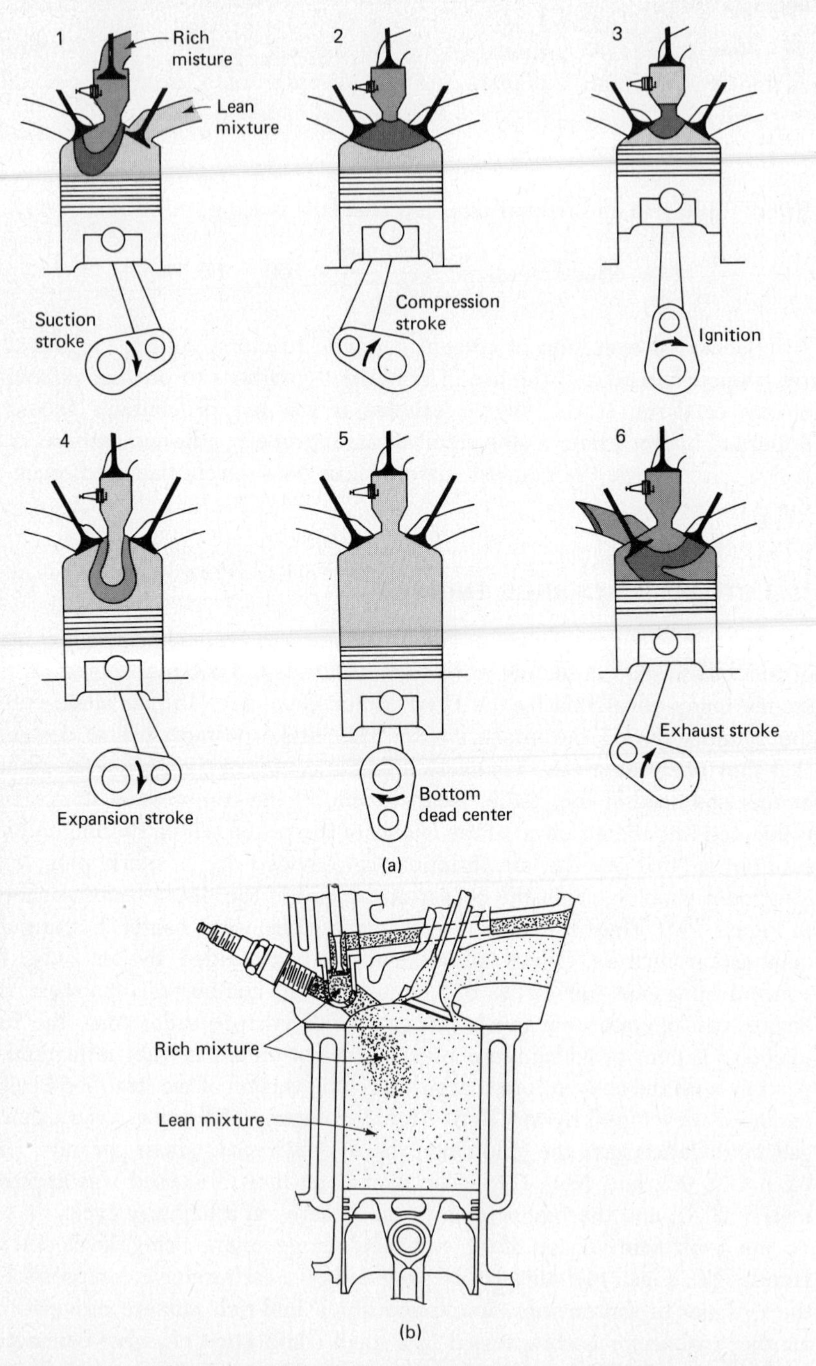

Figure 10-14 (a) Concept of a stratified-charge engine. (b) Practical design of a stratified-charge engine. (SOURCE: Road and Track, February, 1973, p. 119.)

flame propagates into the lean mixture in the cylinder at the beginning of the expansion stroke. The Honda engine is able to operate over a much leaner mixture ratio than does the conventional spark ignition engine. Honda [17] reports the following emissions obtained with a 1955-cm^3 4-cylinder engine: 0.23 g/mi HC, 2.41 g/mi CO, and 0.95 g/mi NO_x. The overall air-fuel ratio varies over the range from 18/1 to 21/1. In one version of the engine the exhaust manifold functions as an external thermal reactor, which reduces the HC and CO emissions.

10-10 ROTARY COMBUSTION ENGINES

Numerous rotary engines have been designed or proposed as possible replacements for the reciprocating engine. The positive displacement rotary piston device actually predates the invention of the reciprocating piston machine, and by 1910 over 2000 specifications of rotary piston devices were on file [18]. The most successful rotary engine to date is that developed by Felix Wankel. In 1954 Wankel found that by rotating an equilateral triangle in a certain manner relative to a containing member, three variable-volume chambers would be formed between the triangle and the containing member. It was not until 1957 that the first Wankel engine was made to run and develop measurable power, but since that time the engine has been developed further. The basic single rotor Wankel engine consists of a chamber shaped like a figure "8" in which a triangular rotor rotates as shown in Figure 10-15. The air-fuel mixture is supplied by a conventional carburetor.

The carbon monoxide and hydrocarbon emissions from a basic rotary combustion engine have been found to be significantly higher than those

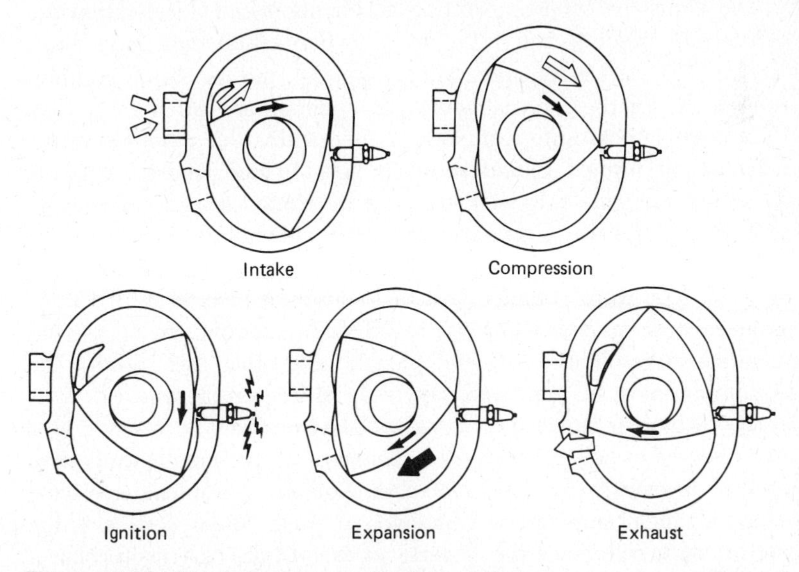

Figure 10-15 Wankel rotary combustion engine.

from a modern conventional reciprocating engine without external emission controls. However, when equipped with a suitable thermal reactor, the Mazda Wankel had emissions of 0.17 g/mi HC, 2.2 g/mi CO, and 0.93 g/mi NO_x. The fuel economy was reported to be 13 mi/gal [19]. The volume of a Wankel engine for a given horsepower output is much less than the volume of a comparable reciprocating engine. Thus, there is sufficient space in the engine compartment of an automobile to accommodate the thermal reactor required to reduce the emissions to an acceptable level. It should be pointed out that the Wankel engine requires a number of complicated seals whose life is unknown at this time.

10-11 ALTERNATIVE VEHICLE POWER SOURCES

At the present time, four power sources with extremely low exhaust emissions appear to have the potential of becoming replacement power sources for automobile propulsion. These include the external combustion engine types—the gas turbine, the Stirling engine, and the Rankine cycle power unit—and some form of electric system. Unfortunately, few complete vehicles powered with these sources have been evaluated under consumer driving conditions to determine drive-ability, economy, maintenance required, service life, and exhaust emission levels.

Under steady-state conditions, the external combustion engines should have a lower emission rate than internal combustion engines since the steady-flow combustion process can be controlled to a higher degree than can the transient combustion processes characteristic of internal combustion engines. The emissions occurring during acceleration and deceleration are more difficult to predict, and little or no experimental data have been published for these modes of operation. In general, the maximum combustion temperatures for the external combustion engines are much lower than those of the internal combustion engines. Consequently, the quantities of the oxides of nitrogen formed should be much less. In one laboratory test a 4-cylinder Stirling engine (Model P-40, built by United Stirling of Sweden and developing 45 hp) was installed in a 3376-lb automobile. When operating on the combined CVS cycle, emissions of 0.43 g/mi CO, 0.12 g/mi HC, and 0.37 g/mi NO_x were obtained at a fuel consumption rate of 19.4 mpg. No external exhaust gas treatment was required. In addition, the engine has been operated on gasoline, diesel fuel, alcohol, and other fuel blends. The fuel consumption must be improved, and some sealing problems exist.

The gas turbine or Brayton cycle power plant has long been envisioned as a suitable source of power for vehicles. Numerous experimental units have been installed in trucks, cars, and race cars. Numerous problems involving gear reduction, high-temperature components, noise, initial cost, and life have been encountered. Since the quantity of exhaust gas of a gas turbine is so much larger than that of an internal combustion engine of comparable horsepower, it is difficult to compare the emissions of the two power plants on the basis of exhaust composition. After studying aircraft gas turbines, Wright [20] concluded that properly designed gas turbines should have emission levels of only 2 to 4 percent those that characterize the untreated gasoline reciprocating engines and only 15 to 30 percent those of reciprocating engines meeting the proposed 1975 emissions standards. Similar trends are reported elsewhere [21]; the Williams Research Corporation [22], for example, confirmed these findings with an experimental car powered by an 80-hp gas turbine.

Nogle [23] reported the results of tests performed on a gas turbine combustor designed for a 123-hp turbine being developed by Chrysler for automotive applications. The tests gave emissions of 0.42 g/mi HC, 3.36 g/mi CO, and 0.40 g/mi NO_x. A wide variety of fuels were used in the testing program. Gas turbine powered buses are undergoing evaluation over routes in several cities.

Various versions of the Rankine cycle "steam engine" have been adapted to vehicle propulsion over the years, yet as of 1979 no vapor-powered vehicle was commercially available. Several organizations [24, 25, 26] are involved in developing Rankine cycle power systems for vehicles. Several major problem areas encountered in the development of suitable power plants are initial cost, large condenser area, maintenance and reliability, fuel economy, lubrication and high freezing temperature when water is employed, and startup time. Notwithstanding the aforementioned difficulties, Bjerklie and Sternlicht [27] concluded, as the result of an extensive study of the potential of the Rankine engine for vehicle use, that in every way the "ideal" fluid Rankine cycle turbine engine offers major advantages over the Otto and steam engines.

In 1970 the state of California and the U.S. Department of Transportation initiated the California Steam Bus Project. Three contractors were selected, and steam power plants for buses were developed by William M. Brobeck and Associates, Lear Motor Corp., and Steam Power Systems [28]. The power unit of the Lear system was a turbine, whereas both the other companies employed reciprocating units. The results of typical tests conducted over bus routes in the city of San Francisco are presented in Table 10-2 [28]. Data from typical diesel-powered buses are included for comparative purposes. From the data presented in Table 10-2 it is evident that although the CO and NO_x emissions are reduced, the fuel consumption is greatly increased. As of 1974, there was nothing to indicate that the California Steam Bus Project would be continued. Additional information on vapor power systems for automobiles may be found in publications of the Society of Automotive Engineers [29].

As in the case of the "steam system," electrical systems were competitors of the gasoline engine in the early years of motor car development. While suitable electric motors and control systems are currently available, no satisfactory battery system exists. Major difficulties are battery weight (2270 to 3400 lb), initial cost ($900 to $12,400), recharging time, and limited range. Most experimental two-passenger cars employing available batteries have

Table 10-2 RESULTS OF CALIFORNIA STEAM BUS TESTS

VEHICLE	EXHAUST EMISSIONS (g/hp·hr) CO	HC	NO_x	FUEL CONSUMPTION (ROUTE MILEAGE)
Brobeck	2.0	1.2	1.2	1.1
Lear	7.7	1.1	1.6	0.7
SPS	2.7	1.6	1.5	0.6
GM diesel (6-cylinder)	4.4	2.5	9.0	3.58
GM diesel (8-cylinder)	7.9	0.9	8.4	3.68
Cummins diesel	2.3	0.5	10.2	3.02

SOURCE: R. A. Renner. "Reviewing the California Steam Bus Project." Soc. Automotive Engineers Paper 730218, January, 1973. Reprinted with permission of the Society of Automotive Engineers, Inc. Copyright, 1973.

ranges of from 40 to 100 mi at speeds of 30 to 40 mi/hr and recharging times of 10 to 12 hr.

A few companies are reported to be developing batteries lighter in weight and with much shorter charging times. One such company, Electric Fuels Propulsion [30], is reported to have developed a Tri-Polar lead cobalt battery having a 30-min charging time. Should this battery or others prove successful, electrical vehicles may become available for short-range use in the future. It should be pointed out that while the electric vehicle is of itself practically free of exhaust emissions, the ultimate source of the required electrical energy for recharging the batteries is the large electric power plant. If the use of electric cars becomes widespread, an enormous additional load will be placed upon the already overloaded electric industry.

10-12 DIESEL ENGINE EMISSIONS

In the basic diesel cycle air alone is compressed to a high pressure and temperature during the compression stroke. Prior to top dead center, fuel is injected into the cylinder and ignition occurs when the fuel vapor and air mixture reaches the auto-ignition temperature. The total quantity of fuel injected is controlled by the load and speed of the engine. The fuel-air ratio of the diesel engine varies from 0.01 to 0.06, which corresponds to air-fuel ratios of from 100 : 1 to 15 : 1. Thus the mixture ratios of the diesel are much leaner than those of the gasoline spark ignition engine. The diesel engine operates on a two-stroke or four-stroke cycle and is either naturally aspirated or supercharged. In addition the engine is classified as direct-injection, D.I., if the fuel is injected directly into the cylinder, or indirect-injection, I.D.I., if the fuel is injected into a precombustion or divided-combustion chamber.

The emissions from diesel engines are (1) odor, (2) smoke, (3) carbon monoxide, (4) unburned hydrocarbons, (5) oxides of nitrogen, and (6) noise. All of the emissions are complex functions of the following basic variables:

1. Combustion chamber design
2. Fuel injector design

3. Fuel composition, including additives
4. Fuel-air ratio
5. Naturally aspirated or supercharged engine
6. Engine condition

Operating and emission characteristics of an automotive-type diesel engine are presented in Figure 10-16 [31, 32].

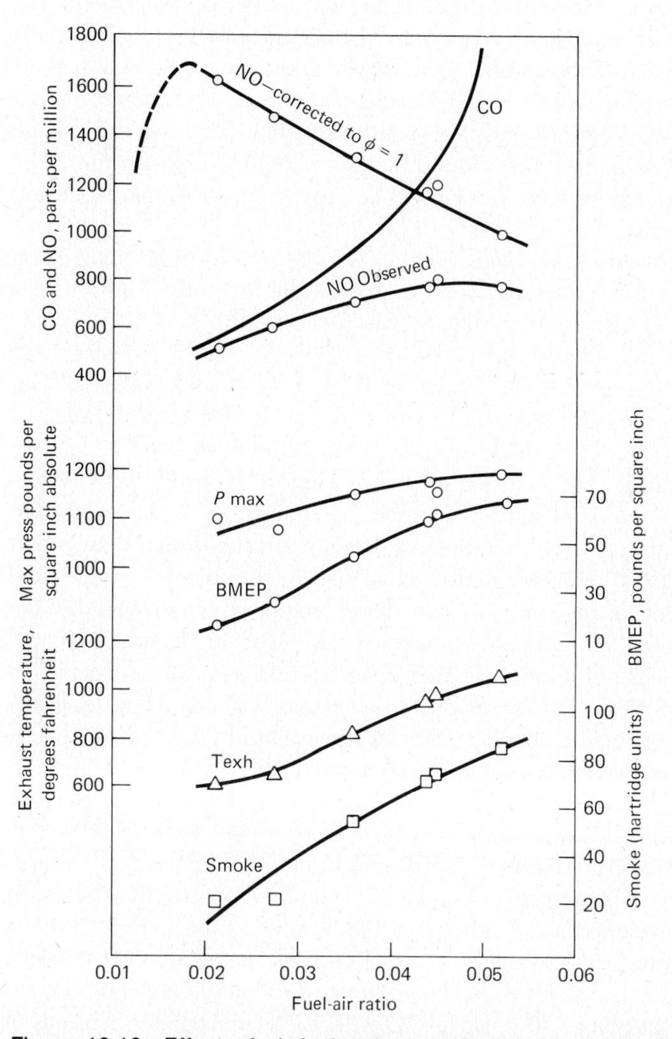

Figure 10-16 Effect of air-fuel ratio on NO and CO emissions and other performance parameters. (SOURCE: *Engine Emissions, Pollutant Formation, and Measurement*. G. S. Springer and D. J. Patterson, ed. New York: Plenum Publishing Corporation, 1973; and E. F. Obert. *Internal Combustion Engines and Air Pollution*. New York: Intext Educational Publishers, 1973.)

The odor of the exhaust gas from a diesel engine, although as late as 1974 still not classified as an air pollutant, is highly unpleasant to some people and has received much study. The source of the odor is attributed to small concentrations of several unburned hydrocarbons. The general topic of odor is presented in Chapter 11 and will not be considered further here. Smoke from a diesel engine is classified as black, white or gray, and blue. Black smoke is primarily unburned carbon which has agglomerated into small particles. On a mass basis, smoke is the smallest of the major emissions from a diesel engine, but it is the most noticeable. Black smoke can be caused by an overload condition; that is, too much fuel is being fed to the engine. White or gray smoke is the residue of the combustion of fuel droplets and represents correct operation. Blue smoke indicates unburned fuel or lubricating oil droplets, which usually means maintenance is required. Many fuel additives have been investigated in an attempt to find suitable smoke suppressants but none have been found that are completely satisfactory. The best method of smoke reduction seems to be good combustion chamber and fuel injector design.

Carbon monoxide, like black smoke, is produced during combustion when the oxygen present is insufficient to oxidize the fuel fully. Thus CO is a direct function of the fuel-air ratio. At a fuel-air ratio of 0.01, the CO concentration in the exhaust gas is less than 200 ppm and it increases to a value of 4000 ppm at a fuel-air ratio of 0.06. Experiments indicate that, unlike concentrations of smoke and carbon monoxide, the concentration of unburned hydrocarbons in diesel exhaust is not directly related to load or engine speed. Rather, hydrocarbon emissions seem to be functions of injection equipment and combustion chamber geometry. It is believed that control of the unburned hydrocarbons depends upon the design details that affect the fuel spray geometry and residual fuel in the injector nozzle [26]. The formation of nitric oxide in the diesel engine is controlled by the chemical reactions and mechanisms already discussed in Section 10-4 and Chapter 8 and does not warrant further consideration here. It is sufficient to point out that to reduce NO formation it is necessary to change at least one of the three major factors which affect the formation of NO: that is, reduce (1) the amount of available oxygen or nitrogen, (2) the peak temperature in zones with sufficient oxygen, or (3) the residence time at temperatures above 3200°R (1800°K).

The California Air Resources Board, the Engine Manufacturers Association, the Society of Automotive Engineers, and the Coordinating Research Council have developed a test procedure [33] known as the CARB 13-mode cycle for evaluating the emissions of diesel engines. Based upon that cycle, the 1974 standards for diesel engines and heavy-duty gasoline engines are 16 g/hp·hr of HC and NO_x, 40 g/hp·hr CO with EPA smokemeter readings of 20 percent during acceleration and 15 percent during lug down. The 1975 California emission standard for HC and NO_x is 5 g/hp·hr. In comparison,

Table 10-3 DIESEL EXHAUST EMISSIONS, CARB CYCLE

ENGINE[a]	EMISSIONS (g/hp·hr) CO	NO_x	HC
416 in.^{3}, 6-cyl., N.A., swirl chamber	2.7	3.1	0.9
416 in.^{3}, 6-cyl., T.C., swirl chamber	2.2	3.4	0.7
360 in.^{3}, 6-cyl., N.A., D.I.	5.8	3.3	3.6
685 in.^{3}, 6-cyl., N.A., D.I.	3.5	15.5	0.5
685 in.^{3}, 6-cyl., 10% recirculation	5.0	9.1	0.7
685 in.^{3}, 0.5:1 water-fuel injection	4.3	11.4	0.5
685 in.^{3}, 1:1 water-fuel injection	4.7	7.8	0.8

SOURCE: C. J. Walder. "Reduction of Emissions from Diesel Engines." Soc. Automotive Engineers Paper 730214, January, 1973. Reprinted with permission from the Society of Automotive Engineers, Inc. Copyright, 1973.

[a] N.A. = naturally aspirated; T.C. = turbocharged; D.I. = direct injection.

the target of the diesel engine manufacturers is 3 g/hp·hr HC, 7.5 g/hp·hr CO, 12.5 g/hp·hr NO, plus the smoke standard. Typical values of exhaust emission of current engines are presented in Table 10-3, taken from data published by Walder [34]. From the data presented in Table 10-3 for the 685-in.^{3} engine it can be seen that by employing either combustion gas recirculation or water injection the emission of oxides of nitrogen can be lowered.

Comparison between data on large truck-type diesel engines and data on engines suitable for automotive installations is difficult. However, references [35] and [36] have presented some comparative data which are summarized in Table 10-4. The most difficult emission control problem for diesel engines is to reduce both the smoke (particulate) and the oxides of nitrogen simultaneously. The proposed emission standards for both NO_x and particulate will be extremely difficult to attain.

Table 10-4 AUTOMOBILE DIESEL EXHAUST EMISSIONS

CAR	DISPLACEMENT (in.^{3})	MILEAGE (mi/gal)	EXHAUST EMISSIONS (g/mi) HC	CO	NO_x
Mercedes 240 D	134	25.9	0.18	1.30	1.01
Mercedes 300 D	183	23.8	0.16	0.85	1.72
Peugot	83	35.9	1.11	1.70	0.67
Perkins	247	25.7	0.72	2.86	1.48
Cutlass Diesel		26.9	0.47	2.00	0.70
Cutlass Gasoline	350	19.9	0.24	2.16	0.85
VW Rabbit					
Gasoline	96.9	31.7	0.23	3.70	0.63
Diesel	89.7	52.2	0.37	0.79	0.87

10-13 TURBOJET ENGINE AND GAS TURBINE EMISSIONS

When compared with emissions from other sources of air pollution, the total of exhaust emissions from turbojet-powered aircraft is relatively small [37, 38]. However, similar to the case of diesel-powered vehicles under load, the smoke from a turbojet engine during landing and takeoff is extremely visible without engine modifications. One reason for this is that the combustors of conventional gas turbines and turbojet engines have been developed over the years to give high performance, reliability, and long life. Until the early 1970s, little consideration was given to the quantity of smoke and exhaust emissions. Public concern subsequently influenced industry to make substantial efforts to reduce smoke emissions. It is also interesting to note that not all of the air pollutants in the vicinity of an airport are emitted by aircraft. Ground servicing equipment and vehicles entering and leaving the airport are also major sources of pollutants. CO emissions from ground equipment frequently are nearly as severe as those due to aircraft movement [39].

In reporting the levels of exhaust emissions from gas turbines and turbojet engines, we must take air-fuel ratios into account when comparing these engines to other sources of air pollutants. The automobile engine normally operates with an air-fuel ratio between 13 and 16. The turbojet engine operates with air-fuel ratios that are extremely lean, that is, 60 : 1 at full load and 150 : 1 at idle [40]. It is common practice to specify the quantity of emissions from a gas turbine by a ratio termed the *emission index*, which is defined as the pounds of a pollutant per thousand pounds of fuel burned (or grams of pollutant per kilogram of fuel). Table 10-5 presents measured emission data for several typical turbojet engines in use in 1970. Note that the largest quantities of pollutants are generated during the idle mode of operation.

Table 10-5 POLLUTANTS FROM AIRCRAFT JET ENGINES

		EMISSION INDEX[a]			
ENGINE		CO	NO_x	HC	PARTICULATE
J-57	Takeoff	6	1	1	14
	Approach	8	4	2	7
	Idle	57	2	4	10
JT 8-D	Takeoff	1.4	5.8	2.5	
	Approach	3.2	6.3	3.5	
	Idle	34	2.7	11.4	
TF-33	Takeoff	0.4	3.1	1.3	1.9
	Approach	2.4	1.7	1.7	1.1
	Idle	28	1.4	43	2.4

SOURCE: J. B. Heywood, J. A. Fay, and L. H. Linden. "Jet Aircraft Air Pollutant Production and Dispersion." AIAA Paper 70-115, January, 1970.

[a] Mass of pollutant per thousand mass units of fuel.

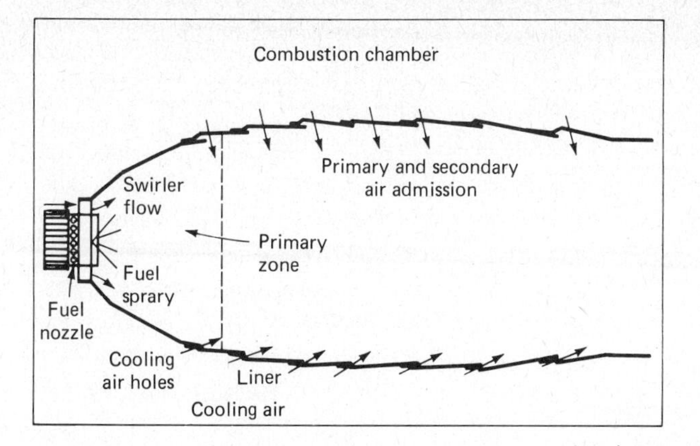

Figure 10-17 Schematic of flow in a gas turbine combustion chamber.

The combustion chamber or combustor of stationary gas turbines, vehicular gas turbines, turbojet engines, and turboprop engines are similar in operation. The combustion phenomenon in a conventional combustor, as shown in Figure 10-17, is divided into zones. In the primary zone, the fuel is mixed with primary air and is partially burned. Both locally fuel-rich and fuel-lean regions exist, owing to the lack of complete homogeneity. Recirculation in the primary zone maintains combustion stability. Additional air enters the secondary zone and completes the combustion reactions. A third portion of the air is fed into the combustion chamber along the inside surface of the liner to act as a coolant, thus protecting the liner. The gas temperature is reduced to the required turbine inlet temperature by dilution with additional quantities of secondary air. The combustion and mixing processes occurring in the combustor are highly complex, and at the present time no completely satisfactory analytical model is available for use by the designer. The formation of the oxides of nitrogen is governed by the phenomena discussed in Chapter 8. From that discussion it will be recalled that the quantities of oxygen and nitrogen, the temperature, and the residence time influence the formation of NO. The quantities of CO and unburned hydrocarbons are likewise complex functions of the localized air-fuel ratio, degree of mixing, and velocity. Experimental data for the concentration of pollutants as functions of location in the combustor as well as the emission indices for various modes of operation prepared by Sawyer [40] are presented in Figure 10-18.

The soot or smoke problem associated with early gas turbine models was the result of high temperatures and high fuel concentrations in the primary zone of the combustor. The soot formation zone typically lies within the envelope of fuel as it is sprayed from the pressure nozzle. There are a

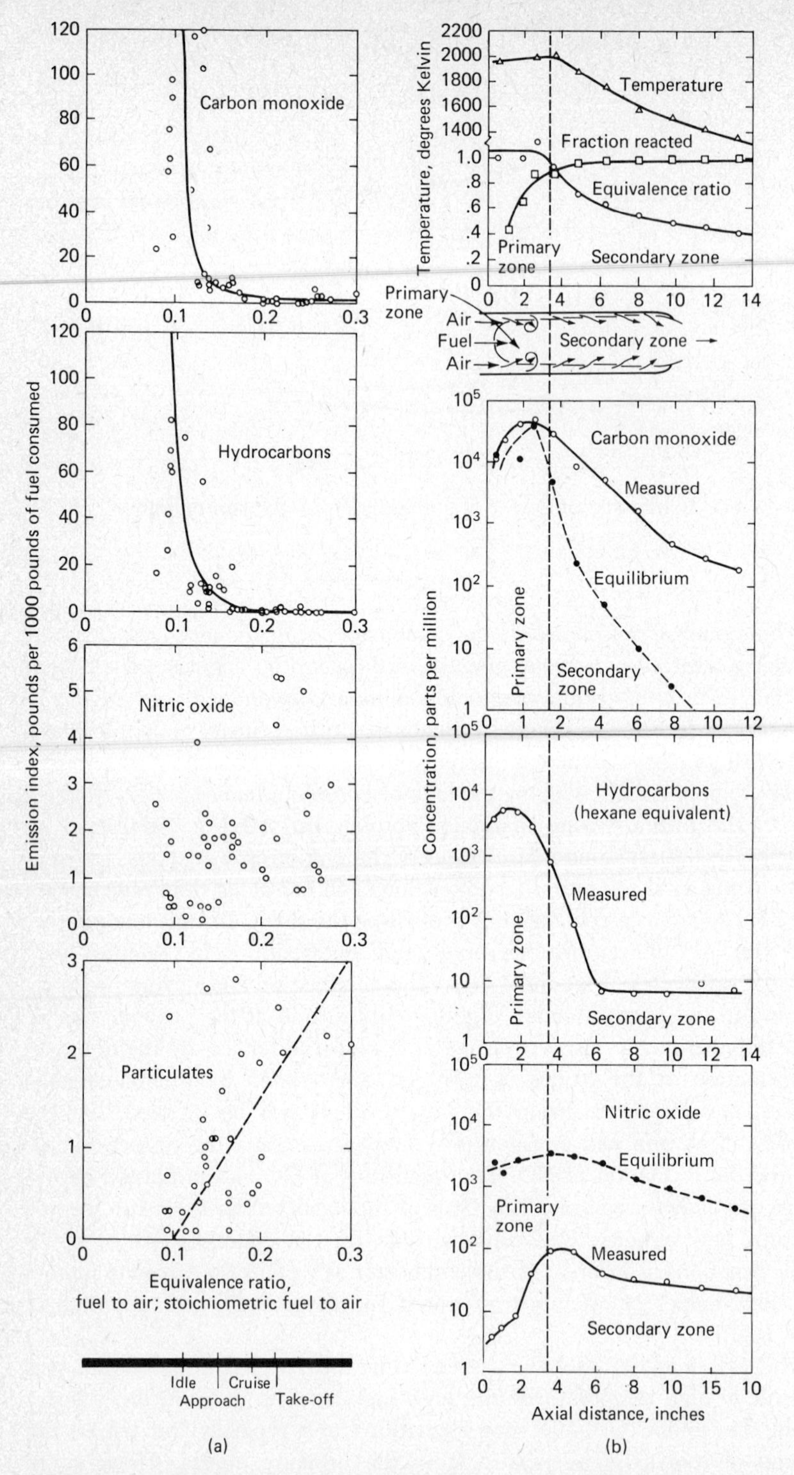

Figure 10-18 Performance characteristics of gas turbine combustors. (a) Operating conditions at which pollutants form. (b) Where pollutants form. (Measurements made along combustor center line.) (SOURCE: R. F. Sawyer. *Astronautics and Aeronautics*, April, 1970.)

number of ways to decrease soot formation in combustors:

1. Increase the fuel-spray cone angle, which increases mixing.
2. Increase the air flow into the primary zone, effectively lowering the temperature and fuel-air ratio in that region. An adverse effect of this method is that it increases the gas velocity in that region, which in turn increases the difficulty of ignition.
3. Increase the liner pressure drop, which effectively increases the turbulence level and promotes mixing. This also increases fuel consumption.
4. Use an airblast atomizer in place of or in conjunction with a conventional pressure atomizer. In this approach, air at high velocity is used to blast or atomize the fuel droplets.
5. Use water injection to reduce the average temperature in the hot zone of the combustor.
6. Make use of fuel additives, which attack the basic chemistry of the combustion process. Barium and manganese compounds have been used in the past, but apparently the effectiveness of such additives is sensitive to the type of engine.

Technique 4 is probably one of the most widely used today to overcome smoke formation. This is especially true with the advent of higher-pressure-ratio engines now in use, since the higher ratio has led to increased smoke problems. For one reason, the higher pressure leads to increased temperatures in the combustion zone, although higher fuel economy also results. The main effect of increased combustor pressure is on the spray pattern of conventional pressure atomizers. Atomization occurs closer to the spray nozzle, with less penetration of the spray into the primary zone because of increased air resistance. In order to take advantage of the increased pressure-ratio effects, such as fuel economy, a different fuel injection system is mandated. The air-blast atomizer is one approach. In its simplest form, fuel is allowed to run along a metal plate and drop or splash off the end. However, a high-velocity air stream is directed at the end of the plate, and this high-energy flow stream shatters the fuel into very small droplets. The air flow may reach a velocity of 400 ft/s. As a rule of thumb, it requires about 3 lb of air per pound of fuel for air-blast atomization, although this ratio may reach 5 or 6. As might be anticipated, the mean droplet size resulting from this operation is inversely proportional to the air-blast velocity.

The reduction in carbon monoxide and hydrocarbon emissions from gas turbine combustors is attained primarily through increased combustion efficiency, which may be achieved in the following ways:

1. Operate the primary zone at an equivalence ratio around 0.8. If the value of ϕ is much higher, the CO will be great simply as a result of

the equilibrium CO formed in conjunction with the CO_2 and O_2 in the gas stream.

2. Improve the fuel atomization, to achieve better burning.
3. Reduce the film cooling air to the combustor. This air cools the combustion process, and as a result the CO does not get oxidized to CO_2.
4. Increase the combustion zone volume, thus providing an increased time in the combustion zone for the gases.

Temperature is the dominant factor in the formation of NO_x in combustors, but residence time is another important variable. Lowering the average temperature of the gas is not effective; it is essential that hot spots in the combustion zone be eliminated as well. A number of techniques may be used to reduce NO_x formation. Among these are:

1. Operate with a lean primary zone. This method of operation tends to lower the temperature but at the same time to increase the CO and hydrocarbon emissions and lower the smoke (see Figure 10-18).
2. Operate with a rich primary zone. This approach reduces the air, and especially the oxygen, available for oxidation of the atmospheric nitrogen. The method is of theoretical interest only, however; it is impractical because of poor fuel economy and increased smoke.
3. Reduce the primary-zone volume, which reduces the time in residence at high temperatures.
4. Increase the liner pressure drop. This approach tends to increase the turbulence within the combustor and to reduce local hot spots. At the same time, however, it increases the fuel consumption.
5. Use water injection. This method lowers the average temperature level in the combustor.

As we review the techniques for achieving improved emission control, it becomes obvious that there are fundamental design conflicts between CO and HC control, on one hand, and NO_x control on the other. For example, carbon monoxide emissions are greatly reduced by operating at relatively low temperatures. There is a fairly narrow band of temperature within which it is possible to meet both CO and NO_x emission limits. To meet EPA standards it is necessary to keep CO emissions below 70 ppm and NO_x below 4.5 ppm. A plot of CO and NO_x levels as a function of primary-zone temperature is shown in Figure 10-19 [41]. The CO and NO_x limits quoted above are also shown. On the basis of these data, we conclude that only in the temperature range of roughly 1610° to 1730°K (2440° to 2650°F) for the primary zone will both pollutant gases meet the desired emission standards. (Another source [42] presents a graph equivalent to Figure 10-19, based on experimental measurements. This source indicates that the required temperature range to meet EPA CO and NO_x standards is 1300° to 1390°K [1870° to 2030°F]. These values are considerably below those quoted above. However, the data

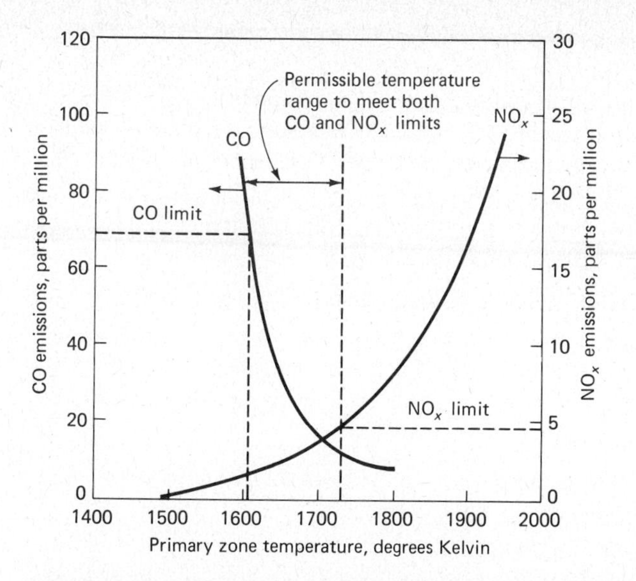

Figure 10-19 CO and NO_x emissions as a function of gas turbine combustor primary-zone temperature, with indication of preferred operating range. (Courtesy of Prof. A. H. LeFebvre, Purdue University, W. Lafayette, Ind.)

may not be in conflict. To ascertain any real difference it is necessary to examine in more detail the method used to measure temperature and concentration in the combustor as well as the position at which the measurements were taken. Nevertheless, both sources lead to the same result, namely, that the band of temperature available to meet both CO and NO_x emissions is extremely narrow.)

Another method of presenting the information shown in Figure 10-19 is given by Figure 10-20(a). In this case the primary-zone temperature is plotted against the overall fuel-air ratio supplied to the combustor. Low fuel-air ratios lead to low primary-zone temperatures and excessive CO formation due to quenching effects. On the other side of the scale, relatively high fuel-air ratios result in high primary-zone temperatures and hence excessive NO_x formation. Thus the low emissions range is typified by a restricted primary-zone temperature or fuel-air ratio range. From a design viewpoint, this suggests that combustors ought to be built so that the gases stay within the suggested ranges as much as possible. One possible method of achieving this would be a variable-geometry combustor. Under this arrangement one could control the air to the primary and secondary zones as fuel requirements change. Another possibility is staging of fuel injection. Several fuel injectors would be employed at various axial positions to maintain the desired temperature and/or fuel-air ratio throughout the combustor, thus achieving reasonable emission levels. Also, variation of the fuel injected from different injectors could take into account the need for a wide variation in fuel-air ratio as the engine operates over the range from idle to full power.

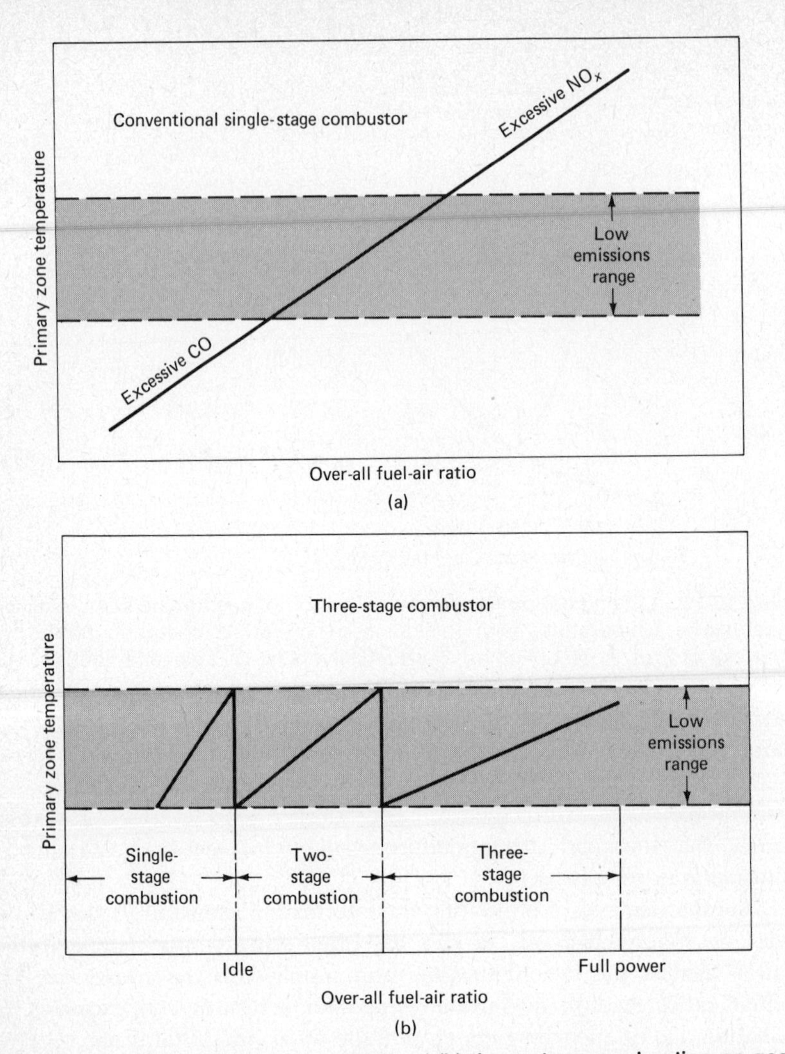

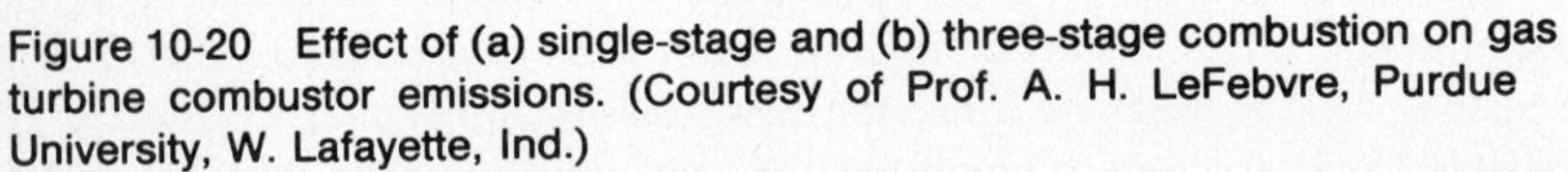
Figure 10-20 Effect of (a) single-stage and (b) three-stage combustion on gas turbine combustor emissions. (Courtesy of Prof. A. H. LeFebvre, Purdue University, W. Lafayette, Ind.)

Figure 10-20(b) illustrates how this might be done for a combustor designed for a maximum of three stages of combustion.

New combustors [39] which are relatively smoke-free have been developed. These combustors provide improved fuel atomization and mixing in the primary zone. Improved utilization of secondary air has reduced the maximum temperature. The influence of the location of secondary air injection holes is illustrated by Figure 10-21, which gives data on the nitrogen oxide emission of a vehicle gas turbine combustor [43]. In modification no. 1, the location of the primary air holes of the standard combustor

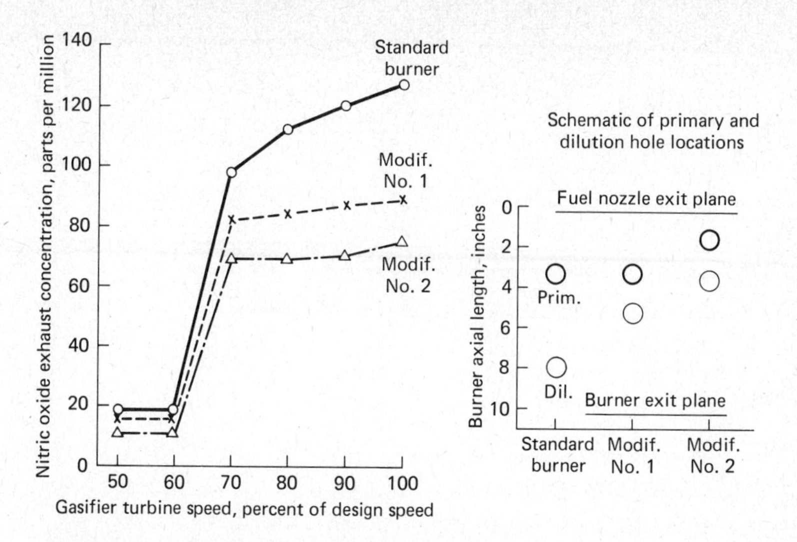

Figure 10-21 Effect of location of secondary air injection holes on NO production in a gas turbine combustor.

was maintained while the secondary air holes were moved forward (toward the fuel nozzle) by $2\frac{3}{4}$ in. This relocation resulted in a decrease of approximately 34 percent in the primary combustion zone. In modification no. 2 both the primary and secondary air holes were moved forward $1\frac{3}{4}$ in., thus reducing the residence time to about 44 percent of that of the standard combustor. Similar results were obtained by keeping the location of the holes the same but changing their size. Water injection into the combustor of gas turbines has also been found to result in a substantial reduction in NO formation [44]. For a more extensive discussion of emissions from continuous combustion systems, the reader is referred to the literature [45].

A major research program currently being conducted by the engine manufacturers and NASA has been undertaken to develop engines with greatly reduced emissions. In order to obtain a more satisfactory comparison of engine emission characteristics, the Environmental Protection Agency (EPA) [46] has defined an emission parameter (EPAP) based upon the rate of emissions from an engine obtained at specified engine thrust levels while operating over a standard takeoff–landing cycle. Thus the EPAP is similar to the emission standard for cars and trucks. The EPAP is obtained by measuring the concentration of the specified pollutant in the exhaust of the engine while the engine is operating at a specified thrust level. A standardized formula developed by EPA is then employed to calculate the emission rate in grams per hour for each mode of engine operation. The flight cycle consists of 19 min of idle outward bound mode, 0.5 min, takeoff mode, 2.5 min, climb-out mode, no descent mode, 4.5 min, approach mode, and 7 min, taxi, idle mode.

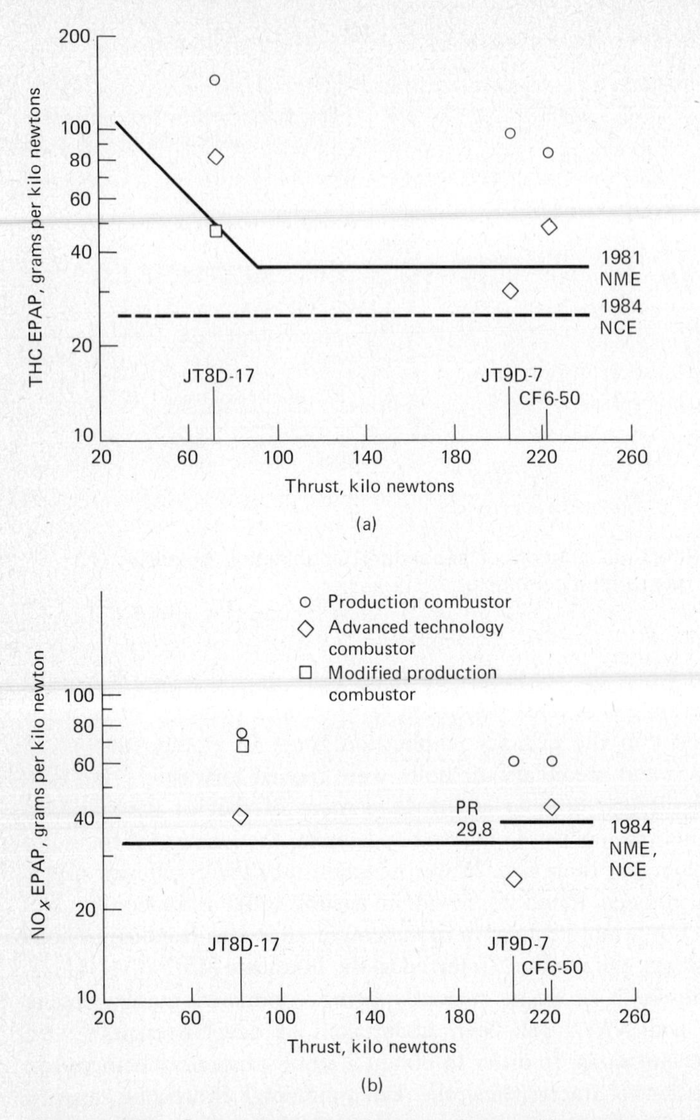

Figure 10-22 Comparison of emissions of several jet engines and the revised EPA standards. (NME—new manufactured engine; NCE—new certified engine.) (a) Total HC emissions and standards. (b) NO_x emissions and standards. (SOURCE: NASA Technical Memorandum 79009, October, 1978.

The time in hours per mode is multiplied by the emission rate to obtain the emissions per mode. These emissions per mode are added to obtain the EPAP emissions in grams of pollutant per kilonewton of thrust per cycle. Figure 10-22 presents comparison of the emissions of several engines and the revised EPA standards. Note that in almost all cases major improvements will be required to attain the EPA goals for emissions.

10-14 ALTERNATIVE FUELS AND THEIR UTILIZATION

In view of the projected decrease in the availability of petroleum based fuels, intensive studies of alternative fuels are being conducted in the United States and throughout the world. The chemical compositions of the major fuels proposed to date are somewhat different from those derived from crude oil and the emission and performance characteristics of engines using those fuels are not completely known. Since the projected costs of the alternative fuels appear to be somewhat higher than the costs of current (1980) fuels and the available supplies lower than what we have been accustomed to, much greater emphasis will be placed upon developing engines that attain minimum emissions and maximum fuel economy. In view of this outlook, it is desirable to examine briefly the alternative fuels and their characteristics.

Alcohols can be produced from coal, petroleum, and biomass feedstocks [47]. Ethyl alcohol (ethanol) can be produced from various biomass feedstocks using a variety of available and developing processes. Methyl alcohol (methanol) is produced primarily from petroleum and coal. The heat of vaporization of alcohol is higher than that for gasoline whereas the heat of combustion is approximately one-half. The alcohols may be blended with gasoline in mixtures of up to about 18–20 percent alcohol or used as straight fuels in spark ignition engines. A mixture of 18–20 percent alcohol and gasoline will separate if a small amount of water is present. Due to the higher octane number (109) for pure alcohol, a much higher compression ratio (up to 14 : 1) can be used in an alcohol engine than is used in current gasoline engines.

Limited experimental results to date, employing gasohol (a mixture of 10 percent methanol and 90 percent gasoline) in engines designed to run on gasoline, indicate little change in either emissions or fuel consumption when compared with gasoline [48]. When either methanol or ethanol were employed as fuels in engines modified for their use (higher compression ratio, correct spark timing, and A/F) it was found that the energy efficiency was increased by 6–10 percent and the emissions without a catalyst were reduced by the following amounts: CO, 60 percent; HC, 78 percent; and NO_x, 55 percent of those for a production gasoline engine [48]. It should be pointed out that while studies report an increase in fuel efficiency when alcohol is used in an engine operating under optimum conditions for that fuel, the fuel consumption (gallons) will be larger than for gasoline due to the much lower heating value.

When methanol is heated it undergoes the following endothermic reaction:

$$CH_3OH(l) \underset{\text{heat}}{\overset{\text{catalyst}}{\rightarrow}} 2H_2(g) + CO(g)$$

The products, termed reformed methanol, can be used as a gaseous fuel in a spark ignition engine. The necessary heat can be obtained from the engine exhaust. In tests conducted by the Nissan Motor Company [48] using a 1952 cm^3 4-cylinder engine, the thermal or energy efficiency was 35 percent and the emissions were low enough to meet the 1982 Japanese standards without additional exhaust gas treatment.

Alcohol is difficult to ignite in a diesel engine and it does not appear advantageous to use it as a straight fuel. However, emulsions of diesel fuel and up to 45 percent ethanol have been used successfully in diesel engines [48]. Alcohols have also been used successfully as fuels for gas turbines, jet engines, and steam generators.

Oil shale and coal may be processed by a variety of methods [49] to yield hydrocarbons that can be refined to produce fuels suitable for use in prime movers. Hydrocarbon material called kerogen is bound in oil shale and can be released by heating the stone to approximately 900°F. The distillate released by this retorting process (about 40 gallons per ton of stone) is similar to petroleum crudes. Coal liquefaction technologies have been known for over 50 years. One process involves pyroloysis, where coal is heated without oxygen to produce a range of molecular weight liquids, gases, and char. Catalytic liquefaction processes bring the coal in direct contact with a catalyst in the presence of hydrogen to produce a range of hydrocarbon liquids. Methanol has been produced from coal in several commercial scale plants outside of the United States using synthetic gas. Mobil Oil has developed, in pilot stage, a methanol to high-octane gasoline process using zeolite catalysts [49].

Although liquid fuels from shale oil and coal can be made compatible with most engine systems, the degree to which they would be able to meet emission constraints depends upon the degree of upgrading that is necessary and practical. In all cases the liquids derived from either oil shale or coal have a high nitrogen content, many trace metals and other contaminants, and low H/C ratios. Automobile engines can tolerate little nitrogen in the fuel if they are to approach the research goal of 0.4 g/mi of oxides of nitrogen. This level is difficult to approach without any fuel-bound nitrogen. Catalytic treatment of exhaust gases is not possible if there are certain metals in the synthetic fuel. Deterioration of catalysts from leaded gasoline is well appreciated. Not so well known is the poisoning of catalysts by phosphorous compounds, arsenic, beryllium, cadmium, lead, mercury, and selenium. All of these latter substances have been found in oil shale. The degree to which they would have to be removed in the refining process is unclear.

Because of the low H/C ratio of the fuels derived from shales and coal without extensive upgrading by hydrogenation (a very costly process in both energy and money), such fuels may prove more suitable for use in diesel type engines. The diesel engine can be developed to burn the heavier hydrocarbon fuels. A potential problem exists with synthetic diesel fuel from shale oil because of the possible low cetane number. In addition, the increased aromatic content of many shale-derived fuels could lead to increased particulate emission from diesel engines and polynuclear aromatics from spark ignition engines [49].

The preceding brief review of the potential alternative fuels indicates that much work must be done to achieve maximum energy conservation, minimum emission of air pollutants, and the conservation of our resources. In conclusion it should be pointed out that some of the fuels may be used in combination in engines that are similar to the stratified engine (PROCO) being developed by the Ford Motor Company.

QUESTIONS

1. At what air-fuel ratio do the untreated exhaust products of a pre-1970 gasoline engine have a minimum amount of unburned hydrocarbons? What is the relative concentration of NO_x at that mixture ratio?
2. Why is it impossible to operate a gasoline engine (no pollution controls) at an air-fuel ratio which will give minimum CO, HC, and NO_x simultaneously?
3. Explain how changes in each of the following items influence the emissions from a gasoline engine: carburetion, spark timing, compression ratio, and combustion chamber surface-to-volume ratio.
4. What are the methods currently being used to reduce evaporative and crankcase emissions?
5. Data presented in Table 10-4 show that marked reduction in emissions can be obtained by changing from gasoline to other types of fuel. What are the possibilities of such a change-over?
6. What are the effects upon output and fuel consumption of combustion gas recirculation and spark retardation methods of reducing exhaust gas emissions?
7. What is the basis of operation of external thermal reactors when used to reduce HC and CO emissions? What are some of the disadvantages of this method of emission reduction?
8. Why does the temperature of exhaust gas entering a thermal reactor need to be higher than that of gas entering a catalytic reactor?
9. What is the effect of misfires upon a thermal reactor?
10. List and discuss the problems encountered in the operation of catalytic oxidizing reactors.
11. Describe how the combustion phenomena in a stratified-charge engine reduce exhaust emissions.
12. Why should the combustion products from an external combustion engine contain fewer air pollutants than those from an internal combustion engine?
13. What was the major disadvantage of the steam systems developed for bus propulsion when compared to a diesel engine?
14. Why should the CO and HC emissions from a diesel engine be inherently less than those from a conventional spark ignition gasoline engine?
15. If the emissions from a diesel engine are so much less than those from a comparable gasoline engine, why are diesel engines not widely used in automobiles?
16. With regard to quantity, what is the major air pollutant emitted by turbojet aircraft? What do most people, on the basis of their own observations, associate with turbojet engines?
17. Under what operational condition does a conventional turbojet engine emit the most air pollutants?

18. Describe the basic approaches available to the designer for reducing the emissions from a turbojet combustor.
19. How can the NO_x emission from a turbojet combustor be reduced?

PROBLEMS

10-1. Determine the stoichiometric air requirement for the combustion of normal octane (C_8H_{18}), in pounds of air per pound of fuel.

10-2. Determine the stoichiometric air requirement for a fuel containing by weight 50 percent *n*-octane (C_8H_{18}), 25 percent 1-heptene (C_7H_{14}), and 25 percent methanol (CH_3OH), in pounds of air per pound of fuel.

10-3. A gasoline engine is operated at an equivalence ratio of (a) 1.10 and (b) 0.95. What is the air-fuel ratio, in pounds of air per pound of fuel?

10-4. Determine the stoichiometric A/F ratio for the following fuels: (a) hydrogen, (b) ammonia (NH_3), (c) methanol (CH_3OH), and (d) natural gas assuming it is methane (CH_4).

10-5. Determine the stoichiometric A/F ratio for (a) diesel fuel (cetane, $C_{16}H_{34}$) and (b) ethyl alcohol (C_2H_5OH).

10-6. A turbojet operates at a equivalence ratio of 0.22 when burning jet fuel ($C_{11}H_{26}$). Determine the A/F and F/A ratios.

10-7. Determine the stoichiometric A/F ratio for fuel mixture of gasoline (C_7H_{13}) and methanol having (a) 10 percent and (b) 20 percent methanol by weight.

10-8. A gasoline C_7H_{13} is burned with 20 percent excess air. (a) What is the corresponding A/F ratio? (b) What is the equivalence ratio?

10-9. Make a table for gasoline, diesel fuel, and jet engine fuel, giving the A/F for (a) a stoichiometric mixture, and (b) an actual mixture when ϕ(gasoline) = 0.98, ϕ(diesel) = 0.7, and ϕ(jet) = 0.22.

10-10. The exhaust gas measured at the exhaust manifold of a gasoline engine has a composition equivalent to 70 g/mi CO and 8 g/mi of NO_x. What is the conversion efficiency of a catalytic converter required to meet the proposed 1985 emission standards of 3.4 g/mi CO and 0.4 g/mi NO_x?

10-11. Consider that the composition of an automotive engine fuel is represented by C_8H_{18}. (a) How many cubic feet per minute of air is required by a 4-stroke engine with a displacement volume of 320 in.^{3}? (b) Now consider that the exhaust emissions contain 2000 ppm of CO, 100 ppm of HC, and 50 ppm of NO. If the stoichiometric amount of air is used for combustion, determine the pounds per minute of CO, HC, and NO emitted. The engine runs at 4000 rpm, and the exhaust gas exits at 1 atm and 1000°F.

10-12. It has been proposed on the basis of experimental data that the quantity of NO_x formed in the combustor of a gas turbine is related to other experimental parameters in the following manner: $[NO_x] \approx V_c P^{1.2} \exp(0.01T_f)/T_f$, where V_c is the combustor volume, P is the absolute pressure, and T_f is the hot-zone temperature in degrees Kelvin. What is the percent change in NO_x concentration when (a) the combustor pressure is raised from 12 to 15 atm, (b) the temperature is raised from 1700° to 1900°K, and (c) the pressure is raised from 10 to 12 atm while the temperature is lowered from 2000° to 1800°K, all other factors being held constant?

10-13. Consider the equation for the NO_x concentration in a gas turbine combustor given in Problem 10-12. Operating conditions are 10 bars and 1750°K. If a new design for the same combustor requires a pressure of 12 bars, estimate what maximum temperature should be used to maintain the same NO_x level as before, all other factors remaining the same.

10-14. A gas turbine combustor operates initially at 8 bars and with a hot-zone temperature of 1725°K. A new design condition of 1775°K is proposed for that pressure. On the basis of the equation given in Problem 10-12 for NO_x concentration, estimate what percent change in combustor volume would be required in order to keep the NO_x emissions at the same level, all other operating variables remaining the same.

References

1. A. J. Haagen-Smit. "Chemistry and Physiology of Los Angeles Smog." *Ind. Engr. Chem.* **44** (June 1952): 1342–1346.
2. D. J. Patterson and N. A. Henein. *Emissions from Combustion Engines and Their Control*. Ann Arbor, Mich.: Ann Arbor Science Publishers, Inc., 1972.
3. E. F. Obert. *Internal Combustion Engines and Air Pollution*. New York: Intext Educational Publishers, 1973.
4. R. C. Lee and D. B. Wimmer. "Exhaust Emission Abatement by Fuel Variations to Produce Lean Combustion." Soc. Automotive Engineers Paper 680769, 1968.
5. M. C. Baxter, G. W. Leck, III, and P. E. Mizelle. "Total Emissions Control Possible with LP-Gas Vehicle." Soc. Automotive Engineers Paper 680529, 1968.
6. R. W. McJones and R. J. Corbeil. "Natural Gas Fueled Vehicles Exhaust Emissions and Operational Characteristics." Soc. Automotive Engineers Paper 700078, and *SAE Journal* **78** (June 1970): 31–34.
7. J. G. Finegold and W. D. Van Vorset. "Engine Performance with Gasoline and Hydrogen, A Comparative Study." Engineering Systems Dept., U. of California, Los Angeles, 1974.
8. J. E. Witzky and J. M. Clark, Jr. "The Third Cycle—Stratified Charge." *Mech. Engr.* **91** (March, 1969): 29–35.
9. W. G. Agnew. *Future Emission-Controlled Spark-Ignition Engines and Their Fuels*. General Motors Corporation Research Publication GMR-880, 1969.
10. J. E. Nicholls, I. S. El Messiri, and H. K. Newhall. "Inlet Manifold Water Injection for Control of Nitrogen Oxides, Theory and Experiment." Soc. Automotive Engineers Paper 690018, 1969.
11. R. C. Stempel and S. W. Martins. "Fuel Economy Trends and Catalytic Devices." Soc. Automotive Engineers, Paper 740594, 1974.
12. D. J. Patterson, R. H. Kadlec, and E. A. Sondreal. "Warmup Limitations on Thermal Reactor Oxidation." Soc. Automotive Engineers Paper 730201, January 1973.
13. R. H. Kadlec, E. A. Sondreal, D. J. Patterson, and M. W. Graves, Jr. "Limiting Factors on Steady-State Thermal Reactor Performance." Soc. Automotive Engineers Paper 730202, January, 1973.
14. W. Glass, D. S. Kim, and B. J. Kraus. "Synchrothermal Reactor System for Control of Automotive Exhaust Emission." Soc. Automotive Engineers Paper 700147, January, 1970.

15. J. Dunn. "Ford's New Smog-Free Engine." *Popular Science Monthly*, May, 1970, pp. 55–57.
16. A. J. Scussel, A. O. Simho, and W. R. Wade. "The Ford PROCO Engine Update." Soc. Automotive Engineers, Paper 780699, 1978.
17. J. L. Bascunana. "Divided Combustion Chamber Gasoline Engines—A Review for Emissions and Efficiency." *J. Air Pollu. Control Assoc.* **24**, no. 7 (July 1974): 674–679.
18. R. F. Ansdale. *The Wankel R.C. Engine*. New York: A. S. Barnes and Company, 1969.
19. *Popular Science*, September, 1973.
20. E. S. Wright. "The Potential of the Gas-Turbine Vehicle in Alleviating Air Pollution." ASME Paper 70-WA/GT-8, August, 1970.
21. *Mech. Engr.* **93**, January, 1971.
22. "Low Emission Automotive Turbine Unveiled." *Gas Turbine International* **12** (January–February 1971): 39–41.
23. T. D. Nogle. "The ERDA/Chrysler Upgraded Automotive Gas Turbine Engine Emission Control System." NATO/CCMS Fourth International Symposium on Automotive Propulsion Systems, Washington, D.C., 1977.
24. *Ind. Research.* **23**, February, 1971.
25. *Trailer Life*, January, 1971, pp. 90–91.
26. *Mech. Engr.* **92** (July 1970): 94–96.
27. J. W. Bjerklie and B. Sternlicht. "Critical Comparison of Low-Emission Otto and Rankine Engine for Automotive Use." Soc. Automotive Engineers Paper 69004, January, 1969.
28. R. A. Renner. "Reviewing the California Steam Bus Project." Soc. Automotive Engineers Paper 730218, January, 1973.
29. Soc. Automotive Engineers Papers 740295, 740296, 740297, and 740298.
30. J. P. Tumda. "New Electrics Make Performance Breakthroughs." *Popular Science* **54**, February, 1971, and *Machine Design*, **42** (June 1970): 19.
31. D. J. Patterson, ed. *Engine Emissions, Pollutant Formation, and Measurement*. New York: Pergamon Press, 1973.
32. E. F. Obert. *Internal Combustion Engines and Air Pollution*. New York: Intext Educational Publishers, 1973.
33. Soc. Automotive Engineers Paper 700671, 1970.
34. C. J. Walder. "Reduction of Emissions from Diesel Engines." Soc. Automotive Engineers Paper 730214, January, 1973.
35. K. J. Springer and T. M. Baines, "Emissions from Diesel Versions of Passenger Cars." Soc. Automotive Engineers Paper 770818, 1977.
36. K. J. Spring and R. C. Stahman. "Diesel Car Emission—Emphasis on Particulate and Sulfate." Soc. Automotive Engineers Paper 770254, 1977.
37. S. Hockheiser and E. R. Lozano. "Air Pollution Emission from Jet Aircraft Operating in New York Metropolitan Area." Soc. Automotive Engineers Paper 680339, 1968.
38. R. E. George, J. A. Verssen, and R. L. Chase. "Jet Aircraft, A Growing Pollution Source." *J. Air Pollu. Control Assoc.* **19** (November, 1969): 847–885.
39. R. E. George, J. S. Nevitt, and J. A. Verssen. "Jet Aircraft Operation: A Threat to the Environment." Paper 71-117, Air Pollution Control Association, June, 1971.

40. R. F. Sawyer. "Reducing Jet Pollution Before It Becomes Serious." *Astronautics and Aeronautics* 8 (April 1970): 62–67.
41. A. H. Lefebvre. Cranfield Institute of Technology, Bedford, England, private communication, 1974.
42. T. F. Nagey et al. "A Low Emission Gas Turbine Passenger Car?" *Mech. Engr.* **96**, no. 1 (1974): 14.
43. W. Cornelius and W. Wade. "The Formation and Control of Nitric Oxide in a Regenerative Gas Turbine Burner." Soc. Automotive Engineers Paper 700708, September, 1970.
44. R. D. Klapatch and T. R. Koblish. "Nitrogen Oxide Control with Water Injection in Gas Turbines." ASME Paper 71-WA/GT-9, Annual Winter Meeting, 1971.
45. W. Cornelius and W. Agnew, eds. "Emissions from Continuous Combustion Systems." *Proceedings of the Symposium on Emissions from Continuous Combustion Systems*. New York: Plenum Press, 1971.
46. National Air Quality, Monitoring, and Emissions Trends Report, 1977, EPA-450/2-78-052, December, 1978.
47. The Report of the Alcohol Fuels Policy Review, U.S. Department of Energy, DOE/PE-0012, June 1979.
48. Alcohol Fuels Technology, Third International Symposium, Vols. I, II, III, Asilomar, California, May 1979.
49. A Discussion Paper on Shale Oil and Coal Liquids and Their Use as Transportation Fuels, Automotive Transportation Center, Purdue University, West Lafayette, Ind., September 1979.

Chapter 11
Odor Control

11-1 INTRODUCTION

Odor is probably the most complex of the air pollution problems. Because of the somewhat nebulous nature of odors, they are classified as noncriteria pollutants by the Environmental Protection Agency. Taste and smell are the two chemical senses, since they appear to be physiological reactions to contact with certain specific substances. In the case of smell, some individuals have the ability to detect very minute quantities of substances—in the range of 1 ppb. Unfortunately, the human nose is the only good measuring device known, and it is notoriously undependable. Moreover, people have mixed reactions to a given odor. What is offensive to one is acceptable to another, as demonstrated by people's varied responses to the different perfumes worn by men and women. Not only is there marked disagreement as to the offensiveness of selected odors, but also two other problems hinder attempts at odor control. First, unfamiliar odor is more easily detected and is more likely to cause complaints than a familiar one. Second, because of a phenomenon known as odor fatigue, given sufficient time a person can become accustomed to almost any odor and be conscious of its only when a change in intensity occurs. Many words are employed to describe odors, the act of "smelling," and the intensity of odors. Unfortunately, these words and phrases are not as definite or as widely accepted as are the terms employed in other areas of air pollution. The science of smell is *osmics*. The word *osmorphoric* indicates an odorous substance. The physical act of smelling is *olfaction*.

11-2 THE SENSE OF SMELL AND THEORIES OF ODOR

The right and left olfactory clefts are located approximately above the juncture of the nasal passages and the top of the throat. The total surface area of each chamber is about 1 in.^{2} in an adult. The long, narrow olfactory cells are located in the olfactory cleft with their length perpendicular to the plane of the nasal cavity. The olfactory nerve carries the "smell impulse" from the olfactory cleft to the olfactory bulb of the brain. A complete description of the olfactory system is given by Moncrieff in reference 1.

An extensive review of the theories of odor lead Moncrieff [1] to conclude that there are three basic theories of odor. Moncrieff has developed a theory of his own which states that in order to be odorous, a substance must be (1) volatile, so that it continually loses molecules to the atmosphere for transportation to the olfactory apparatus, (2) capable of being absorbed on the sensitive surface of the olfactory epithelium, and (3) customarily absent from the olfactory region. "Customarily absent" means that the substance must be one which is not already present on the olfactory epithelium, so that when it arrives there, carried by the inspired air, it brings about a change. Only change causes sensation.

The mechanism of olfaction would seem to be accomplished in the following stages:

1. The odorous material is volatile and is continually losing molecules to the atmosphere.
2. Some of these molecules are inspired through the nasal passages, and, though usually not aided by a sniff, they are directed to the olfactory receptors.
3. They find the olfactory receptors and lodge there; that is, they are adsorbed on sites of suitable dimensions on the olfactory receptors.
4. The lodgment is accompanied by an energy change, adsorption being an exothermic process.
5. This energy change causes electric impulses to pass up the olfactory nerve to the brain.
6. Brain processes result in the sensation of smell.

Finally, Moncrieff [1] suggests that odor perception may involve several processes in the nose, some physical and some chemical. Perhaps in olfaction the physical vibratory system is present in the olfactory epithelium and the chemicals come from outside, that is, from the odorants.

Numerous efforts have been made to classify odors and to relate the chemical composition and molecular structure of the odorous substances to their odors. To date no completely satisfactory relationship has been established. The view has prevailed that odor is *not* dependent on chemical constitution, but on the physical differences brought about by the arrangement of the constitutive groups. After an extensive study of odorous substances, Moncrieff [1] concluded that there is no simple and consistent relation between odor and constitution. What must be recognized is that the dependence of odor characteristics upon molecular configuration is only half the story; the other half is to be found in the receptors and brain of the individual who is doing the smelling. Moncrieff was able to compile 62 general principles relating odor to chemical composition. We list several of these principles here for illustrative purposes.

1. Strong odor is often found accompanied by volatility, and chemical reactivity and unsaturation run parallel to odor.
2. The main factor in determining odor is the architectural type of the molecule.
3. In ring compounds, the number of ring members often determines the odor.
   a. Five- to six-member rings: bitter almonds and methol odor
   b. Six- to nine-member rings: transitional odors
   c. Nine- to twelve-member rings: camphor or mint odor
   d. Thirteen-member rings: woody or cedar odors
   e. Fourteen- to sixteen-member rings: musk and peach odors
   f. Seventeen- and eighteen-member rings: civet odors
   g. Rings of more than eighteen members: faint or no odor

Smells often carry intense emotional value, depending on a given individual's personal experiences. A person's favorable reaction to an odor, or lack thereof, can sway judgment. Molecular configuration and physiology of the receptors and the nervous pathway must be considered together.

11-3 PHYSICAL PROPERTIES OF ODOROUS SUBSTANCES

As in the case of the chemistry of odorous substances, many studies have been conducted to determine relationships between the physical properties of odorous substances and their odors. Some success has been achieved in this area.

VAPOR PRESSURE

If it is postulated that molecules or particles of substances must come in contact with the olfactory sensors in order to be odorous, then those substances with higher vapor pressures would provide increased numbers of molecules. For example, ether, chloroform, and gasoline have stronger odors than do less volatile liquids. Similarly, volatile solids like camphor and moth balls have stronger odors than glass. Some exceptions should be noted. For example, civetone and musk xylol are highly odorous yet have very low vapor pressure—between 10^{-3} and 10^{-4} mm Hg. Water, having a vapor pressure of 17.5 mm Hg, is odorless.

SOLUBILITY

It is known that some of the most highly odorous materials are soluble in water and fat. This is understandable since such substances are able to penetrate the watery mucus surrounding the flagella and then through the fatty flagella themselves. A notable exception is glycol, which is readily soluble in fat or water but is considered odorless.

INFRARED ABSORPTION

There is some evidence that odorous substances strongly absorb infrared radiation. The phenomenon was at one time attributed to the presence of moisture, but it has been shown that dry odorous substances do absorb infrared rays. The underlying principle is that just as absorption bands in the visible spectrum determine color, absorption bands in the infrared (or the ultraviolet) may determine odor. The existence of an absorption band indicates that the substance has an intramolecular vibration of the same frequency as that of the absorbed light, since the absorption depends on interference between the vibrations of the molecule and of the light. However, there is no apparent reason why such absorption bands should be more noticeable in the infrared than in the visible or the ultraviolet spectrum. Exceptions are paraffin and carbon disulfide, which do smell but are transparent in the infrared range.

ULTRAVIOLET IRRADIATION

Solutions of odorous materials such as eugenol, safrol, and so on, were made in glycerine or paraffin or water, and it was found that after exposure, they exhibited a Tyndall effect; that is, a beam of ultraviolet light passing through the solution was scattered by the particles of solute and so became opalescent. This phenomenon is a strong function of the solution temperature; as the temperature rises, the solution becomes clearer. Ultraviolet lamps produce ozone in an oxygen-laden gas. Deodorizing is then accomplished since ozone oxidizes most odorant materials.

RAMAN EFFECT

When monochromatic light—for example, the green light from a mercury vapor lamp—is scattered by a pure substance, the scattered light is no longer homogeneous but, when examined with a spectrometer, is found to consist of light of both longer and shorter wavelengths than the monochromatic light used. The differences in wavelength are the Raman shift. The Raman spectrum is characteristic of any pure compound. It is related to the infrared absorption spectrum. If, for example, green mercury lamp light having a wavelength of 4358 Å falls on benzene, some light of wavelength 5000 Å is reflected. The shift is thought to be due to a resonance effect, derived from the energy of the molecules of the scattering substance. This is evidently a method of gaining information with respect to the molecular vibrations.

It is clear, then, that the Raman shift is a measure of the intramolecular vibrations of a substance; since odor commonly is also supposed to be dependent on the intramolecular vibrations, we might expect to find a close relation between Raman shift and odor. The alkyl mercaptans R-SH are a homologous series with powerful, objectionable odors, and the odors of the first members of the series are very similar to each other. Comparison of the Raman spectra of methyl, ethyl, propyl, butyl, and amyl mercaptans shows that they all have a Raman shift of 2567 to 2580 cm^{-1}. No other compounds have this shift, and no other compounds have the mercaptan odor and, presumably, the mercaptan molecular vibrations. Exceptions exist. Glycerine is odorless, but it has a vapor pressure and some fat solubility. It also has shifts in its Raman spectrum at 466-2, 880-2, 955-3, and 202 cm^{-1}, all of which should excite olfaction.

ADSORPTION

Activated carbon, such as coconut charcoal, will adsorb large quantities of odorous substances. It will take up at least half its own weight of benzene. It will adsorb chloroform and similar substances. It picks up ammonia well, but adsorbs gases like hydrogen, nitrogen, oxygen, carbon monoxide, and carbon dioxide only in small quantities. Since gases that excite smell are usually adsorbed well by carbon and other surface agents, it is not unlikely that adsorption in the olfactory apparatus precedes the sensation of smell. Adsorption of odorous molecules on solid surfaces may take place in three ways: (1) through the condensed layer of moisture usually present on all surfaces, (2) by direct adsorption to the clean solid surface from which the moisture film has been removed, and (3) by direct attraction of charged particles by an oppositely (electrically) charged surface.

11-4 ODOR MEASUREMENT TECHNIQUES

The human nose assisted by selected devices or prescribed procedures is the basic system for evaluating odors. In general, odor determinations are made by panels of from 2 to 15 trained individuals subjected to given concentrations of odorous substances. The measurements of odors fall into two general categories: determining the threshold concentration of odorous substances and establishing the type and intensity of atmospheric odors. A possible third category would be the tracing or "tracking" of a given odor in the atmosphere to its source. The first category involves primarily "pure" odors, that is, the odor of a single substance. In the second category, the odors of several substances may combine in the atmosphere to give the impression of a single odor. The ability to discriminate may be extremely important in this category.

The four attributes [2, 3] of the olfactory sense applicable to measurement are listed and defined as follows:

ODOR ATTRIBUTE	DEFINITION
Intensity	Magnitude of perceived sensation
Pervasiveness	Change of magnitude or acceptability on dilution. Detectability
Quality	Similarity of odor sensation, chemical nature, or functional grouping. Characteristic properties.
Acceptability	Degree of like or dislike of odor sensation

Intensity is some numerical or verbal indication of the strength of the odor. A gradual increase in intensity is readily detected, although persons may become fatigued by odor.

Pervasiveness, according to Nader [4], is sometimes referred to as odor potential ratio or threshold dilution ratio. These essentially are measures of the ability of an odor to pervade a large volume of dilution air and continue to possess a detectable intensity. A pervasive odor such as might result from mercaptans and decomposed proteins will tend to spread in all directions over a community.

Quality describes the characteristics of odors in terms of association with a familiar odorant, such as coffee or onions, or by associating a familiar odor with an unfamiliar odorant by analogy.

Acceptability—the degree of like or dislike of the odor sensation—depends to a large degree upon the experiences of the person doing the odor evaluation. The association of odor with events can result in the odor being considered as pleasant or disagreeable.

With the exception of odor quality, the above attributes are evaluated by reference to subjective scales with varying numbers of points. A commonly used scale has five points:

0	No perception
1	Very faint perception
2	Faint perception
3	Easily noticeable
4	Strong
5	Overpowering perception

Nine-point scales and four-point scales have also been employed. The reproducibility of odor ratings decreases sharply if more than a ten-point scale is employed.

It will be recalled from Section 11-3 that several chemical and physical properties of odorants are involved in the olfactory process. Correlations between odor and odorant properties are summarized below [3]:

Intensity	Concentration Volatility Fat or water solubility
Pervasiveness	Concentration Chemical nature
Quality	Molecular size and shape Infrared or Raman spectrum Chemical nature or functional grouping
Acceptability	—

The generally accepted relationship between odor intensity and odorous contaminant concentration, as illustrated in Figure 11-1, is given by the Weber-Fechner equation:

$$P = K \log S \tag{11-1}$$

where P is the magnitude of sensory response, or odor intensity; K is a constant; and S is the magnitude of stimulus, or odor concentration. Values of K have been found to vary from 0.3 to 0.6 [5]. A similar equation known as the psychophysical power law is given by the relation

$$P = K_2 S^n \tag{11-2}$$

where K_2 and n are constants peculiar to each odorant and are determined experimentally [6]. The value of n usually ranges from 0.15 to 0.8.

Through use of the sensory attribute of pervasiveness, the problem of measuring odorants in a manner similar to, and at concentrations detectable by, the human olfactory system is neatly avoided. All such measurements involve the determination of the number of dilutions of a sample of the odorous gas required to render it odorless. The odor threshold is then expressed not in terms of odorant concentration, but rather in such subjective units as "threshold," "odor units," or "dilutions." Thus the relative strength of the odor is determined, with stronger odors requiring greater

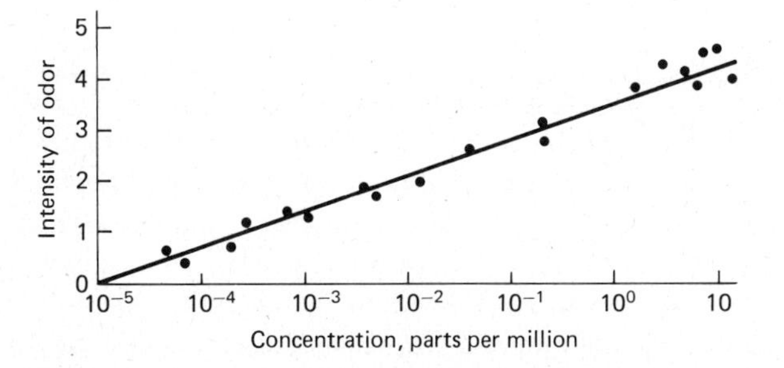

Figure 11-1 The results of several tests of odor intensity.

numbers of dilutions with odor-free air to bring the odorous gas to the threshold of odor detection.

The syringe-dilution method adopted in ASTM Procedure 1391-57 has received considerable attention. A sample of the gas whose odor is to be measured is diluted with odorless air until a dilution is achieved in which the odor can be barely perceived. The ratio of the total volume of this diluted mixture (original sample volume plus volume of diluting air) to the volume of the original sample is a measure of the concentration of odor in the original sample. The diluted gas sample can be injected directly into the nostril from a syringe made for that purpose. Another method currently in vogue, known as the Scentometer, allows for odor evaluation directly in the field. This instrument has the widest dilution range (up to 128) of all vapor-solution instruments designed for field use. The unit consists of two telescoping metal tubes. Holes in the inner tube are closed or opened by adjusting the relative position of the two tubes, thus varying the ratios of contaminated air to clean air.

Two primary sources of error in threshold measurements are presentation and control of the stimulus and changes in the state of the odorant in collection and transport of samples. The latter errors are introduced by adsorption on collecting apparatus and temperature effects on odorants. Evaluations of samples with a high moisture content or with a high dew point are especially unreliable because of temperature and condensation effects. The errors are introduced because repeatability of response is often difficult. Portable devices, such as the above, are limited in their range of dilution (to 128) and are severely hampered because a single organoleptic sensor (observer) is used. Underlying all of the threshold methods is the assumption that any odorant in the atmosphere that produces a positive sensation of odor is undesirable. It seems that the purpose of malodor control is the elimination of odorants that produce objectionable odor response, not just any response. Accordingly, threshold measurements would best be directed toward determining the objectionability threshold, not the intensity threshold. This could be done with exactly the same devices previously described but using an objectionability scale rather than an intensity scale. Superthreshold measurement, as it is called, is simply the direct application of the rating scales previously described to odorant concentrations clearly above the sensory threshold.

A possible means of converting standard analytical instruments or methods into odor-measuring devices is to use the superthreshold rating techniques to determine the appropriate value K in the Weber-Fechner equation. In so doing, the problem of relative insensitivity of analytical methods as compared to the human olfactory sense is avoided by evaluating the odor intensity of odorant concentrations within the analytical range. Of course, this can be done only for odorant emissions that are consistent in composition and for which a physiochemical property can be correlated with subjective odor attributes. Emissions from many industrial processes appear to meet these requirements. If values of K can be determined for various malodorous effluents, analytical instruments can be used to measure odor intensity directly and establish the degree of control required. These measurements can be made in the field, thus precluding the need to transport samples and to make allowances for fatigue and adaptation of the human olfactory sense.

An instrument for measuring the intensity of diesel exhaust odor has been developed by Arthur D. Little, Inc. [7]. An odor panel was employed in the determination of the chemical species of diesel exhaust which contribute to the characteristic burnt-oily odor. After those species were identified, values of the constants in an equation similar to the Weber-Fechner equation relating odor intensity to chemical concentration of the odorous species were determined. The instrument measures the concentration of the hydrocarbon species which are responsible for the diesel exhaust odor and by means of the intensity-concentration equation relates the measured concentration to an intensity readout scale.

11-5 ODOR THRESHOLD VALUES

As has been mentioned previously, the concentrations of odorous substances that can be detected by the human nose vary by many orders of magnitude. Examples of odor threshold concentrations for selected substances are presented in Table 11-1. An indication of the range of odor threshold concentrations of similar species is presented in Table 11-2 [8]. A more extensive list of the threshold concentrations of 100 petrochemicals using sensory methods has been published by Hellman and Small [9]. The magnitude of the odor control problem can be seen when we compare detection of acetone versus

Table 11-1 ODOR THRESHOLDS IN AIR

CHEMICAL	ODOR THRESHOLD (ppm)	ODOR DESCRIPTION
Acetic acid	1.0	Sour
Acetone	100.0	Chemically sweet
Amine monomethyl	0.021	Fishy, pungent
Amine trimethyl	0.0021	Fishy, pungent
Ammonia	46.8	Pungent
Carbon disulfide	0.21	Vegetable sulfide
Chlorine	0.314	Bleach, pungent
Diphenyl sulfide	0.0047	Burnt, rubbery
Formaldehyde	1.0	Hay or strawlike
Hydrogen sulfide	0.00047	Eggy
Methanol	100.	Sweet
Methylene chloride	214.	
Phenol	0.047	Medicinal

SOURCE: G. Leonardos, D. Kendall, and N. Bernard. "Odor Threshold Determinations of 53 Odorant Chemicals." *J. Air Pollu. Control Assoc.* 19, no. 2 (1969): 91–95.

Table 11-2 ODOR THRESHOLDS FOR SOME NITROGENOUS COMPOUNDS

COMPOUND	CONCENTRATION (ppm)
Trimethyl amine	0.00021
Nitrobenzene	0.0047
Pyridine	0.021
Aniline	1.0
Ammonia	46.8
Diethyl formanide	100.0

SOURCE: G. Leonardos, D. Kendall, and N. Bernard. "Odor Threshold Determinations of 53 Odorant Chemicals." *J. Air Pollu. Control Assoc.* **19**, no. 2 (1969): 91–95.

hydrogen sulfide. In the case of acetone a concentration of 80 ppm would not be detected, since its threshold value is 100 ppm. However, a concentration of 0.0008 ppm of H_2S (five orders of magnitude less than that of acetone) would constitute a possible odor problem, because the threshold value for H_2S is only 0.00047 ppm. Hence a value of 0.0008 ppm for H_2S is readily detected.

11-6 APPLICATIONS OF ODOR MEASUREMENTS

The design of odor control systems must be based partially on odor threshold values. If odorous substances cannot be completely removed from a waste gas by incineration or other techniques, then it is necessary to disperse the pollutant into the atmosphere. Dispersion must be effective enough to assure that odor threshold values are not exceeded at ground level at any position downwind from the stack where humans might normally live, work, or play. Example 11-1 illustrates a typical dispersion calculation for an odorous material.

Example 11-1

Assume that carbon disulfide gas is being discharged from a stack 30 m tall. The rate of discharge is 400 lb of CS_2 per hour. On a clear day in June when the wind speed is 10 mi/hr, will the odor be detected in a populous region 1000 m downwind?

SOLUTION

We will use the dispersion equation,

$$C = \frac{Q}{\pi u \sigma_y \sigma_z} \exp\left(\frac{-H^2}{2\sigma_z{}^2}\right) \qquad \text{(4-13)}$$

developed in Chapter 4, where u is the wind speed, Q is the emission rate, H

is the stack height, and σ_y and σ_z are the dispersion standard deviations. In this particular case,

$$u = 10 \text{ mi/hr} = 4.47 \text{ m/s}$$

$$Q = 400 \text{ lb/hr} = 50.4 \times 10^6 \mu\text{g/s}$$

$$H = 30 \text{ m}$$

and, from Figures 4-6 and 4-7,

$$\sigma_y = 160 \text{ m}$$

$$\sigma_z = 120 \text{ m}$$

Substituting these values into Equation (4-13),

$$C = \frac{50.4 \times 10^6}{3.14(4.47)(160)(120)} \exp\left[-\frac{1}{2}\left(\frac{30}{120}\right)^2\right] = 180 \ \mu\text{g/m}^3$$

Equation (1-1) may now be used to determine the concentration in parts per million. Since the molar mass of carbon disulfide is 76 g/g·mole,

$$C = \frac{180(24.5)}{76(10^3)} = 0.058 \text{ ppm}$$

The calculated concentration of 0.058 ppm is much less than the threshold value of 0.21 ppm. Therefore no carbon disulfide odor should be detected in the atmosphere at 1000 m downwind.

Figure 11-2 presents the calculated concentrations of carbon disulfide as a function of downwind distance at the centerline for the conditions of the preceding example. Note that no odor of CS_2 should be detected at a distance greater than approximately 500 m.

It is often convenient in odor control problems to express the quantity or concentration of the odorous substance in terms of odor units. One odor unit is that quantity of odorous substance necessary to contaminate 1 ft^3 of clean air to the median threshold odor detection response in humans. For example, it would require 1/(0.00000021) ft^3 of CS_2 to contaminate 1 ft^3 of clean air to the 0.21 ppm odor threshold level. In the case of methanol, which has an odor threshold concentration of 100 ppm, one odor unit would be 1/(10,000) ft^3 of methanol vapor since that quantity would contaminate 1 ft^3 of air to the odor threshold. Expressed in a different way, 1 ft^3 cubic foot of methanol vapor would contaminate 10,000 ft^3 of clean air to the threshold level. Thus 1 ft^3 of pure methanol vapor is equivalent to 10,000 odor units. In like manner, 1 ft^3 of pure CS_2 can contaminate 1/(0.00000021) ft^3 or 4,750,000 ft^3 of clean air and thus is equivalent to 4,750,000 odor units. The rate of discharge of an odorous substance expressed in odor units can be employed in odor dispersion calculations.

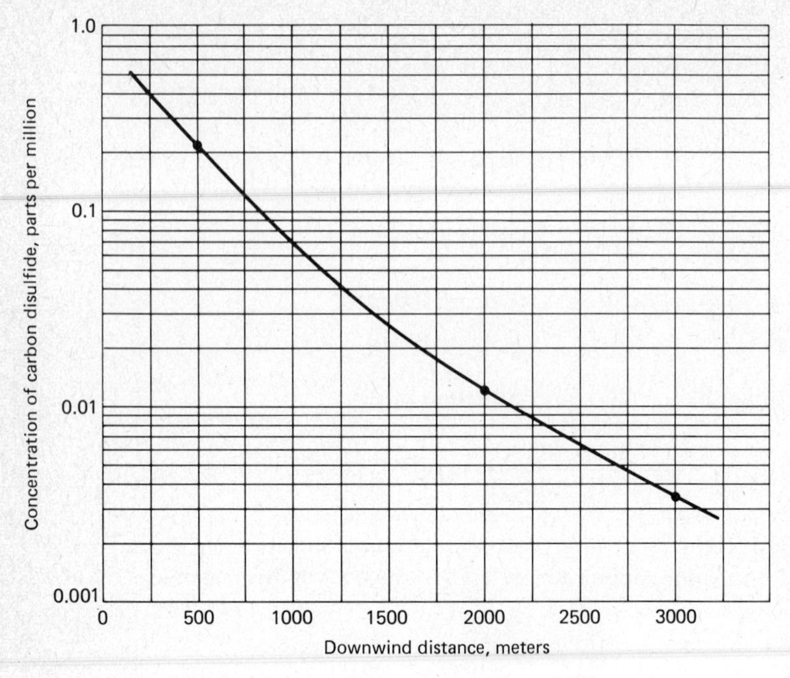

Figure 11-2 Calculated downwind concentration of CS_2.

For example, the volumetric rate of discharge of carbon disulfide in the preceding dispersion example is

$$\text{vol. rate} = \frac{400 \text{ lb/hr}}{76 \text{ lb/lb·mole}}\left(\frac{379 \text{ ft}^3}{\text{lb·mole}}\right) = 2000 \text{ ft}^3/\text{hr}$$

Since 1 ft^3 of CS_2 is equivalent to 4,750,000 odor units, the rate of emission of odor units of carbon disulfide is simply $2000 \times 4{,}750{,}000 = 9.5 \times 10^9$ odor units/hr, or 2.64×10^6 odor units/s. Substitution of this value for the rate of discharge Q into the diffusion equation used earlier gives a concentration at 1000 m of

$$C = \frac{2.64 \times 10^6}{3.14(4.47)(160)(120)} = 9.6 \text{ odor units/m}^3 = 0.28 \text{ odor units/ft}^3$$

Since an odor threshold of 0.21 ppm is, by definition, equal to 1 odor unit/ft^3, a value of 0.28 odor unit/ft^3 cannot be detected.

From the preceding example we can conclude that the phenomenon of dilution or dispersion could be employed to establish regulations governing the allowable emission concentration of chemically identifiable substances [10]. Either the dispersion equation previously discussed or some other equally appropriate analytical method of calculating the concentration could be used. Feldstein et al. [10] employ a log-normal distribution and observed standard deviation data for SO_2 in the San Francisco Bay area to obtain an allowable emission concentration of 100 times the odor threshold for typical stack heights. Thus in the case of phenol having an odor threshold concentration of 0.047 ppm, the allowable emission concentration would be 4.7 or 5 ppm.

11-7 ODOR CONTROL METHODS

The two general approaches to odor control are (1) to reduce concentration so that the smell is less intense and therefore less objectionable; and (2) to change or mask the quality of the odorant so that the smell becomes more pleasant and acceptable to the populace. In the first approach, odor control is achieved by reducing the source of the odor; diluting the odor by greater dispersal; removing the odor from a gas stream by adsorption, absorption, or oxidation; or chemically converting the odorous product to one less odorous. In the second approach, attempts are made to hide the original odor by introducing a more powerful odor having a pleasing scent or by modifying the original odor.

11-7-A Ventilation

Confining odorous air by means of forced-air hood and duct systems is still the most commonly employed method of odor control. Discharging the odorous air by means of a stack tall enough to let natural atmospheric dispersion take place is the simplest method of odor dissemination. When odors are to be dispersed from an elevated source, such as a stack, the ground-level concentrations can be calculated as functions of stack geometry, odorant concentration in the emitted gas stream, and meteorological conditions just as in the treatment of diffusion of other air pollutants. An example of the procedure to be followed is presented in Section 11-6. It must be pointed out that such procedures give predicted average concentrations over a specific time period. In the case of a foul odor, even low concentrations may be objectionable to some people. In addition, Turk [11] notes that odorants may not always disperse in the atmosphere as gaseous matter. Instead, the odorous material may be absorbed on particles and thereby reach the nose of the subject in a form different from what we would have been exposed to had the odorous substance been entirely in the gaseous state. In support of his argument, Turk [12] cites evidence showing that the removal of particulate matter from diesel exhaust by thermal precipitation effects a marked reduction in odor intensity. If the nature of the odor is such that dispersion in the atmosphere is not a satisfactory method of control, then other methods must be examined.

11-7-B Adsorption

Activated carbon adsorption may be applied to deodorize (1) a gas stream before being discharged to the atmosphere, (2) an odorous outside atmosphere air before it is used in a building, or (3) an indoor atmosphere being recirculated. Conventional face velocities of 375 to 500 ft/min are used and bed thicknesses may vary from 1 to 3 in. (thin beds) to 1 to 3 ft (thick beds). Care must be taken to ascertain whether or not nonodorous substances that are easily adsorbed by the charcoal are present in the odorous gas stream. These nonodorous substances may saturate the activated carbon bed reducing or preventing the bed functioning as an odor control device. Example 11-2 illustrates the application of an equation based upon a simplified adsorption model for the determination of maximum adsorption filter life. (Also, see the general discussion in Section 6-3.)

Example 11-2

A charcoal filter element with dimensions of $24 \times 24 \times 8\frac{3}{4}$ in. thick and containing 45 lb of activated charcoal pellets is to be used to remove methanol vapor from an air stream with a flow rate of 300 scf/min and a vapor concentration of 300 ppm. It is desired to estimate the filter life on the basis that the charcoal can adsorb up to 50 percent of its initial weight of methanol vapor and the filter efficiency is 90 percent. The maximum flow rate of gas for the specified filter element is 1000 scf/min.

SOLUTION

In lieu of specific information, we assume a vertical adsorption wave (see Section 6-3) and use the simplified equation proposed by Turk. The Turk equation is

$$t = \frac{WS}{EQC}$$

where S is the proportionate maximum saturation of adsorbent (a fraction); W is the weight of the adsorbent; E is the efficiency of the filter (a fraction near unity); Q is the gas flow rate; and C is the concentration of odorous material.

Since the rated capacity of the filter element is 1000 scf/min, which is larger than the required flow rate, only one filter will be used in the preliminary calculation. The values of the independent variables in the Turk equation, in a consistent set of units, are:

$$C = \frac{(\text{ppm})(\text{molar mass})(10^3)}{24.5} = \frac{300(32)(10^3)}{24.5} = 392 \times 10^3\ \mu\text{g/m}^3$$

$$= 392 \times 10^{-3}\ \text{g/m}^3$$

$$W = 45\ \text{lb} = 20{,}400\ \text{g}$$

$$S = 0.5$$

$$Q = 300 \text{ ft}^3/\text{min} \times 0.0283 \text{ m}^3/\text{ft}^3 = 8.49 \text{ m}^3/\text{min}$$
$$E = 0.90 \text{ (90 percent)}$$

Substituting the above values into the Turk equation yields

$$t = \frac{0.5(20{,}400)\text{ g}}{0.9\,(8.49 \text{ m}^3/\text{min})\left[(392 \times 10^{-3}\text{ g})/\text{m}^3\right]} = 3400 \text{ min (or 56.7 hr)}$$

This answer represents the maximum time the filter can be used without having vapor pass through the element unadsorbed. To account for a nonvertical adsorption wave, a more conservative design would call for two filters to be used in series or a reduction in the time of operation.

11-7-C Absorption

Air scrubbers or washers may be employed when the odorant is (1) soluble in the scrubbing liquid, (2) condensable at the temperature of the scrubbing liquid, which could be water, or (3) capable of becoming attached to particulate in the scrubber. Numerous types of scrubbers employing a wide range of liquids are available [13], the choice dependent upon the odor to be removed. Plain water or water with added chemical reactants is widely used because of the low cost involved. Additional discussion pertaining to absorption is presented in Sections 6-5 through 6-7.

11-7-D Flame Oxidation

As already pointed out, incineration by direct flame at temperatures from 1100° to 1500°F (850° to 1100°K) has been found effective for removing odors from gas streams. In most cases, natural gas is selected as the fuel, since the major products of combustion are CO_2 and water vapor. Care must be exercised to see that partial oxidation does not occur, since an increase in odor could result, as exemplified by the conversion of alcohols to highly odorous carboxylic acid. In addition to natural gas, propane and low sulfur oil may be used as fuels. Whenever possible, heat recovery systems should be used to conserve fuel. Section 6-12 includes additional discussion of flame oxidation.

11-7-E Catalytic Oxidation

Catalytic incineration may be employed to remove odors by oxidation. The required temperature is from 500° to 800°F lower than that required for flame incineration. In addition, the required equipment is simpler in nature. Unfortunately, the life of the catalyst can be reduced by (1) poisoning, (2) obstruction of the catalytic surface by carbonaceous deposits, or (3) loss of catalytic surface owing to abrasion by solid particles in the gas stream. (Also see Section 6-12-C.)

11-7-F Chemical Oxidation

Chemical conversion of odorous substances may be accomplished by oxidizing agents such as ozone, permanganates, hypochlorites, chlorine, and chlorine dioxide [14]. Ozone converts organic matter by oxidative degradation, usually to aldehydes, ketones, and acids. Ozone must be used in a confined and unoccupied space. Potassium permanganate solutions have been found to deodorize sulfur compounds, amines, phenols, and unsaturated compounds like styrene and acrolein. An example of odor reduction by chemical oxidation is

$$CH_3SH + \text{oxidant} \rightarrow SO_2 + CO_2 + H_2O$$

The odor threshold of CH_3SH is only 1 ppb and is easily detected. Of the three products of oxidation, two of these (CO_2 and H_2O) are odorless, and SO_2 has an odor threshold of 1 ppm, which is three orders of magnitude greater than that of CH_3SH. Hence the possibility of detection is greatly reduced.

11-7-G Counteractions and Masking

Unpleasant odors may be controlled by modifying their quality by means of admixture with more pleasing odors without any chemical change in the original odorants themselves. We have only to recall the number of "deodorants"—perfumes, lotions, creams, and so on—available today to observe the widespread use of such substances to control body odors. The early Romans, Egyptians, and others have been said to have practiced air pollution control when they employed large quantities of perfumes in their homes and places of worship. Observations show that when two odorous substances of given concentrations are mixed in a given ratio, the resulting odor of the mixture may be much less intense than that of the separate components. This phenomenon has been termed odor modification. Many industrial odor modifiers are available. Care must be taken to ensure that the odorous substance for which masking is being considered does not constitute a health hazard at any concentration. Several companies provide consulting services in this area [15].

QUESTIONS

1. Why is it necessary to use descriptive words in defining odors?
2. Why is it difficult to establish levels or concentrations of unpleasant odors which can be used as air quality standards in the manner in which standards are established for air pollutants like CO, for example?

3. Explain the difficulty in obtaining a satisfactory relationship between some physical or chemical characteristic of a molecule and the odor of the substance.
4. How are odor intensities determined? What scale is used?
5. What is odor fatigue? What is its importance?
6. Conduct an experiment in which three persons attempt to describe to others the odor of some substance that they have not experienced.
7. What is the range of sensitivity of the average nose to odor intensity?
8. What general systems are available for odor control?
9. It has been argued that the use of tall stacks to reduce the ground-level concentration of sulfur dioxide is not a satisfactory method of control since the quantity of sulfur dioxide emitted into the atmosphere has not been reduced. Does this same argument apply to the use of stacks for odor control?

PROBLEMS

11-1. Phenol vapor, part of a waste gas stream, is discharged at 5.0 kg/hr from an effective stack 40 m tall. The wind speed at the top of the stack is 5 m/s, and the atmosphere is in a class *C* stability condition. Determine whether phenol vapor, C_6H_5OH, can be detected by the average person anywhere downwind at ground level.

11-2. Consider the data of Problem 11-1. What emission rate of phenol in kilograms per hour, is permissible so that detection by humans is just barely possible and at what distance will this occur, in kilometers?

11-3. Consider the data of Problem 11-1, but use an effective stack height of (a) 25 m and (b) 15 m. On the average, how many kilometers downwind at ground level can the gas be detected if the discharge rate is now 20.4 kg/hr?

11-4. A waste-gas stream containing formaldehyde is discharged from a stack with an effective height of 30 m. The formaldehyde (CH_2O) discharge rate is 240 kg/hr, and the wind speed at the top of the stack is 6 m/s. For a class *D* stability condition, will the formaldehyde be detected by humans at ground level anywhere downwind?

11-5. Consider the data of Problem 11-4. What formaldehyde emission rate, in kilograms per hour, makes human detection of the odor just barely possible, and at what distance, in kilometers, will this occur?

11-6. Consider the data of Problem 11-4. On the average, how many kilometers downwind at ground level can the gas be detected by humans?

11-7. With reference to Problem 11-1, what is the rate of emission of phenol in odor units per second? Assume that the specific volume of the pollutant is 24.4 m^3/kg·mole.

11-8. With reference to Problem 11-4, what is the rate of emission of formaldehyde in odor units per second? Assume that the atmospheric temperature of the pollutant is 20°C at a pressure of 1 bar.

References

1. R. W. Moncrieff. *The Chemical Senses*. London: Leonard Hill Book Company, 1967.

2. G. Leonardos. "A Critical Review of Regulations for the Control of Odors." *J. Air Pollu. Control Assoc.* **24** (1974): 456.
3. R. A. Duffee. "Appraisal of Odor Measurement Techniques." *J. Air Pollu. Control Assoc.* **18** (1968): 472–474.
4. J. S. Nader. "Current Techniques of Odor Measurement." *AMA Archives of Ind. Health* **17**, no. 5 (1958).
5. H. Henning. *Der Geruch*. Leipzig, 1916.
6. J. J. Franz. Report on Proposed Odor Terminology. TT-4 Odor Committee, Air Pollution Control Association, March, 1977.
7. P. L. Levins et al. "Chemical Analysis of Diesel Exhaust Odor Species." *Trans. Soc. Automotive Engineers*, no. 730216, 1974.
8. G. Leonardos, D. Kendall, and N. Barnard. "Odor Threshold Determinations of 53 Odorant Chemicals." *J. Air Pollu. Control Assoc.* **19**, no. 2 (1969): 91–95.
9. T. M. Hellman and F. H. Small. "Characterization of the Odor Properties of 101 Petrochemicals Using Sensory Methods." *J. Air Pollu. Control Assoc.* **24**, no. 10 (October 1974): 979–982.
10. M. Feldstein, D. A. Levaggi, and R. Thuillier. "Odor Regulation by Emission Limitation." Paper 73-273, Annual Meeting, Air Pollution Control Association, June, 1973.
11. A. Turk. "Measuring and Controlling Odors." *Heating, Piping, and Air Conditioning* **40** (1968): 207–210.
12. A. Turk. "Industrial Odor Control and Its Problems." *Chem. Engr.* **73** (1966): 70–78.
13. B. N. Murthy. "Odor Control by Wet Scrubbing, Selection of Aqueous Reagents." Paper 73-275, Annual Meeting, Air Pollution Control Association, June, 1973.
14. D. A. Lundgren, L. W. Reese, and L. D. Lehman. "Odor Control by Chemical Oxidation—Cost, Efficiency, and Bases of Selection." Paper 72-116, Annual Meeting, Air Pollution Control Association, June, 1972.
15. O. B. Lauren. "Odor Modification Concepts and Applications." 12th Annual Purdue Air Quality Conference, November, 1973.

Appendix A Instrumentation

A-1 INTRODUCTION

The determination of the quantity of a given air pollutant in an exhaust-gas stream or in the ambient atmosphere requires care and the use of sensitive instrumentation since, in either case, the concentration of the pollutant of interest is small. Because of the similarity in the basic requirements for accurate gas sampling and the types of instrumentation employed, we discuss both areas of measurement together. It should be pointed out, however, that whereas the concentration of the species of interest may be only a few parts per million or billion in the ambient atmosphere, the concentrations in the exhaust-gas streams may be several percent.

A-2 SAMPLING TRAIN

The determination of the components of a flowing gas stream can be no more accurate than the degree to which the sample obtained represents the flowing gas. For meaningful results, the sample must represent the varying component concentration and composition of the actual gas stream. When large stacks or gas ducts are involved, a representative gas sample requires that sampling traverses of the gas passage be made.

Sampling devices and procedures must satisfy the following general requirements:

1. All components of the sampling train must be chemically inert with respect to the gases being sampled. This requirement may be difficult to meet since stainless steel and many plastics cause some chemical reactions between the air pollutants. Care must be taken to prevent the absorption of the pollutants on the surfaces of the sampling train.

2. The temperature in the sampling probe and gas lines must be maintained above the dew point of the condensable vapors in the gas sample. In addition, some installations require that the temperature of the probe and line be above or below specific chemical reaction temperatures to prevent unwanted changes in the chemical composition of the sample gas.

3. The total volume flow rate of the gas stream being sampled must be known. The volume flow rate of the sample gas stream must also be known.

4. The sample must be obtained isokinetically. That is, the velocity of the gas entering the sampling probe must have the same magnitude and direction as the gas in the main stream.

5. The quantity of gas collected must be sufficient to satisfy the requirements of the methods or devices used in the analysis of the sample.

6. The location of the sampling station should be in a straight section of duct or stack at least 15 pipe diameters downstream of a bend or obstruction. There should be no bends or obstructions for at least 10 pipe diameters downstream of the sampling station.

7. In sampling the ambient atmosphere, the sampling time period should be long enough to satisfy the time required by the analytical method selected.

8. The temperature and static pressure of the sample gas stream must be measured so that the volume of gas at standard conditions can be calculated.

The various procedures and devices available for analysis of the gas sample once it has been obtained will be discussed in the following sections. Specific details for individual instruments are available from the manufacturers of the equipment or may be found in suitable reference publications [1, 2].

A-3 PARTICULATE ANALYSIS

Most devices employed in sampling for particulate air pollutants are merely collectors. Analyzing the particulate and determining its quantity must be done separately. Since the particulate can be in both the solid and liquid phases, several types of collectors may be required. In many cases, the different collectors may be used in series. Care must be exercised to prevent collection of the particulate on the walls of the sampling probe and connecting tubing. It should also be pointed out that if particulate collectors precede

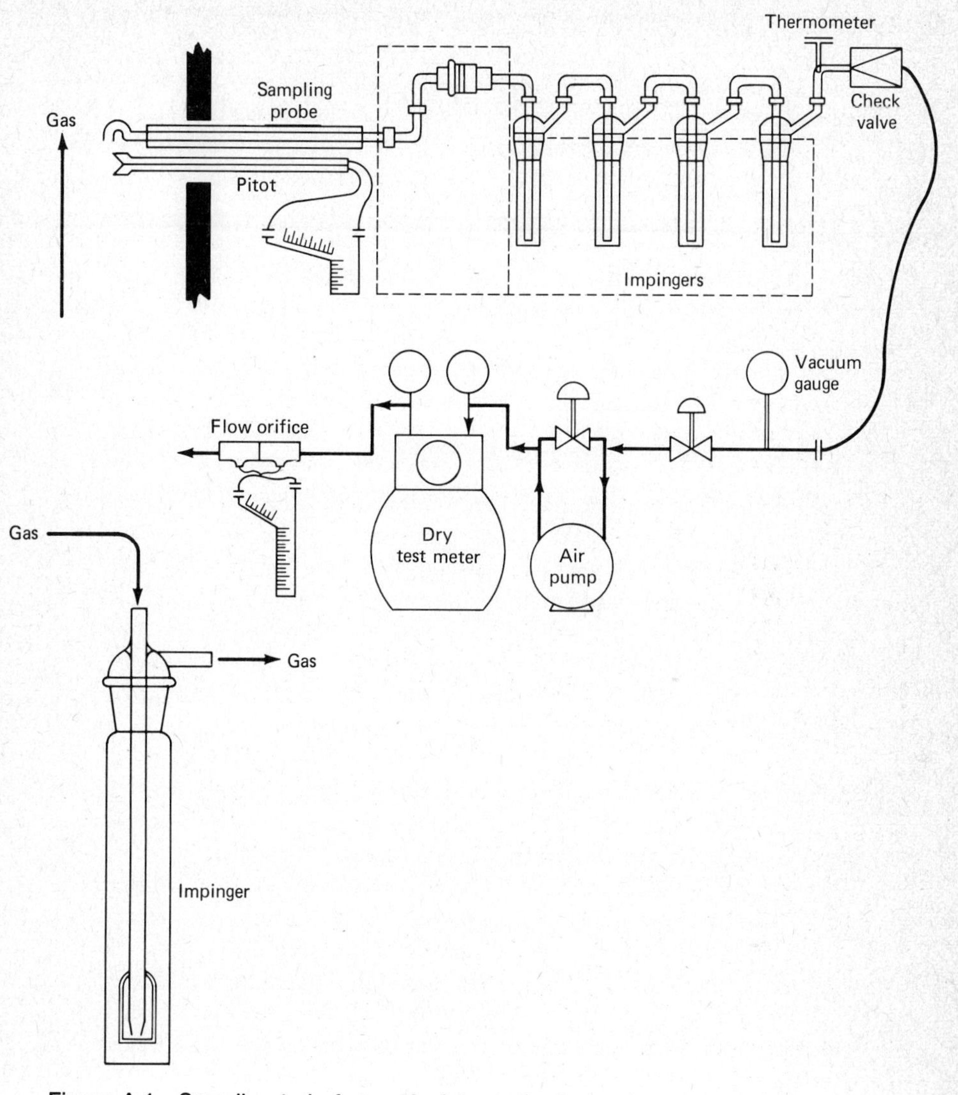

Figure A-1 Sampling train for particulate matter in stack gas.

gas samplers, the gases of interest may be absorbed or may undergo chemical reaction on the surfaces of the particulate collected. This undesirable situation must be prevented. Figure A-1 illustrates a sampling train for particulates.

The basic mechanisms employed in the operation of these collectors are the same as those used in large particulate control devices; therefore, these collectors may be considered miniature collectors. Probably the most widely used collector is the filter. Paper disks, pleated paper, cotton, wool or

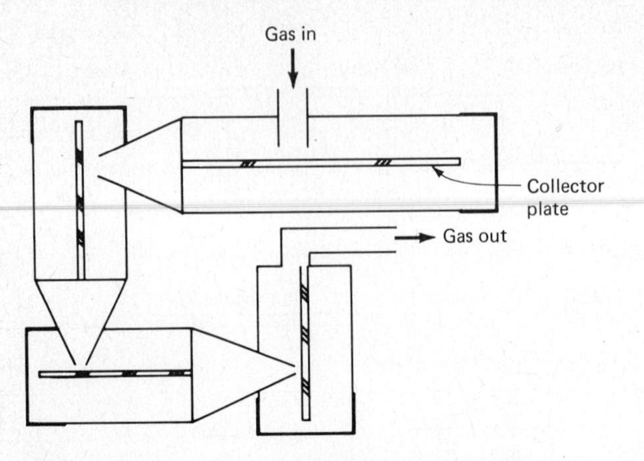

Figure A-2 Impingers for collecting particulate.

asbestos bags, glass or wool fiber, wire screen, and so on, are all available as basic filtering elements. Small wetted packed towers, simple bubbling devices, and venturi scrubbers are used as scrubbing or washing collectors. Electrical precipitators and small cyclone separators are also available. Dry impactors are widely used. In these devices (Figure A-2), the contaminated gas flows through a small orifice and impinges upon a microscope slide or the surface of a liquid. Some particulate collect on the surface. In most cases, a series of orifices and collecting surfaces are arranged so that the size of the orifice is progressively reduced. Thus smaller and smaller particles are collected, with each microscope slide containing a limited range of particle sizes. It has been reported that when sonic velocities are employed, particles as small as 0.1 μm are collected by wet impingers.

Graduated series of metal-cloth screens or a microscope may be used to measure the size of the particulate collected. Wet chemistry may be employed to determine the chemical composition of the particulate. Devices and procedures are currently being developed that use various laser beams to measure air-borne particulate. The intensity of the beam is attenuated in proportion to the concentration and size of the particulate in the beam. It is predicted that lasers will find wide application in the future.

A-4 GAS ANALYSIS

For years, skilled laboratory technicians have performed wet chemical analyses to determine the quantities of gaseous air pollutants in the atmosphere or in sampled gas streams. Because of the recent explosive growth in the air pollution control field, however, the demand for air sample measurements can no longer be met by such procedures. Often the need arises for continuous or on-line analysis of air samples at remote locations, and such needs cannot be satisfied by normal chemical laboratory procedures. The demand for instruments designed to aid in vehicle emission analyses has also caused major changes in available instrumentation. It has been estimated that well over 15,000 air sampling and analyzing devices are available from manufacturers. We make no attempt to discuss specific devices, but we describe selected systems to indicate the basic principles involved. Excellent descriptions of the wet chemical methods may be found in reference 3.

A-4-A Colorimetric Method

Air pollutants are being continuously measured by what might be called the first generation of air pollution instruments. These instruments were adapted from instruments or procedures in use in the chemical process industry. Typically, they operate on the following principle: solution of the gas to be determined in an aqueous medium; reaction with a color-forming reagent in that medium; and measurement of the color formed by photoelectric means. Such analyzers are used to measure atmospheric concentrations of SO_2, NO, NO_2, and total oxidants; source emissions of NO_x; and vehicle emissions of CO. These instruments require considerable attention, and their measurements are likely to be affected by interfering substances found in polluted air.

A modification of the basic colorimetric method is used in the paper-tape analyzer. The basis for the detection of SO_2, NO_x, or CO is a chemical reaction which takes place on a test paper strip that has been impregnated with suitable chemicals. The result of the reaction is a colored stain that is monitored photoelectrically. The test paper is in the form of a continuous motor-driven reel of paper tape so that continuous monitoring is possible.

A-4-B Chemical-sensing Electrodes and Electrochemical Cells

In the chemical-sensing analyzer, a known volume of gas is brought into contact with an absorbing solution having a reference pH value. The solution, containing the dissolved gaseous pollutant (SO_2 or NO_x), then passes an ion-selective electrode where the ion concentration proportional to the pollutant absorbed is measured electronically.

Electrochemical cells are employed to measure the current induced by the electrochemical reaction of selected air pollutants at a sensing electrode. The gas to be detected—SO_2, NO_x, or CO—diffuses through a semipermeable membrane into the cell. The rate of diffusion is proportional to the concentration of the species of interest. When an oxidizing electrolyte is employed in the cell, electrons are released at the sensing electrode by the electrochemical oxidation reaction. The production of electrons causes this electrode to be at a lower potential relative to the counterelectrode. An electron current will flow that can be amplified by suitable electronic equipment. Selectivity of the cell is determined by the semipermeable membrane, the electrolyte, the electrode materials, and the retarding potential.

A-4-C Nondispersive Infrared (NDIR)

Physical phenomena rather than chemical characteristics may be used to infer the concentration of gaseous air pollutants. An example is the nondispersive infrared instrument used to measure gases that absorb infrared radiation, such as carbon monoxide. In this device the species being measured are used to detect themselves. The method of measurement is based upon the principle of selective absorption. A particular wavelength of infrared energy peculiar to a given gas will be absorbed by that gas while other wavelengths will be transmitted. The absorption band for carbon monoxide, for example, is between 4.5 and 5 μm.

The detector consists of two chambers, as shown in Figure A-3, having equal volumes separated by a flexible metallic diaphragm and a stationary metallic button. The metallic button and diaphragm form a capacitor. Both lower chambers in the figure contain equal quantities of the gas of interest. Two identical infrared sources direct radiation through the two separate identical cells or chambers shown in the middle of the figure. One chamber, termed the reference cell, is filled with an inert gas, usually nitrogen. The

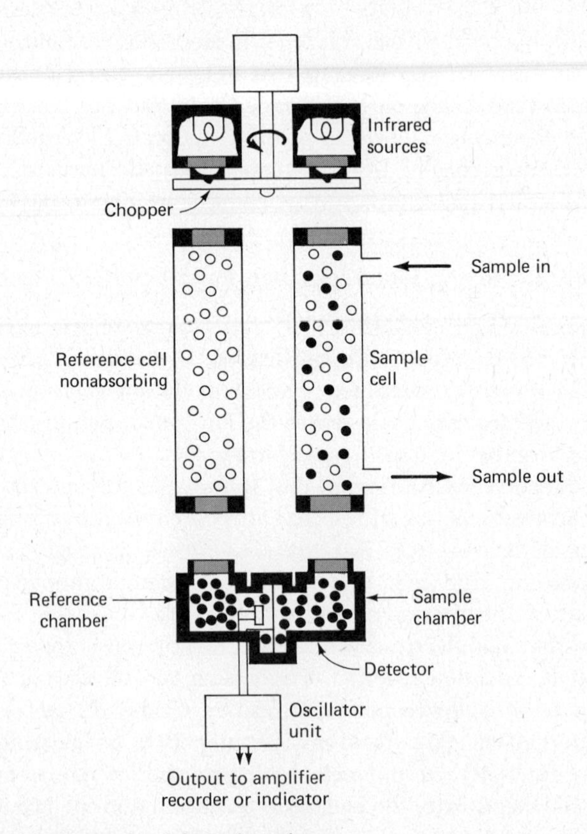

Figure A-3 Nondispersive infrared analyzer.

other cell, called the sample cell, is a tube through which the gas to be analyzed flows. When the gas species of interest is present, it absorbs the infrared radiation in an amount directly proportional to the molecular concentration of the component of interest. No infrared absorption occurs in the reference cell. After passing through either the reference cell or the sample cell, the radiant energy is absorbed by the gas in the detector cells. This absorption of radiant energy causes the gas in the reference detector cell to be heated to a higher temperature than the gas in the sample detector cell. Since the two detector cells are sealed, the pressure becomes higher in the reference detector cell than in the sample detector cell, and the diaphragm will be deflected. The resulting variation in capacitance is directly proportional to the concentration of the species of interest in the sample cell.

A chopper is placed between the infrared sources and the sample and reference cells to provide an ac signal to improve the characteristics of the required electronic circuitry. Filters are employed to block out all wavelengths of infrared radiation except those in the absorption range of the gas of interest. Thus different instruments must be used if two or more gases which absorb infrared radiation are together in the sample gas stream. Instruments employing nondispersive infrared radiation are available to measure SO_2, NO_x, CO, and HC from stationary sources; CO in the ambient atmosphere; and CO, NO_x, and HC in vehicle exhausts. Other characteristics of the NDIR instruments may be found in reference 4. Instruments employing ultraviolet and visible nondispersive absorption are also available.

A-4-D Gas Chromatography

Gas chromatography is employed to separate the different species of interest that are contained in a sample of gas. Basically, gas chromatography is a collective name for a group of methods mainly employed to separate volatile substances for analysis. By means of some suitable device, normally a small glass syringe, the sample is introduced into one end of a long narrow chromatographic column. This column or tube (3 to 10 ft long and from 0.01 to 2 in. in diameter) contains a nonvolatile substance, termed the stationary phase, which acts as a selective retardant. The stationary phase may be either an absorbent (solid) or an adsorbent (liquid). In packed columns the liquid is distributed over an inert solid support, while in capillary columns it coats the inner wall of the tube.

An inert gas that is not retained by the stationary phase flows at a constant rate through the column and is termed the carrier gas. Its function is to transport the sample or solute molecules through the column. The effluent carrier gas leaving the column carries with it sample constituents, emerging at different times depending upon the retention time in the column. A suitable detector is used to indicate the presence of the components in the carrier gas. The output of the detector is fed into a strip-chart

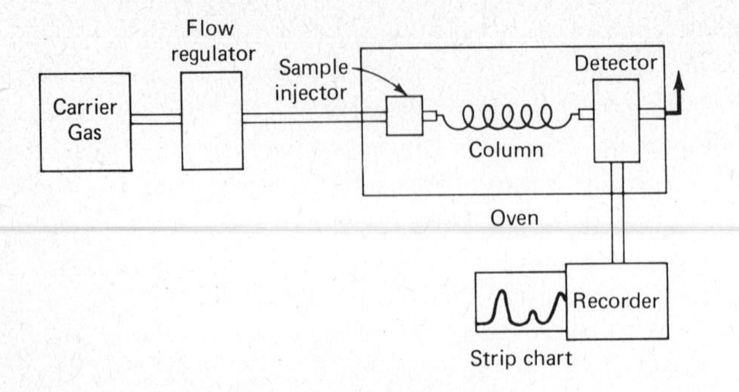

Figure A-4 Gas chromatography.

recorder. The graphic representation, called a chromatogram, usually has a straight-line base on which are superimposed a series of spikes. The location or time of the spikes indicates components of the sample gas, and the height or area of the spikes gives a measure of the quantity of the component present in the sample. Figure A-4 illustrates the system. A constant flow rate of the carrier gas must be maintained during gas analysis. The column may be heated or cooled to accomplish the desired degree of separation of given species. It should also be pointed out that the column must be calibrated so that the time of emergence of each species is known.

A wide variety of detectors, including flame ionization detectors and thermal conductivity cells, are available for use with gas chromatographs. References 5 and 6 may be consulted for more detailed information.

A-4-E Flame Ionization Detector (FID)

In the flame ionization detector, the sample gas is injected into the flame which is created by burning hydrogen in either air or oxygen. The detector is shown in Figure A-5. The hydrogen fuel flows through a metallic capillary tube, and the air flows around it. The flame is positioned between electrodes which have a voltage drop of a few hundred volts between them. When the hydrogen alone is burning, very few ions are formed. When the sample gas containing hydrocarbons is injected into the hydrogen stream, ions are formed in the flame and go to the positive collector electrode. The resulting dc signal produced is proportional to the number of ions formed. The number of ions formed is in turn proportional to the number of carbon atoms in the flame. The FID is sensitive only to the total quantity of hydrocarbons present and does not differentiate between species. When the quantity of an individual species of hydrocarbon is of interest, gas chromatography may be employed to separate the species.

A-4-F Chemiluminescence

Chemiluminescent analyzers are employed to measure the quantities of the oxides of nitrogen or oxidants measured as ozone. When NO reacts with

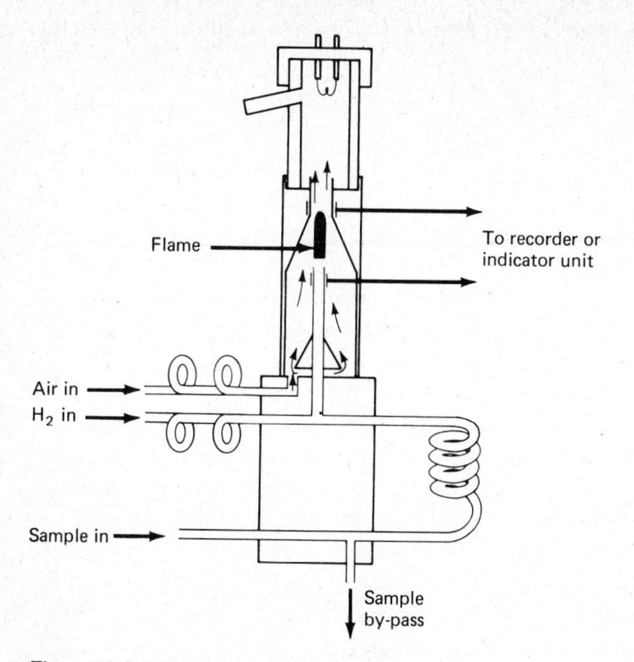

Figure A-5 Flame ionization detector.

ozone, a quantity of excited NO_2 species is produced which then proceeds to the ground state with the emission of radiation energy. The radiant energy is measured by a photomultiplier tube whose output is amplified and fed to suitable readout devices. The chemical reactions are:

$$NO + O_3 \rightarrow NO_2 + O_2$$
$$NO + O_3 \rightarrow NO_2^* + O_2$$

About 5 to 10 percent of the total quantity of NO_2 is produced by the second reaction. Then

$$NO_2^* \rightarrow h\nu + NO_2$$

The radiation $h\nu$ is received by the photomultiplier tube. The intensity of radiation is proportional to the original quantity of NO in the sample gas.

When NO_2 as well as NO is present in the sample gas, the gas is first passed through a heated stainless steel tube before being introduced into the chemiluminescent reaction chamber. The chemical reaction is

$$NO_2 \xrightarrow[1350°F]{\text{stainless steel}} NO + O_2$$

Thus when the sample gas contains both NO and NO_2, a two-step process is necessary. The gas is first passed through the stainless steel converter and then reacted with ozone in the chemiluminescent chamber to obtain total NO_x values. In the second step, the sample gas is passed by the converter and reacted with ozone to obtain the NO value. The quantity of NO_2 in the sample gas is then determined by difference.

Ozone when allowed to react on a surface (organic dye on silica gel) also produces chemiluminescence. The resultant emission is detected by a phototube, and the current generated is directly related to the mass of ozone per unit time flowing over the surface.

A-4-G Mass Spectrometer

Although originally thought of as a laboratory research device, the mass spectrometer is now being widely applied to analyze air pollutants. Basically the mass spectrometer is an instrument used to separate and identify atoms and molecules. The particles of interest are first ionized, accelerated by an electric field, and directed across a magnetic field which deflects their flight into a curved path, as shown in Figure A-6. The heavier of two ions will have a path with a larger radius of curvature and thus will strike a target or detector at a different location than will the lighter ion. Since the velocity of the particles, the strength of the magnetic field, and the paths of flight can be readily determined, the masses of various ions (atoms and molecules) may be calculated precisely.

For singly charged ions of mass m accelerated through a voltage potential V, the kinetic energy acquired is eV. The ions then enter a magnetic field with a discrete velocity v given by the equation

$$eV = \tfrac{1}{2}mv^2 \tag{A-1}$$

where e is the electron charge. For positive ions having charge states n ($n = 1, 2, 3, \ldots$), with n denoting the number of electrons stripped from neutral atoms, the more general expression is

$$neV = \tfrac{1}{2}mv^2 \tag{A-2}$$

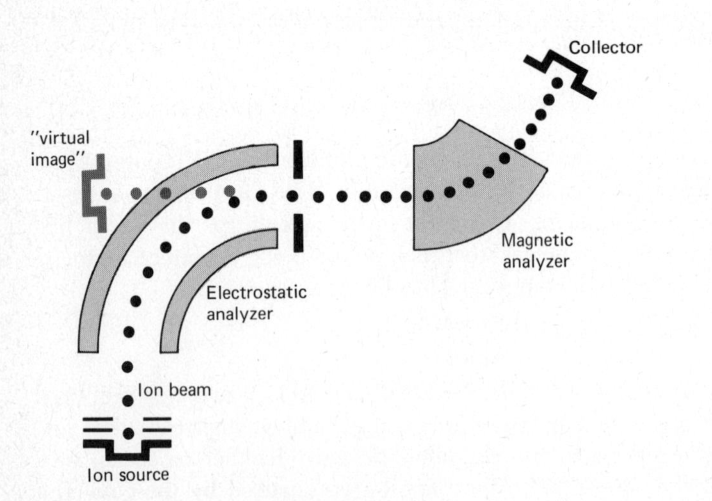

Figure A-6 Mass spectometer operation.

For a magnetic field of strength B perpendicular to the velocity vector of the ions, the ions will be deflected into a circular path with a balancing of centrifugal and centripetal forces. Thus

$$Bnev = \frac{mv^2}{R} \quad \text{or} \quad Bne = \frac{mv}{R} \tag{A-3}$$

where R is the radius of the ion path.

Eliminating v between Equations (A-2) and (A-3) gives

$$R = \frac{1}{B}\left(\frac{2mV}{ne}\right)^{1/2} \tag{A-4}$$

The differential form of Equation (A-4) is

$$\frac{2\,\Delta R}{R} = \frac{\Delta m}{m} + \frac{\Delta v}{v} - \frac{2\,\Delta B}{B} \tag{A-5}$$

For a homogeneous magnetic field and ions having a negligible energy spread, the last two terms of Equation (A-5) will vanish, leaving

$$\left(\frac{\Delta m}{m}\right)R = 2(\Delta R) \tag{A-6}$$

The left-hand term gives a measure of the mass dispersion or separation of ions along the focal plane. Thus

$$\left(\frac{m_2 - m_1}{m_1}\right)R_1 = 2(R_2 - R_1) \tag{A-7}$$

Many modifications of the mass spectrometer are available and are classified according to the total arc of curvature from 60 through 360 degrees. Numerous devices are available for producing ions and serving as detectors.

A similar device with a different measurement principle is the time-of-flight or pulsed-mass spectrometer. In this device the ion is accelerated by a pulsed electric field and directed axially through a long chamber (approximately 170 cm in length) in which a good vacuum is maintained. The velocity of the ion is obtained by solving Equation (A-2). Thus

$$v_t = \left(\frac{2neV}{m_i}\right)^{1/2} \tag{A-8}$$

where i indicates the species. If the drift region (tube length) is of length L, the time of drift is

$$t_i = \frac{Lm_i^{\,1/2}}{(2neV)^{1/2}} \tag{A-9}$$

The difference in drift times of two ions having masses m_1 and m_2 is

$$\Delta t = \frac{L\left(m_1{}^{1/2} - m_2{}^{1/2}\right)}{\left(2neV\right)^{1/2}} \tag{A-10}$$

Electronic devices are employed to measure Δt values.

Additional information pertinent to the calibration and use of mass spectrometers may be found in reference 7 and similar sources.

A-5 MONITORING OF CARBON MONOXIDE AND HYDROCARBONS

The measurement of CO in the ambient environment and from mobile and stationary sources is most frequently accomplished by employing the principles of nondispersive infrared absorption (NDIR), gas chromatography combined with a flame ionization detector (GC-FID), and catalytic oxidation. In catalytic oxidation, the usual technique is to measure the temperature difference between a cell where oxidation occurs and a similar cell where no oxidation occurs. The magnitude of the temperature difference, sensed by the thermocouples, is an indication of the CO concentration in the gas stream. For oxidation of CO, the catalyst is usually Hopcalite, which is a mixture of MnO_2 and CoU, although other catalysts may be used.

The measuring technique recommended by EPA for establishing ambient CO levels is nondispersive infrared spectroscopy. The primary ambient standard on a 1-hr averaging time is 40 $\mu g/m^3$ (35 ppm), while the value is 10 $\mu g/m^3$ (9 ppm) for an 8-hr averaging time (see Table 2-1). The majority of vehicle emission or stationary source monitors for CO are either NDIR or catalytic oxidation types, although GC-FID instruments are also offered by a number of manufacturers.

The major detection device for hydrocarbons is based on flame ionization. It is used alone or preceded by gas chromatography. Although not applicable to ambient air monitoring, nondispersive infrared absorption is applicable to hydrocarbon detection from stationary and mobile sources. An FID instrument monitors total hydrocarbons; it does not separate individual hydrocarbon species. When it is necessary to know individual species concentrations or to separate out the methane component, a gas chromatographic unit precedes the FID unit. This method of operation requires intermittent sampling rather than continuous sampling. The EPA measurement method is the GC-FID technique. Gas chromatographic separation is necessary in this situation, since the 3-hr (6 to 9 A.M.) primary ambient standard of 160 $\mu g/m^3$ (0.24 ppm) is based on hydrocarbon level corrected for methane.

A-6 MONITORING METHODS FOR SULFUR DIOXIDE

It has been estimated that over 100 commercial instruments for the continuous monitoring of SO_2 are on the market. These instruments are based on a number of principles. These include conductivity, colorimetry, nondispersive infrared, coulometry, spectrophotometry, flame emission spectrometry, electrochemistry, and condensation nuclei. Among the most important features required of any of these instruments is sensitivity, since the SO_2 level in most nonurban areas is around 0.02 ppm or less. Many commercial units have a lower detection limit of about 0.01 ppm. Another major requirement for any detection device is noninterference by other substances in the air sample. Carbon dioxide, for example, provides a positive interference in conductimetric instruments, while certain substances such as ozone, nitrogen oxides, some organic compounds, and even fine particulate matter are potential problems in the use of coulometric devices. A flame photometric detector responds to the total of all sulfur compounds present rather than SO_2 alone. One method of overcoming such difficulties is to combine two analytical techniques in series, such as a gas chromatographic–flame photometric detector. Sulfur compounds such as SO_2, H_2S, and CH_3SH are separated on a chromatographic column and then measured individually by flame photometry. Detectability as low as 2 ppb has been reported in this case. Although the advantages of prior separation are apparent, it is to be anticipated that additional operational and maintenance problems may exist for such systems. A rather detailed review [10] of commercially available SO_2 analyzers for continuous monitoring duty appeared in 1973.

Manual methods for periodic sampling of SO_2 have been described in a government publication [11]. The recommended procedures in this case are the West-Gaeke technique and the hydrogen peroxide method. The West-Gaeke method has been chosen by EPA as the standardized technique in the United States, as indicated earlier in Table 2-1. In the table it is called the pararosaniline method, and is a colorimetric measurement. The test involves absorbing sulfur dioxide in a dilute aqueous sodium tetracholoromercurate solution to form the nonvolatile dichlorosulfitomercurate ion. A subsequent reaction of this ion with formaldehyde and bleached pararosaniline leads to the formation of red-purple pararosaniline methylsulfonic acid. The color intensity, measured at 0.56 μm, is proportional to the SO_2 concentration. The technique is applicable for the concentration range from 0.002 to 5 ppm. As indicated above, levels of 0.01 to 0.1 ppm are of primary interest with respect to ambient measurements. The annual primary standard is set at 0.03 ppm. The West-Gaeke method is specific to SO_2 and sulfite salts. However, the presence of nitrogen dioxide and ozone may reduce the reading from its true value. The hydrogen peroxide method is an acid titration technique, and is used commonly in Europe. The air sample is passed through a $0.03N$ hydrogen peroxide solution with a pH of 5. Oxidation of SO_2 to H_2SO_4 occurs, with the resulting acid then titrated with a standard alkali. The titration indicator used is methyl-red bromocresol green, which is green above pH 5 and red below pH 5. Since the method is based on acid equivalents, the presence in the air sample of other acidic or alkaline gases will lead to high or low results, respectively. It is not uncommon for these methods and others to show different results for the same sample, owing to the different interferences frequently found in ambient air samples.

A-7 MONITORING FOR OXIDES OF NITROGEN

A number of basic principles are used in instruments for the detection of NO or NO_2 in the ambient air or from stationary or mobile sources. Included among these are colorimetry, coulometry, electrochemistry, chemiluminescence, and nondispersive infrared, visible, and ultraviolet absorption.

In the middle latitudes the background levels of NO and NO_2 are approximately 2 ppb and 4 ppb, respectively. However, in urban areas the annual mean concentration typically may range from 30 to 100 ppb (0.03 to 0.10 ppm), with hourly levels reaching from 0.10 to 1.0 ppm. The federal annual primary and secondary ambient standard for NO_2 is 0.05 ppm. On the other hand, NO_x emissions from combustion processes may reach 300 to 1000 ppm. Hence the detection range required for ambient air monitoring is quite different from that needed for stationary source emissions. Nondispersive infrared absorption is used for monitoring stationary and mobile sources emissions, but not for ambient air. However, chemiluminescence is a useful technique in all three situations. Ambient air measurement is also carried out by colorimetric and coulometric devices. Stationary source monitors also make effective use of the electrochemical cell as well as nondispersive ultraviolet and visible absorption.

The laboratory or manual methods commonly used to determine NO_2 levels are the Griess-Saltzman and Jacobs-Hochheiser techniques. The former is used in conjunction with California standards, while the latter is that adopted by EPA. The Saltzman colorimetric method employs the reaction of NO_2 in an air sample with sulfanilic acid to form a diazonium salt. This is then coupled with *N*-(1-naphthyl)-ethylenediamine dihydrochloride so that a deep red-violet color is formed. The absorbence of the dye is then measured spectrophotometrically at 0.55 μm and compared to data from carefully calibrated measurements. The method normally is valid for NO_2 levels in air between 0.02 and 0.75 ppm (40 to 1500 $\mu g/m^3$).

In the EPA method the air sample is passed through an aqueous sodium hydroxide solution so that the nitrite ion is formed from the NO_2. Treatment with hydrogen peroxide (H_2O_2) removes SO_2 interference, and the solution is then acidified. At this juncture the nitrous acid produced is measured by the diazotizing coupling discussed above for the Griess-Saltzman technique, with the exception that sulfanilamide replaces sulfanilic acid as a reagent. This method can be used to measure both NO and NO_2 levels. For example, the NO_2 concentration can be measured first. Then in a second sample the NO is oxidized to NO_2 and the total NO_x determined. The NO level is then established by difference.

It should be noted that considerable discrepancy has arisen in the past when NO_x measurements based on different instruments have been compared for the same sample. The federal government continues to play an active role in establishing ground rules for the use of various instruments in pollutant detection. The validity and accuracy of any measurement technique should be clearly established before it is used as a primary method for detecting ambient or emission levels.

A-8 MONITORING OF PHOTOCHEMICAL OXIDANTS

Chemiluminescence is the major method of monitoring photochemical oxidants in the ambient atmosphere, although some commercial instruments are based on colorimetric or coulometric principles. Oxidant measurements must be corrected for the presence of NO_2 and SO_2, and the resulting values are primarily an indication of ozone concentration. One convenient technique for the continuous air monitoring of ozone is based on the chemiluminescence resulting from the reaction between ozone and ethylene at atmospheric pressure [12, 13]. Either atmospheric ozone or nitric oxide can be monitored by a low-pressure chemiluminescent reaction between nitric oxide and ozone [14]. This latter technique has been expanded [15] to include the measurement of total oxides of nitrogen as well as NO_x plus ammonia.

Peroxyacetyl nitrate (PAN) occurs in photochemical smog with concentrations as high as 30 ppb. It is usually monitored in the atmosphere by gas chromatography on a batch basis. A method for continuous monitoring of PAN has been suggested [16]. In this case, chemiluminescence spectra resulting from the reaction of PAN and ozone with triethylamine vapor have proved capable of detecting PAN at concentrations as low as 6 ppb, which is within the range of interest.

References

1. *Methods of Air Sampling and Analysis*. Air Pollution Control Association, Pittsburgh, 1972.
2. D. F. Adorns, ed. *Air Pollution Instrumentation*. Instrument Society of America, Pittsburgh, 1966.
3. A. C. Stern, ed. *Air Pollution*. Vol. II, 2nd ed. New York: Academic Press, 1968.
4. N. Henein and D. Patterson. *Emissions from Combustion Engines and Their Control*. Ann Arbor, Mich.: Ann Arbor Science Publishers, Inc., 1972.
5. H. M. McNair and E. J. Bonelli. *Basic Gas Chromatography*. Walnut Creek, Calif.: Varian Aerograph, 1969.
6. D. Abbott and R. S. Andrews. *An Introduction to Chromatography*. New York: Houghton Mifflin, 1969.
7. F. A. White. *Mass Spectrometry in Science and Technology*. New York: Wiley, 1968.

8. R. K. Stevens and W. F. Herget, eds. *Analytical Methods Applied to Air Pollution Measurements*. Ann Arbor, Mich.: Ann Arbor Science Publishers, Inc., 1974.
9. J. N. Driscoll. *Flue Gas Monitoring Techniques*. Ann Arbor, Mich.: Ann Arbor Science Publishers, Inc., 1974.
10. C. D. Hollowell, G. Y. Gee, and R. D. McLaughlin. *Anal. Chem.* **45** (1973): 63A.
11. *Methods of Measuring and Monitoring Atmospheric Sulfur Dioxide*. PHS Publication No. 999-AP-6, 1964. PB 168-865.
12. C. W. Nederbragt, A. Van der Horst, and T. Van Duijn. *Nature* **206** (1965): 87.
13. G. J. Warren and F. Babcock. *Rev. Sci. Instrum.* **41** (1970): 280.
14. A. Fontijn et al. *J. Anal. Chem.* **42** (1970): 575.
15. J. A. Hodgeson et al. AIAA Paper No. 71-1067, November, 1971.
16. J. N. Pitts, Jr., et al. *Environ. Sci. Tech.* **7**, no. 6 (1973): 550.

Appendix B
Measurement Quantities

B-1 CONVERSION FACTORS

Length $1\ m = 10^6\ \mu m = 10^{10}\ \text{Å} = 3.28\ ft = 39.37\ in.$

Mass $1\ lb_m = 453.6\ g = 7000\ gr = 0.4544\ kg = 0.0005\ ton$

Force $1\ lb_f = 32.174\ lb_m{\cdot}ft/s^2 = 4.448\ N$
$1\ N = 1\ kg{\cdot}m/s^2$

Pressure $1\ atm = 1.013\ bars = 14.7\ lb_f/in.^2 = 760\ mm\ Hg$
$1\ in.\ water = 0.0361\ lb_f/in.^2 = 2.49\ mbar$

Volume $1\ l = 0.0353\ ft^3 = 0.2642\ gal = 61.025\ in.^3$
$1\ ft^3 = 0.02833\ m^3 = 28{,}320\ cm^3 = 7.481\ gal = 28.32\ l = 1728\ in.^3$

Density $1\ lb_m/ft^3 = 0.01602\ g/cm^3$
$1\ gr/ft^3 = 2.29\ g/m^3$

Energy $1\ kJ = 737.5\ ft{\cdot}lb_f = 0.948\ Btu = 239.0\ cal$
$1\ Btu = 252.2\ cal = 778.16\ ft{\cdot}lb_f$
$1\ cal/g{\cdot}mole = 1.80\ Btu/lb{\cdot}mole$

Power $1\ \text{kW} = 3413\ \text{Btu/hr} = 1\ \text{kJ/s} = 0.00134\ \text{hp}$
$1\ \text{hp} = 746\ \text{W} = 33{,}000\ \text{ft·lb}_f/\text{min} = 2545\ \text{Btu/hr}$

Speed $1\ \text{mi/hr} = 0.447\ \text{m/s} = 1.467\ \text{ft/s}$

Viscosity $1\ \text{poise} = 1\ \text{g/cm·s} = 242\ \text{lb}_m/\text{ft·hr} = 361\ \text{kg/m·hr}$

B-2 UNIVERSAL GAS CONSTANT AND GRAVITATIONAL ACCELERATION

$$R_u = 8.315\ \text{kJ/kg·mole·°K} = 0.08315\ \text{bar·m}^3/\text{kg·mole·°K}$$
$$= 1545\ \text{ft·lb}_f/\text{lb·mole·°R} = 10.73\ \text{psia·ft}^3/\text{lb·mole·°R}$$

$$g(\text{at sea level}) = 32.174\ \text{ft/s}^2 = 9.806\ \text{m/s}^2$$

B-3 PROPERTIES OF AIR

Density: at 25°C (77°F) and 1 atm, $\rho = 0.0739\ \text{lb/ft}^3 = 1.183\ \text{kg/m}^3$

Viscosity:

T, °K	300	350	400	450	500	550
μ, kg/m·hr	0.0666	0.0748	0.0825	0.0900	0.0969	0.1031
T, °F	80	170	260	350	440	530
μ, lb/ft·hr	0.0447	0.0503	0.0554	0.0605	0.0651	0.0693

Gas constant: $R = 287.1\ \text{m}^2/\text{s}^2\text{·°K} = 0.2871\ \text{kJ/kg·°K}$
$= 1716\ \text{ft}^2/\text{s}^2\text{·°R} = 53.35\ \text{ft·lb}_f/\text{lb}_m\text{·°R}$

Specific heat: at 25°C (77°F), $c_p = 1.005\ \text{kJ/kg·°K} = 0.240\ \text{Btu/lb·°F}$
$c_v = 0.718\ \text{kJ/kg·°K} = 0.171\ \text{Btu/lb·°F}$

B-4 MOLAR MASSES OF VARIOUS SUBSTANCES, *M*

SUBSTANCE	M	SUBSTANCE	M
Air	28.96	Oxygen (O_2)	32.00
Carbon	12.0	Water (H_2O)	18.02
Methane (CH_4)	16.04	Nitrogen (N_2)	28.01
Carbon monoxide (CO)	28.01	Nitric oxide (NO)	30.00
Carbon dioxide (CO_2)	44.01	Nitrogen dioxide (NO_2)	46.00
Hydrogen (H_2)	2.02	Ammonia (NH_3)	17.03

B-5 VALUES OF THE ERROR FUNCTION erf x

x	erf x	x	erf x	x	erf x	x	erf x
0.	0.	0.42	0.4475	0.79	0.7361	1.32	0.9381
0.02	0.02256	0.43	0.4569	0.80	0.7421	1.34	0.9419
0.04	0.04511	0.44	0.4662	0.81	0.7480	1.36	0.9456
0.06	0.06762	0.45	0.4755	0.82	0.7538	1.38	0.9490
0.08	0.09008	0.46	0.4847	0.83	0.7595	1.40	0.9523
0.10	0.1125	0.47	0.4937	0.84	0.7651	1.42	0.9554
0.11	0.1236	0.48	0.5027	0.85	0.7707	1.44	0.9583
0.12	0.1348	0.49	0.5117	0.86	0.7761	1.46	0.9611
0.13	0.1459	0.50	0.5205	0.87	0.7814	1.48	0.9637
0.14	0.1569	0.51	0.5292	0.88	0.7867	1.50	0.9661
0.15	0.1680	0.52	0.5379	0.89	0.7918	1.52	0.9684
0.16	0.1790	0.53	0.5469	0.90	0.7969	1.54	0.9706
0.17	0.1900	0.54	0.5549	0.91	0.8019	1.56	0.9726
0.18	0.2009	0.55	0.5633	0.92	0.8068	1.58	0.9745
0.19	0.2118	0.56	0.5716	0.93	0.8116	1.60	0.9763
0.20	0.2227	0.57	0.5798	0.94	0.8163	1.62	0.9780
0.21	0.2335	0.58	0.5879	0.95	0.8209	1.64	0.9796
0.22	0.2443	0.59	0.5959	0.96	0.8254	1.66	0.9811
0.23	0.2550	0.60	0.6039	0.97	0.8299	1.68	0.9825
0.24	0.2657	0.61	0.6117	0.98	0.8342	1.70	0.9838
0.25	0.2763	0.62	0.6194	0.99	0.8385	1.72	0.9850
0.26	0.2869	0.63	0.6270	1.00	0.8427	1.74	0.9861
0.27	0.2974	0.64	0.6346	1.02	0.8508	1.76	0.9872
0.28	0.3079	0.65	0.6420	1.04	0.8586	1.78	0.9882
0.29	0.3183	0.66	0.6494	1.06	0.8661	1.80	0.9891
0.30	0.3286	0.67	0.6566	1.08	0.8733	1.82	0.9899
0.31	0.3389	0.68	0.6638	1.10	0.8802	1.84	0.9907
0.32	0.3491	0.69	0.6708	1.12	0.8868	1.86	0.9915
0.33	0.3593	0.70	0.6778	1.14	0.8931	1.88	0.9922
0.34	0.3694	0.71	0.6847	1.16	0.8991	1.90	0.9928
0.35	0.3794	0.72	0.6914	1.18	0.9048	1.92	0.9934
0.36	0.3893	0.73	0.6981	1.20	0.9103	1.94	0.9939
0.37	0.3992	0.74	0.7047	1.22	0.9155	1.96	0.9944
0.38	0.4090	0.75	0.7112	1.24	0.9205	1.98	0.9949
0.39	0.4187	0.76	0.7175	1.26	0.9252	2.00	0.9953
0.40	0.4284	0.77	0.7238	1.28	0.9297		
0.41	0.4380	0.78	0.7300	1.30	0.9340		

B-6 IDEAL-GAS ENTHALPY OF AIR

METRIC DATA: h, kJ/kg; T, °K.

T	h	T	h	T	h	T	h
280	280.1	440	441.6	600	607.0	760	778.2
290	290.1	450	451.8	610	617.5	780	800.0
300	300.2	460	460.0	620	628.1	800	822.0
310	310.2	470	472.2	630	638.6	820	844.0
320	320.3	480	482.5	640	649.2	840	866.1
330	330.3	490	492.7	650	659.8	860	888.3
340	340.4	500	503.0	660	670.5	880	910.6
350	350.5	510	513.3	670	681.1	900	932.9
360	360.6	520	523.6	680	691.8	920	955.4
370	370.7	530	534.0	690	702.5	940	977.9
380	380.8	540	544.4	700	713.3	960	1000.6
390	390.9	550	554.7	710	724.0	980	1023.3
400	401.0	560	565.2	720	734.8	1000	1046.0
410	411.1	570	575.6	730	745.6	1020	1068.9
420	421.3	580	586.0	740	756.4	1040	1091.9
430	431.4	590	596.5	750	767.3	1060	1114.9

ENGLISH DATA: h, Btu/lb; T, °R.

T	h	T	h	T	h	T	h
500	119.5	760	182.1	1160	281.1	1600	395.7
520	124.3	780	186.9	1200	291.3	1620	401.1
530	126.7	800	191.8	1240	301.5	1640	406.5
537	128.1	820	196.7	1280	311.8	1660	411.8
540	129.1	840	201.6	1320	322.1	1680	417.2
550	131.5	860	206.5	1360	332.5	1700	422.6
560	133.9	880	211.4	1400	342.9	1720	428.0
580	138.7	900	216.3	1420	348.1	1740	433.4
600	143.5	920	221.2	1440	353.4	1760	438.8
620	148.3	940	226.1	1460	358.6	1780	444.3
640	153.1	960	231.1	1480	363.9	1800	449.7
660	157.9	980	236.0	1500	369.2	1820	455.2
680	162.7	1000	241.0	1520	374.5	1840	460.6
700	167.6	1040	251.0	1540	379.8	1860	466.1
720	172.4	1080	261.0	1560	385.1	1880	471.6
740	177.2	1120	271.0	1580	390.4	1900	477.1

SOURCE: Adapted or abridged from J. H. Keenan, and J. Kaye, *Gas Tables*. New York: Wiley, 1980.

Answers to Odd-Numbered Problems

1-1	11.4 mg/m^3
1-3	0.94×10^{-6} gr/ft^3
1-5	457 mg/m^3
1-7	(a) 6.99×10^{-5} gr/ft^3; (b) 22.9×10^6 $\mu g/m^3$
1-9	(a) 32.4; (b) 54.3; (c) 85.9 percent
1-11	(a) 2.53×10^{-3} m^{-1}; (b) 2.08; (c) 1.54 km
1-13	68 percent
1-17	(a) 0.87 ppm; (b) 0.274 ppm; (c) 0.10 ppm
1-19	(a) 7200 m; (b) 16.75; (c) 2.33
1-21	(a) 0.552 ft/s; (b) 3970 ft
1-23	33 percent
2-1	(a) 0.63 percent; (b) 0.69 percent
2-3	(a) 588 g/s; (b) 374 g/s
2-5	(a) 135, 620, 52,500 lb/hr; (b) 3500, 6450, 7290 lb/hr; (c) 3.6, 13,630, 11,080 lb/hr
2-7	Yes, 87.5 percent
2-9	99.7 percent
2-11	0.36 lb/10^6 Btu
2-13	84.0 percent
2-15	90 percent
3-1	(a) Stable; (b) slightly stable; (c) unstable; (d) neutral; (e) unstable; (f) weakly stable

3-3 (a) 544.7°R; (b) 531.8°R; (c) 517.9°R; (d) 294.4°K; (e)297.4°K; (f) 307.4°K
3-5 (a)12.9°F/1000 ft; (b) 0.96°F/1000 ft; (c) -1.0°C/100 m; (d) 0°C/100 m
3-7 13.9 mi/hr
3-9 (a) 2.78, 3.06, 3.64, 5.37 m/s; (b) 2.91, 3.24, 3.92, 6.14 m/s
3-11 6900 ft; (b) 4830 ft
3-13 (a) 2590 m; (b) 793 m
3-15 (a) -9.26×10^{-3}°K/m; (b) 378 m; (c) 287°K
3-17 (a) 119.7, 218.8, 319.9, 483, 1020, 1311, 1582, 2884, 3165 m

4-1 (a) $x = 0.5$, $f(x) = 0.484$; (b) $x = 1$, $f(x) = 0.242$; (c) $x = 1$, $f(x) = 0.176$
4-3 (a) 0.044, 0.011, 0.002
4-5 (a) 130, 440, 570, 560, 400, 170, 50; (b) 36, 200, 310, 400, 320, 150, 45
4-7 1.5 km, 750 $\mu g/m^3$
4-9 1.2 km, 828 $\mu g/m^3$
4-11 675 $\mu g/m^3$, 1.46 km
4-13 45, 250, 420, 500, 485, 335, 130, 50
4-19 540 $\mu g/m^3$, 2.8 km
4-23 $x = 1250$ to 3000 m, $y = \pm 50$ m
4-25 (a) 56 m; (b) 65 m; (c) 77 m
4-27 480 m, 0.63 ppm
4-29 (a) 90 m; (b) 65 m
4-31 (a) 3900 g/s; (b) 5060 g/s
4-33 0.31 g/s
4-35 (a) 71; (b) 95; (c) 200; (d) 148 m
4-39 125 m, 185 m, 214 m, 88 m
4-41 (a) 181; (b) 260; (c) 282; (d) 190; (e) 427 m
4-43 (a) -33 percent; (b) 12 percent; (c) 28 percent
4-45 155 $\mu g/m^3$

5-1 No
5-3 97.9 percent
5-5 96.6 percent
5-7 (a) 96.6 percent; (b) 81.6 percent; (c) 90.2 percent
5-9 1.25 gr/ft^3
5-11 92 percent
5-13 2.14 gr/ft^3
5-15 80 percent
5-17 0.614 kg/s
5-19 (a) 5000 m^2/kg; (b) 500 m^2/kg
5-21 (a) 1.59×10^{15}; (b) 1.59×10^{12}
5-23 (a) 4.56×10^9; (b) 0.0873 kg; (c) 133.6 m^2/kg
5-27 61.4 percent
5-29 (a) 30; (b) 1.35; (c) 31.3, 9.64; (d) 22.9, 1.35, 24.0, 7.35 μm
5-31 (a) 26 μm; (b) 3.1
5-33 (a) Yes; (b) 5.4, 1.65 μm; (c) 83 percent
5-35 (a) 0.51, 1.00; (b) $5.06 \times 10^{-4}/1$; (c) 1260 ft/min; (d) 0.636/1

5-37 (a) 468 m/s; (b) 6.73×10^{-8} m; (c) 1.28
5-39 (a) 19.8 μm/s; (b) 4.86, 7.46 μs
5-41 (a) 19.7 μm/s; (b) 4.54, 7.0 μs
5-43 6.31×10^{-5}, 1.18×10^{-5} g/m²·s
5-45 3.82 μg/m²·s
5-47 (a) 1.2×10^{-11}; (b) 1.4×10^{-9}; (c) 2.9×10^{-7}; (d) 4.4×10^{-3} g/m²·s
5-49 466 μg/m³
5-51 (a) (2) 5000 m; (b) (2) 4000 m; (c) (2) 1500 m; (d) (2) 1100 m
5-53 2.37/1
5-55 0.5 m/s
5-57 (a) 2160; (b) 0.20 m; (c) 0.70, 0.45, 0.25, 0.11, 0.03
5-59 (a) 0.25 m; (b) 3185; (c) 0.67, 0.39, 0.17, 0.04
5-61 0.35 m/s
5-63 88 μm
5-65 51 μm
5-67 83 percent
5-69 (a) 81 percent; (b) 62.5 percent; (c) 72 percent; (d) 75 percent
5-73 (a) 10 μm; (b) 14.2 μm
5-75 (a) 13.9 m/s; (b) 9.97 μm; (c) 4.46 μm, (d) 14.1 μm
5-77 (a) 0.13 in. H_2O; (b) 0.12 in. H_2O
5-79 (b) 0.22, 0.50, 0.80, 0.95
5-81 (a) 9.73×10^{-4} m, 3.89; (b) 9.73×10^{-4} m, 1.95; (c) 9.73×10^{-4} m, 0.97
5-83 (a) 8.4×10^{-4} m, 4.2; (b) 8.4×10^{-4} m, 1.68; (c) 8.4×10^{-4} m, 0.84
5-85 (a) 23 percent; (b) 74 percent; (c) 88 percent
5-87 (a) 0.012, 0.39, 0.50, 0.43, 0.29, 0.10, 0.02;
(b) 0.54, 0.60, 0.50, 0.36, 0.20, 0.04, 0.005; (c) 0, 0, 0.33, 0.42, 0.35, 0.18
5-89 (a) 65.4; (b) 40.1 in. H_2O
5-91 (a) 46 m/s; (b) 36.1 cm H_2O; (c) 15.6 cm H_2O
5-93 (a) 0.56; (b) 0.23; (c) 0.043; (d) 0.026; (e) 0.016
5-95 (a) 56 m/s; (b) 31.8 cm H_2O; (c) 15.4 cm H_2O
5-97 (a) 0.60; (b) 0.33; (c) 0.15; (d) 0.074
5-99 4.1×10^{-12} m²
5-101 (a) 4.6×10^{-4} in.·lb_m/in H_2O·min²; (b) 6.4×10^{-11} ft²
5-103 (a) 180 N/m², 1.80 mbar; (b) 1.08 kg/m²; (c) 634 N/m², 6.34 mbar
5-105 (a) 30,000 kg/m²·s; (b) 104,300 s^{-1}
5-107 6.7 hr
5-109 74.4 hr
5-111 (a) 8.77×10^{-13} m²; (b) 19,180 s^{-1}
5-113 (a) 0.0267; (b) 0.0464; (c) 0.205; (d) 0.404; (e) 1.98 m/s
5-115 (a) 86.9; (b) 95.2; (c) 98.3 percent
5-117 (a) 99; (b) 99.9; (c) 99.99 percent
5-119 (a) 95.5; (b) 46.3; (c) 11.7 percent
5-121 (a) 100 percent; (b) 76.6 percent
5-123 (a) 98.9; (b) 93.8; (c) 50.1 percent
5-125 (a) 178; (b) 148 percent
5-127 (a) 15.6 percent; (b) 28 percent
5-129 $1.66/ft²/min
5-131 (a) 97 percent; (b) 97.9, 75 percent; 99.4, 50 percent; (c) 12.9 percent

6-1 (a) 61.7 cm^3, 23.1; (b) 0.69; (c) 0.90
6-3 (a) 0.0151 m; (b) 7.7×10^{-5} m/s; (c) 2.8 hr
6-5 (a) 0.0172 m; (b) 1.14×10^{-5} m/s; (c) 4.4 hr
6-7 0.79 hr
6-9 (a) −64 percent; (b) −32 percent
6-11 5.5 hr
6-13 2.3 hr
6-15 1.2 hr
6-17 (a) 1.60 ft, ~0.62 in. H_2O/ft
6-19 (a) (1) 1.63; (2) 0.73; (3) 125 percent; (4) ~5.2
6-21 (a) (1) 1.20 ft; (2) 0.82 ft; (3) 1.58 ft; (4) ~8.2 ft
6-23 (a) 1.54; (b) ~7
6-25 2.16 m^3/min
6-27 (a) 3350 lb/hr; (b) 29 in.; (c) 7.7 or 9.6 in. H_2O; (d) ~9; (e) ~21 ft
6-29 (a) 30.2; (b) 389
6-31 (a) 1.27; (b) 89; (c) 0.74, 0.0106
6-33 (a) 111,000 kJ/kg·mole; (b) 2.9×10^{12} cm^3/g·mole·s
6-35 (a) 114,400 kJ/kg·mole; (b) 2.5×10^{9} cm/g·mole·s
6-37 CO = 0.028, O_2 = 0.064
6-39 (a) 215 ppm; (b) 180 ppm
6-41 (a) 96.2 percent; (b) 82.1 percent
6-43 (a) 130 ppm; (b) 55 ppm
6-45 (a) 155/1; (b) 618°K; (c) −61 percent
6-47 (a) 164/1; (b) −62 percent; (c) 645°F

7-1 78 percent
7-3 63 percent
7-5 (a) 3.12; (b) 248; (c) 1220

8-1 0.375 tons/day
8-3 (a) 740 ppm; (b) 295 ppm
8-5 (a) 875 ppm
8-7 (a) 0.50, 0.81, 0.99; (b) 2410°K
8-9 (a) 0.14, 0.49, 0.83; (b) 2325°K
8-11 (a) 250 ppm; (b) 420 ppm
8-13 (a) 0.11, (b) 0.57

9-1 0.211 μm
9-3 (a) 2.8×10^{6}; (b) 4.67×10^{-16}

10-1 15.05
10-3 (a) 13.2; (b) 15.3
10-5 (a) 1489/1; (b) 8.07/1
10-7 (a) 13.69; (b) 12.88

10-9 (a) 14.5, 14.89, 15.2; (b) 14.79, 21.3, 69
10-11 (a) 370 ft^3/min; (b) CO = 0.019, HC = 0.0040, NO = 0.00052 lb/min
10-13 1725°K

11-1 0.0065 ppm, No
11-3 (a) ~0.15 to 0.45 km; (b) ~0.05 to 0.5 km
11-5 190–220 kg/hr
11-7 0.256 units/s

Index

82 83 84 9 8 7 6 5 4 3